# NOISE AND VIBRATION ANALYSIS

# NOISE AND VIBRATION ANALYSIS

## SIGNAL ANALYSIS AND EXPERIMENTAL PROCEDURES

**Anders Brandt**

*Department of Industrial and Civil Engineering*
*University of Southern Denmark*

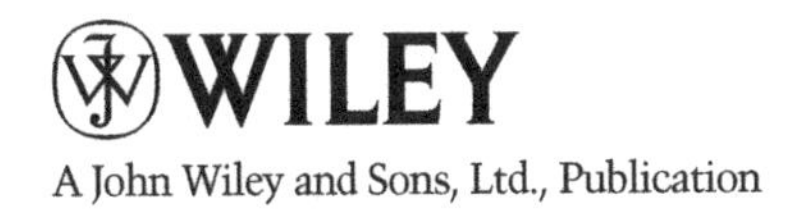
A John Wiley and Sons, Ltd., Publication

This edition first published 2011


*Registered office*
John Wiley & Sons Ltd, The Atrium, Southern Gate, Chichester, West Sussex, PO19 8SQ, United Kingdom

For details of our global editorial offices, for customer services and for information about how to apply for permission to reuse the copyright material in this book please see our website at www.wiley.com.



*Library of Congress Cataloguing-in-Publication Data*

Brandt, Anders.
Noise and vibration analysis : signal analysis and experimental procedures / Anders Brandt.
p. cm.
Includes bibliographical references and index.
ISBN 978-0-470-74644-8 (hardback)
1. Vibration–Mathematical models. 2. Noise–Mathematical models. 3. Acoustical engineering. 4. Stochastic analysis. 5. Signal processing. I. Title.
TA355.B674 2011
620.3–dc22

2010039788

A catalogue record for this book is available from the British Library.

Print ISBN: 9780470746448
E-PDF ISBN: 9780470978177
O-Book ISBN: 9780470978160
E-Pub ISBN: 9780470978115

Typeset in 9/11pt Times by Aptara Inc., New Delhi, India.

# Contents


**About the Author** xv

**Preface** xvii

**Acknowledgements** xxi

**List of Abbreviations** xxiii

**Notation** xxv

**1 Introduction** **1**
1.1 Noise and Vibration 1
1.2 Noise and Vibration Analysis 2
1.3 Application Areas 3
1.4 Analysis of Noise and Vibrations 3
*1.4.1 Experimental Analysis* 4
1.5 Standards 4
1.6 Becoming a Noise and Vibration Analysis Expert 4
*1.6.1 The Virtue of Simulation* 4
*1.6.2 Learning Tools and the Format of this Book* 5

**2 Dynamic Signals and Systems** **7**
2.1 Introduction 7
2.2 Periodic Signals 8
*2.2.1 Sine Waves* 8
*2.2.2 Complex Sines* 10
*2.2.3 Interacting Sines* 11
*2.2.4 Orthogonality of Sines* 12
2.3 Random Signals 13
2.4 Transient Signals 14
2.5 RMS Value and Power 15
2.6 Linear Systems 16
*2.6.1 The Laplace Transform* 17
*2.6.2 The Transfer Function* 20
*2.6.3 The Impulse Response* 21
*2.6.4 Convolution* 22
2.7 The Continuous Fourier Transform 25
*2.7.1 Characteristics of the Fourier Transform* 27
*2.7.2 The Frequency Response* 29

| | | | |
|---|---|---|---|
| | *2.7.3* | *Relationship between the Laplace and Frequency Domains* | 29 |
| | *2.7.4* | *Transient versus Steady-state Response* | 30 |
| 2.8 | Chapter Summary | | 31 |
| 2.9 | Problems | | 32 |
| | References | | 33 |
| **3** | **Time Data Analysis** | | **35** |
| 3.1 | Introduction to Discrete Signals | | 35 |
| 3.2 | The Sampling Theorem | | 35 |
| | *3.2.1* | *Aliasing* | 37 |
| | *3.2.2* | *Discrete Representation of Analog Signals* | 38 |
| | *3.2.3* | *Interpolation and Resampling* | 40 |
| 3.3 | Filters | | 42 |
| | *3.3.1* | *Analog Filters* | 43 |
| | *3.3.2* | *Digital Filters* | 45 |
| | *3.3.3* | *Smoothing Filters* | 46 |
| | *3.3.4* | *Acoustic Octave Filters* | 47 |
| | *3.3.5* | *Analog RMS Integration* | 49 |
| | *3.3.6* | *Frequency Weighting Filters* | 49 |
| 3.4 | Time Series Analysis | | 51 |
| | *3.4.1* | *Min- and Max-analysis* | 51 |
| | *3.4.2* | *Time Data Integration* | 51 |
| | *3.4.3* | *Time Data Differentiation* | 55 |
| | *3.4.4* | *FFT-based Processing* | 58 |
| 3.5 | Chapter Summary | | 58 |
| 3.6 | Problems | | 59 |
| | References | | 60 |
| **4** | **Statistics and Random Processes** | | **63** |
| 4.1 | Introduction to the Use of Statistics | | 63 |
| | *4.1.1* | *Ensemble and Time Averages* | 64 |
| | *4.1.2* | *Stationarity and Ergodicity* | 64 |
| 4.2 | Random Theory | | 65 |
| | *4.2.1* | *Expected Value* | 65 |
| | *4.2.2* | *Errors in Estimates* | 65 |
| | *4.2.3* | *Probability Distribution* | 66 |
| | *4.2.4* | *Probability Density* | 66 |
| | *4.2.5* | *Histogram* | 67 |
| | *4.2.6* | *Sample Probability Density Estimate* | 68 |
| | *4.2.7* | *Average Value and Variance* | 68 |
| | *4.2.8* | *Central Moments* | 70 |
| | *4.2.9* | *Skewness* | 70 |
| | *4.2.10* | *Kurtosis* | 70 |
| | *4.2.11* | *Crest Factor* | 71 |
| | *4.2.12* | *Correlation Functions* | 71 |
| | *4.2.13* | *The Gaussian Probability Distribution* | 72 |
| 4.3 | Statistical Methods | | 74 |
| | *4.3.1* | *Hypothesis Tests* | 74 |
| | *4.3.2* | *Test of Normality* | 77 |
| | *4.3.3* | *Test of Stationarity* | 77 |

4.4 Quality Assessment of Measured Signals 81
4.5 Chapter Summary 84
4.6 Problems 85
References 86

**5 Fundamental Mechanics 87**
5.1 Newton's Laws 87
5.2 The Single Degree-of-freedom System (SDOF) 88
*5.2.1 The Transfer Function* 88
*5.2.2 The Impulse Response* 89
*5.2.3 The Frequency Response* 91
*5.2.4 The Q-factor* 94
*5.2.5 SDOF Forced Response* 95
5.3 Alternative Quantities for Describing Motion 95
5.4 Frequency Response Plot Formats 97
*5.4.1 Magnitude and Phase* 97
*5.4.2 Real and Imaginary Parts* 100
*5.4.3 The Nyquist Plot – Imaginary vs. Real Part* 100
5.5 Determining Natural Frequency and Damping 103
*5.5.1 Peak in the Magnitude of FRF* 103
*5.5.2 Peak in the Imaginary Part of FRF* 103
*5.5.3 Resonance Bandwidth (3 dB Bandwidth)* 104
*5.5.4 Circle in the Nyquist Plot* 104
5.6 Rotating Mass 104
5.7 Some Comments on Damping 106
*5.7.1 Hysteretic Damping* 106
5.8 Models Based on SDOF Approximations 107
*5.8.1 Vibration Isolation* 107
*5.8.2 Resonance Frequency and Stiffness Approximations* 110
5.9 The Two-degree-of-freedom System (2DOF) 110
5.10 The Tuned Damper 113
5.11 Chapter Summary 115
5.12 Problems 115
References 116

**6 Modal Analysis Theory 119**
6.1 Waves on a String 119
6.2 Matrix Formulations 120
*6.2.1 Degree-of-freedom* 121
6.3 Eigenvalues and Eigenvectors 122
*6.3.1 Undamped System* 122
*6.3.2 Mode Shape Orthogonality* 125
*6.3.3 Modal Coordinates* 127
*6.3.4 Proportional Damping* 128
*6.3.5 General Damping* 130
6.4 Frequency Response of MDOF Systems 133
*6.4.1 Frequency Response from* $[M]$, $[C]$, $[K]$ 133
*6.4.2 Frequency Response from Modal Parameters* 134
*6.4.3 Frequency Response from* $[M]$, $[K]$, *and* $\zeta$ *– Modal Damping* 138
*6.4.4 Mode Shape Scaling* 138

| | | |
|---|---|---|
| | *6.4.5 The Effect of Node Lines on FRFs* | 139 |
| | *6.4.6 Antiresonance* | 140 |
| | *6.4.7 Impulse Response of MDOF Systems* | 141 |
| 6.5 | Time Domain Simulation of Forced Response | 141 |
| 6.6 | Chapter Summary | 143 |
| 6.7 | Problems | 144 |
| | References | 145 |
| **7** | **Transducers for Noise and Vibration Analysis** | **147** |
| 7.1 | The Piezoelectric Effect | 147 |
| 7.2 | The Charge Amplifier | 148 |
| 7.3 | Transducers with Built-In Impedance Converters, 'IEPE' | 149 |
| | *7.3.1 Low-frequency Characteristics* | 150 |
| | *7.3.2 High-frequency Characteristics* | 151 |
| | *7.3.3 Transducer Electronic Data Sheet, TEDS* | 152 |
| 7.4 | The Piezoelectric Accelerometer | 152 |
| | *7.4.1 Frequency Characteristics* | 153 |
| | *7.4.2 Mounting Accelerometers* | 155 |
| | *7.4.3 Electrical Noise* | 155 |
| | *7.4.4 Choosing an Accelerometer* | 155 |
| 7.5 | The Piezoelectric Force Transducer | 157 |
| 7.6 | The Impedance Head | 158 |
| 7.7 | The Impulse Hammer | 159 |
| 7.8 | Accelerometer Calibration | 159 |
| 7.9 | Measurement Microphones | 161 |
| 7.10 | Microphone Calibration | 162 |
| 7.11 | Shakers for Structure Excitation | 162 |
| 7.12 | Some Comments on Measurement Procedures | 163 |
| 7.13 | Problems | 164 |
| | References | 165 |
| **8** | **Frequency Analysis Theory** | **167** |
| 8.1 | Periodic Signals – The Fourier Series | 167 |
| 8.2 | Spectra of Periodic Signals | 169 |
| | *8.2.1 Frequency and Time* | 170 |
| 8.3 | Random Processes | 170 |
| | *8.3.1 Spectra of Random Processes* | 171 |
| 8.4 | Transient Signals | 173 |
| 8.5 | Interpretation of spectra | 173 |
| 8.6 | Chapter Summary | 175 |
| 8.7 | Problems | 175 |
| | References | 176 |
| **9** | **Experimental Frequency Analysis** | **177** |
| 9.1 | Frequency Analysis Principles | 177 |
| | *9.1.1 Nonparametric Frequency Analysis* | 178 |
| 9.2 | Octave and Third-octave Band Spectra | 179 |
| | *9.2.1 Time Constants* | 179 |
| | *9.2.2 Real-time versus Serial Measurements* | 179 |

9.3 The Discrete Fourier Transform (DFT) 180
*9.3.1 The Fast Fourier Transform, FFT* 181
*9.3.2 The DFT in Short* 182
*9.3.3 The Basis of the DFT* 183
*9.3.4 Periodicity of the DFT* 183
*9.3.5 Properties of the DFT* 186
*9.3.6 Relation between DFT and Continuous Spectrum* 186
*9.3.7 Leakage* 187
*9.3.8 The Picket-fence Effect* 189
*9.3.9 Time Windows for Periodic Signals* 191
*9.3.10 Time Windows for Random Signals* 198
*9.3.11 Oversampling in FFT Analysis* 199
*9.3.12 Circular Convolution and Aliasing* 199
*9.3.13 Zero Padding* 200
*9.3.14 Zoom FFT* 201
9.4 Chapter Summary 202
9.5 Problems 203
References 204

**10 Spectrum and Correlation Estimates Using the DFT 205**
10.1 Averaging 205
10.2 Spectrum Estimators for Periodic Signals 206
*10.2.1 The Autopower Spectrum* 207
*10.2.2 Linear Spectrum* 208
*10.2.3 Phase Spectrum* 208
10.3 Estimators for PSD and CSD 209
*10.3.1 The Periodogram* 209
*10.3.2 Welch's Method* 211
*10.3.3 Window Correction for Welch Estimates* 211
*10.3.4 Bias Error in Welch Estimates* 212
*10.3.5 Random Error in Welch Estimates* 217
*10.3.6 The Smoothed Periodogram Estimator* 221
*10.3.7 Bias Error in Smoothed Periodogram Estimates* 223
*10.3.8 Random Error in Smoothed Periodogram Estimates* 224
10.4 Estimator for Correlation Functions 224
10.5 Estimators for Transient Signals 226
*10.5.1 Windows for Transient Signals* 227
10.6 Spectrum Estimation in Practice 228
*10.6.1 Linear Spectrum Versus PSD* 228
*10.6.2 Example of a Spectrum of a Periodic Signal* 229
*10.6.3 Practical PSD Estimation* 231
*10.6.4 Spectrum of Mixed Property Signal* 233
*10.6.5 Calculating RMS Values in Practice* 234
*10.6.6 RMS From Linear Spectrum of Periodic Signal* 234
*10.6.7 RMS from PSD* 236
*10.6.8 Weighted RMS Values* 236
*10.6.9 Integration and Differentiation in the Frequency Domain* 238
10.7 Multi-channel Spectral Analysis 238
*10.7.1 Matrix Notation for MIMO Spectral Analysis* 239
*10.7.2 Arranging Spectral Matrices in MATLAB/Octave* 240

10.8 Chapter Summary 240
10.9 Problems 241
References 242

**11 Measurement and Analysis Systems 245**
11.1 Principal Design 246
11.2 Hardware for Noise and Vibration Analysis 246
*11.2.1 Signal Conditioning* 247
*11.2.2 Analog-to-digital Conversion, ADC* 247
*11.2.3 Practical Issues* 253
*11.2.4 Hardware Specifications* 255
*11.2.5 Transient (Shock) Recording* 257
11.3 FFT Analysis Software 257
*11.3.1 Block Processing* 258
*11.3.2 Data Scaling* 259
*11.3.3 Triggering* 259
*11.3.4 Averaging* 260
*11.3.5 FFT Setup Parameters* 261
11.4 Chapter Summary 261
11.5 Problems 261
References 262

**12 Rotating Machinery Analysis 263**
12.1 Vibrations in Rotating Machines 263
12.2 Understanding Time–Frequency Analysis 264
12.3 Rotational Speed Signals (Tachometer Signals) 265
12.4 RPM Maps 267
*12.4.1 The Waterfall Plot* 268
*12.4.2 The Color Map Plot* 268
12.5 Smearing 269
12.6 Order Tracks 272
12.7 Synchronous Sampling 272
*12.7.1 DFT Parameters after Resampling* 276
12.8 Averaging Rotation-speed-dependent Signals 276
12.9 Adding Change in RMS with Time 277
12.10 Parametric Methods 281
12.11 Chapter Summary 282
12.12 Problems 282
References 283

**13 Single-input Frequency Response Measurements 285**
13.1 Linear Systems 286
13.2 Determining Frequency Response Experimentally 286
*13.2.1 Method 1 – the $H_1$ Estimator* 286
*13.2.2 Method 2 – the $H_2$ Estimator* 288
*13.2.3 Method 3 – the $H_c$ Estimator* 289
13.3 Important Relationships for Linear Systems 290
13.4 The Coherence Function 291
13.5 Errors in Determining the Frequency Response 291
*13.5.1 Bias Error in FRF Estimates* 292

| | | | |
|---|---|---|---|
| | *13.5.2* | *Random Error in FRF Estimates* | 293 |
| | *13.5.3* | *Bias and Random Error Trade-offs* | 295 |
| 13.6 | Coherent Output Power | | 295 |
| 13.7 | The Coherence Function in Practice | | 296 |
| | *13.7.1* | *Non-random Excitation* | 297 |
| 13.8 | Impact Excitation | | 297 |
| | *13.8.1* | *The Force Signal* | 298 |
| | *13.8.2* | *The Response Signal and Exponential Window* | 300 |
| | *13.8.3* | *Impact Testing Software* | 300 |
| | *13.8.4* | *Compensating for the Influence of the Exponential Window* | 303 |
| | *13.8.5* | *Sources of Error* | 305 |
| | *13.8.6* | *Improving Impact Testing by Alternative Processing* | 306 |
| 13.9 | Shaker Excitation | | 306 |
| | *13.9.1* | *Signal-to-noise Ratio Comparison* | 307 |
| | *13.9.2* | *Pure Random Noise* | 308 |
| | *13.9.3* | *Burst Random Noise* | 310 |
| | *13.9.4* | *Pseudo-random Noise* | 310 |
| | *13.9.5* | *Periodic Chirp* | 311 |
| | *13.9.6* | *Stepped-sine Excitation* | 311 |
| 13.10 | Examples of FRF Estimation – No Extraneous Noise | | 312 |
| | *13.10.1* | *Pure Random Excitation* | 312 |
| | *13.10.2* | *Burst Random Excitation* | 312 |
| | *13.10.3* | *Periodic Excitation* | 314 |
| 13.11 | Example of FRF Estimation – with Output Noise | | 315 |
| 13.12 | Examples of FRF Estimation – with Input and Output Noise | | 316 |
| | *13.12.1* | *Sources of Error during Shaker Excitation* | 318 |
| | *13.12.2* | *Checking the Shaker Attachment* | 318 |
| | *13.12.3* | *Other Sources of Error* | 319 |
| 13.13 | Chapter Summary | | 319 |
| 13.14 | Problems | | 321 |
| | References | | 321 |
| | | | |
| **14** | **Multiple-input Frequency Response Measurement** | | **323** |
| 14.1 | Multiple-input Systems | | 323 |
| | *14.1.1* | *The 2-input/1-output System* | 324 |
| | *14.1.2* | *The 2-input/1-output System – matrix notation* | 325 |
| | *14.1.3* | *The $H_1$ Estimator for MIMO* | 326 |
| | *14.1.4* | *Multiple Coherence* | 327 |
| | *14.1.5* | *Computation Considerations for Multiple-input System* | 329 |
| | *14.1.6* | *The $H_v$ Estimator* | 329 |
| | *14.1.7* | *Other MIMO FRF Estimators* | 330 |
| 14.2 | Conditioned Input Signals | | 331 |
| | *14.2.1* | *Conditioned Output Signals* | 333 |
| | *14.2.2* | *Partial Coherence* | 333 |
| | *14.2.3* | *Ordering Signals Prior to Conditioning* | 334 |
| | *14.2.4* | *Partial Coherent Output Power Spectra* | 334 |
| | *14.2.5* | *Backtracking the H-systems* | 335 |
| | *14.2.6* | *General Conditioned Systems* | 336 |
| 14.3 | Bias and Random Errors for Multiple-input Systems | | 336 |

14.4 Excitation Signals for MIMO Analysis 337
*14.4.1 Pure Random Noise* 337
*14.4.2 Burst Random Noise* 338
*14.4.3 Periodic Random Noise* 338
*14.4.4 The Multiphase Stepped-sine Method (MPSS)* 338
14.5 Data Synthesis and Simulation Examples 339
*14.5.1 Burst Random – Output Noise* 339
*14.5.2 Burst and Periodic Random – Input Noise* 342
*14.5.3 Periodic Random – Input and Output Noise* 342
14.6 Real MIMO Data Case 345
14.7 Chapter Summary 348
14.8 Problems 349
References 350

**15 Orthogonalization of Signals 351**
15.1 Principal Components 351
*15.1.1 Principal Components Used to Find Number of Sources* 353
*15.1.2 Principal Components Used for Data Reduction* 355
15.2 Virtual Signals 360
*15.2.1 Virtual Input Coherence* 361
*15.2.2 Virtual Input/Output Coherence* 364
*15.2.3 Virtual Coherent Output Power* 364
15.3 Noise Source Identification (NSI) 367
*15.3.1 Multiple Source Example* 367
*15.3.2 Automotive Example* 370
15.4 Chapter Summary 372
15.5 Problems 373
References 373

**16 Advanced Analysis Methods 375**
16.1 Shock Response Spectrum 375
16.2 The Hilbert Transform 378
*16.2.1 Computation of the Hilbert Transform* 379
*16.2.2 Envelope Detection by the Hilbert Transform* 379
*16.2.3 Relating Real and Imaginary Parts of Frequency Response Functions* 380
16.3 Cepstrum Analysis 384
*16.3.1 Power Cepstrum* 385
*16.3.2 Complex Cepstrum* 387
*16.3.3 Inverse Cepstrum* 387
16.4 The Envelope Spectrum 388
16.5 Creating Random Signals with Known Spectral Density 390
16.6 Operational Deflection Shapes – ODS 391
*16.6.1 Multiple-Reference ODS* 392
16.7 Introduction to Experimental Modal Analysis 393
*16.7.1 Main Steps in EMA* 393
*16.7.2 Data Checks* 394
*16.7.3 Mode Indicator Functions* 395
*16.7.4 The MAC Matrix* 397
*16.7.5 Modal Parameter Extraction* 398

16.8 Chapter Summary 399
16.9 Problems 400
References 400

**Appendix A Complex Numbers 403**

**Appendix B Logarithmic Diagrams 407**

**Appendix C Decibels 411**

**Appendix D Some Elementary Matrix Algebra 413**
Reference 415

**Appendix E Eigenvalues and the SVD 417**
E.1 Eigenvalues and Complex Matrices 417
E.2 The Singular Value Decomposition (SVD) 418
Reference 419

**Appendix F Organizations and Resources 421**

**Bibliography 423**

**Index 429**

# About the Author

Anders Brandt has more than twenty years experience as a consultant and short-course instructor in experimental vibration analysis. During his entire career, he has worked on providing increased understanding of the measurement and analysis procedures used in experimental vibration analysis. Currently, Anders Brandt is an Associate Professor of Experimental Dynamics and Signal Processing at the University of Southern Denmark, where his main research interests are in applied signal processing and operational modal analysis. Anders is a popular short-course instructor and lecturer on the topics covered by this book.

# Preface

The material in this book has been developing in my mind for more than twenty years of teaching. During these years I have been teaching over 200 shortcourses for engineers in the industry on techniques for experimental noise and vibration analysis and also on how to use commercial measurement and analysis systems. In addition, in the late 1990s I developed and taught three master's level courses in experimental analysis of vibrations at Blekinge Institute of Technology in Sweden. Noise and vibration analysis is an interdisciplinary field, incorporating diverse subjects such as mechanical dynamics, sensor technology, statistics, and signal processing. Whereas there are many excellent and comprehensive books in each of these disciplines, there has been a lack of introductory material for the engineering student who first starts to make noise and/or vibration measurements, or the engineer who needs a reference in his or her daily life. In addition, there are few textbooks in this field presenting the techniques as they are actually used in practice. This book is an attempt to fill this void.

My aim for this book is for it to serve both as a course book and as supplementary reading in university courses, as well as providing a handbook for engineers or researchers who measure and analyze acoustic or vibration signals. The level of the book makes it appropriate both for undergraduate and graduate levels, with a proper selection of the content. In addition the book should be a good reference for analysts who use experimental results and need to interpret them. To satisfy these rather different purposes, for some of the topics in the book I have included more detail than would be necessary for an introductory text. To facilitate its use as a handbook, I have also included a short summary at the end of each chapter where some of the key points of the chapter are repeated.

This book contains background theory explaining the majority of analysis methods used in modern commercial software for noise and vibration measurement and analysis, with one exception: experimental modal analysis is only briefly introduced, as this is a specialized field with some excellent textbooks already available. This book also includes a number of tools which are usually not found in commercial systems, but which are still useful for the practitioner. With modern computer-based software, it is easy to export data to, e.g., MATLAB/Octave (see below), and apply the techniques there.

Since it is an introductory text, most of the content of this book is of course available in more specialized textbooks and scientific papers. A few parts, however, include some improvements of existing techniques. I will mention these points in the descriptions of the appropriate chapters below.

Signal analysis is traditionally a field within electrical engineering, whereas most engineers and students pursuing noise and vibration measurements are mechanical or civil engineers. The aim has therefore been to make the material accessible, particularly to students and engineers of these latter disciplines. For this reason I have included introductions to the Laplace and Fourier transforms – both essential tools for understanding, analyzing and solving problems in dynamics. Electrical engineering students and practitioners should still find many of the topics in the book interesting.

Signal analysis is a subject which is best learned by practicing the theories (as, perhaps, all subjects are). I have therefore incorporated numerous examples using MATLAB or GNU Octave throughout the book. Further examples and an accompanying toolbox which can be used with either MATLAB or GNU

Octave can be downloaded from Internet. More information about this is located in Section 1.6. I strongly recommend the use of these tools as a complement to reading this book, regardless of whether you are a student, a researcher or an industry practitioner.

Chapter 2 introduces dynamic signals and systems with the aim of being an introduction particularly for mechanical and civil engineering students. In this chapter the classification of signals into periodic, random and transient signals is introduced. The chapter also includes linear system theory and a comprehensive introduction to the Laplace and Fourier transforms, both important tools for understanding and analyzing dynamic systems.

In Chapter 3 some fundamental concepts of sampled signals are presented. Starting with the sampling theorem and continuing with digital filter theory, this chapter presents some important applications of digital filters for fractional octave analysis and for integrating and differentiating measured signals.

Chapter 4 introduces some applied statistics and random process theory from a practical perspective. It includes an introduction to hypothesis testing as this tool is sometimes used for testing normality and stationarity of data. This chapter also gives an introduction to the application of statistics for data quality assessment, which is becoming more important with the large amounts of data collected in many applications of noise and vibration analysis.

Chapters 5 and 6 provide an introduction to the theory of mechanical vibrations. I anticipate that the contents of these two chapters will already be known to many readers, but I have found it important to include them because my presentation focuses on the experimental implications of the theory, unlike the presentation in most mechanical vibration textbooks, and because some later chapters in the book need a foundation with a common nomenclature. Chapter 6 also includes an accurate and fast method for computing forced response of linear systems in the time domain which is very attractive, e.g., to produce known experimental signals for testing out signal analysis procedures. This method, developed by Professor Kjell Ahlin, has been presented at conferences, but deserves better dissemination.

In Chapter 7 the most important transducers used for measurements of noise and vibration signals are presented; specifically the accelerometer, the force sensor and the microphone. Because piezoelectric sensors with built-in signal conditioning (so-called IEPE sensors) are widely used today, this technology is presented in some depth. In this chapter I also present some personal ideas on how to become a good experimentalist.

The analysis techniques mostly used in this field are based on the Discrete Fourier Transform (DFT), computed by the FFT. Spectrum analysis is therefore an important part of this book and Chapters 8 through 10 are spent on this topic. Chapter 8 introduces basic frequency analysis theory by presenting the different signal classes, and the different spectra used to describe the frequency content of these signals.

In Chapter 9 the DFT and some other techniques used to experimentally determine the frequency content of signals are presented. The properties of the DFT, which are very important to understand when interpreting experimental frequency spectra, are presented relatively comprehensively.

Chapter 10 includes a comprehensive presentation of how spectra from periodic, random and transient signals, and mixes of these signal classes, should be estimated in practice. Also, I mention a convenient technique for removing harmonics in spectral density estimates using the smoothed periodogram method; which, to my knowledge, has never been presented before. Chapter 10 also includes a comprehensive explanation of Welch's method for PSD estimation, including overlap processing, as this is the method used in virtually all commercial software. The treatment of practical spectral analysis in this chapter should also be of use to engineers outside the field of acoustics and vibrations who want to calculate and/or interpret spectra by using the FFT.

In Chapter 11 the design of modern data acquisition and measurement systems is described from a user perspective. In this chapter both hardware and software issues are penetrated.

Chapter 12 addresses order tracking, which is a common technique for analysis of rotating machinery equipment. The chapter describes the most common techniques used to measure such signals both with fixed sampling frequency and with synchronous sampling.

Frequency response functions are important measurement functions in experimental noise and vibration analysis and are used, for example, in experimental modal analysis. Chapter 13 therefore covers techniques for measuring frequency responses for single-input/single-output (SISO) systems. Both impact excitation and shaker excitation techniques are presented in detail. In Chapter 14 the techniques are extended to multiple-input/multiple-output (MIMO) systems. In Chapters 13 and 14 I also present a technique which has not, to the best of my knowledge, been presented before. Using well-known periodic excitation signals, I show that the bias error in frequency response estimates with extraneous noise present in both input and output signals can be eliminated by time domain averaging, for single-input as well as multiple-input systems.

Chapter 15 presents some relatively advanced techniques used for multichannel analysis, namely principal components and virtual signals. These techniques are commonly used for noise path analysis and noise source identification in many of the sophisticated software packages available commercially. I present these concepts in some depth, since they are not readily available in other textbooks.

In Chapter 16 I have collected a number of more advanced techniques that engineers in this field should be acquainted with. This chapter presents, in order, the shock response spectrum, the Hilbert transform with applications, the cepstrum and envelope spectrum, how to produce Gaussian time signals with known spectral density, and finally two very important tools: operational deflection shapes, and experimental modal analysis. The latter is a comprehensive technique and only briefly introduced.

In the Appendix section I have included some fundamentals on complex numbers, logarithmic diagrams and the decibel unit, matrix theory, and eigenvalues and the singular value decomposition. The reader who does not feel confident with some of these concepts will hopefully find enough theory in these appendices to follow the text in this book. The last appendix contains some references to good sources for more information within the noise and vibration community. I hope the newcomer to this field can benefit from this list.

# Acknowledgements

This book is inspired partly by class notes I wrote for two classes at Blekinge Institute of Technology, BTH. I am especially grateful to Professor Ingvar Claesson and the Department of Signal Processing at BTH for supporting me in writing these early texts. Also, Timothy Samuels did a great job translating an early manuscript from Swedish to English.

My most sincere appreciation goes to Professor Kjell Ahlin, my colleague and friend for many years. Our many long discussions have strongly contributed to my understanding of this subject and I am grateful for the data provided by Professor Ahlin for examples in Chapter 16.

Dr Per-Olof Sturesson and the noise and vibration group at SAAB Automobile AB have been invaluable resources of feedback and have provided data for Chapters 12 and 15. For this, and many ideas and discussions, I am very grateful. Special thanks also goes to Mats Berggren.

My thanks extend to Professor Jiri Tuma for supporting me with data for Chapter 12 and for kind support through times.

Svend Gade and Brüel and Kjær A/S are acknowledged, along with Niels Thrane, for allowing me to reuse an illustration and an overview description of the Discrete Fourier Transform from an old B & K Technical Review, which I find is of great value for presenting the DFT.

I have always found the many participants at the International Modal Analysis Conference (IMAC), organized by Society for Experimental Mechanics (SEM), an invaluable source of inspiration and knowledge. Special thanks to Tom Proulx, Al Wicks, Dave Brown, and Randall Allemang for their outstanding support and encouragement and continuous willingness to give from their wealth of knowledge.

This book would not be what it is without the professional staff at Wiley, who have been of great help throughout the work. My thanks extend particularly to Debbie Cox and Nicky Skinner who have both been of great help.

Particularly I also wish to thank Dr Julius S. Bendat, Professor Rune Brincker, Knut Bertelsen (in memoriam), and Professor Bo Håkansson for their willingness to always share their knowledge and for inspiring me, to Claus Vaarning and Soma Tayamon for reading parts of the manuscript and offering many good comments, and to all the professional people I have had the opportunity of learning from during my career.

Finally I am, of course, thankful to a great number of people who have inspired and supported me, and to all my students and short-course participants over the years who have taught me so much. And to my family for having endured a long time without seeing very much of me.

# List of Abbreviations

| | |
|---|---|
| 2DOF | Two degrees-of-freedom system |
| AC | Alternating current |
| ADC | Analog-to-digital converter |
| BT | Bandwidth-time (product) |
| CSD | Cross-spectral density function |
| DAC | Digital-to-analog converter |
| DC | Direct current |
| DFT | Discrete Fourier transform |
| DOF | Degree-of-freedom (point and direction) |
| ESD | Energy spectral density |
| FE | Finite element |
| FEM | Finite element method |
| FFT | Fast Fourier transform |
| FIR | Finite impulse response (filter) |
| FRF | Frequency response function |
| HF | High frequency |
| HP | Highpass |
| IDFT | Inverse discrete Fourier transform |
| IEPE | Integrated electronics piezoelectric (sensor) |
| IFFT | Inverse fast Fourier transform |
| IIR | Infinite impulse response (filter) |
| IRF | Impulse response function |
| LF | Low frequency |
| ISO | International standardization organization |
| MDOF | Multiple degrees-of-freedom |
| MIF | Mode indicator function |
| MIMO | Multiple-input/multiple-output |
| MISO | Multiple-input/single-output |
| MPSS | Multi-phase stepped sine |
| MrMIF | Modified real mode indicator function |
| MvMIF | Multivariate mode indicator function |
| NSI | Noise source identification |
| NSR | Noise-to-signal ratio |
| ODS | Operating deflection shape |
| PDF | Probability density function |
| PSD | Power spectral density |
| RMS | Root mean square |

| | |
|---|---|
| RPM | Revolutions per minute |
| SDOF | Single degree-of-freedom |
| SIMO | Single-input/multiple-output |
| SISO | Single-input/single-output |
| SNR | Signal-to-noise ratio |
| SRS | Shock response spectrum |
| SVD | Singular value decomposition |
| TEDS | Transducer electronic data sheet |

# Notation

$< x >$ Average of $x$  
$\mathcal{F}[\,]$ Fourier transform of [ ]  
$\mathcal{H}[\,]$ Hilbert transform of [ ]  
$\mathcal{L}[\,]$ Laplace transform of [ ]  
$\mathrm{E}[\,]$ Expected value  
$a, a(t)$ Vibration acceleration  
$A_{pqr}$ Residue of mode $r$, between points $p$ and $q$  
$A_{xx}$ Autopower spectrum of $x$  
$B$ Bandwidth in [Hz]  
$B_e$ Equivalent (statistical) bandwidth in Hz  
$B_{en}$ Normalized equivalent bandwidth (dimensionless)  
$B_r$ Resonance bandwidth in Hz  
$c_p$ Power cepstrum  
$c_r$ Modal (viscous) damping of mode $r$  
$\delta(t)$ Dirac's unit impulse  
$\Delta f$ Frequency increment of discrete Fourier transform  
$\Delta t$ Time increment in [s]  
$\varepsilon$ Normalized error  
$f$ Frequency in [Hz]  
$f_n, f_r$ Undamped natural frequency  
$g^2(f)$ Virtual coherence function  
$\gamma_{yx}^2$ Coherence function between $x$ (input) and $y$ (output)  
$\gamma_{y:x}^2$ Multiple coherence of $y$ (output) with all $x_q$ (inputs)  
$G_{xx}(f)$ Single-sided autospectral density of $x$  
$G'_{xx}$ Principal component  
$[G_{xx}]$ Single-sided input cross-spectral matrix  
$G_{yx}(f)$ Single-sided cross-spectral density between $x$ (input) and $y$ (output)  
$\left[G_{yx}\right]$ Single-sided input/output cross-spectral matrix  
$h(n)$ Discrete impulse response  
$h(t)$ Analog impulse response  
$H(f)$ Analog frequency response function  
$H(k)$ Discrete frequency response function  
$H(s)$ Transfer function  
$\mathrm{Im}[\,]$ Imaginary part of [ ]  
$\mathrm{j}$ Imaginary number, $\sqrt{-1}$  
$k$ Discrete (dimensionless) frequency variable  
$k_r$ Modal stiffness of mode $r$

| | |
|---|---|
| $K_x$ | Kurtosis of $x$ |
| $\lambda$ | Eigenvalue |
| $\mu_x$ | (Theoretical) mean value of $x$ |
| $m_r$ | Modal mass of mode $r$ |
| $M_n$ | $N$th statistical (central) moment |
| $n$ | Discrete (dimensionless) time variable |
| $\phi$ | Phase, general random variable |
| $p_x(x)$ | Probability density of $x$ |
| $P(x)$ | Probability distribution of $x$ |
| $\{\psi\}_r$ | Mode shape vector of mode $r$ |
| $[\Psi]_r$ | Mode shape matrix of mode $r$ |
| $Q$ | Quality factor ($Q$-factor) |
| $Q_r$ | Modal scale constant of mode $r$ |
| $R_{xx}(\tau)$ | Autocorrelation of $x$ |
| $R_{yx}(\tau)$ | Cross-correlation between $x$ (input) and $y$ (output) |
| Re[ ] | Real part of [ ] |
| $s$ | Laplace operator (in [rad/s]) |
| $s_r$ | Pole, root to characteristic polynomial |
| $\sigma_x$ | Standard deviation of $x$ |
| $S_x$ | Skewness of $x$ |
| $S_{xx}(f)$ | Double-sided autospectral density of $x$ |
| $S_{yx}(f)$ | Double-sided cross-spectral density between $x$ (input) and $y$ (output) |
| $[G_{yx}]$ | Single-sided input/output cross-spectral matrix |
| $t$ | Analog time |
| $T$ | Measurement time |
| $\tau$ | Time delay, time lag variable for correlation functions |
| $T_x(k)$ | Discrete transient spectrum of $x$ |
| $u, u(t)$ | Vibration displacement |
| $v, v(t)$ | Vibration velocity |
| $w(n)$ | Discrete time window |
| $x(n)$ | Discrete/sampled (input) signal |
| $x(t)$ | Analog (input) signal |
| $\tilde{x}(t)$ | Hilbert transform of $x(t)$ |
| $X(f)$ | (Continuous) Fourier transform of $x(t)$ |
| $X'$ | Spectrum of virtual signal |
| $X(k)$ | Discrete Fourier transform of $x(n)$ |
| $X_L(k)$ | Linear (RMS) spectrum of $x(n)$ |
| $y(n)$ | Discrete/sampled (output) signal |
| $y(t)$ | Analog (output) signal |
| $\omega$ | Angular frequency in [radians/s] |
| $\zeta_r$ | Relative (viscous) damping |

# 1

# Introduction

This chapter provides a short introduction to the field of noise and vibration analysis. Its main objective is to show new students in this field the wide range of applications and engineering fields where noise and vibration issues are of interest. If you are a researcher or an engineer who wants to use this book as a reference source, you may want to skim this chapter. If you decide to do so, I would recommend you to read Section 1.6, in which I present some personal ideas on how to use this book, as well as on how to go about becoming a good experimentalist – the ultimate goal after reading this book.

I want to show you not only the width of disciplines where noise and vibrations are found. I also want to show you that noise and vibration *analysis*, the particular topic of this book, is truly a fascinating and challenging discipline. One of the reasons I personally find noise and vibration analysis so fascinating is the interdisciplinary character of this field. Because of this interdisciplinary character, becoming an expert in this area is indeed a real challenge, regardless of which engineering field you come from. If you are a student just entering this field, I can only congratulate you for selecting (which I hope you do!) this field as yours for a lifetime. You will find that you will never cease learning, and that every day offers new challenges.

## 1.1 Noise and Vibration

Noise and vibration are constantly present in our high-tech society. Noise causes serious problems both at home and in the workplace, and the task of reducing community noise is a subject currently focused on by authorities in many countries. Similarly, manufacturers of mechanical products with vibrations causing acoustic noise, more and more find themselves forced to compete on the noise levels of their products. Such competition has so far occurred predominantly in the automotive industry, where the issues with sound and noise have long attracted attention, but, at least in Europe, e.g., domestic appliances are increasingly marketed stressing low noise levels.

Let us list some examples of reasons why vibration is of interest.

- Vibration can cause *injuries and disease* in humans, with 'white fingers' due to long-term exposure to vibration, and back injuries due to severe shocks, as examples.
- Vibration can cause *discomfort*, such as sickness feelings in high-rise buildings during storms, or in trains or other vehicles, if vibration control is not successful.
- Vibration can cause *fatigue*, i.e., products break after being submitted to vibrations for a long (or sometimes not so long) time.

*Noise and Vibration Analysis: Signal Analysis and Experimental Procedures*, First Edition. Anders Brandt.

- Vibration can cause *dysfunction* in both humans and things we manufacture, such as bad vision if the eye is subjected to vibration, or a radar on a ship performing poorly due to vibration of the radar antenna.
- Vibration can be used for cleaning, etc.
- Vibration can cause noise, i.e., unpleasant sound, which causes annoyance as well as disease and discomfort.

To follow up on the last point in the list above, once noise is created by vibrations, noise is of interest, e.g., for the following reasons.

- Excessive noise can cause hearing impairment.
- Noise can cause discomfort.
- Noise can (probably) cause disease, such as increased risk of cardiac disease, and stress.
- Noise can be used for burglar alarms and in weapons (by disabling human ability to concentrate or to cope with the situation).

The lists above are examples, meant to show that vibrations and noise are indeed interesting for a wide variety of reasons, not only to protect ourselves and our products, but also because vibration can cause good things.

Besides simply reducing sound levels, much work is currently being carried out within many application areas concerning the concept of *sound quality*. This concept involves making a psychoacoustic judgment of how a particular sound is experienced by a human being. Harley Davidson is an often-cited example of a company that considers the sound from its product so important that it tried to protect that sound by trademark, although the application was eventually withdrawn.

Besides generating noise, vibrations can cause mechanical fatigue. Now and then we read in the newspaper that a car manufacturer is forced to recall thousands of cars in order to exchange a component. In those cases it is sometimes mechanical fatigue that has occurred, resulting in cracks initiating after the car has being driven a long distance. When these cracks grow they can cause component breakdown and, as a consequence, accidents.

## 1.2 Noise and Vibration Analysis

This book is about the *analysis methods* for analyzing noise and vibrations, rather than the mechanisms causing them. In order to identify the sources of vibrations and noise, extensive analysis of measured signals from different tests is often necessary. The measurement techniques used to carry out such analyses are well developed, and in universities as well as in industry, advanced equipment is often used to investigate noise and vibration signals in laboratory and in field environments.

The area of experimental noise and vibration analysis is an intriguing field, as I hope this book will reveal. It is so partly because this field is multidisciplinary, and partly because dynamics (including vibrations) is a complicated field where the most surprising things can happen. Using measurement and analysis equipment often requires a good understanding of mechanics, sensor technology, electronic measurement techniques, and signal analysis.

Vibrations and noise are found in many disciplines in the academic arena. Perhaps we first think of mechanics, with engines, vehicles, and pumps, etc. However, vibrations are also found also in civil engineering, in bridges, buildings, etc. Many of the measurement instruments and sensors we use in the field of analyzing vibrations and noise are, of course, electrical, and so the field of electrical engineering is heavily involved. This makes the initial study of noise and vibration analysis difficult, perhaps, because you are forced to get into some of the other fields of academia. Hopefully, this book can help bridge some of the gaps between disciplines.

If many academic disciplines are involved with noise and vibrations, the variety in industry is perhaps even more overwhelming. Noise and vibration are important in, for example, military, automotive, and aerospace industries, in power plants, home appliances, industrial production, hand-held tools, robotics, the medical field, electronics production, bridges and roads, etc.

## 1.3 Application Areas

As evident from the first sections of this chapter, noise and vibration are important for many reasons, and in many different disciplines. Within the field of noise and vibration, there are also many different, more specialized, disciplines. We need to describe some of these a little more.

*Structural dynamics* is a field which describes phenomena such as resonance in structures, how connecting structures together affect the resonances, etc. Often, vibration problems occur because, as you probably already know, resonances amplify vibrations – sometimes to very high levels.

*Environmental engineering* is a field in which environmental effects (not to be confused with the 'green environment') from such diverse phenomena as heat, corrosion, and vibration, etc., are studied. As far as vibrations are concerned, *vibration testing* is a large industrial discipline within environmental engineering. This field is concerned with a particular product's ability to sustain the vibration environment it will encounter during its lifetime. Sensitive products such as mobile phones and other electronic products are usually tested in a laboratory to ensure they can sustain the vibrations they will be exposed to during their lifetime. Producing standardized tests which are equivalent to the product's real-life vibration environment, is often a great challenge. Transportation testing of packaging is a closely related field, in which the interest is that, for example, the new video camera you buy arrives in one piece when you unpack the box, even if the ship that delivered it encountered a storm at sea.

*Fatigue analysis* is a field closely related to environmental engineering. However, the discipline of fatigue analysis is usually more involved with measuring the stresses on a product and, through mathematical models such as Wöhler curves etc., trying to predict the lifetime of the product, e.g., before fatigue cracks will appear. From the perspective of experiments, this practically means it is more common to measure with strain gauges rather than accelerometers.

*Vibration monitoring* is another field, where the aim is to try to predict when machines and pumps, for example, will fail, by studying (among many things) the vibration levels during their lifetime. In civil engineering, a somewhat related field, *structural health monitoring* attempts to assess the health of buildings and bridges after earthquakes as well as after aging and other deteriorating effects on the structure, based on measurements of (among many things) vibrations in the structures.

*Acoustics* is a discipline close to noise and vibration analysis, of course, as the cause of acoustic noise is often vibrations (but sometimes not, such as, for example, when turbulent air is causing the noise).

## 1.4 Analysis of Noise and Vibrations

There are several ways of analyzing noise and vibrations. We shall start with a brief discussion of some of the methods which this book is *not* aimed at, but which are crucial for the total picture of noise and vibration analysis, and which is often the reason for making experimental measurements.

*Analytical analysis* of vibrations is most commonly done using the *finite element method*, FEM, through normal mode analysis, etc. In order to successfully model vibrations, usually models with much greater detail (finer grid meshes, correctly selected element types, etc.) need to be used, compared with the models sufficient for static analysis. Also, dynamic analysis using FEM requires good knowledge of boundary conditions etc. For many of these inputs to the FEM software, experiments can help refine the model. This is a main cause of much experimental analysis of vibrations today.

For *acoustic analysis*, acoustic FEM can be used as long as the noise (or sound) is contained in a cavity. For radiation problems, the *boundary element method*, BEM, is increasingly used. With this method, known vibration patterns, for example from a FEM analysis, can be used to model how the sound radiates and builds up an acoustic field.

FEM and BEM are usually restricted to low frequencies, where the mode density is low. For higher frequencies, *statistical energy analysis*, SEA, can be used. As the name implies, this method deals with the mode density in a statistical manner, and is used to compute average effects.

### *1.4.1 Experimental Analysis*

In many cases it is necessary to measure vibrations or sound pressure, etc., to solve vibration problems, because the complexity of such problems often make them impossible to foresee through analytical models such as FEM. This is often referred to as trouble-shooting. Another important reason to measure and analyze vibrations is to provide input data to refine analytical models. Particularly, damping is an entity which is usually impossible to estimate through models – it needs to be assessed by experiment.

Experimental analysis of noise and vibrations is usually done by measuring accelerations or sound pressures, although other entities can be measured, as we will see in Chapter 7. In order to analyze vibrations, the most common method is by *frequency analysis*, which is due to the nature of linear systems, as we will discuss in Chapter 2. Frequency analysis is a part of the discipline of *signal analysis*, which also incorporates filtering signals, etc. The main tool for frequency analysis is the FFT (fast Fourier transform) which is today readily available through software such as MATLAB and Octave (see Section 1.6), or by the many dedicated commercial systems for noise and vibration analysis. Methods using the FFT will take up the main part of this book.

Some of the analysis necessary to solve many noise and vibration problems needs to be done in the time domain. Examples of such analysis is fatigue analysis, which incorporates, e.g., cycle counting, and data quality analysis, to assess the quality of measured signals. For a long time, the tools for noise and vibration analysis were focused on frequency analysis, partly due to the limited computer performance and cost of memory. Today, however, sophisticated time domain analysis can be performed at a low cost, and we will present many such techniques in Chapters 3 and 4.

## 1.5 Standards

Due to the complexity of many noise and vibration measurements, international standards form an important part of vibration measurements as well as of acoustics and noise measurements. Acoustics and vibration standards are published by the main standardization organizations, ISO (International Standardization Organization), IEC (International Electrical Committee), and, in the U.S., by ANSI (American National Standards Institute). The general recommendation from many acoustics and vibration experts is that, if there is a standard for your particular application – use it. It is outside the scope of this book, and practically impossible, to summarize all the standards available. Some of the many standards for signal analysis methods used in vibration analysis are, however, cited in this book.

## 1.6 Becoming a Noise and Vibration Analysis Expert

The main emphasis in this book is on the signal analysis methods and procedures used to solve noise and vibration problems. To be successful in this, it is necessary to become a good experimentalist. Unfortunately, this is not something which can be (at least solely) learned from a book, but I want to make some recommendations on how to enter a road which leads in the right direction.

### *1.6.1 The Virtue of Simulation*

As many of the theories of dynamics, as well as those of signal analysis, are very complex, a vital tool for understanding dynamic systems and analysis procedures, is to simulate simplified, isolated, cases, where the outcome can be understood without the complicating presence of disturbance noise, complexity of structures, non-ideal sensors, etc. I have therefore incorporated numerous examples in this book which use simulated measurement data with known properties. A practical method to create such signals is presented in Section 6.5. The importance of this cannot be overrated. Before making a measurement of noise or vibrations, it is crucial to know what a correct measurement signal should look like, for example. The hidden pitfalls in, particularly, vibration measurements are overwhelming for the beginner (and sometimes even for more experienced engineers). The road to successful vibration measurements therefore goes through careful, thought-through simulations.

Another important aspect of good experiments, is to make constant checks of the equipment. In Section 7.12 I present some ideas of things to check for in vibration measurements. In Section 7.8 I also present a by no means new technique, but nevertheless a simple and efficient one (mass calibration, if you already know it) to verify that accelerometers are working correctly. These devices are, like many sensors, sensitive and can easily break, and unfortunately, they often break in such a way that it can be hard to discover without a proper procedure to verify the sensors on a known signal. Single-frequency calibration, which is common for absolute calibration of accelerometers, usually completely fails to discover the faults present after an accelerometer has been dropped on a hard floor.

Having written this, I want to stress that good vibration measurements are performed every day in industry and universities. So, the intention is, of course, not to discourage you from this discipline, but simply to stress the importance of taking it slowly, and making sure every part of the experiment is under your control, and not under the control of the errors.

### *1.6.2 Learning Tools and the Format of this Book*

If you anticipated finding a book with numerous data examples from the field by which you would learn how to make the best vibration measurements, you will be disappointed by this book. The main reasons for this are twofold; (i) for the reasons just given in the preceding section, real vibration measurements are usually full of artifacts from disturbance noise, complicated structures, etc.; and (ii) each structure or machine or whatever is measured, has its own vibration profile, which makes 'typical examples' very narrow. If you work with cars, or airplanes, or sewing machines, or hydraulic pumps, or whatever, your vibration signals will look rather different from signals from those other products.

I have instead based most examples in this book on simplified simulations, where the key idea of discussion is easily seen. These examples will, hopefully, provide much deeper insights into the fundamental signal analysis ideas we discuss in each part of the book. They are also easily repeated on your own computer, which leads us to the next important point.

I believe that signal analysis (like, perhaps, all subjects) is far too mathematically complicated to understand through reading about it. Instead, I believe strongly in simulation, and application of the theories by your own hands. I have therefore throughout the book given numerous examples using the best tool I know of – MATLAB. This software is, in my opinion, the best available tool for signal analysis and therefore also for the vibration analysis methods we are concerned with in this book. If you do not already know MATLAB, you will soon learn by working through the examples.

The drawback of MATLAB may be that it is commercial software, and therefore costs money. If you find this to be an obstacle you cannot overcome, you can instead use GNU Octave, which is free software published under the GNU General Public License (GPL) and can be freely downloaded from http:///www.gnu.org/software/octave/. Octave is to a large extent compatible with MATLAB in the sense that MATLAB code, with some minor tweaks, can run under Octave. I have made sure that all examples in this book run under both MATLAB and Octave, so you are free to choose whichever of the two software tools.

In addition to the examples in this book, there will be a free accompanying toolbox for MATLAB/Octave made available by me to aid your learning. There will also be more examples than could fit this book. More information about this toolbox and examples for instructors, etc., can be found at the book website at www.wiley.com/go/brandt.

# 2

# Dynamic Signals and Systems

Vibration analysis, and indeed the field of mechanical dynamics in general, deals with *dynamic events*, i.e., for example forces and displacements which are *functions of time*. This chapter aims to introduce many of the concepts typical for dynamic systems, particularly for mechanical and civil engineering students who may have little theory at their disposal for understanding this subject. We will start with some rather simple signals, and later in this chapter introduce some important concepts and fundamental properties of dynamic signals and systems. This chapter also covers basic introductions to the Laplace and Fourier transforms – two very important mathematical tools to describe and understand dynamic signals and systems.

This chapter deals with *continuous signals*, as most of our understanding of engineering principles is based on the theory of continuous signals and differential calculus. In Chapter 3 we will introduce experimental signals, i.e., sampled signals as we find them in measurement systems. Before that, however, we need to have a general understanding of what characterizes dynamic signals and systems.

## 2.1 Introduction

In this book, we will call any physical entity that changes over time a *signal*, regardless of whether it is a measured signal, or an analytical (mathematical) 'signal'. Some examples of signals are thus

- the force acting on a car suspension (in a particular direction) as we drive the car on a road, or
- the sound pressure at the ear of an operator of some machine, or
- the displacement of a point (in a particular direction) on a vibrating handle on a hand-held machine such as a pneumatic drilling machine.

The analysis of (dynamic) signals is often called *signal analysis* or sometimes *signal processing*. I make the distinction, along with some, but not all authors, that *signal analysis* is the process of *extracting and interpreting* useful information in a signal, for example by a frequency spectrum, some statistical numbers, etc. By *signal processing*, on the other hand, I mean the actual process (usually a mathematical algorithm or similar) used in processing a signal from one form to another form. With this distinction, signal analysis will often include some signal processing, but not the other way around. This book deals with the signal analysis procedures used to understand signals that describe mechanical vibrations and acoustic noise, and many of the methods we use throughout the book will include signal processing procedures. There are many excellent books that include a more in-depth coverage of the topics discussed

*Noise and Vibration Analysis: Signal Analysis and Experimental Procedures*, First Edition. Anders Brandt.

in this chapter, for example (Oppenheim *et al.* 1999; Proakis and Manolakis 2006) for general signal analysis, and (Haykin 2003) for systems analysis.

A *dynamic system* is a physical entity that has one or more outputs (responses), caused by one or more inputs, and where both input(s) and output(s) are dynamic, i.e., they change over time. In this book, the most common system will be a mechanical system, or sometimes a vibro-acoustic system. The former includes inputs in the form of forces and torques, and outputs in the form of some time derivative of motion, i.e., displacement, velocity, or acceleration. The latter is a combined system where the outputs, in addition to motion responses, can be acoustic (sound) pressure or some other acoustical entity. In a sense, a system can be thought of as a 'black box', with the inputs and outputs, and the relationships that relate the outputs to the inputs. The simplest system we will use is the mechanical single-degree-of-freedom, SDOF, system we will introduce in Chapter 5.

In terms of the frequency content of signals, we often separate signals into three different *signal classes*, namely

- periodic signals
  signals which repeat themselves with a period, $T_p$
- random signals (stochastic processes)
  signals which at each time instant are independent of values at other instants, and
- transient signals
  signals which have limited length, usually they die out after a certain time.

Determining to which such class a particular signal belongs is often called *signal classification*, a field particularly important when damaging effects of vibrations are of interest, such as in fatigue analysis and in environmental testing. We will describe some important fundamental properties of each of these classes in this chapter. Another way of classifying signals is into stationary and nonstationary signals, see Chapter 4, or into deterministic versus nondeterministic signals. A deterministic signal is a signal for which there is a closed form expression so that from a part of the signal, the entire signal for all times, past and present, can be expressed mathematically. Periodic signals and most transient signals belong to the class of deterministic signals whereas *random signals* (noise) belong to the other class, the nondeterministic signals, which cannot be described in the past or present based on a shorter observation, as their values are random at each instant in time. In practice, of course, we often encounter signals which are mixed combinations of the 'pure' signal classes described here, for example periodic signals with background noise. The interpretation of such signals can sometimes be difficult and will be discussed with respect to frequency analysis in Chapter 10.

As we will see in later chapters, random and transient signals have continuous spectral content, as opposed to periodic signals which have discrete spectra (with only some frequencies present). Because of this fundamental difference, we will introduce different types of spectral scaling in Chapter 8 for describing the different types of signals.

## 2.2 Periodic Signals

Periodic vibrations occur whenever we have repeating phenomena such as a reciprocating engine running at constant RPM or a rotating device such as a turbine, for example. The simplest periodic signal is the sine wave which we start the discussion with.

### *2.2.1 Sine Waves*

One of the most fundamental dynamical signals is the *sinusoid*, or sine wave, which has some very interesting properties that we will discuss in this and subsequent sections. A sine signal is defined

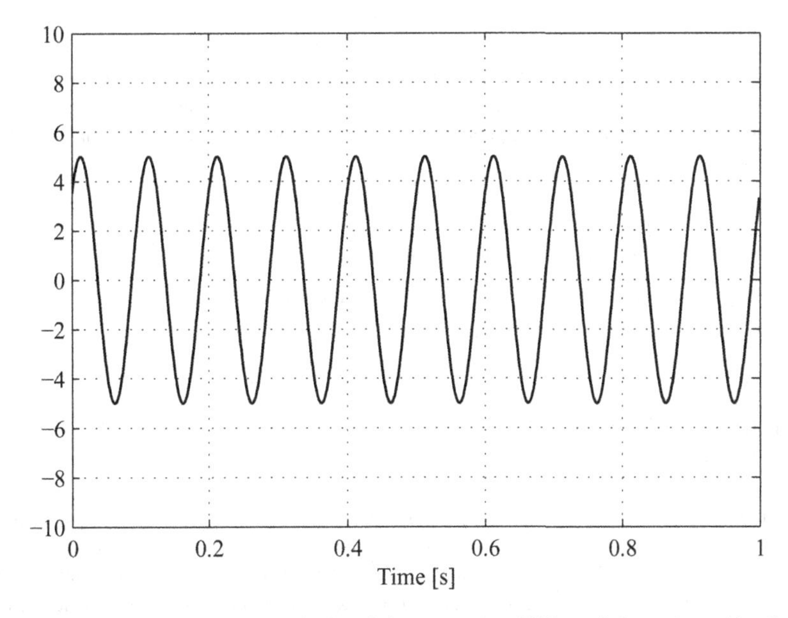


**Figure 2.1** Sine wave with amplitude $A = 5$, frequency $f = 10$ Hz, and phase, $\phi = \pi/4$ radians

by three parameters; the *amplitude*, $A$, the *angular frequency*, $\omega$, and the *phase angle*, $\phi$. With these parameters defined, the time dependent sine is defined by

$$x(t) = A \sin(\omega t + \phi). \tag{2.1}$$

The amplitude, $A$, defines the maximum of the sine, since $-1 \leq \sin(\omega t + \phi) \leq 1$ for all angles $\omega t + \phi$. The angular frequency in [rad/s] is often replaced by the (cyclic) frequency in [Hz] , $f$, defined by the relationship $\omega = 2\pi f$. The phase, $\phi$ of the sine, finally, defines a shift along the time axis and can be calculated from the function value at time zero, i.e., $x(0) = A \sin(\phi)$. A sine with amplitude $A = 5$, frequency $f = 10$ [Hz] and phase angle $\phi = \pi/4$ radians, is plotted in Figure 2.1. The *period*, $T_p$ of the sine (or of any periodic signal) is the time for one complete cycle, which for the sine is related to the frequency by

$$T_p = \frac{1}{f}. \tag{2.2}$$

The *cosine* is similar to the sine, and in this text we will often refer to both the sine and cosine as 'sines'. The relationship between the sine and cosine is

$$\cos(\phi) = \sin(\phi + \pi/2) \tag{2.3}$$

i.e., the cosine lags behind the sine by $90^\circ$, or $\pi/2$ radians.

There are many reasons why sines are important in vibration analysis. The most fundamental reason is perhaps that a sine represents a *single frequency*, and as we will see in Section 2.6.1, for linear systems, sinusoidal inputs result in sinusoidal outputs. This is often referred to as *harmonic response*. Another important reason for using sines is that through the theory of Fourier series, we know that all periodic signals are composed of a sum of sines, see Section 8.1. A third reason why sines and cosines are important is that they are *orthogonal*, see Section 2.7.1 and that they are used as the so-called *basis functions* in the Discrete Fourier transform, see Chapter 9.

### 2.2.2 Complex Sines

A common approach when dealing with periodic signals is to use *complex sines*. It is essential to understand how this is used and we will therefore discuss complex sines in some depth. If you are not familiar with complex numbers, Appendix A gives an overview. Assume first that we have a real, time-dependent signal,

$$y(t) = A\cos(\omega t + \phi). \tag{2.4}$$

A corresponding complex sine, $\tilde{y}(t)$ is now defined as

$$\tilde{y}(t) = Ae^{j(\omega t+\phi)} = Ce^{j\omega t} = C\left[\cos(\omega t) + \mathrm{j}\sin(\omega t)\right] \tag{2.5}$$

where

$$C = Ae^{\mathrm{j}\phi}. \tag{2.6}$$

Using this notation, our actual (original) signal is

$$y = \mathrm{Re}\,[\tilde{y}] \tag{2.7}$$

By introducing the complex signal, $\tilde{y}(t)$, we are now able to easily change both the amplitude and phase of our signal, for example, by passing the complex sine through a frequency response (see Section 2.7.2), i.e., some physical process that affects the amplitude and phase. The resulting, true signal is then obtained by taking the real part of the complex signal, which follows from the orthogonality between the real and imaginary parts. We achieve the same result as if we had calculated the result using trigonometric functions for addition and multiplication, but in a usually much easier way. In some applications the imaginary part of the complex signal also has interpretations, which we shall not discuss here, but in general it can be said that the imaginary part is simply 'following along' as a 'complement' in the calculations.

**Example 2.2.1** *As an example of using a complex sine, assume that we have a sinusoidal force with amplitude 30 N and frequency 100 Hz. The force acts on an SDOF system with a resonance* $f_0 = 100$ *Hz, where the frequency response between force input and acceleration output is* $0.1\angle 90^\circ$ *[(m/s2)/N]. We let the phase of our force be the reference, that is,* $0^\circ$*. What is the resulting acceleration?*

*Note: This example is by necessity a little premature, as we will not present frequency responses until later in this chapter, see Section 2.7.2. However, at the moment it is sufficient to know that the output of a linear system, at each frequency, is the product of the input (force in our example) and the frequency response at that frequency. The frequency response is a frequency-dependent function which at each frequency is a complex number describing amplitude gain factor and phase effect if described in polar form, so the example illustrates how the complex sine formulation simplifies the calculation when we multiply two complex values.*

*Our force signal,* $F(t)$*, can be written in complex form as*

$$F(t) = Ce^{\mathrm{j}2\pi f_0 t} \tag{2.8}$$

*where* $C = 30\angle 0^\circ$ *[N] and* $f_0 = 100$ *[Hz]. Furthermore, the frequency response at 100 Hz is*

$$H(100) = 0.1\angle 90^\circ. \tag{2.9}$$

*We thus obtain that the resulting acceleration is*

$$A = F \cdot H = 30 \cdot 0.1\angle\left(0 + (90^\circ)\right) = 3\angle 90^\circ \mathrm{m/s^2} \tag{2.10}$$

*or if we write the actual, real acceleration, that is, the real part of Equation (2.10), then*

$$a\,(t) = 3\cos\left(2\pi \cdot 100t + \pi/2\right)\mathrm{m/s^2}. \tag{2.11}$$

*End of example.*

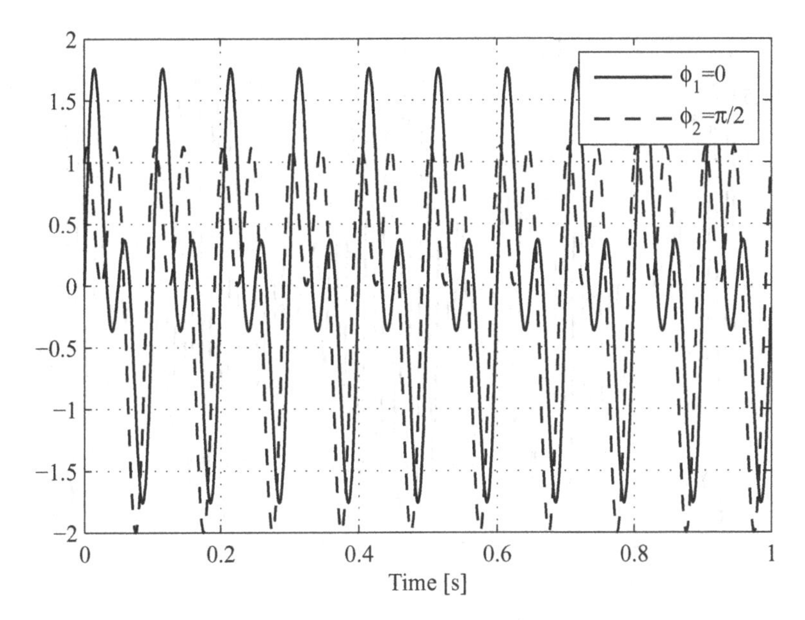


**Figure 2.2** Sum of two sines with frequencies 10 and 20 Hz, respectively. Two cases of phase difference are shown; solid: both signals in phase (phase angles $\phi = 0$); dashed: phase angle of 20 Hz sine $\phi_2 = \pi/2$. The sum signal has a period of $T = 0.1$ s, which corresponds to one period of the 10 Hz sine and 2 periods of the 20 Hz sine

### 2.2.3 Interacting Sines

Next we will study effects of summing and multiplying sines with different frequencies. When two sines with different frequencies are combined, the result depends on the frequencies and phase angles of the two sines. Assume that we sum two sines with frequencies $f_1$ and $f_2$ Hz. The resulting signal

$$y(t) = \sin(2\pi f_1 t) + \sin(2\pi f_2 t) \tag{2.12}$$

will be a periodic signal if there is a time $T_p$ for which both sines make integer number of periods. This will be the case if $f_1$ and $f_2$ are both rational numbers, or if they are related so that their ratio is a rational number. An example is illustrated in Figure 2.2 where the result of the sum of a sine with frequency $f_1 = 10$ Hz and a sine with frequency $f_2 = 20$ Hz is plotted. In the figure, the sum is shown for two cases; the two signals in phase (both have a phase of 0 radians), and with the phase of the second sine being $\pi/2$ relative to the phase of the first sine. In both cases the period will be $T_p = 1/f_1$ as the second frequency is exactly twice the first one.

As seen in Figure 2.2 the resulting sum of the two sines is a periodic signal. As is evident from the two signals in the plot, the actual shape of the signal depends on the phase between the two sines. Another important observation from the plot is that there is no well-defined amplitude, since the maximum value of each of the signals is different! Amplitude is a useful concept only for single sines, not for signals containing several sines.

A special effect of the combination of two sines, *beating*, occurs when two sines with frequencies relatively near each other are summed, as seen in Figure 2.3. In the figure the sum of a sine with frequency $f_1 = 100$ Hz and a sine with frequency $f_2 = 90$ Hz is plotted. As is seen in the figure, the result shows a 'high-frequency' sine with a 'slowly' varying amplitude and it can be seen that the amplitude varies with a frequency of 10 Hz (from the period defined between two of the instances where the amplitude is, for example, zero).

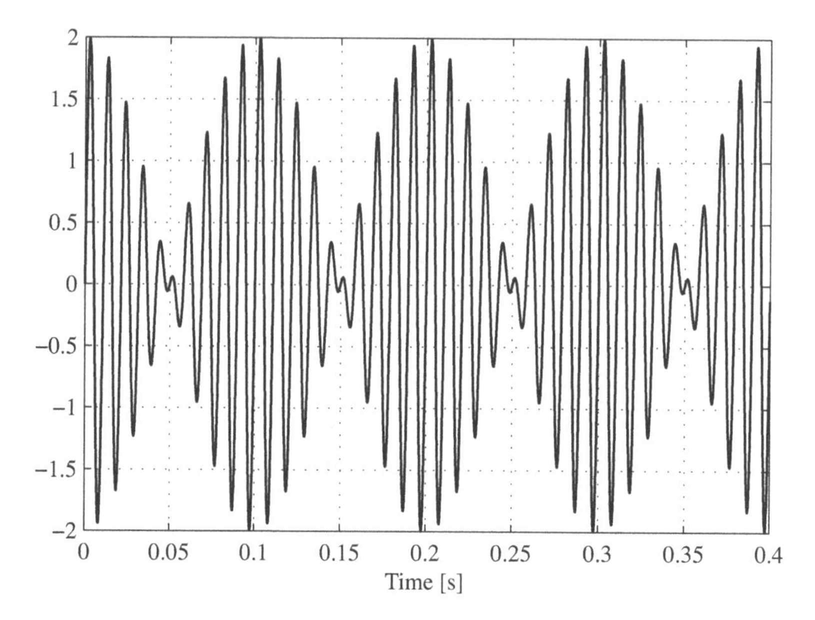


**Figure 2.3** Sum of two sines with beating. The signal in the figure is the sum of a sine with frequency $f_1 = 100$ Hz and another sine with frequency $f_2 = 90$ Hz, both with amplitudes of unity. The sum signal has a periodic beating with a frequency of 10 Hz, corresponding to the difference between the frequencies, $f_1 - f_2$

From basic trigonometry we have the formula for the sum of two sines

$$\sin(u) + \sin(v) = 2\sin\left(\frac{u+v}{2}\right)\cos\left(\frac{u-v}{2}\right) \tag{2.13}$$

which shows one of the relationships between the sum of two sines and multiplication of two sines (or a sine and a cosine to be exact). From this relationship we see that the effect of summing two sines is equal to multiplying the mean and half the difference frequencies. The beating effect thus occurs either when two sines with close frequencies are *summed*, or when two sines with largely different frequencies (typically a high frequency and a much lower frequency) are *multiplied*.

The effect of beating is important in noise and vibration applications, not the least because our human hearing is sensitive to amplitude fluctuations. Naturally two sines with close frequencies can often occur and they are often causes of unwanted noise effects, particularly from rotating machines, see also Chapter 12.

### 2.2.4 *Orthogonality of Sines*

The concept of *orthogonal signals* is very important in signal analysis. For example, we will use the concept of orthogonality between general signals in Chapter 15. The definition of orthogonality between any two signals $u(t)$ and $v(t)$ is that the integral of the product of the two signals is zero, i.e.,

$$\int_{-\infty}^{\infty} u(t) \cdot v(t) \mathrm{d}t = 0 \tag{2.14}$$

It should be noted that if the integral in Equation (2.14) is fulfilled, then the *mean* (average value) of the product of the two signals is also zero since the mean equals the integral divided by the time of integration. Often it is easier to think of the mean of a signal rather than its integral.

In this section we will discuss specifically the concept of orthogonality between sines and cosines, which is essential (among other things) to understand the Fourier transform. For two rational frequencies $f_1$ and $f_2$, the product between two sines and/or cosines gives a new periodic signal as we discussed in Section 2.2.3. If we let the period of the new signal be $T_p$, then we have the orthogonality relationships

$$\frac{1}{T_p}\int_0^{T_p} \cos(2\pi f_1 t)\cdot\cos(2\pi f_2 t) = \begin{cases} 0, & f_1 \neq f_2 \\ 1/2, & f_1 = f_2 \end{cases} \tag{2.15}$$

which is also valid if the cosines are replaced by sines, and

$$\frac{1}{T_p}\int_0^{T_p} \cos(2\pi f_1 t)\cdot\sin(2\pi f_2 t) = 0, \text{ for all } f_1, f_2. \tag{2.16}$$

Equation (2.15) states that in order for the average of the product of two cosines (or sines if both cosines in the equation are replaced) to be non-zero, then the signals must have the same frequency, whereas Equation (2.16) states that the product of a sine and a cosine always has zero mean, even if the frequencies of the sine and cosine are the same. There is a limitation to when this is mathematically true, and that is that the frequencies $f_1$ and $f_2$ have to be rational numbers, so that there is a common period $T_p$ over which the two sines/cosines each have an integer number of periods, as otherwise the integral cannot be calculated as stated in the equations. If one or both of the frequencies are not rational numbers, there will not be any period over which one of the integrals in Equations (2.15) and (2.16) will be exactly zero. However, the product inside the integral will still be a signal with an 'apparent' zero mean so from a practical standpoint the effect is a 'roundoff error', see Problem 2.3. Signals that have this property are sometimes called *almost-periodic* signals.

## 2.3 Random Signals

As mentioned in the chapter introduction, signals can be either deterministic or random. Random vibrations typically occur when the forces are caused by many independent contributions, such as the roughness of a road producing random force inputs to the tires of a car, or the sound produced by turbulent air coming out of a ventilation system, etc. Random signals are mathematically described by *stochastic processes*, which we will discuss in Chapter 4. In this section, we will limit the discussion to some fundamental aspects of random signals.

A random signal $x(t)$, is a signal for which the function values at different instances $t$ and $t+\tau$, i.e., $x(t)$ and $x(t+\tau)$ are independent. Thus knowing (recording) $x(t)$ for any amount of time does not help at all to predict future values. Since most random signals we find in vibration applications have some causing mechanism behind them which has some particular 'pattern', the random signals will have some resulting pattern. We may for example drive a car at constant speed over a type of asphalt road which has a certain surface 'shape', which causes the sound produced by the road to sound 'constant' in some way. If this is the case the random signal has constant statistical values such as RMS value (see Section 2.5), spectrum (see Section 8.3.1) etc., and we refer to the random signal as a *stationary* random signal. Note however, that over a long enough time, most random signals are not stationary, as for example the asphalt type will change after a while, or the wind speed for wind-induced vibrations, etc.

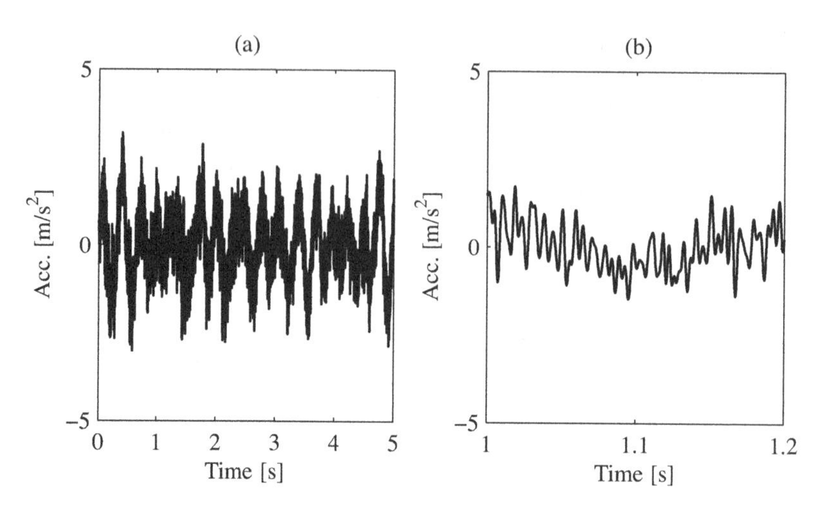


**Figure 2.4** Example of random signal. The figure shows the acceleration on the frame of a truck driving on a rough road. In (a) the acceleration over five seconds is displayed and in (b) the same acceleration signal is zoomed in to show a small part of the data. See text for discussion

An example of a random signal is shown in Figure 2.4. The example is taken from an accelerometer (see Section 7.4) measuring the acceleration on the frame of a truck driving on a rough road. In the plot in Figure 2.4 (a), a 5-second frame of data is plotted, which shows random variations. The plot in Figure 2.4 (b) shows a small part of the data, from 1 to 1.2 s, which reveals the random 'ringing' of the signal. A word is appropriate here about the nature of the signal in Figure 2.4. You may see a seemingly periodic behavior of the signal, with a period of approx. 0.3 s. How can we be sure this signal is random and not periodic? The answer is, that we cannot determine this at all from the figure. Indeed, this question, although so apparently simple, turns out to be very difficult in practice. For now, we leave the discussion on this difficult issue to Chapters 4 and 10 where it belongs.

## 2.4 Transient Signals

The third fundamental signal class is the class of *transient signals*. A transient signal is a signal which has a limited duration, i.e., it dies out after a while. Examples of such signals are for example the vibrations when we cross a railroad with our car, or the sound of a car door closing. Transient signals are usually, but not always, deterministic; for example, burst random noise that is described in Section 13.9.3 is a, rare, exception. If the signal is deterministic, it means that the same signal is repeated if the event is repeated, for example we can imagine each sound from a gunshot producing the same sound pressure at a particular location relative to the gun barrel. This is of course an idealized example, which does not take into account any statistical spread between each gunshot, etc. We will say more about spread in measurements, etc. in Section 4.1.

An example of a transient signal is shown in Figure 2.5 in the form of an exponentially decaying sine. A characteristic that separates transients from periodic and random signals is that because the transients die out, it is not relevant to discuss the *power* of the transient (remember, power is defined as energy per time unit). Instead of power, we can relate to the *energy* of the transient, or sometimes simply the sum (integral) of it. If the measured entity is a force, for example, we can relate to the integral of the force, which we know as the *impulse* of the force.

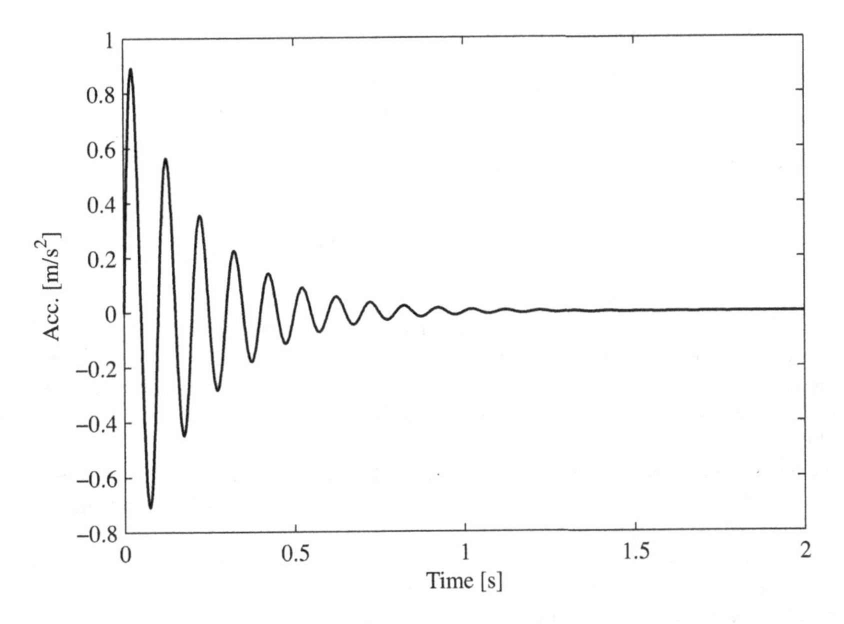


**Figure 2.5** Example of a transient signal; an exponentially decaying sine

## 2.5 RMS Value and Power

From the discussions in the previous sections in this chapter it should be apparent that the properties of dynamic signals in general cannot be summarized by a single value. Often it is, however, useful to be able to compare two dynamic signals and distinguish which one is 'larger'. The most common measure used in this respect is the *root mean square*, or RMS value. The RMS value of a signal $x(t)$, based on an averaging time of $\tau$ s, which we can denote $x_{RMS}$, is defined by

$$x_{RMS} = \sqrt{\frac{1}{\tau}\int_{0}^{\tau} x^2(t)\mathrm{d}t} \tag{2.17}$$

that is, the RMS value is the square root of the *mean square* of the signal. The 'origin' of the RMS value is based on a simple electrical circuit as illustrated in Figure 2.6. In such a circuit the instantaneous

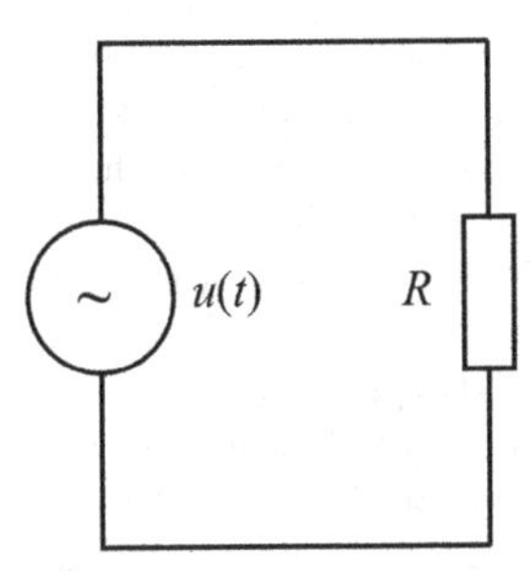


**Figure 2.6** A simple electrical circuit with an AC voltage source and a resistor

power dissipating through the resistor is

$$P_u(t) = \frac{u^2(t)}{R}. \tag{2.18}$$

The average power, which we denote $< P_u >$, based on $\tau$ seconds of $u(t)$ is now

$$< P_u > = \frac{1}{R\tau} \int_0^\tau u^2(t)\mathrm{d}t \tag{2.19}$$

where $R$ is the resistance. Equation (2.19) is the mean square value of the voltage $u(t)$, divided by the resistance, $R$. This means that if we replace the dynamic (AC) voltage $u(t)$ with a DC voltage $u_{DC} = u_{\mathrm{RMS}}$ from Equation (2.17), the mean power will be equivalent. This in turn means that the heat dissipated by the resistance (or the light emitted if $R$ is a light bulb) will be the same. This is the essence of the RMS value.

In noise and vibration applications, the RMS value is often relevant. For instance, the ear is essentially sensitive to the RMS value of the sound pressure in the ear canal. Sound level meters therefore measure RMS values, as will be discussed in Section 3.3.5.

The RMS value is the most common value used when a single value is wanted to describe the level of a dynamic signal. It is, however, by no means the only one. And it should be emphasized that the only thing the RMS level tells us is what the square root of the mean square of the signal is. In Chapter 4 we will discuss several more statistical values such as, for example, skewness and kurtosis, that are also often used to describe the characteristics of dynamic signals.

From the discussion of the simple electrical circuit above, it is clear that the electrical power is proportional to the *square* of the voltage. It is very common in signal analysis and vibration analysis, to refer to all squared units as 'power', although in many cases the actual signal squared may not be directly proportional to the actual power in units of watts [W]. For example, if we measure an acceleration, the square of the acceleration will be referred to as the 'power of the acceleration' although for mechanical systems, the power is actually related to the square of the *velocity* (as the kinetic energy is $mv^2/2$ which should be well known from mechanical dynamics). It is important to realize this somewhat 'sloppy' use of the term 'power' in order not to be confused later in this book (or in your professional career, for that matter).

## 2.6 Linear Systems

As we mentioned in the introduction of this chapter, a system is an entity which has one or more *inputs* causing one or more *outputs*. A dynamic system is often defined (rather theoretically) as a *linear system* if it can be described by linear differential equations. If it is not linear, it is called a *nonlinear system*. In this section we will show what implications this theoretical definition has, and we will discuss briefly when we can expect a system to be linear. In Chapters 13 and 14 we will discuss how to identify linear systems from measurements of input(s) and output(s), and then we will also discuss practical methods of testing if the estimated system is linear or nonlinear.

A particularly interesting class of linear systems is the class of *time invariant* systems. Such a system is a linear system for which all parameters are constant (independent of time). In mechanical systems, this means that the masses, springs and dampers are not changing with time. This is often a reasonable assumption during, for example, the time over which we measure a system, but over a long enough time span, very few real systems are time invariant. The characteristics of a bridge, for example, can change due to the temperature changing between day and night, or its characteristics (on a more long-term span) can change due to aging or fatigue of the structure.

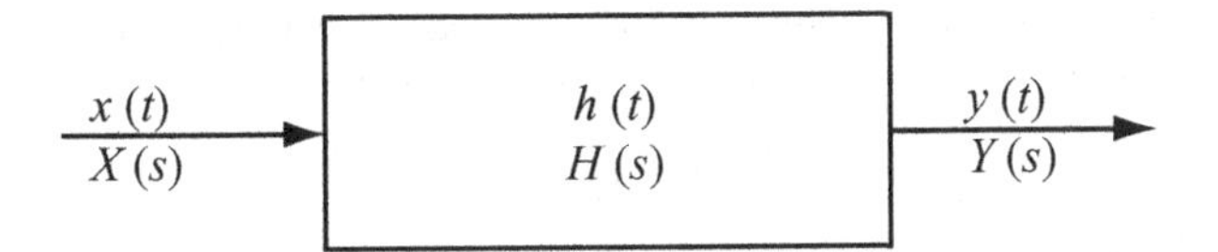


**Figure 2.7** Linear system as a 'black box' with time signals and Laplace domain equivalents

In principal a system can be thought of as a 'black box' relating the inputs and the outputs caused by those inputs as illustrated for a single-input/single-output system in Figure 2.7. In the remainder of this section we will look into how the input and output of such systems are related when the system is a time invariant, linear system. The main theory we will use for describing the linear system, is the *Laplace transform*. If you feel you have a good understanding of the Laplace transform, I still recommend you read the following subsections at least briefly, as the treatment here is probably less abstract than you have seen in math classes, and may reveal one or two pieces of information you have not thought about before. If you have never seen the Laplace transform before, the following is meant to serve as an introduction sufficient to follow the discussions in the rest of this book.

## 2.6.1 *The Laplace Transform*

The Laplace transform is a mathematical tool that can be used (among other things) to solve systems described by linear differential equations. The feasibility of the Laplace transform theory for our purpose, is a result of the fact that it is very general, and is easily related to experimentally available entities such as time signals and frequency spectra (Fourier transforms).

If we have a signal $x(t)$ we define its Laplace transform, $\mathcal{L}[x(t)] = X(s)$ by

$$\mathcal{L}[x(t)] = X(s) = \int_{0^-}^{\infty} x(t)e^{-st}dt \tag{2.20}$$

where the complex variable $s$ is the *Laplace operator* which we will sometimes divide into its real and imaginary parts as

$$s = \sigma + j\omega. \tag{2.21}$$

The Laplace transform, $X(s)$ is an algebraic expression; in our cases with differential equations it is usually a polynomial. We often refer to the Laplace variable, $s$, and the function $X(s)$ in the Laplace s-plane as belonging to the *Laplace domain*, whereas the original time signal, $x(t)$, is in the *time domain*. We can thus transform signals from one domain to the other with the forward or inverse Laplace transform. Later, in Section 2.7 we will introduce the similar *frequency domain* for the Fourier transform.

Note that the integral in Equation (2.20) starts at '$0^-$' which ensures that we will include any Dirac impulse functions at time zero, see Section 2.6.3 below. If we have a Laplace transform $X(s)$, we can use the *inverse Laplace transform* denoted $\mathcal{L}^{-1}[X(s)]$ to go backwards to get the time function, $x(t)$, i.e.,

$$x(t) = \frac{1}{2\pi j} \lim_{T\to\infty} \int_{\beta-jT}^{\beta+jT} X(s)e^{st}ds. \tag{2.22}$$

In order to understand Equation (2.22), you will need to know complex calculus, which we will leave out here. The important Laplace transform pairs, i.e., time functions and their Laplace domain counterparts which we need will be presented later in this section.

The Laplace transform has some important properties related to our application of it, which we will now present. The Laplace transform is a *linear transform* which means that

$$\mathcal{L}[a_1 x_1(t) + a_2 x_2(t)] = a_1 \mathcal{L}[x_1(t)] + a_2 \mathcal{L}[x_2(t)] \tag{2.23}$$

for any real scalar constants $a_1$ and $a_2$.

Further, what makes the Laplace transform particularly useful to solve differential equations, is that it transforms linear differential equations into polynomials in $s$. This is a fact because the Laplace transform of the n-th derivative of $x(t)$, i.e., $\mathcal{L}\left[x^{(n)}(t)\right]$ is

$$\mathcal{L}\left[x^{(n)}(t)\right] = s^n X(s) - s^{n-1}x(0) - s^{n-2}x^{(1)}(0) - \ldots - x^{(n-1)}(0) \tag{2.24}$$

where $x(0)$, $x^{(1)}(0)$ etc. are the *initial conditions* of the differential equation. Note the difference between the n-th *power* of $s$, $s^n$, and the $n$th *derivative* of $x(t)$, where we use parentheses, $x^{(n)}(t)$. Equation (2.24) means that the Laplace transform of the first derivative, $\dot{x}(t)$, of $x(t)$, is

$$\mathcal{L}[\dot{x}(t)] = sX(s) - x(0) \tag{2.25}$$

and the Laplace transform of the second derivative, $\ddot{x}(t)$ is

$$\mathcal{L}[\ddot{x}(t)] = s^2 X(s) - sx(0) - \dot{x}(0). \tag{2.26}$$

The initial conditions in the previous equations are necessary to solve differential equations with arbitrary initial conditions. However, there is an important principle that we will utilize later, namely the principle of *superposition*, which says that, if an input $x_1(t)$ causes an output $y_1(t)$, and another input $x_2(t)$ causes an output $y_2(t)$, then for a linear system, the input signal $x_1(t) + x_2(t)$ will be $y_1(t) + y_2(t)$. Thus, if the initial conditions are not zero, we can always calculate the contribution due to a particular input signal (or change in the input signal) under the assumption of zero initial conditions, and then add the vibrations that were there before.

You should remember from your calculus class that the solution to a linear differential equation generally consists of two parts, the *homogeneous* (transient) solution, and the *particular* (forced, or steady-state) solution. The total solution is the sum of those two solutions. You should note that when using the Laplace transform to solve a linear differential equation, we get both those solutions. This adds to the wide applicability of the Laplace transform. Also see Section 2.7.4 for a discussion on transient versus steady-state response.

Some common Laplace transform pairs are given in Table 2.1. For more comprehensive tables of Laplace transform pairs, any standard mathematical reference book can be used, for example (Zwillinger 2002).

**Table 2.1** Common Laplace transform pairs. Note that pairs 1 and 2 are for the special case where all initial conditions are zero, see the text for details

| # | Description | $x(t)$ | $X(s)$ |
|---|---|---|---|
| 1 | Differentiation | $\dot{x}(t)$ | $sX(s)$ |
| 2 | Integration | $\int x(t)\mathrm{d}t$ | $\frac{1}{s}X(s)$ |
| 3 | Dirac pulse | $\delta(t)$ | 1 |
| 4 | Exponential | $\mathrm{e}^{at}$ | $\frac{1}{s-a}$ |
| 5 | Time delay | $x(t-\tau)$ | $\mathrm{e}^{-s\tau}X(s)$ |
| 6 | Convolution | $\int_{-\infty}^{\infty} x(u)y(t-u)\mathrm{d}u$ | $X(s)Y(s)$ |

An important theorem we will use extensively later in this book is the theorem of *partial fraction expansion*. This theorem applies to any function $H(s)$ which is a ratio of two polynomials $P(s)$ and $Q(s)$, i.e.,

$$H(s) = \frac{P(s)}{Q(s)} = \frac{P(s)}{(s - s_1)(s - s_2)\cdots(s - s_{N_q})} \tag{2.27}$$

and for which the polynomial order of $Q$, $N_q$, is at least one more than the order of $P$, $N_p$, i.e., $N_q > N_p$. If those conditions are met, the theorem says that the function $H(s)$ can be divided into a sum

$$H(s) = \sum_{r=1}^{N_q} \frac{A_r}{s - s_r}. \tag{2.28}$$

where $s_r$ is the $r$th *root* of $Q(s)$, i.e., a solution $Q(s_r) = 0$.

The variables $s_r$ in Equation (2.28) are called the *poles* of $H(s)$, and the variables $A_r$ are called the *residues*. To calculate the residues, we can use the so-called *hand-over method* (sometimes called the *cover method*), which says that

$$A_r = (s - s_r)\frac{P(s)}{Q(s)}\bigg|_{s=s_r}. \tag{2.29}$$

This method is called the hand-over method because Equation (2.29) says that to calculate the residue for pole $r$, $A_r$, then from an expansion of the denominator,

$$H(s) = \frac{P(s)(s - s_r)}{(s - s_1)(s - s_2)...(s - s_r)...(s - s_{N_q})}\bigg|_{s=s_r} \tag{2.30}$$

we see that the factors $(s - s_r)$ in the numerator and denominator cancel out. Therefore, without going to the length of Equation (2.30), we can instead 'hold our hand' over the factor $s - s_r$ in Equation (2.27), and insert $s = s_r$ in the remaining equation. The hand-over method does not work if there is a repeated pole, that is two or more values of $s_r$ are coinciding. The partial fraction expansion still applies, however, and as we are interested in using the Laplace transform as a tool to explain the principle of systems theory, we leave this special case out here.

**Example 2.6.1** *As an example of using the Laplace transform to solve differential equations, let us solve the differential equation*

$$\ddot{y} + 3\dot{y} + 2y = x(t) \tag{2.31}$$

*for an input signal $x(t)$ which is*

$$x(t) = 10e^{-3t} \tag{2.32}$$

*and, for simplicity, initial conditions*

$$y(0) = 0 \;\; \dot{y}(0) = 0. \tag{2.33}$$

*Laplace transformation of Equation (2.31) gives*

$$[s^2 + 3s + 2]Y(s) = \frac{10}{s + 3} \tag{2.34}$$

*where we have used transform pair number 4 in Table 2.1. We divide left- and right-hand sides of the equation by the polynomial in s on the left-hand side, which gives us*

$$Y(s) = \frac{10}{(s + 3)(s^2 + 3s + 2)} = \frac{10}{(s + 3)(s - s_1)(s - s_2)}. \tag{2.35}$$

*where $s_1$ and $s_2$ are the roots of the polynomial $s^2 + 3s + 2$, i.e., $s_1 = -2$ and $s_2 = -1$.*

*Next, we use partial fraction expansion on Equation (2.35) which yields*

$$Y(s) = 10\left[\frac{A_1}{s+3} + \frac{A_2}{s+2} + \frac{A_3}{s+1}\right] \tag{2.36}$$

*where the residues, $A_n$, can be found by applying the hand-over method on Equation (2.35), which gives $A_1 = 0.5$, $A_2 = -1$, $A_3 = 0.5$. Thus we have a solution in the s-plane*

$$Y(s) = \frac{5}{s+3} - \frac{10}{s+2} + \frac{5}{s+1}. \tag{2.37}$$

*We now go to Table 2.1 to find the inverse solution*

$$y(t) = 5e^{-3t} - 10e^{-2t} + 5e^{-t} \tag{2.38}$$

*which is our end result. You should also look at Problems 2.9 and 2.10 to learn how to use MATLAB/Octave to solve this problem.*

*End of example.*

## 2.6.2 The Transfer Function

From the definition of the Laplace transform in Section 2.6.1 the *transfer function*, $H(s)$, of a system follows straightforwardly. For any linear (single-input/single-output) system with a Laplace transform of the input $X(s)$ and of the output $Y(s)$, the transfer function is defined as the ratio of the output and input Laplace transforms, i.e.,

$$H(s) = \frac{Y(s)}{X(s)}. \tag{2.39}$$

The transfer function for any linear system has a unique expression, i.e., it is independent of the input and output signals; for any input signal, the output signal Laplace transform will be such that the ratio $Y(s)/X(s) = H(s)$.

The practical use of the transfer function is that the output can be calculated for an arbitrary input by multiplying the transfer function with the input, which follows directly from rewriting Equation (2.39) into

$$Y(s) = X(s)H(s). \tag{2.40}$$

The output time signal (the solution to the linear differential equation for a particular forcing function), is found be applying the inverse Laplace transform on Equation (2.40), i.e.,

$$y(t) = \mathcal{L}^{-1}[Y(s)] = \mathcal{L}^{-1}[X(s)H(s)] \tag{2.41}$$

where the inverse Laplace transform is usually found by table lookup, after some algebraic manipulation to yield Laplace expressions that can be found in the Laplace table.

An important concept related to the transfer function is the concept of *poles*. The poles of a transfer function (or, as we often say, the poles of the *system* $H(s)$) are the roots of the denominator of $H(s)$. We will look at the poles of a simple mechanical system, which we will describe in more detail in Chapter 5. Let us assume we have a second order differential equation

$$m\ddot{y} + c\dot{y} + ky = x(t) \tag{2.42}$$

where for the moment we only assume the constants $m$, $c$, and $k$ are time invariant constants and the initial conditions are zero (i.e., $x(0) = 0$ and $\dot{x}(0) = 0$). By Laplace transforming this equation we first obtain

$$\left(ms^2 + cs + k\right)Y(s) = X(s) \tag{2.43}$$

after which we form the transfer function

$$H(s) = \frac{Y(s)}{X(s)} = \frac{1}{ms^2 + cs + k} = \frac{1/m}{s^2 + c/ms + k/m} \tag{2.44}$$

where we have rearranged the denominator so that the highest order is free of any constant. The poles of $H(s)$ are now the roots of the denominator in Equation (2.44). From Equation (2.43) we can see that the roots of the polynomial are in fact the non-trivial solutions to

$$\left(ms^2 + cs + k\right) Y(s) = 0 \tag{2.45}$$

i.e., for mechanical systems, which we are particularly interested in here, the equation for *free vibrations*. Thus, *the poles are the solutions (frequencies) which give us free vibrations for a mechanical system.* (Remember that the nontrivial solutions are the solutions of Equation (2.45) for which $y(t) \neq 0$. They are called nontrivial, of course, because if indeed $y(t) = 0$, for a mechanical system, we have no vibrations, and that is surely a trivial solution.)

Roots of the numerator polynomial in $H(s)$, i.e., values of the Laplace operator $s$ for which $H(s) = 0$, are called *zeros* of $H(s)$ (or zeros of the *system*). The poles and zeros are important properties because they build up the transfer function. In Section 2.7.2 we will look more into these important properties. First, however, we will now look at the inverse transform of the transfer function.

**Example 2.6.2** *Define the transfer function for the system described by the differential equation in Example 2.6.1. Find the poles of the system.*

*We already have a Laplace transform of the differential equation, so the transfer function simply becomes*

$$H(s) = \frac{Y(s)}{X(s)} = \frac{1}{s^2 + 3s + 2} \tag{2.46}$$

*and the poles are the roots of the denominator polynomial, which were already calculated in Example 2.6.1, i.e.,* $s_1 = -2$, *and* $s_2 = -1$ .

*End of example.*

## 2.6.3 The Impulse Response

A very important Laplace transform pair, is the one that specifies that multiplication in the Laplace domain corresponds to *convolution* in the time domain, and vice versa. Thus, the Laplace domain relationship in Equation (2.40) is equivalent in the time domain to the convolution integral

$$y(t) = \int_{-\infty}^{\infty} x(u)h(t-u)\mathrm{d}u. \tag{2.47}$$

The function $h(t)$ in Equation (2.47) is the inverse Laplace transform of the transfer function $H(s)$ and is called the *impulse response* of the system. The name implies that the function $h(t)$ is obtained as the output of the system, when the input of the system, $x(t)$ is an ideal impulse, a so-called *Dirac unit impulse function*, $\delta(t)$.

The Dirac unit impulse (or simply the Dirac impulse), $\delta(t)$ is an idealized function with the special properties

$$\begin{aligned} &\delta(t) = 0, t \neq 0 \\ &\int_{0^-}^{0^+} \delta(t)\mathrm{d}t = 1. \end{aligned} \tag{2.48}$$

These properties imply that the Dirac impulse is infinitely narrow, and infinitely high, so that the area under it is unity. The Laplace transform of the Dirac impulse is

$$\mathcal{L}\left[\delta(t)\right] = 1 \tag{2.49}$$

so that, as mentioned above, if $x(t) = \delta(t)$, then

$$y(t) = \int_{-\infty}^{\infty} \delta(u)h(t-u)\mathrm{d}u = h(t). \tag{2.50}$$

Naturally, we could also obtain this relationship by using the Laplace transform, i.e.,

$$y(t) = \mathcal{L}^{-1}\left[1 \cdot H(s)\right] = h(t). \tag{2.51}$$

**Example 2.6.3** *Find the impulse response of the system from Example 2.6.2.*

*By applying the hand-over method on the transfer function in Example 2.6.2 we find that the transfer function can be written as*

$$H(s) = \frac{1}{s+1} - \frac{1}{s+2}. \tag{2.52}$$

*Thus the impulse response becomes*

$$h(t) = e^{-t} - e^{-2t}. \tag{2.53}$$

*End of example.*

The impulse response, and the convolution integral so closely related to it through Equation (2.47), are very important concepts. The convolution integral is in fact so important that we will devote a whole section to it.

## 2.6.4 *Convolution*

As we saw in the previous section, convolution of two signals is important for understanding linear systems, as the output of a linear system is the convolution result between the input time signal and the impulse response. The convolution is also important for understanding many aspects of frequency analysis, as multiplication in the time domain corresponds to convolution in the frequency domain, and conversely, multiplication in the frequency domain corresponds to convolution in the time domain, see Section 2.7.

In my experience, the convolution process is perceived as rather difficult by many students. Yet it is essential to understand, in order to grasp the nature of linear systems and of frequency analysis, and we will therefore describe the convolution process in some depth here. Although the convolution process is certainly complex (in the sense of complexity) and the results of the convolution of most pairs of signals impossible to foresee without actually computing it, the principle of convolution is rather simple, as we will see henceforth.

The convolution is often denoted by an asterisk, $*$, and the convolution result, $y(t)$, between two time signals $x(t)$ and $h(t)$, is defined as

$$y(t) = x(t) * h(t) = \int_{-\infty}^{\infty} x(u)h(t-u)\mathrm{d}u \tag{2.54}$$

where the integration time variable is substituted by the variable $u$. It is rather easy to show by variable substitution (see Problem 2.8) that changing the order of the signals that are convolved, does not change

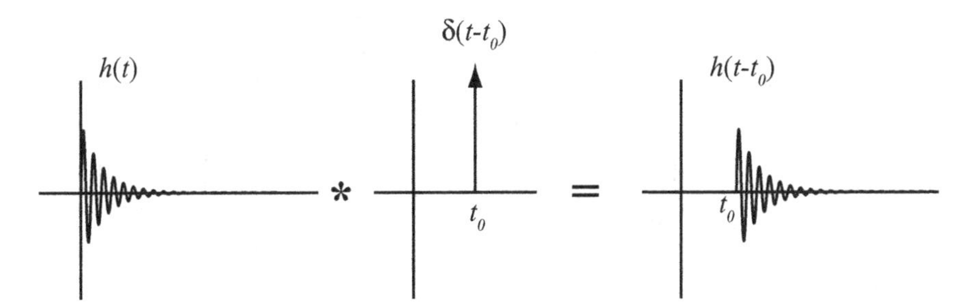


**Figure 2.8** Illustration of the result that convolving a signal (here an impulse response $h(t)$) by a Dirac impulse located at $t_0$, i.e., $\delta(t-t_0)$ results in shifting the impulse response to start at point $t = t_0$, i.e., the result is $h(t-t_0)$

the result, i.e.,

$$h(t) * x(t) = x(t) * h(t). \tag{2.55}$$

A case of special interest is the convolution between a function and the *Dirac unit impulse*, $\delta(t)$, or a translated unit impulse, $\delta(t-t_0)$, as defined in Section 2.6.3. Since the Dirac function $\delta(t-t_0)$ is nonzero only at $t = t_0$, it is relatively easy to show that

$$h(t) * \delta(t-t_0) = h(t-t_0) \tag{2.56}$$

for any delay $t_0$. Obviously then, convolving any function by the Dirac function $\delta(t-t_0)$ translates the convolved function in such a way that the function value at the origin moves from $t = 0$ to $t = t_0$, as illustrated in Figure 2.8.

As mentioned above, the actual result of the convolution of two signals is in general impossible to foresee simply by knowing the two signals, because of the effective 'mixing' of the signals through the convolution process. A property of the convolution process is, however, that, loosely speaking, the resulting function will have the 'qualities' of both signals. If, for example, one of the signals is rippling whereas the other signal is smooth, then the result will likely have some ripple.

In order to understand the convolution integral, we will split the convolution process up into a few pieces, as illustrated in Figure 2.9. First, the impulse response $h(t)$ is usually of limited length, which is illustrated in Figure 2.9(a). Now, the integral in Equation (2.54) has a factor $h(t-u)$ rather than $h(u)$. We start by observing, in Figure 2.9(b), that $h(-u)$ is the function $h(u)$ reversed along the $u$-axis ($x$-axis). In Figure 2.9(c), $h(t-u)$ is obtained by shifting the function $h(-u)$ to the 'start point' $u = t$, which, if $t$ is positive as in the figure, is positive along the $u$-axis. (Think of the point $h(0)$ which corresponds to $h(t-u)$ where $u = t$.)

Now that we have seen that $h(t-u)$ is $h(u)$ time-reversed and shifted to time $u = t$, we can go to the next 'step' in the convolution integral in Equation (2.54), which is the multiplication of $h(t-u)$ by $x(u)$ and then taking the time integral of this product, as indicated in Figure 2.9(d) and Figure 2.9(e). In Figure 2.9(f), finally, the integral of the product is marked by the shaded area where, of course, negative area under the $x$-axis is subtracted from the positive area above the $x$-axis. The complication of the convolution process is that *the entire process thus described, is calculating one, single, value of* $y(t)$. In order to calculate the convolution value of, say, $y(t_2)$, then $h(-u)$ is shifted to $u = t_2$ and the multiplication and subsequent integration is recalculated. This process is then repeated again and again, as $t$ passes all values along the time axis for which we want to calculate the convolution result $y(t)$.

From the above discussion it follows that to calculate the entire function $y(t)$, the time-reversed impulse response $h(t-u)$ is sliding along the time axis. This means that the impulse response acts as

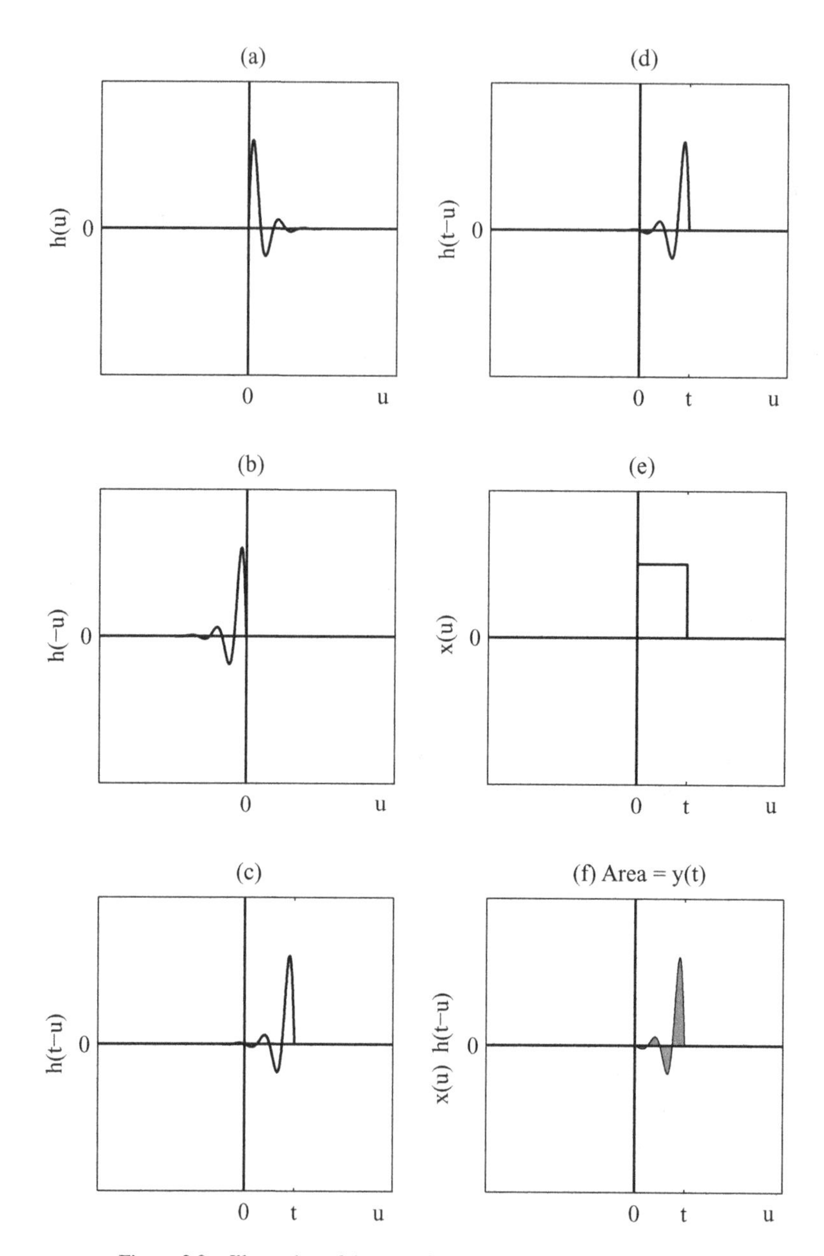


**Figure 2.9** Illustration of the convolution process; see text for details

a weighting function, where the output $y(t)$ at each $t$ is the input signal $x(t)$ weighted 'backwards' in time by the impulse response $h(t)$, as illustrated in Figure 2.10. This leads us to the important concept of *causality*.

A *causal* physical system, is a system for which the outputs at every instant in time only depend on past inputs. Such a system does not 'foresee the future' which is an important concept in physics. Naturally, any system we observe in nature will be causal. From the understanding we now have of the

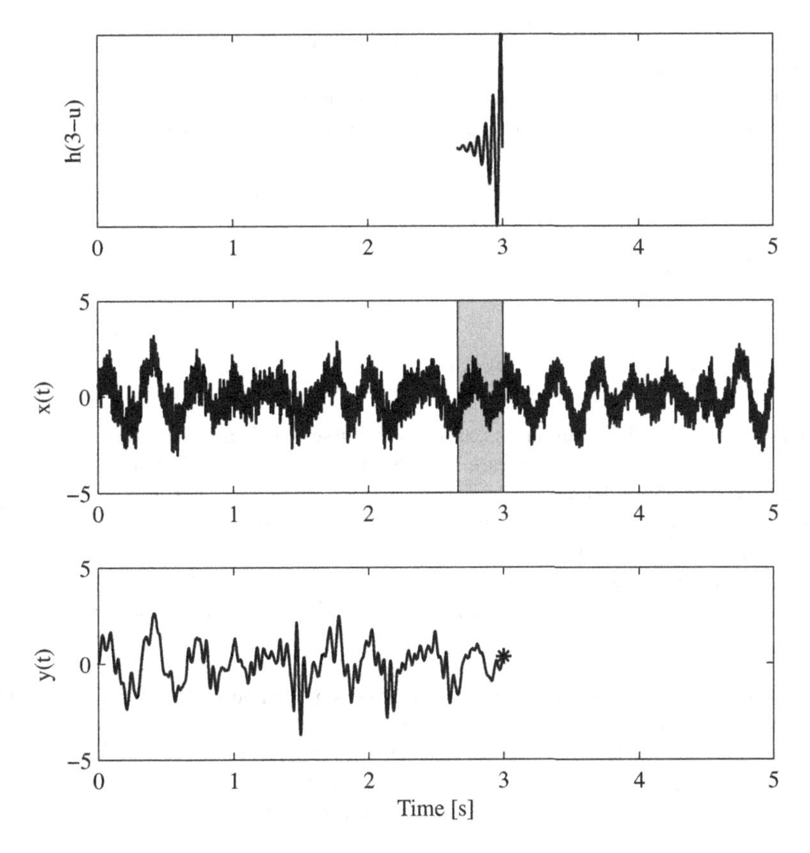


**Figure 2.10** Illustration of the impulse response 'sliding' through time to obtain the convolution output result. The figure illustrates the moment when the impulse response has 'moved' up to $t = 3$ seconds, and the sample for $y(3)$, which is the result of the integral of the product of the impulse response as shown and the part of $x(t)$ which coincides in time with $h(3 - u)$, is marked by an asterisk. The shaded area is illustrating the part of $x(t)$ which is coinciding with the impulse response $h(3 - u)$

convolution process and the definition of the impulse response, it is straightforward to conclude that any causal system will have an impulse response obeying

$$h(t) = 0, t < 0 \tag{2.57}$$

as otherwise it would mean that at an arbitrary time instant $t = t_0$, the convolution process would act on 'future' values of $x(t)$, where $t > t_0$.

Convolution is a very basic mathematical operation. Polynomial multiplication, for example, is an example where the polynomial coefficients are convolved to find the resulting polynomial, although we are then talking about discrete convolution instead of continuous convolution as we have so far talked about. See Problem 2.10 for more details.

## 2.7 The Continuous Fourier Transform

We will soon continue to explain the natural sequel of the transfer function and the impulse response, namely the frequency response. Before we can do that, however, we need a short introduction to the continuous Fourier transform. This transform is the basis for frequency analysis of aperiodic signals, i.e., random and transient signals. The Fourier transform is also useful because it relates the abstract transfer function with the more 'experimentally friendly' frequency response.

The Fourier transform, $X(f)$, of a time signal, $x(t)$, is defined by the *Fourier transform integral*

$$\mathcal{F}[x(t)] = X(f) = \int_{-\infty}^{\infty} x(t) e^{-j2\pi f t} dt. \tag{2.58}$$

The time signal can be calculated from the Fourier transform, $X(f)$, by the *inverse Fourier transform* defined by

$$\mathcal{F}^{-1}[X(f)] = x(t) = \int_{-\infty}^{\infty} X(f) e^{j2\pi f t} df. \tag{2.59}$$

You may not recognize the above definitions, as in maths classes, the Fourier transform is usually defined using the angular frequency $\omega$, which means there has to be a factor of $1/2\pi$ in the definition of the inverse transform. The above definitions, however, are more physically meaningful and appropriate for the signal analysis that we are going to use in this book. Similar to the Laplace transform, as mentioned before, we often refer to the Fourier transform functions as being in the *frequency domain*.

Like the Laplace transform, the Fourier transform is a linear transform, so, if $x_1(t)$ and $x_2(t)$ have Fourier transforms $X_1(f)$ and $X_2(f)$, respectively, and $a_1$ and $a_2$ are real constants, then

$$X(f) = \mathcal{F}[a_1 x_1(t) + a_2 x_2(t)] = a_1 \mathcal{F}\{x_1(t)\} + a_2 \mathcal{F}[x_2(t)]. \tag{2.60}$$

It can be helpful to understand some basic characteristics of the Fourier transform by studying some transform pairs. Therefore, in Table 2.2 some of the most common transform pairs are presented. The description to the left relates to the time signal, except for pair number 8, frequency translation.

The transform pairs numbers 1 and 2 show that differentiation and integration in the time domain is equivalent to multiplication and division by $j\omega = j2\pi f$, respectively. Transform pairs 3 and 4 show that a

**Table 2.2** Common Fourier transform pairs

| # | Description | $x(t)$ | $X(f)$ |
|---|---|---|---|
| 1 | Differentiation | $\dot{x}(t)$ | $j\omega X(f)$ |
| 2 | Integration | $\int x(t) dt$ | $\frac{X(f)}{j\omega}$ |
| 3 | Constant | 1 | $\delta(f)$ |
| 4 | Dirac pulse | $\delta(t)$ | 1 |
| 5 | Gaussian pulse | $e^{-\pi t^2}$ | $e^{-\pi f^2}$ |
| 6 | Symmetry | $X(t)$ | $x(-f)$ |
| 7 | Time translation (delay) | $x(t-\tau)$ | $e^{-j2\pi f \tau} X(f)$ |
| 8 | Frequency translation | $x(t) e^{j2\pi a t}$ | $X(f-a)$ |
| 9 | Complex conjugation | $x^*(t)$ | $X^*(-f)$ |
| 10 | Rectangular window | $1, 0 \le t \le T$ | $T \frac{\sin(\pi f T)}{(\pi f T)} e^{-j\pi f T}$ |
| 11 | Multiplication | $x(t) y(t)$ | $\int_{-\infty}^{\infty} X(u) Y(f-u) du$ |
| 12 | Convolution | $\int_{-\infty}^{\infty} x(u) y(t-u) du$ | $X(f) Y(f)$ |
| 13 | Parseval's theorem | $\int_{-\infty}^{\infty} x(t) y^*(t) dt$ | $\int_{-\infty}^{\infty} X(f) Y^*(f) df$ |

constant in one domain corresponds to a Dirac pulse in the other domain. Transform pair 5 is a somewhat interesting case, showing that the Fourier transform of a Gaussian pulse is a Gaussian pulse. Although not of any great importance for us as such (because we rarely have vibrations corresponding to Gaussian pulses), this fact gives some insight into the Fourier transform. It can be generalized loosely to say that a function which is 'narrower' than a Gaussian pulse in one domain, is 'wider' than the Gaussian pulse in the other domain, and vice versa.

The symmetry property in transform pair 6 shows that if we know a transform pair in 'one direction', we can use the same transform pair in the other direction by replacing the frequency variable $f$ by $-f$. A special result of this is, that if we perform four forward transforms, we end up with the original function, because we get

$$x(t)\xrightarrow{\mathcal{F}}X(f)\xrightarrow{\mathcal{F}}x(-t)\xrightarrow{\mathcal{F}}X(-f)\xrightarrow{\mathcal{F}}x(t). \tag{2.61}$$

Time translation as in pair 7, which is the same as a time *delay* if $\tau$ is positive, is a common effect in filters, and in some mechanical systems. The effect in the frequency domain is to add a linear phase component $\phi(f) = -2\pi\tau \cdot f$. Frequency translation as in transform pair 8 has several important uses in vibration analysis, the most important perhaps that of *zoom FFT* as described in Section 9.3.14. It is also known as amplitude modulation, which we will briefly discuss in Section 16.4.

Complex conjugation in transform pair 9 will be commented in Section 2.7.1 below. Transform pair 10 is very important, as we very often multiply functions with a rectangular window, for example to limit the measurement time. A comment on the form on the right-hand side is motivated; the reason there is a factor $e^{-j\pi fT}$ in the frequency domain is that the pulse in the time domain is not symmetric from $-T/2$ to $T/2$. This corresponds to a time translation of $T/2$ and therefore the Fourier transform will include an exponential term, as shown by pair 7. If the pulse had been symmetric in the time domain, the exponential term would disappear in the frequency domain.

Transform pairs 11 and 12 show a very important property of linear systems, that a multiplication in one domain corresponds to convolution in the other domain. We have included both directions here to make this statement explicitly. Finally, transform pair 13, Parselval's theorem, is particularly interesting when $x(t) = y(t)$. The transform pair then says that the mean square value of the time signal is equal to the frequency summation of $|X(f)|^2$. In other words; instead of calculating an RMS value in the time domain, we can integrate the magnitude squared of the Fourier transform in the frequency domain. The scaling factors will be dealt with in Chapter 9 where we will use Parseval's theorem for discrete spectra.

### *2.7.1 Characteristics of the Fourier Transform*

To appreciate some of the characteristics of the Fourier transform, it is necessary to understand a few fundamental mathematical principles. First of all, the Fourier transform is based on the orthogonality between sines and cosines as described in Section 2.2.4. The integral of the transformed signal multiplied by the complex sine $e^{-j2\pi ft}$ essentially extracts the mean of the product over the whole time interval. If there is some signal content around the frequency $f$, then the integral will result in a nonzero value, otherwise not.

Furthermore, the Fourier transform can obviously be regarded as two integrals

$$X(f) = \mathcal{F}\left[x(t)\right] = \int_{-\infty}^{\infty} x(t)\cos(2\pi ft)\mathrm{d}t - \mathrm{j}\int_{-\infty}^{\infty} x(t)\sin(2\pi ft)\mathrm{d}t. \tag{2.62}$$

In Equation (2.62) it is seen that the real part of the Fourier transform comes from a multiplication of the signal $x(t)$ by a cosine, and the imaginary part from a multiplication by a sine. In order to understand

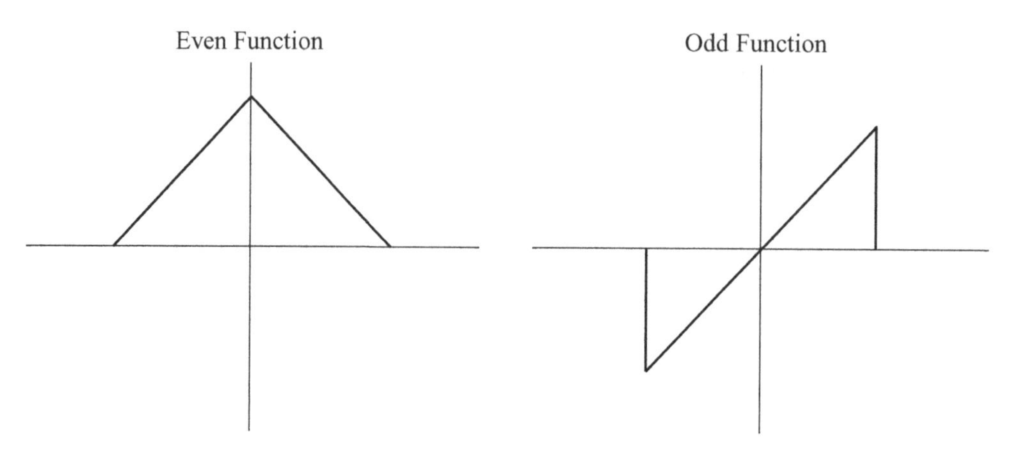


**Figure 2.11** Illustration of even and odd functions

the implications of this, we need to look at the properties of even and odd functions. An even function $x_e(n)$ is a function for which

$$x_e(-t) = x_e(t) \tag{2.63}$$

and an odd function $x_o(n)$ is a function for which

$$x_o(-t) = -x_o(t). \tag{2.64}$$

An even and an odd function are illustrated in Figure 2.11. All real functions can be split into an even and an odd function, which easily follows from the evident relationships

$$x_e(t) = \frac{1}{2}[x(t) + x(-t)] \tag{2.65}$$

and

$$x_o(t) = \frac{1}{2}[x(t) - x(-t)] \tag{2.66}$$

and by observing that the sum of the even and odd function is the original signal $x(t)$.

Some important properties related to even and odd functions, which we present here without proof, are

- The product of an even and an odd function is an odd function.
- The product of two even or two odd functions is an even function.
- A symmetric integral of an odd function is zero.

Because of the last item in the list, even and odd functions are particularly useful when we integrate functions symmetrically around the $x$-axis, for example, the Fourier transform time integral from negative to positive infinity. Obviously the cosine is an even function, and the sine is an odd function. From the above properties it then follows that for a real time signal, $x(t) = x_e(t) + x_o(t)$, the real part of the Fourier transform, $X(f)$, will be

$$\text{Re}[X(f)] = \int_{-\infty}^{\infty} [x_e(t) + x_o(t)] \cos(2\pi f t) \text{d}t = \int_{-\infty}^{\infty} x_e(t) \cos(2\pi f t) \text{d}t \tag{2.67}$$

since the product of the even cosine and $x_o(t)$ will be an odd function, and thus the integral value will be zero. Similarly for the imaginary part

$$-\mathrm{Im}[X(f)] = \int_{-\infty}^{\infty} [x_e(t) + x_o(t)] \sin(2\pi f t) \mathrm{d}t = \int_{-\infty}^{\infty} x_o(t) \sin(2\pi f t) \mathrm{d}t. \tag{2.68}$$

Furthermore, because $cos(-2\pi ft) = cos(2\pi ft)$, it follows directly from Equation (2.67) that

$$\mathrm{Re}\,[X(f)] = \mathrm{Re}\,[X(-f)] = \mathcal{F}\,[x_e(t)] \tag{2.69}$$

Similarly because $sin(-2\pi ft) = -sin(2\pi ft)$ it follows from Equation (2.68)

$$\mathrm{Im}\,[X(f)] = -\mathrm{Im}\,[X(-f)] = \mathcal{F}\,[x_o(t)]\,. \tag{2.70}$$

Hence, the real part of the Fourier transform is an even function and the imaginary part an odd function, and each part consists of the frequency information of the even and odd parts of $x(t)$, respectively.

It should be noted that the results in Equations (2.69) and (2.70) could also be concluded from transform pair 9 in Table 2.2, by noting that the time signal $x(t)$ is real and therefore its complex conjugate equals the function itself. From the Fourier transform pair it then follows that the Fourier transform $X(f)$ of a real signal $x(t)$ must satisfy the relationships in the left-hand sides of Equations (2.69) and (2.70).

### 2.7.2 *The Frequency Response*

In Section 2.6.3 we showed that the impulse response, $h(t)$ is the inverse Laplace transform of the transfer function, $H(s)$. If we use the Fourier transform to transform the impulse response into the frequency domain we obtain the *frequency response function*, $H(f)$, (more often referred to as simply 'frequency response', or FRF). By the Fourier transform relationship number 12 in Table 2.2, and replacing $y(t)$ with the impulse response $h(t)$ and rearranging, it is obvious that the frequency response is

$$H(f) = \frac{Y(f)}{X(f)}. \tag{2.71}$$

The most intuitive interpretation of the frequency response is, that it is the ratio of a sinusoidal output, and a corresponding sinusoidal input. $H(f)$ at each frequency is a complex number and the magnitude of it is the ratio of the two amplitudes, and the phase of $H(f)$ is the phase difference $\angle(Y(f)) - \angle(X(f))$.

While the transfer function is a mathematical, abstract entity which we can use as a tool for solving differential equations, the frequency response is an entity which we can measure experimentally, as we will discuss in Chapters 13 and 14. The frequency response can also be calculated directly from mathematical models in the form of differential equations, or for example from finite element models. It is therefore a very common analysis function for dynamic systems, both analytically and experimentally. In Figure 5.3 on page 93 an example of an FRF of a simple mechanical system with one mass, spring and damper, is shown.

### 2.7.3 *Relationship between the Laplace and Frequency Domains*

The frequency domain and Laplace domain are conceptually quite different. While the frequency domain contains a real frequency axis, the Laplace domain contains a more abstract complex operator, $s$. For physical systems there is, however, a direct relationship between the two which applies to the transfer function. The frequency response can be obtained by evaluating the transfer function on the imaginary axis, or

$$H(\mathrm{j}\omega) = H(s)|_{s=\mathrm{j}\omega}. \tag{2.72}$$

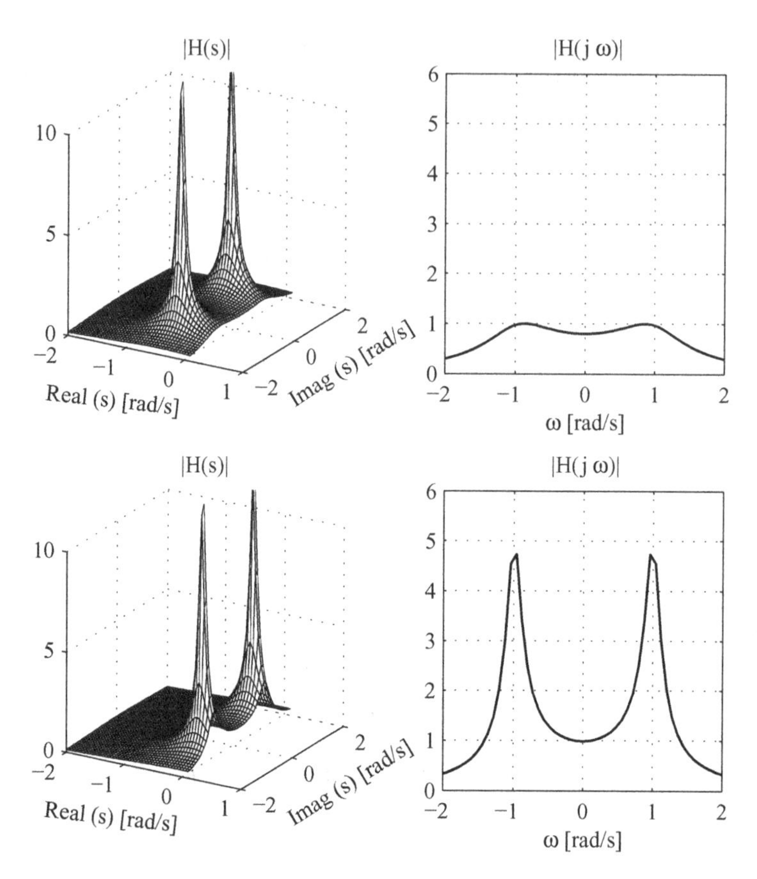


**Figure 2.12** Illustration of the relationship between the Laplace domain and the frequency domain. The frequency response is obtained by evaluating the transfer function on the imaginary axis in the Laplace domain, i.e., where $s = j\omega$. Upper left: magnitude of Laplace transfer function with poles $s_{1,2} = -0.5 \pm j$; upper right: corresponding frequency response; lower left: magnitude of Laplace transfer function with poles $s_{1,2} = -0.1 \pm j$; lower right: corresponding frequency response. From the figure it is clear that the closer to the imaginary axis the poles are located, the larger the peaks in the frequency response

The effect of Equation (2.72) is illustrated in Figure 2.12, for the magnitudes $|H(s)|$ and $|H(j\omega)|$. The transfer function $H(s)$, or at least its magnitude, can be visualized as a surface above the $s$-plane as in Figure 2.12. In the figure, the transfer function has a pair of complex conjugate poles, as typically results from a second-order linear differential equation. When evaluating $H(s)$ on the imaginary axis $s = j\omega$, it is obvious that the closer the poles are to the imaginary axis, the larger are the peaks in the frequency response. In Chapter 5 we will show that this corresponds to lower damping in the system.

## 2.7.4 *Transient versus Steady-state Response*

In Section 2.6.1 we mentioned the transient and forced parts of the solution to differential equations of linear systems. The forced part of the solution is often referred to as the *steady-state* solution, as it is the solution which remains after the transient response has died out. The practical implications of these two properties are very important, particularly how they relate to the Laplace and Fourier transforms.

If we have a linear system, described by its transfer function, $H(s)$, we can solve both the transient and the steady-state response by using the Laplace transform approach, as described in Section 2.6.1. If, instead, we use the frequency response, and multiply it by the Fourier transform of the input signal, we will get only the steady-state response, as the frequency domain only relates to the steady-state response, as we look at a particular frequency at the time.

This often leads to a misunderstanding, however, in the use we have of *measured* frequency responses. Such a function can very well be used to find the transient as well as the steady-state response. If we inverse Fourier transform the frequency response, and use the *impulse response* instead, the convolution of the input signal with the impulse response, will include both the transient and the steady-state response. So, even if we measure the frequency response by a steady-state signal, such as broadband noise (see Chapter 13 on how to measure FRFs), it *includes the information necessary for both the transient and steady-state response.* It is only in the frequency domain we do not have access to the transient part.

## 2.8 Chapter Summary

In this chapter we have introduced some basic concepts about dynamic signals and systems. We started by defining the most fundamental dynamic signal, the *sine wave* $x(t)$ by its amplitude, $A$, its angular frequency (in radians/s), $\omega$, and the initial phase angle, $\phi$, or in equation form

$$x(t) = A \sin(\omega t + \phi) \tag{2.73}$$

and we noted that the circular frequency (in Hz) is $f = \omega/(2\pi)$.

We then introduced the concept of the three fundamental signal classes, periodic signals, random signals, and transient signals. These three classes of signals often have to be separated in signal analysis, as they have very different properties.

Linear systems theory was then introduced. The most important concept of systems theory is to apply a 'black box' kind of approach where the output of the system to a known input can be calculated, or the system can simply be described by one if its important system functions, the *transfer function*, $H(s)$, the *impulse response*, $h(t)$, or the *frequency response*, $H(f)$.

The transfer function $H(s)$ is defined as the ratio $Y(s)/X(s)$, where $Y(s)$ is the Laplace transform of the output signal $y(t)$, and $X(s)$ is the Laplace transform of the input signal $x(t)$, and it defines the properties of the dynamic system. The *poles* of the system, i.e., the roots of the denominator in $H(s)$ are particularly interesting because they give the frequencies where the system has free vibrations (if it is a transfer function of a mechanical system). The transfer function can (in theoretical analysis) be used to find the solutions (responses) for any input (force) by using the inverse Laplace transform, or

$$y(t) = \mathcal{L}^{-1}\left[X(s)H(s)\right]. \tag{2.74}$$

The impulse response, the inverse Laplace transform of the transfer function, can alternatively be used to calculate the output by the convolution integral

$$y(t) = \int_{-\infty}^{\infty} x(u)h(t-u)du. \tag{2.75}$$

The impulse response is experimentally obtainable, although usually indirectly by first estimating the frequency response, $H(f)$, and then use the inverse Fourier transform to calculate $h(t)$.

The frequency response function (FRF), finally, is probably the most commonly used description of dynamic systems, as it is easily estimated from measurements of (for example) forces and acceleration

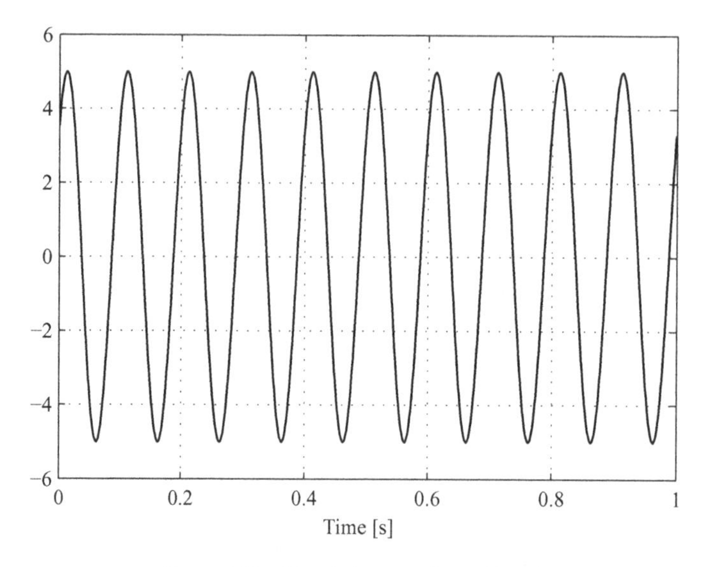


**Figure 2.13** Plot of sine wave for problem 2.1

signals, see Chapters 13 and 14. The FRF of a system can be obtained theoretically from the transfer function by evaluating $H(s)$ on the imaginary axis $s = j\omega$. If it is determined experimentally, it is essentially defined in the frequency domain by

$$H(f) = \frac{Y(f)}{X(f)}. \tag{2.76}$$

## 2.9 Problems

Many of the problems following are supported by the accompanying ABRAVIBE toolbox for MATLAB/Octave and further examples which can be downloaded with the toolbox. If you have not already done so, please read Section 1.6, and follow the instructions to download this toolbox together with example files.

**Problem 2.1** *Determine the expression of the sine wave $x(t) = A\sin(\omega t + \phi)$, which is plotted in Figure 2.13. Answer the following questions:*

*a) What is the amplitude?*
*b) What is the (circular) frequency (in Hz)?*
*c) What is the initial phase angle?*
*d) Derive an expression of a complex sine $\tilde{x}(t)$ describing $x(t)$ so that* $\text{Re}[\tilde{x}] = x(t)$.

**Problem 2.2** *Calculate the root mean square (RMS) value of a signal $x(t)$,*

$$x(t) = 4\cos(4t + \pi/8). \tag{2.77}$$

**Problem 2.3** *Write a MATLAB/Octave function*

```
function m = multsines(f1,f2)
```

*which calculates the mean, m, of the product of the two sines with frequencies* $f_1$ *and* $f_2$*, and plots the product in a figure. Choose a sampling frequency of 10 times the highest frequency of* $f_1$, $f_2$*, and a total time of 10 times the largest period.*

*Use this function to experiment with different frequencies to see for which frequencies the mean does not become zero. Also look at the plots and investigate what happens when the difference between the two frequencies becomes small.*

**Problem 2.4** *Find the poles of the transfer function* $H(s)$*,*

$$H(s) = \frac{1}{s^2 + 2s + 4}. \tag{2.78}$$

**Problem 2.5** *Find the partial fraction expansion of the transfer function in Problem 2.4.*

**Problem 2.6** *Calculate the impulse response* $h(t)$ *of the system described by the transfer function in Problem 2.4. Write a MATLAB/Octave script that plots the function.*

**Problem 2.7** *Calculate the frequency response* $H(\mathrm{j}\omega)$ *of the system described by the transfer function in Problem 2.4. Write a MATLAB/Octave script that plots the function.*

**Problem 2.8** *Prove that changing the order of convolution does not change the result of the convolution, i.e., that* $h(t) * x(t) = x(t) * h(t)$.

**Problem 2.9** *Investigate the MATLAB/Octave commands*

```
poly
roots
residue
```

*to find how to solve a differential equation such as the one in Example 2.6.1 on page 19–20.*

**Problem 2.10** *Investigate the MATLAB or Octave command*

```
conv
```

*to find how to calculate the total polynomial in the numerator in Example 2.6.1 (see Section 2.6.4).*

*(Hint: Try convolving the two polynomials* $s - s_1$ *and* $s - s_2$ *with numbers replacing* $s_1$ *and* $s_2$*. In MATLAB/Octave these polynomials are defined as the vectors V1 = [1, − S1] and V2 = [1, − S2], where you replace variables S1 and S2 by numbers. Compare the results with what you obtain manually!)*

## References

Haykin S 2003 *Signals and Systems* 2nd edn. John Wiley & Sons, Ltd.

Oppenheim AV, Schafer RW and Buck JR 1999 *Discrete-Time Signal Processing*. Pearson Education.

Proakis JG and Manolakis DG 2006 *Digital Signal Processing: Principles, Algorithms, and Applications* 4th edn. Prentice Hall.

Zwillinger D (ed.) 2002 *CRC Standard Mathematical Tables and Formulae* 31 edn. Chapman & Hall.

# 3

# Time Data Analysis

In modern measurement and analysis systems for noise and vibration analysis, it is usually possible to record time signals for later analysis. In many software packages, the time data can then be processed with functions such as filtering, statistics, etc. The aim of this chapter is to introduce some fundamental aspects of digital signals and time domain processing of time discrete signals, necessary to understand in order to correctly analyze time signals. Such processing is often referred to as *time series analysis*.

This chapter will by necessity be brief and introductory in nature. Still, I hope the information will be useful for the non-expert practitioner or for example mechanical and civil engineering students, enabling them to use the information provided to perform some important analysis tasks such as resampling, and octave filtering, in the time domain. Electrical engineering students may also find some otherwise not so readily available information on applications of signal processing. For example, this chapter includes some filters for integration and differentiation of time signals not readily available in most textbooks.

Practical issues regarding data acquisition of time signals are left for Chapter 11, where, for example, the most common type of analog-to-digital converters (ADC), the sigma–delta ADC, will be presented, together with practical considerations of discretization, etc. In the present chapter we will assume that we can somehow produce a sampled signal and we will discuss how to process it to obtain good quality results.

## 3.1 Introduction to Discrete Signals

All signals we measure on mechanical systems are of course analog, that is they are defined in continuous time. When we record signals, they are converted to *time discrete* signals by analog-to-digital conversion (ADC). If an analog signal $x(t)$ is sampled with *sampling frequency* $f_s = 1/\Delta t$, we denote the new, time discrete signal by $x(n\Delta t)$ or, for simplicity, $x(n)$ or sometimes $x_n$. We call $\Delta t$ the *sampling increment*.

## 3.2 The Sampling Theorem

The *sampling theorem*, which is fundamental for all digital signal analysis, was first formulated by Nyquist in 1928, in a paper that is reprinted in Nyquist (2002). The paper by Nyquist did not receive much attention at the time, however. It was not until Shannon added important contributions in a paper in 1949, reprinted in Shannon (1998), that the sampling theorem and its interpretations were more widely spread. Actually, Shannon stated in his paper that the sampling theorem was ‘common knowledge in the

*Noise and Vibration Analysis: Signal Analysis and Experimental Procedures,* First Edition. Anders Brandt.

communication art,' but he is widely acknowledged for formalizing the mathematics of it in a precise and accessible way.

The sampling theorem can be formulated several ways. The formulation we will use here is given in Equation (3.2). Assume the frequency spectrum of a signal $x(t)$ is zero outside a frequency interval, $B = (f_1, f_2)$, the *bandwidth* of the signal, i.e., if we denote the Fourier transform by $\mathcal{F}$, then

$$|X(f)| = |\mathcal{F}[x(t)]| = 0, \ f \notin B. \tag{3.1}$$

The sampling theorem says that the analog signal $x(t)$ can be *uniquely represented* by its discrete samples, if (and only if) it is sampled using a sampling frequency larger than twice the bandwidth, $B$, i.e., $f_s > 2 \cdot (f_2 - f_1)$. If this is fulfilled, then the analog signal can be reconstructed using Equation (3.2)

$$x(t) = \sum_{n=-\infty}^{\infty} x(n)\text{sinc}[f_s(t - n\Delta t)] \tag{3.2}$$

where

$$\text{sinc}(x) = \frac{\sin(\pi x)}{\pi x}. \tag{3.3}$$

Half of the sampling frequency, $f_s/2$, is generally called the *Nyquist frequency* , being named so by Shannon. It is important to understand that what the sampling theorem means is that if the sampling theorem is fulfilled before sampling an analog signal, then the sampled signal, $x(n)$, is an exact representation of the analog signal. In other words the sampled signal contains all information in the analog signal. It is shown in Equation (3.2) by observing that *any* value of the analog signal, $x(t)$, can be calculated using the samples in the discrete signal, $x(n)$.

Why the sampling theorem holds is not easily understood intuitively (mind you it was 'revealed' only as late as 1928, or perhaps 1949). However, it follows from the fact that sampling can be understood as multiplication of the analog signal by a pulse train $s_1(t)$, as illustrated in Figure 3.1. As we know from Chapter 2, multiplication in the time domain is equivalent to convolution in the frequency

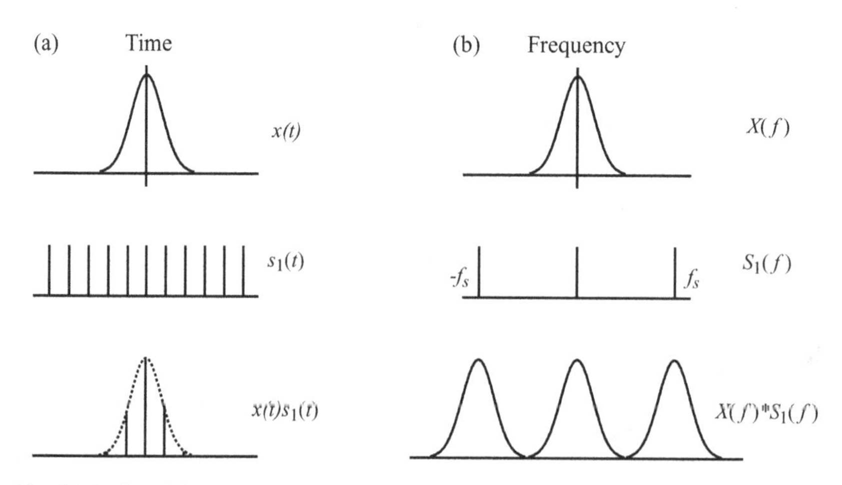


**Figure 3.1** Illustration of the sampling process as a multiplication of the analog signal $x(t)$ by a pulse train $s_1(t)$. (a) the time domain process (b) the frequency domain equivalent, the convolution of the frequency spectra $X(f)$ and $S_1(f)$. The time signal used here is a Gaussian pulse, which has a Fourier transform, also a Gaussian pulse. This is chosen only for the simplicity of the plot. In most cases, of course, the time signal will be some continuous signal. This will be discussed more in Chapter 9

domain. The Fourier transform of a time domain 'pulse train' such as the signal $s_1$ is another 'pulse train' in the frequency domain, with a distance between the pulses equal to the sampling frequency, $f_s$. The convolution of the two spectra thus results in a repetition of the spectrum of the sampled signal $x(t)$ around each multiple of the sampling frequency $f_s$. Thus, in order for the convolution not to mix the spectrum content, the bandwidth of the sampled signal $x(t)$ must be less than $f_s/2$, the Nyquist frequency. By fulfilling the sampling theorem we therefore make sure that the spectrum of the sampled signal, within $|B| < f_s/2$, is the same as the original spectrum of $x(t)$. Thus the original signal can be retrieved back by bandpass filtering the sampled signal $x(n)$. Proofs of the sampling theorem can be found in standard textbooks on digital signal processing, for example (Oppenheim *et al.* 1999; Proakis and Manolakis 2006).

Although not of much practical interest for noise and vibration analysis, it should be noted that the sampling theorem implies that the signal has to be sampled at more than twice the *bandwidth* of the signal and not twice the highest frequency of it, as is often, but incorrectly, said. A band-limited signal, where the frequency content of the signal does not start at 0 Hz and go up to the bandwidth, $B$, but rather lies in some frequency range where $f_{max} \gg B$, can thus be sampled at a sampling frequency much lower than the frequency $f_{max}$. This is frequently used in some applications such as in modern cell phones, where signals at, e.g., the 1.8 GHz band of the GSM network, are sampled at a few MHz.

### 3.2.1 Aliasing

If we do not fulfill the sampling theorem, a phenomenon called *aliasing* or *frequency-folding* will occur. These names, which both refer to the same phenomenon, come from two different ways of illustrating the phenomenon, illustrated in Figure 3.2. Aliasing as illustrated in Figure 3.2(a) occurs if we sample a sine signal with a sampling frequency less than twice the frequency of the sine, which results in a sine signal of a different frequency. For example, the frequencies $0.4f_s$ and $0.6f_s$ will, after sampling, produce the same signal. The same is true for the frequencies $1.1f_s$ and $0.1f_s$. We thus observe that all frequencies are mirrored in the Nyquist frequency, $f_s/2$.

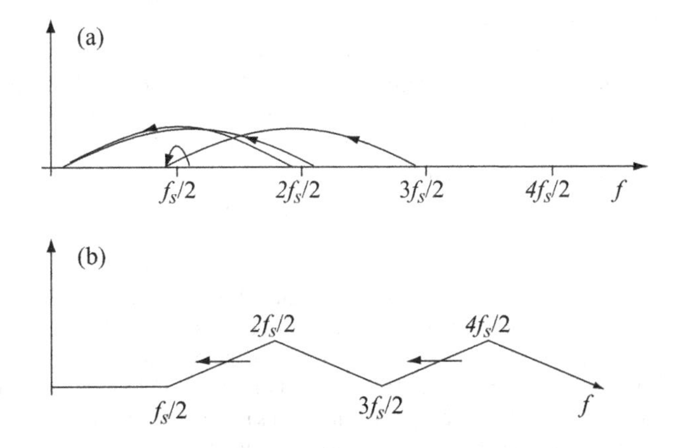


**Figure 3.2** Illustration of (a) aliasing; (b) folding. In (a) a signal (sinusoid) which lies some small amount above half the sampling frequency, behaves, after sampling, as if it were a sinusoid lying the same amount under half the sampling frequency. It thus behaves like a signal with a frequency other than that which it really has, therefore the name aliasing. In (b) the same phenomenon is illustrated through so-called frequency folding which name arises from observing that the frequency axis is folded at all multiples of half the sampling frequency, as in the figure. After the folding of the frequency axis is completed, the entire frequency axis will go 'back and forth' between 0 and $f_s/2$

Another way of illustrating the same phenomenon is found by cutting the frequency axis at each multiple of the Nyquist frequency, $f_s/2$, and folding the frequency axis like an accordion around these points, as illustrated in Figure 3.2(b). Thus the name folding. For broadband signals (signals with continuous frequency content, i.e., random and transient signals) aliasing will still occur, of course, but in a more complicated way.

Perhaps you notice a contradiction here. In Section 3.2 we said that the sampling theorem relied on the *bandwidth* of the signal. But the bandwidth of a sine is really zero, so the discussion on aliasing/folding here must surely be mushy? Indeed, if we know *a priori* that a signal is a sine, we can actually sample it at any (low) sampling rate and reproduce it according to the sampling theorem. But that relies on the fact that *we have to know it is a sine.* The illustration of aliasing here, is an effect of the periodicity in the spectrum illustrated in Figure 3.1. It says that, *if we do not band-limit the signal* prior to sampling it, we will not be able to tell whether a peak in the spectrum belongs to a frequency component at that frequency, or from a frequency component at that frequency plus some multiple (positive or negative) of the sampling frequency. By not band-limiting the signal prior to sampling, we indirectly assume that all frequency components of the signal are in the frequency region between 0 and $f_s/2$. So there is no contradiction here. Actually, aliasing is very easy to illustrate, see Problem 3.1.

A direct implication of the sampling theorem is that we must make sure before sampling a signal, that it has no frequency content above half the sampling frequency. This is done by an analog *antialias filter* before the A/D converter (ADC) and will be discussed in Chapter 11. The antialias filter must have a cutoff frequency below $f_s/2$, as the slope of any analog filter above the cutoff frequency is finite. The ratio between the sampling frequency and the cutoff frequency of the antialias filter is called the *oversampling ratio* (sometimes oversampling factor), and the steeper the slope of the antialias filter, the lower the oversampling factor needs to be. Traditionally the oversampling factor was always 2.56 in FFT analyzers, but in some modern analyzers with sigma–delta ADCs, the oversampling factor has been reduced, see Chapter 11.

### 3.2.2 Discrete Representation of Analog Signals

An effect of the limited oversampling ratio used in most measurement systems designed for noise and vibration signals is that time signals do not appear correct when plotted. In Figure 3.3 a signal is plotted with 2.56 times oversampling, and with 20 times oversampling. It is evident from a careful study of the figure that the signal with low oversampling ratio is not accurately describing the signal. The signal seems to 'jump' between the sampling points, which is of course impossible. It is important to understand that the signal with 2.56 times oversampling still includes all *information* in the signal. This means, among other things, that a spectrum computed from this signal will be correct. Only the time domain representation of the signal is limited due to the low oversampling ratio. This will be further discussed in Section 3.4.1.

Another effect often encountered when measuring noise and vibration signals is the effect on the time domain representation, of reducing the bandwidth of a signal. In Figure 3.4(a), a pulse from an impact test (see Chapter 13) using a (high) sampling frequency of 10 kHz is shown. This results in a bandwidth of the signal of approximately 4 kHz (10 kHz divided by 2.56, the oversampling ratio). The time signal is well described as an approximate half-sine (half a period of a sine). In Figure 3.4(b), the same signal is displayed with a bandwidth of 400 Hz, corresponding to a sampling frequency of approx. 1 kHz at an oversampling ratio of 2.56. Note how the signal with low bandwidth oscillates before and after the pulse. This effet, known as the Gibbs phenomenon, is strictly due to the limited bandwidth, and does not mean that there is anything wrong with the signal (unless, of course, we want to accurately describe it, for example by its peak and width, which are obviously not described well in Figure 3.4(b)). It is important to understand this effect in order not to misinterpret signals similar to the one in Figure 3.4(b) to indicate an error.

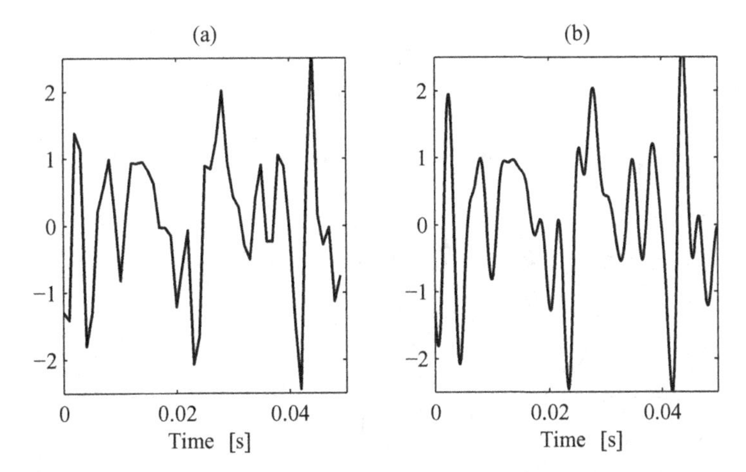


**Figure 3.3** Illustration of oversampling ratio. A signal measured by a noise and vibration measurement system is usually sampled using an oversampling ratio of 2.56 or some similar number, as illustrated in (a). This signal is an inaccurate description of the analog signal because of the low oversampling ratio. Note the abrupt change of the signal with sharp edges between the samples. If this is, for example, an acceleration signal, it would mean that the structure (point where the sensor is located) would have a 'violent' behavior with rapidly changing acceleration. That is, of course, not physical. Increasing the oversampling ratio to 20 times, as in (b), the true signal behavior, with smooth variations, is revealed. The signal samples in (a) are, however, enough to represent the information in the original, analog signal. As we will see later in this chapter and in later chapters of this book, the signal in (b) can be produced from the signal in (a), and spectra can accurately be computed from the signal in (a). However, some time domain information cannot be directly extracted from the signal in (a), for example min- and max-values, etc., see Section 3.4.1

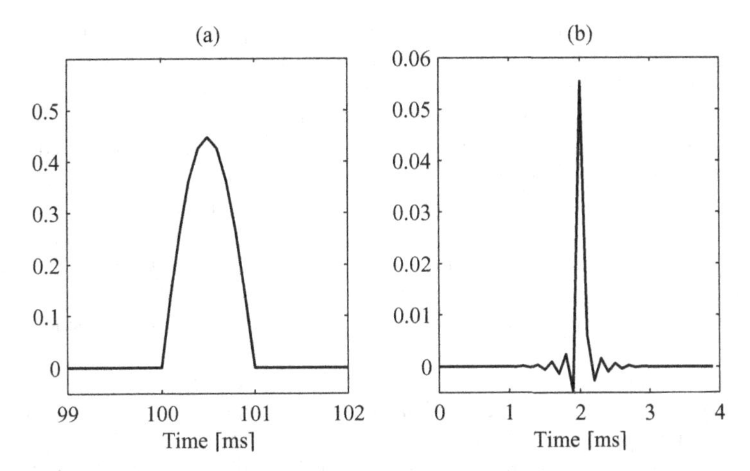


**Figure 3.4** Illustration of the effect of reducing the bandwidth of a signal. In (a) a half-sine pulse sampled with a sampling frequency of 10 kHz and an oversampling ratio of 2.5 is shown. In (b), the same signal is shown after sampling by a sampling frequency of only 1 kHz, and a bandwidth of 400 Hz. Note how oscillations appear both before and after the pulse, and the pulse amplitude is highly affected. Due to the nature of linear systems, where each frequency is independent of all other frequencies (see Section 2.6), the signal in (b) is entirely valid in the frequency range up to 400 Hz, however, as we will illustrate in Section 13.8, where we will discuss impact testing, a common method for exciting structures where the phenomenon illustrated here often appears

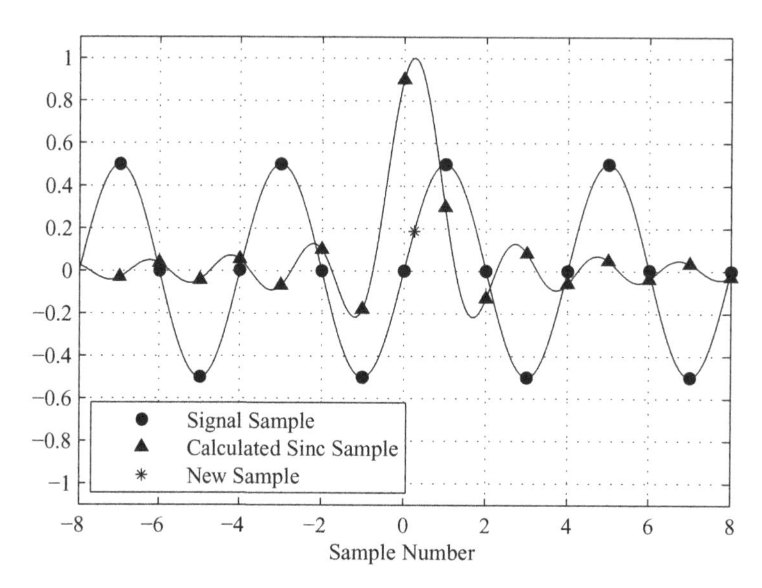


**Figure 3.5** Illustration of exact (sinc) interpolation according to Equation (3.2). The sinc function is centered at the time value to be calculated, then it is sampled at the time points which coincide with the sampled function, and finally the product of corresponding samples is summed to produce the new sample value. Note that only a few values around the point of interpolation are shown for clarity, but that more points outside the figure where used to produce the accurate new sample value. See text for comments

## *3.2.3 Interpolation and Resampling*

The formula in Equation (3.2) can be used for the purpose of *interpolation*, that is, to compute a value of the analog signal, at any value of $t$ between the samples. The interpolation principle according to Equation (3.2) is illustrated in Figure 3.5. The sinc function in Equation (3.2) is centered at the time $t$, where we wish to calculate a new sample as is seen in the figure. Then the sinc function is calculated at every time increment $n\Delta t$ where we have our samples. The two functions are multiplied together, and summed to form the new sample value $x(t)$.

In practice, we do not have an infinite number of samples. This produces a truncation error when interpolating data, reducing the accuracy of interpolation that can be performed after sampling. The sinc function falls off by $1/t$ away from the center point, which is a rather slow decay. To obtain a high accuracy when interpolating signals, rather long interpolation filters should therefore be used. In fact, in order to produce the accurate new sample in Figure 3.5, several hundred samples outside the plot range were used.

Another important error occurs at the ends of the data where there is limited (or no) data on one side of the interpolation point $t$ since the interpolation formula in Equation (3.2) is symmetric around the point of interpolation. Some data at the ends should therefore preferably be excluded after resampling; a hundred samples are normally sufficient for time domain analysis. If very accurate data are needed, for example in order to perform input/output analysis in the frequency domain, many thousand points may have to be discarded at each end.

**Example 3.2.1** *In many software packages there are efficient algorithms to interpolate a signal. In MATLAB/Octave, there are several ways, one of the best being the* **resample** *command. In this example we will illustrate the process of resampling a signal with a different sampling frequency which is often*

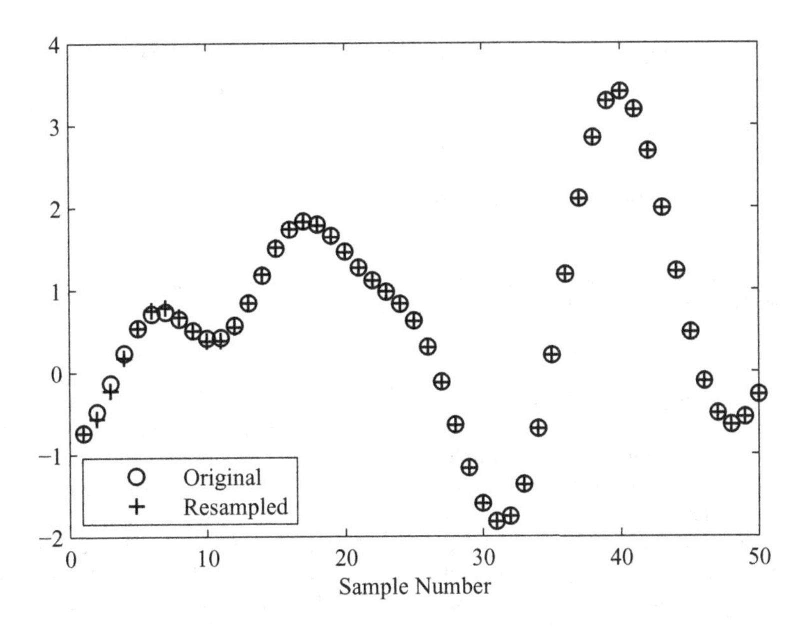


**Figure 3.6** Plot for Example 3.2.1. The original signal samples are indicated by rings, and the recreated samples obtained by interpolation of the decimated signal are indicated by plus signs. There is apparently an error for the first samples, due to the lack of symmetry at the end of the data

*of great use for different analysis processes. The following MATLAB/Octave code produces a random signal with 10 times oversampling which is then decimated by a factor 4 by taking every 4th sample. The thus decimated signal is then interpolated (resampled) up to the original sampling frequency again, and the samples are plotted of both signals with rings and plus signs, respectively.*

```
N=1000;                    % Number of samples to start with
x=randn(N,1);              % Gaussian noise, oversampling ratio is 2
x=resample(x,5,1);         % Resample to 10 times oversampling
x25=x(1:4:end);            % Oversampling ratio of x25 is 2.5
xr=resample(x25,4,1);      % oversampling ratio of xr is 10
% Plot 'original' data x, and the resampled xr
plot(1:50,x(1:50),'ok',1:50,xr(1:50),'+k')
legend('Original','Resampled')
xlabel('Sample Number')
```

*The result of the plot is shown in Figure 3.6 where it can be seen that except for a few samples at the beginning of the signal, the interpolation reproduces the same samples that were formerly removed by the decimation. Note that the use of a random signal on the second line will result in a different signal every time you run the code above.*

*End of example.*

If we have sampled a signal at a particular sampling rate, we can also use interpolation to calculate the signal corresponding to another sampling rate. This is usually referred to as *resampling* the signal, as was used by the command **resample** in Example 3.2.1. When resampling a signal, it is essential, first of all, to differentiate between *upsampling* and *downsampling*, respectively. In the first case, upsampling a signal

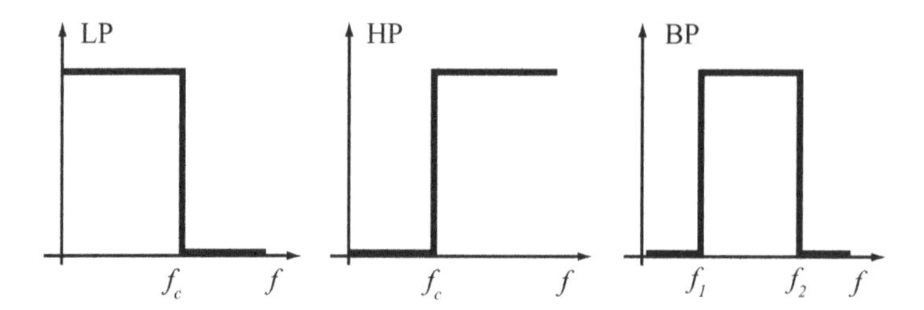


**Figure 3.7** Ideal filter characteristics. 'LP' indicates the characteristic of an ideal lowpass filter, 'HP' the ideal highpass filter characteristic, and 'BP' the ideal bandpass characteristic

means increasing its sampling frequency. This can be done by directly implementing the interpolation formula in Equation (3.2) at suitable locations for a new sampling frequency, usually at some fractional of the original sampling frequency. The **resample** command, for example, has a syntax **resample(x,P,Q)** which resamples the signal at a new sampling frequency of $f_{s,\text{new}} = P/Qf_{s,\text{old}}$, and the ratio $P/Q$ can be larger or smaller than one.

If downsampling is required, however, it is a completely different case, as this will potentially produce aliasing if the bandwidth of the signal is not taken into consideration. A lowpass filter must then be applied prior to decimating data (removing samples is called decimation in this context). The resample command in MATLAB/Octave is applying such a lowpass filter when needed, and provides a good way of ensuring accurate resampling for upsampling as well as downsampling.

Finally the potential of using the **resample** command for lowpass filtering should be stressed. If the purpose is to increase the oversampling ratio, instead of designing a lowpass filter and having to elaborate on things such as time delay, phase linearity, filter slopes, etc., in most cases it is better to use the **resample** command. To reduce the bandwidth of the signal, you first resample down to a lower sampling frequency, twice the bandwidth you want, and then up again, as was done in Example 3.2.1. The new signal will be equivalent to the original signal filtered by a very sharp lowpass filter (with near ideal filter characteristics as indicated in Figure 3.7, see next section).

## 3.3 Filters

The design of suitable filters for various purposes is a large part of the field of signal processing which requires a deep understanding of many aspects of signal processing to be applied correctly. In this section we will discuss some basic principles, that will allow you to use for example MATLAB/Octave for some filtering operations that are often required for noise and vibration signals.

A filter is often described by its frequency response, which for filters is often referred to as the *filter characteristic*. Very often we want to filter a signal with an *ideal* filter characteristic, whereas, due to physical limitations, we have to do with some compromise between computational efficiency and filter characteristic. The three most common types of filters are the *lowpass, highpass* and *bandpass*, filters illustrated (as ideal filters) in Figure 3.7. As the names imply, the lowpass filters let low frequencies pass, and consequently high frequencies are blocked, or filtered away. Similarly, the highpass filter is used to filter away low frequencies, whereas for the bandpass filter all frequencies, except those in a certain *passband region*, are filtered away.

The ideal filters in Figure 3.7 cannot be physically realized. With digital filter designs, however, it is possible to get arbitrarily close to ideal characteristics, at the expense of two sorts, namely computational cost, and time delay. This will be discussed in Section 3.3.2. The common way to describe filter characteristics, is by the *asymptotic behavior* of the filter. A lowpass filter, for example, is then described by two parameters; its *cutoff frequency*, $f_c$, and the *slope of the filter* above that cutoff frequency. It is particularly important to understand that the cutoff frequency is almost always defined as the *−3 dB frequency*, where the gain function of the filter has decreased by 3 dB, which means that, for a lowpass filter,

$$|H(f_c)| = \frac{|H(f \ll f_c)|}{\sqrt{2}}. \tag{3.4}$$

In measurement applications this means that, expressed as an error, the error at the cutoff frequency is approximately 30 %; a rather significant number. If we want to estimate the RMS value of a sine, for example, it is therefore essential to carefully evaluate the effects of filter cutoff frequencies.

### 3.3.1 Analog Filters

Much of the filtering theory still used goes back to the days of analog filters, before digital signal processing was common. Thus a few words on some common analog filter characteristics is motivated. An analog filter can be characterized by its gain and phase functions, which is the magnitude and phase of the frequency response of the filter.

The simplest analog filter is a so-called *first-order filter* which for a lowpass version has a filter characteristic of

$$H(j\omega) = \frac{1}{1 + (j\omega/\omega_c)} \tag{3.5}$$

where $\omega_c$ is the filter cutoff frequency. The first-order lowpass filter is common in electronics, although in vibration equipment, the first-order highpass filter is perhaps even more common, as it is included in many sensors and signal conditioning units, see Chapter 7.

The first-order highpass filter has a characteristic

$$H(j\omega) = \frac{(j\omega/\omega_c)}{1 + (j\omega/\omega_c)} \tag{3.6}$$

which is plotted in Figure 3.8 in amplitude and phase versus frequency. The gain of the first-order highpass filter is

$$|H(j\omega)| = \frac{(\omega/\omega_c)}{\sqrt{1 + (\omega/\omega_c)^2}} \tag{3.7}$$

and some useful numbers for measurement applications is found in Table 3.1.

One of the most common general filters is the *Butterworth filter*, which (for a lowpass filter) has a filter gain of

$$|H_b(j\omega)| = \frac{1}{\sqrt{1 + (\omega/\omega_c)^{2n}}} \tag{3.8}$$

where the integer $n$ is called the *filter order*. A comparison of Equation (3.8) with the magnitude of Equation (3.5) shows that a first-order lowpass filter is identical to a first-order Butterworth filter with $n = 1$. The Butterworth filter is a useful filter because it has maximum flat gain characteristic, and the phase characteristic is relatively linear. It is therefore a commonly used filter in many applications. For example, the filters used for standardized octave and third-octave filtering are third order Butterworth filters, see Section 3.3.4 where we will also look into how MATLAB/Octave can be used to define digital versions of Butterworth filters.

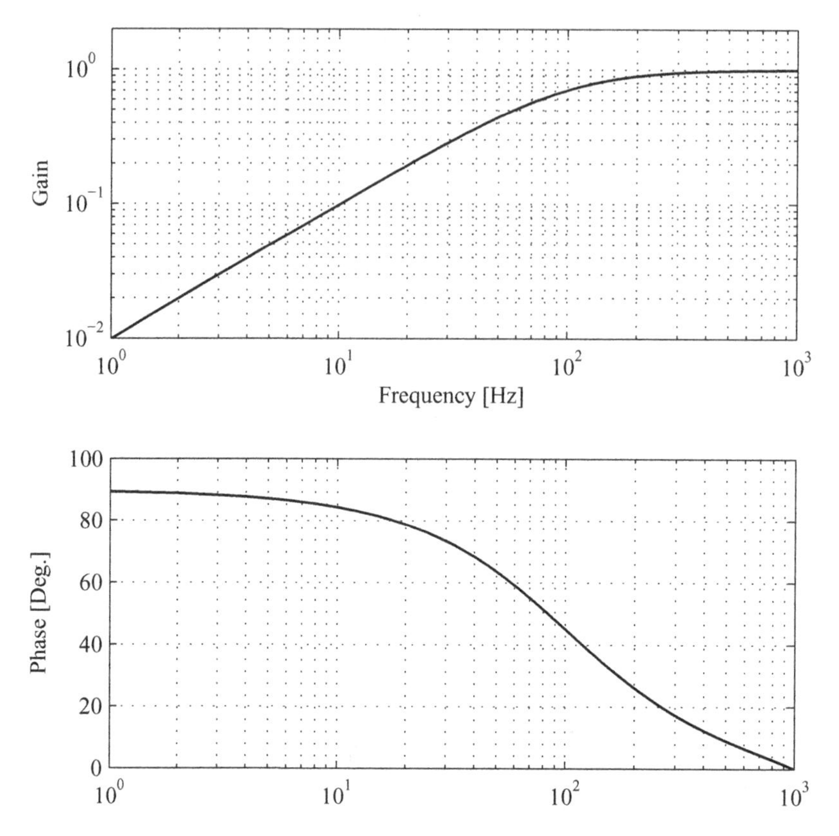


**Figure 3.8** Filter characteristics of a first-order highpass filter with a cutoff frequency of $f_c = 100$ Hz. It is common to plot filter gains with a scale in dB. However, for physical interpretation of actual numbers, we have chosen to show the gain characteristic on a logarithmic scale here

The filter order determines the *asymptotic slope* of the filter, which is approached at frequencies higher than the cutoff frequency (for a lowpass filter). It is easy to determine that the asymptotic slope is −20 dB/decade or −6 dB/octave per order of the filter. A decade is an increase in frequency by 10 times, whereas an octave is a doubling of frequency. For a third-order filter, for example, the slope is thus −60 dB/decade.

**Table 3.1** Values of gain and phase characteristics of a first-order highpass filter

| $f/f_c$ [–] | $\lvert H(f)\rvert$ [–] | $\angle H(f)$ [°] |
|---|---|---|
| 0.1 | 0.0995 | 84.3 |
| 0.9 | 0.669 | 48.0 |
| 1 | 0.707 | 45.0 |
| 2 | 0.894 | 26.5 |
| 10 | 0.995 | 5.71 |
| 20 | 0.999 | 2.86 |

### 3.3.2 Digital Filters

Digital filters are often used when analyzing noise and vibration data, for example to reduce the bandwidth of a signal prior to performing frequency analysis or for acoustic octave and third octave filtering, human whole body vibration (comfort) filters, or shock response spectra. Digital filters and their design is a discipline of its own in the field of digital signal processing, and here we will touch only on some simple facts that are important to understand from a user perspective.

A rather general definition of a digital filter between an input $x_n$ and an output $y_n$ is shown in Equation (3.9)

$$a_0 y_n = \sum_{k=0}^{N_b} b_k x_{n-k} - \sum_{l=1}^{N_a} a_l y_{n-l} \tag{3.9}$$

where the filter coefficients $a_l$ and $b_k$ are defining the filter characteristics. If $N_a$ is zero, i.e., the filter is using only old input values to compute the output, then the filter is called a Finite Impulse Response (FIR) filter, as it will have a finite impulse response of length $N_b + 1$. The impulse response is the output for an impulse input $x_0 = 1$ and $x_n = 0$ for all remaining $n$. If, on the other hand, there are nonzero $a_l$ coefficients in the filter, it is called an *Infinite Impulse Response* (IIR) filter, or sometimes a *recursive* filter. IIR filters are in general more time efficient than FIR filters, i.e., more powerful filter characteristics can be accomplished using fewer filter coefficients. In some special cases, however, FIR filters are preferred, especially if linear-phase characteristics are wanted, as will be discussed below.

Often we would like to create a digital filter with characteristics equivalent to a specified analog filter, because most filter theory was developed in the analog era. There is no exact such transformation, however, and thus the science of digital filters deals to a great extent with how to make the digital filter that best approximates the analog filter characteristics, in some respect. Here, it is sufficient to mention a few basic facts about digital filter performance that a user must be aware of.

It is important to know that the digital filter approximation of an analog filter performs worse and worse the closer one gets to the Nyquist frequency. So, if the filter characteristics are defined in the analog frequency domain, such as for octave filters, and whole body filters, one should use sufficient oversampling prior to filtering the signal. In the IEC/ANSI standards for octave filters, (IEC 61260 1995) in Europe and (ANSI S1.11 2004) in United states, for example, a minimum oversampling of 5 times the highest center frequency is recommended, see Section 3.3.4. In general, it is recommended to use at least 10 times oversampling when performing digital filtering of data unless you are sure some other factor is better.

The two most common transforms used to convert an analog filter to a corresponding digital filter are the *bilinear* and the *impulse invariant* transforms, although there is a large variety of other transforms available, all with different advantages and disadvantages. To thoroughly understand these transforms it is convenient to understand the $z$-transform, which we will not introduce here. It is, however, possible to compute the frequency response of the digital filter directly from the coefficients $a_l$ and $b_k$ in Equation (3.9), so a $z$-transform model is not needed, although it certainly aids in the understanding of digital filters (and discrete systems in general). In MATLAB/Octave the frequency response can be computed from the digital filter $a_l$ and $b_k$ coefficients by the command **freqz**.

An important issue with filters is the *time delay*. Most digital filters delay the signal by some number of samples, sometimes an integer number samples, and sometimes a fractional number of samples. To calculate the delay of a filter, it is useful to define the *group delay* of the filter which is defined as

$$\tau_g(\omega) = -\frac{\mathrm{d}}{\mathrm{d}\omega} \angle(H(\mathrm{j}\omega)) \tag{3.10}$$

and thus (in most cases) is a frequency dependent number. The group delay of a digital filter is scaled in samples if no sampling frequency (or, actually, a sampling frequency of 1 Hz) is used. In MATLAB/Octave, the command to calculate it is **grpdelay**. For data analysis, it is sometimes necessary, or at least convenient, that the time delay of the filter be an integer number of samples, so that the data at the output of the filter are still synchronously sampled with the input data.

In addition to the delay of the filter it is also important to understand that a filter can have a *transient response,* just like a physical system. From Equation (3.9) it follows that, in the simplest case with a FIR filter, it will take $N_b$ samples before the input 'fills' the filter so that the output is actually calculated using the full length of $N_b$ samples. The filter transient can be several times longer than this length. However, just like for physical systems, transient effects are longer the less damping the filter has. For the types of filters presented in this chapter, the damping is high, and transient effects thus less disturbing.

An important concept relating to filters is the notion of *phase distortion.* If we ask ourselves what the phase characteristic of a filter should be, in order for a certain signal to pass the filter with the same relative phase between different frequencies, it turns out that the answer is that the phase should be linear with frequency, see Problem 3.2. Of course zero phase characteristic for all frequencies could also be a solution, but this is unfortunately impossible to achieve if the filter is to have a gain characteristic other than a constant.

The easiest way to produce a filter with linear phase is by designing a FIR filter with symmetric coefficients. For a linear-phase filter the group delay is constant and for the most common type of linear-phase filters, FIR filters with symmetric coefficients (see Section 3.4) of length $2N + 1$, the time delay is always equal to $N$.

We have mentioned that the oversampling factor usually needs to be at lest 5–10 times. The oversampling factor should, however, in general not be too large, as this produces larger filters (more filter coefficients), and potential numerical truncation problems.

A useful trick to produce linear phase characteristics, even for an IIR filter, can be achieved by filtering the data first in the normal (time) direction, and then running the filter backwards in time. The scaling of the filter has to be considered when using this method. In MATLAB/Octave, there is a command, **filtfilt**, which performs this type of filtering, including scaling the output so that the efficient (gain) filter characteristic is the same as for the IIR filter used in the 'normal' way.

It is rather common to use several filters connected in so-called *cascade coupling*, which means that one filter is following another, practically meaning we first filter data with one filter, and then use that output as input to a new filter. Due to the fact that polynomial multiplication is equivalent to convolution of the two polynomials, convolving the filter coefficients of the numerator and denominator, respectively, of each filter, produces the total digital filter parameters. This will be illustrated later in this chapter, in Example 3.3.4.

### *3.3.3 Smoothing Filters*

A common filter for averaging several adjacent values in a signal together is the *smoothing filter.* The simplest digital smoothing filter is a filter with the same weighting factors for all filter coefficients. Such filters are frequently used in order tracking applications (see Chapter 12), and many other applications. To obtain linear phase characteristics, the **filtfilt** function in MATLAB/Octave is preferably used. We illustrate this with an example.

**Example 3.3.1** *Filter a signal in variable x with a smoothing filter of length L* $= 10$ *samples.*

*We define the digital filter a and b coefficients and subsequently filter the content of variable x by the MATLAB/Octave code*

```
L=10;     % Smoothing filter length
a=1;
b=1/L*ones(1,L);
xfiltered=filtfilt(b,a,x);
```

*End of example.*

**Table 3.2** Center frequencies for standardized 1/1-and 1/3-octave filters as specified in (ANSI S1.11 2004; IEC 61260 1995) for some typical frequencies in the audio band. The frequency series continues with the same numbers multiplied or divided by 10, 100 and so on, for frequencies higher and lower, respectively, than those specified in the table

| | Center frequency [Hz] | | | Center frequency [Hz] | |
|---|---|---|---|---|---|
| Band | 1/1-octave | 1/3-octave | Band | 1/1-octave | 1/3-octave |
| 15 | 31.5 | 31.5 | 23 | | 200 |
| 16 | | 40 | 24 | 250 | 250 |
| 17 | | 50 | 25 | | 315 |
| 18 | 63 | 63 | 26 | | 400 |
| 19 | | 80 | 27 | 500 | 500 |
| 20 | | 100 | 28 | | 630 |
| 21 | 125 | 125 | 29 | | 800 |
| 22 | | 160 | 30 | 1000 | 1000 |

### 3.3.4 *Acoustic Octave Filters*

In acoustics (particularly) it is very common to analyze the frequency content of signals by means of a set of parallel bandpass filters, whose time domain output signals are analyzed various ways, as we will discuss in Section 9.2. The bandwidth of the filters are usually either a whole octave (1/1-octaves) or sometimes one-third of an octave (1/3-octaves), and center frequencies and filter characteristics are standardized in (IEC 61260 1995) for Europe, and in (ANSI S1.11 2004) for U.S. Those two standards are compatible so that center frequencies and filter shapes are identical.

The standardized center frequencies for whole-octave and third-octave filters are tabulated in Table 3.2.

Octave and third-octave filters are two examples of a more general set of $1/n$ octave bands, where the integer, $n$, is usually (but not necessarily) 1, 3, 6, 12, and 24. The higher fractional octave bands were used more in the past when narrowband analysis using FFT was not as readily available as it is today. Today there is little use for those filter types, but they are still sometimes used. To define the center frequencies and bandwidths of fractional octave bands, the standard specifies an octave ratio, $G$, defined by

$$G = 10^{3/10} \approx 1.9953 \tag{3.11}$$

which is then used to calculate the exact center frequencies of each filter by the formulas

$$\begin{aligned} f_c &= 1000G^{(x-30)/n}, \; b \text{ odd} \\ f_c &= 1000G^{(2x-59)/n}, \; b \text{ even} \end{aligned} \tag{3.12}$$

where $x$ is a positive or negative integer corresponding to the band number.

The filters for $1/n$ octave bands are specified as third-order Butterworth filters with the center frequencies from Equation (3.12) and with lower and upper edge bands, $f_l$ and $f_h$ according to

$$\begin{aligned} f_l &= f_c G^{-1/2n} \\ f_h &= f_c G^{1/2n}. \end{aligned} \tag{3.13}$$

Using Equation (3.13) it is easy to verify that $f_c = \sqrt{f_l f_h}$ and that

$$f_h - f_l = f_c \left[ G^{1/2n} - G^{-1/2n} \right] \tag{3.14}$$

which, for example, for a third-octave band is approximately $0.23 f_c$ or 23%.

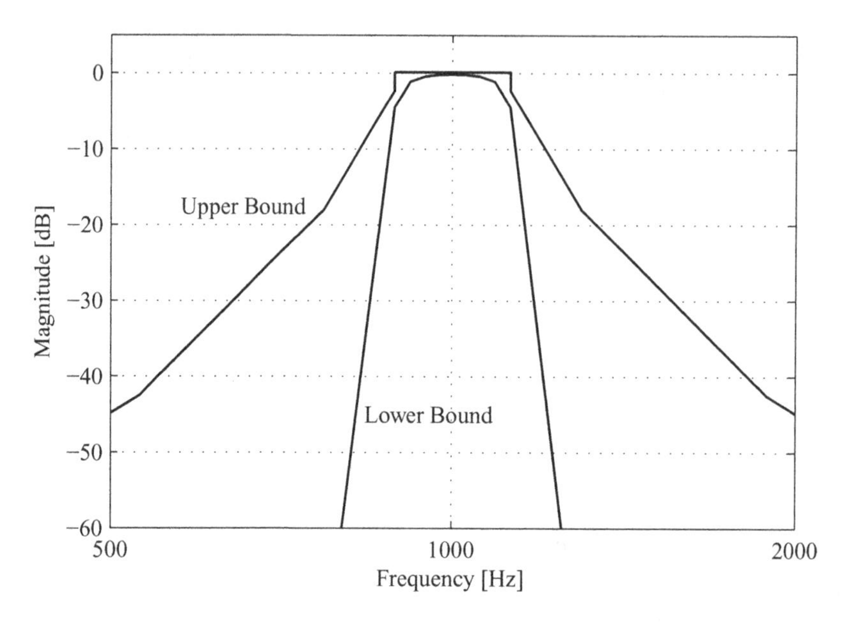


**Figure 3.9** Octave filter limits for an example 1/3-octave filter with center frequency 500 Hz, according to ANSI and IEC standards. See text for details

The filters are allowed to vary within certain limits, as plotted in Figure 3.9 for an example of a 1/3-octave band with center frequency of 1000 Hz. It is easy to make a MATLAB/Octave function that calculates the upper and lower bounds for a particular $1/n$ octave filter so that one can verify that the filter is within the specified bounds.

To define a digital filter which is in accordance with the standards mentioned above, one can use the MATLAB/Octave **butter** command. It is however essential to check the filter shape obtained by this command, as for a certain ratio of center frequency to sampling frequency, the limits specified by the standard will not be met. This is easily done by the **freqz** command mentioned above, as we will show in Example 3.3.2.

**Example 3.3.2** *In this example we will show how to create a fractional filter corresponding to a 1/3-octave filter for band 30 (center frequency 1000 Hz) in MATLAB/Octave. By changing the parameter n in the code below, any 1/n octave filter can be produced.*

```
G=10^0.3;
n=3;
fc=1000;                    % 1/3-octave center freq.
flo = fc/G^(1/2/n);         % low cutoff freq. (definition)
fhi = fc*G^(1/2/n);         % high cutoff freq. (definition)
N=8*1024;                   % Number of frequency lines for H
fs=10000;                   % Sampling frequency
[b,a] = butter(3,[flo/(fs/2) fhi/(fs/2)]);
[H,f]=freqz(b,a,N,fs);    % Filter frequency response
% Perform the acoustic filtering on signal in vector x
y=filter(b,a,x);
```

*End of example.*

### 3.3.5 Analog RMS Integration

In acoustic analysis it is very common that the time signal is analyzed with a running RMS value with a particular integration time, corresponding to the output of an analog integrator as was used in the past. This is what is done in a sound level meter (SLM), where the RMS value of the integrated data is simply converted to a sound pressure level in dB SPL (sound pressure level), see Appendix C. Such a filter is easily designed as a first-order Butterworth lowpass filter, noticing that the time constant (integration time), $\tau$, of an analog filter obeys the relationship

$$\tau = \frac{1}{\omega_c} = \frac{1}{2\pi f_c} \tag{3.15}$$

where $\omega_c$ is the filter cutoff frequency in rad/s and $f_c$ is the cutoff frequency in Hz. In order to get the right scaling of the filter we need to refer back to Equation (3.5) and compare it with the true integrator frequency response of $H(\mathrm{j}\omega) = 1/\mathrm{j}\omega$. We realize we need to divide the filter characteristic of the first-order lowpass filter by $\omega_c$ to get an asymptotic behavior of the filter of $1/\mathrm{j}\omega$. The principle is most easily illustrated by an example.

**Example 3.3.3** *In this example we illustrate how to make an analog integration with a so-called* fast *time constant as specified in many acoustical applications. This means that the time constant should be* $\tau = 1/8$ *[s] (another common time constant in acoustics, the* slow *time constant, is 1 s). We assume the data in vector x are scaled in Pa as coming from a microphone. We recall from Section 3.3.1 that the first-order lowpass filter is identical to a first-order Butterworth filter. For sound, the sound pressure level,* $L_p$*, is the 'instantaneous' integrated RMS value with a particular time constant, in dB relative to 20* $\mu$*Pa. The equivalent sound pressure level,* $L_{eq}$*, is the total RMS level in dB with the same reference. The following MATLAB/Octave code designs an integration filter and produces* $L_p$ *and* $L_{eq}$*, after filtering the signal through an analog integrator filter.*

```
fs=44100;                      % Our sampling frequency
tau=1/8;                       % Time constant in s
fc=1/(2*pi*tau);               % Cutoff freq. in Hz
[b,a]=butter(1,fc/(fs/2));     % Integrator filter
y=filter(b,a,x.^2);            % Filter squared signal
y=y/(2*pi*fc);                 % Scaled, integrated square
y=sqrt(y);                     % Root mean square complete
Lp=20*log10(y/2e-5);           % Sound pressure level in dB
Leq=20*log10(std(y)/2e-5);     % Leq level in dB
```

### 3.3.6 Frequency Weighting Filters

Many noise and vibration applications include some form of frequency weighting, such as acoustic A or C weighting, to account for the frequency dependence of the human ear, as specified in (IEC 61672-1 2005), or various frequency weightings for human comfort analysis to account for the sensory perceptions of vibrations, as specified in (ISO 2631-1: 1997; ISO 2631-5: 2004; ISO 8041: 2005). Such weightings can often be applied more efficiently in the frequency domain as we will discuss in Section 10.6.8. In some cases, however, it is necessary to apply the weightings in the time domain, for example to produce values equivalent to those of a sound level meter.

For space reasons we will limit the discussion here to acoustic A and C filters, these being the most common filters necessary for many acoustic applications. A good source for implementing weighting filters for vibration effects on humans is found in (Rimell and Mansfield 2007), and background information in (Mansfield 2005).

The filter characteristics for the A and C weighting filters are currently defined by the standard (IEC 61672-1 2005) (previously IEC651). The C filter is defined in the Laplace plane by two poles located at 20.6 Hz and two poles located at 12200 Hz, i.e., at $s = -2\pi \cdot 20.6$ and $s = -2\pi \cdot 12200$, since $s$ is defined in rad/s. Thus if we denote the two pole locations by $\omega_1$ and $\omega_2$, respectively, the transfer function is

$$H_C(s) = \frac{C_c s^2}{(s + \omega_1)^2 (s + \omega_2)^2} \tag{3.16}$$

where $C_c$ is a scaling constant to provide a frequency response of 1 at 1000 Hz, as specified by the standard. The $s^2$ factor in the numerator is necessary to yield the correct shape as we will see shortly.

To produce a digital filter that approximates the transfer function in Equation (3.16), it is recommended to use the bilinear transform as mentioned in Section 3.3.2. However, the bilinear transform does not behave nicely when poles are spread apart by as much as they are in this filter. Thus it is more appropriate to define the C-weighting filter as two filters in cascade; one highpass filter

$$H_{HPc} = \frac{C_1 s^2}{(s + \omega_1)^2} \tag{3.17}$$

and a lowpass filter

$$H_{LPc} = \frac{C_1}{(s + \omega_2)} \tag{3.18}$$

which are transformed separately into digital filters which are in turn combined using convolution between the coefficients (or you could run **filter** once for each filter).

The A weighting filter is defined by adding two poles at 107.7 Hz and 737.9 Hz to the C weighting filter. We illustrate the process with an example. It should be noted that to ensure that the digital filters perform as a close approximation to the analog filters, 10 times oversampling should preferably be used.

**Example 3.3.4** *Compute filter coefficients B and A for a digital filter for C weighting when the sampling frequency of the data to be filtered is 44100 Hz, using MATLAB/Octave.*

*We start by defining the denominator and numerator polynomials for the separate HP and LP filters in s. The standard defines the frequency response at 1000 Hz to 1 (0 dB), so we ensure a scaling to achieve that. Then we use the bilinear transform to compute the digital filter coefficients for each filter, and finally convolve the A and B coefficients separately into coefficients for the total (cascade coupled) filter. The entire (MATLAB, see below) code becomes*

```
w1=2*pi*20.6;                          % First (double) pole
D1=conv([1 w1],[1 w1]);                % Denominator of HP filter
jw1k=j*2*pi*1000;                      % Value at 1000 Hz
C1=abs(jw1k+w1)^2/abs(jw1k)^2;         % To make H(1000)=1
w2=2*pi*12200;                         % Second (double pole)
D2=conv([1 w2],[1 w2]);                % Denominator of LP filter
C2=abs(jw1k+w2)^2;                     % Const. to make H(1000)=1
[B1,A1]=bilinear([C1 0 0],D1,fs);      % Digital HP filter
[B2,A2]=bilinear(C2,D2,fs);            % Digital LP filter
B=conv(B1,B2);                         % Total filter, B coeff.
A=conv(A1,A2); % Total filter, A coeff.
```

*With this example it should be easy to add the two poles to produce a similar filter for A weighting. An important note is required here; the* **bilinear** *command in Octave should have* $1/f_s$ *as the third parameter, as the syntax is different from the MATLAB syntax.*

*End of example.*

## 3.4 Time Series Analysis

In this section we will discuss some fundamental time series analysis procedures. Integration and differentiation of vibration signals is very common and natural because of the close relationship between vibration displacement, velocity, and acceleration. It is thus somewhat surprising that so little has been published on best practices to integrate and differentiate vibration signals. The presentation here will be rather practical since we do not have the tools ($z$-transform) to analyze digital filters in detail. Still, it will be useful and hopefully illuminating for some 'best practice' procedures that are not easily found in literature for the non-specialist.

### *3.4.1 Min- and Max-analysis*

In cases where minimum and maximum values in the time signal are to be estimated in an analysis procedure, it is important to consider several factors which we will present here. Such time domain analysis is perhaps most commonly used in fatigue analysis, where different forms of reduction processes are used for cycle counting, such as range pair and rainflow reductions. Transient analysis, for example in pyroshock or drop table applications, is also a common example where time domain analysis is used, and shock response spectrum is another related example.

First of all, peak values in data are heavily influenced by the bandwidth of the analysis, as is evident from Figure 3.4. Thus it is important to select a high enough sampling frequency when recording the signal. If no *a priori* information on the bandwidth of the data is available, this has to be established by increasing the sampling rate until peaks do not increase anymore.

For min- and max-analysis it is also essential to ensure that data are sampled correctly, i.e., with linear-phase filters. This will be discussed in more detail in Section 11.2.2. It should be reminded here, however, that some current measurement systems for noise and vibration analysis do not have linear phase antialias filters, and that care must therefore be taken when this type of time domain analysis is to be performed.

With the normally low oversampling ratio of 2.56 or slightly less, as used in FFT analyzers, min and max values will be seriously wrong. There is, to my knowledge, no general formula for setting an acceptable oversampling ratio in peak analysis. For narrowband data such as from shock response analysis (see Section 16.1), however, it is recommended to use 10 times oversampling, which yields less than 10% error in min and max estimates on vibration levels from SDOF systems, (Wise 1983). From the discussion on interpolation in Section 3.2.3, it is clear that 10 times oversampling can be obtained by resampling the data immediately prior to the time domain analysis. If storage space is limited, data with an oversampling of approximately 2.56 (whatever the hardware manufacturer has implemented) can be stored without losing any significant quality in the data. This oversampling ratio is sufficient to allow for accurate resampling when so needed as we showed in Example 3.2.1.

### *3.4.2 Time Data Integration*

As accelerometers are the most common sensors for vibration measurements, it is common to want to integrate such signals into vibration velocity or displacement. This can seem like a simple task, but as we will show in this section, it is not as easy as one might imagine to integrate time signals accurately. One well-known problem of numerical integration is the problem of the integration constant, or more practically formulated, the problem of DC or low-frequency variations in the signal. DC errors are very common in measurement signals due to offsets in the data acquisition electronics. In many cases this can be easily corrected for by simply removing the mean from the entire time signal prior to integration.

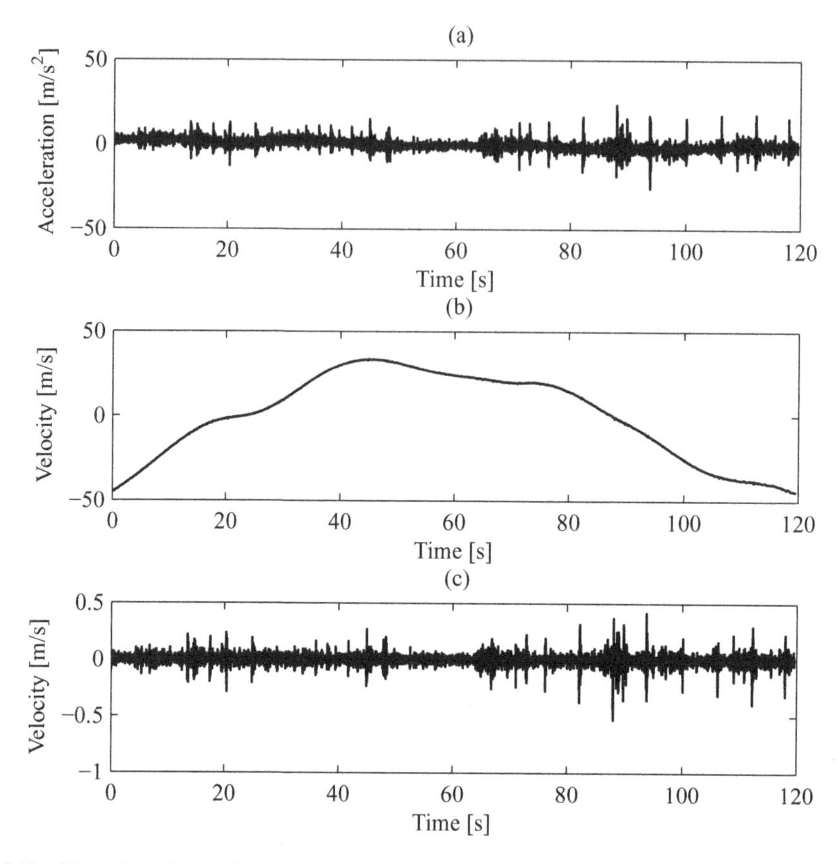


**Figure 3.10** Illustration of the problem of integrating an acceleration signal. In (a) a time record of an acceleration signal from a truck driving on a road is plotted. In (b) the signal is integrated by a precision filter presented later in the section (an accurate IIR filter). The integration reveals that apparently there is some low-frequency variations in the signal in (a) that overshadows the vibrations when integrated. These are likely low acceleration frequencies arising from the road surface variations. In (c) the original signal has been lowpass filtered with a cutoff frequency of 5 Hz prior to integration, which produces the expected vibration velocity

A more serious issue is if there is low-frequency drift in the signal, as this can often cause a low-frequency variation many times larger than the actual vibration signal, as is illustrated in Figure 3.10. Low-frequency drift is also very common in microphone signals and sometimes occurs in accelerometer signals due to temperature variations. If integration of a time signal results in these problems, the only solution is often to include a highpass filter in the integration process, to eliminate frequencies below a certain frequency.

Another problem often encountered is the problem of integrating to absolute displacement. Due to the same reasons already mentioned, it is in most cases not possible to integrate acceleration signals twice and produce absolute displacement. Furthermore, in order to be able to obtain the absolute displacement, the acceleration signal must have a frequency response down to DC (static) acceleration, which excludes the most common piezoelectric signals (see Chapter 7). This does not mean that it is always impossible to obtain absolute position from acceleration measurements, but it is a much more difficult problem than is perhaps first thought.

We will now look at how time domain integration can be performed and what the errors are. As mentioned above, quite surprisingly the problem is not particularly commonly studied in the signal processing literature or in the noise and vibration analysis literature. For that reason we will cover different integrators in some depth to show what performance commonly used algorithms have in practice. (Pintelon and Schoukens 1990) includes a short overview, and they also summarize that the known methods from the field of numerical analysis (trapezoidal rule, Newton–Raphson, etc.) perform poorly compared with what can be done by modern signal processing techniques. The presentation here will end with a very useful high-performance integrator presented in (Pintelon and Schoukens 1990) which today seems to be one of the best choices for integration in time domain.

Integration in the time domain corresponds to division by the factor $j\omega$ in the frequency domain. Ideally, therefore, we would like to filter the data with a filter with the 'true' frequency response, $H_t(\omega)$,

$$H_t(\omega) = \frac{1}{j\omega} \tag{3.19}$$

where the subindex $t$ stands for 'true'. If instead we filter the data with a filter with a frequency response $\hat{H}$, along with (Pintelon and Schoukens 1990), we define a frequency dependent error, $\delta$, as

$$\delta(\omega) = \left| \frac{H_t - \hat{H}}{H_t} \right| \tag{3.20}$$

where we drop the frequency variable $\omega$ on the right-hand side for simplicity. We will thus evaluate different integration filters by comparing their error $\delta$.

A very intuitive, but not very good, filter for integration is obtained by approximating the integral of a time signal $x(t)$ by

$$\int_0^{t_k} x(t)\mathrm{d}t \approx \Delta t \sum_{n=0}^{k} x(n) \tag{3.21}$$

which is easily converted to a recursive difference equation

$$y_n = y_{n-1} + \Delta t \cdot x(n) \tag{3.22}$$

which corresponds to a filter in MATLAB/Octave with $A = [1, -1]$ and $B = \Delta t$. An alternative way of computing it is to use the command **cumsum** in MATLAB/Octave. It can be shown that the frequency response of this simple integrator is

$$H_1 = \frac{\Delta t}{1 - e^{-j2\pi r}} \tag{3.23}$$

where we denote the normalized frequency by $r = f/f_s$. The frequency response in Equation (3.23) is a very bad approximation of true integration, and not worth plotting together with the other integrators. Instead we leave this for Problem 3.3.

The relative error of this integrator, which we denote $\delta_1$ for later comparison, is plotted in Figure 3.12, together with the errors for the integrators we are going to present later in this section. As is seen in the plot, this simple integrator is a rather poor choice, with an error at $r = 0.1$, which corresponds to 10 times oversampling, of a mere $-10$ dB, which corresponds to approximately 30% error. You should apparently avoid integrating using this simple method.

A next step to achieve a more accurate integration can be found by using the bilinear transform on the ideal transfer function $H(s) = 1/s$. This results in an integrator with a difference equation

$$y_n = y_{n-1} + \frac{\Delta t}{2} x_n + \frac{\Delta t}{2} x_{n-1} \tag{3.24}$$

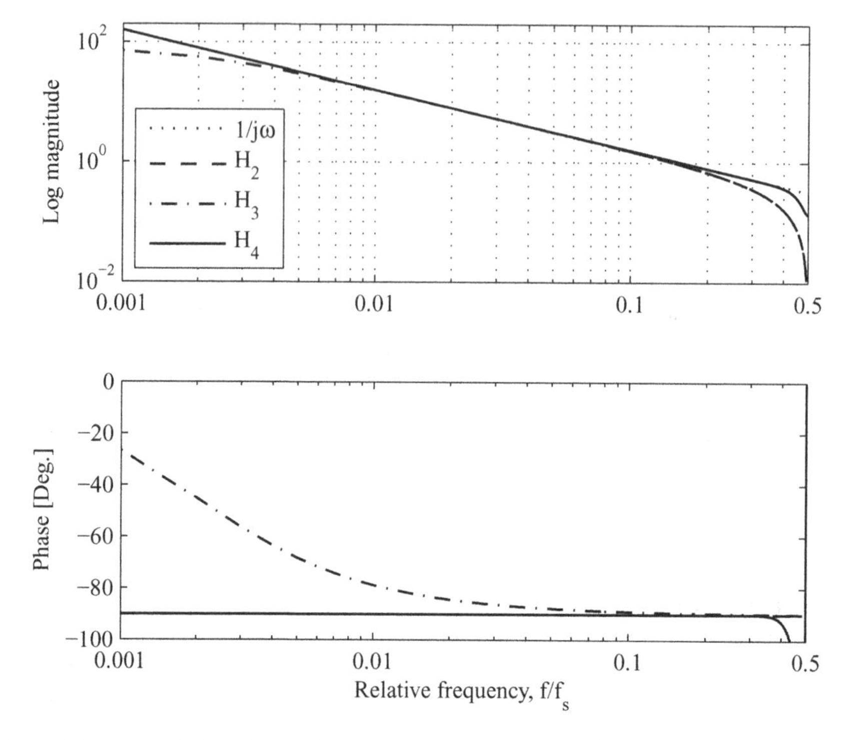


**Figure 3.11** Filter gain and phase characteristics of some common integrator filters. $1/j\omega$ is the ideal integrator, $H_2$ is the trapezoidal rule integrator, $H_3$, the first-order lowpass filter integrator, with a cutoff frequency $f_c = f_s/1000$, and $H_4$ is the combined filter characteristics of a 32 order FIR highpass filter with a cutoff frequency $f_c = f_s/1000$, and the 8-th order IIR filter defined in (Pintelon and Schoukens 1990)

which is easily implemented as a digital filter with $A = [1, -1]$ and $B = 1/2f_s[1, 1]$. This integrator is well known in numerical analysis and is referred to as the 'trapezoidal rule'. It has a frequency response, $H_2$, as shown in Figure 3.11 with a phase which is identical to $-90°$ except at DC. The error, as shown in the comparison plot in Figure 3.12 is less than $-30$ dB, corresponding to approximately 3% error, for oversampling ratios over 10. This may seem like a small error, but compared with the dynamic range of most vibration measurement systems (in excess of 100 dB), it is a poor integrator.

A third integrator which is common in vibration applications is to use a first-order lowpass filter with a low cutoff frequency, as we discussed in Section 3.3.5. This filter has the advantage of including a cutoff frequency (at some low frequency) below which no integration is done. However, as seen in Figure 3.11, where the frequency response, $H_3$, of this type of integrator with a cutoff frequency of $f_s/1000$ is plotted, the phase of this integrator performs rather poorly at frequencies several hundred times the cutoff frequency. The relative error, $\delta_3$, of the integrator is also seen to be poorer than the trapezoidal rule integrator at most relative frequencies.

For high-performance integration, Pintelon and Schoukens (1990) presented two different IIR filter integrators which were developed using advanced digital filter optimization methods. One of their integrators, referred to as the eighth-order integrator, has an integer sample delay, which is preferable for data analysis as data can be shifted to be synchronous with the data that are not integrated. A slight drawback with this filter is that it is somewhat unstable. However, the performance can easily be improved by adding a linear-phase FIR highpass filter prior to the IIR filter. The frequency response,

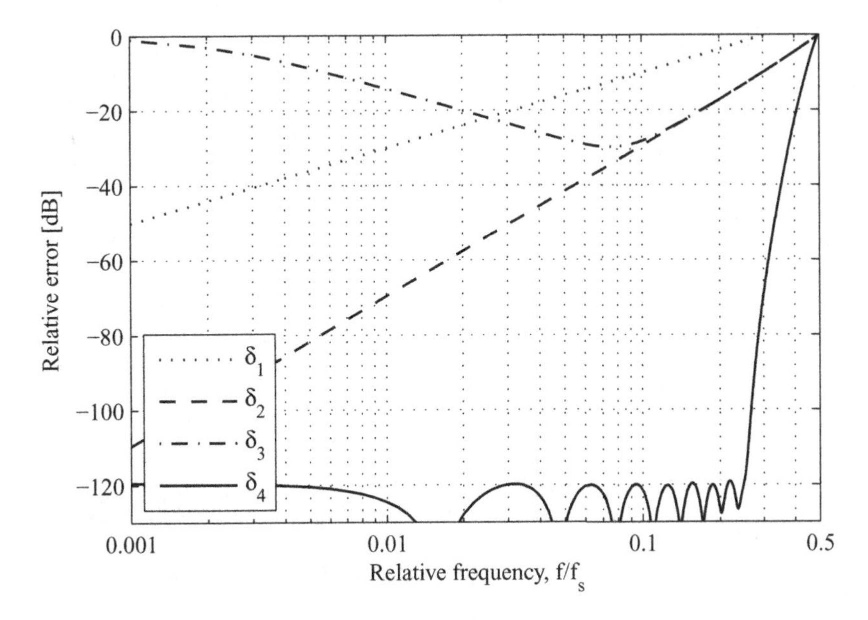


**Figure 3.12** Comparison of relative error for different integrators. $\delta_1$ is the simple integrator using the command **cumsum**, $\delta_2$ is the trapezoidal rule integrator, $\delta_3$, the first-order lowpass filter integrator, with a cutoff frequency $f_c = f_s/1000$, and $\delta_4$ is the combined filter characteristics of a 32 order FIR highpass filter with a cutoff frequency $f_c = f_s/1000$, and the eighth-order IIR filter defined in Pintelon and Schoukens (1990)

$H_4$, and relative error, $\delta_4$, of such an implementation, with the highpass cutoff at $f_s/1000$, are plotted in Figure 3.11 and Figure 3.12, respectively. As can be seen in the second figure, the relative error $\delta_4$ is below $-120$ dB for all frequencies up to $r = 0.25$. This means that 4 times oversampling is sufficient for this type of integration filter, and the dynamic range of the integrator is outstanding compared with the other types of integrators. The computational expense is negligible for most data analysis cases with the performance of modern PCs. As of this writing, integrating a data vector with 1 million samples with the filter implementation including a highpass FIR filter of order 32 and the IIR filters of Pintelon and Schoukens, takes approximately 0.2 s on my computer, compared with approximately 0.03 s for the simple integration using **cumsum**.

### *3.4.3 Time Data Differentiation*

Like integration, differentiating time signals is of rather common interest in many noise and vibration applications. The literature on differentiation is extensive compared with what is written about integration. This is probably largely due to the fact that differentiation poses fewer problems than integration, due to the fact the there is a pole at DC in the ideal differentiator, which has, of course, a transfer function of $H(s) = s$. This means that FIR filters perform very well for differentiation, whereas they are poor for integration, as a pole at DC cannot be implemented by a FIR filter.

For comparison, and perhaps as a warning example, we will start by the perhaps most intuitive and simple differentiator given by the digital filter calculating $y_n$, an estimate of the derivative of $x_n$, by

$$y_n = \Delta t\,[x_n - x_{n-1}] \tag{3.25}$$

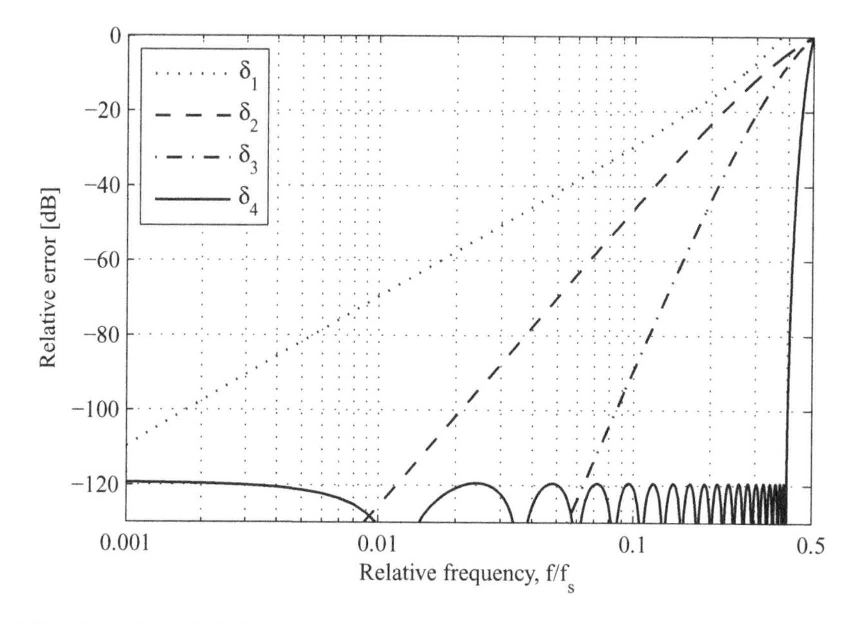


**Figure 3.13** Comparison of relative error for different differentiation filters. The error denoted $\delta_1$ is the error of the simple differentiator obtained by $\Delta t$ times the command **diff**, $\delta_2$ is the error for the maximum flat differentiator according to Equation (3.26) with $N = 2$, $\delta_3$ the error of the same type of differentiator with $N = 4$, and finally $\delta_4$ is the error of an optimal FIR differentiator filter using the Parks–McClellan/Remez method with $N = 40$. It can be seen that the last differentiator performs extremely well all the way up to a relative frequency of 0.4, corresponding to the normal oversampling ratio of noise and vibration measurement systems. Using this differentiator, there is no need for upsampling the signal

which can be computed using the MATLAB/Octave command **diff** times the time increment, $\Delta t$. The relative error, which we defined in Section 3.4.2, of this simple differentiator is a mere $-10$ dB, corresponding to an amplitude error of approximately 30%. We leave the computation of the error to Problem 3.6, but we emphasize that this is not a good differentiator.

The next step could be to use the bilinear transform of the ideal differentiator $H(s) = s$ to calculate digital filter coefficients, as we did for the integrator. This leads to a digital filter with $A = [1, 1]$ and $B = 2f_s[1, -1]$. The relative error for this differentiator, denoted $\delta_1$ is plotted in Figure 3.13 where it can be seen that the error is less than approximately $-30$ dB for frequencies below $f_s/10$. Although perhaps this does not seem too bad, it is a relatively large error compared to the approximately 100 dB dynamic range of most sensors and data acquisition equipment in use. For precise integration we need a better differentiator, as we will present below.

Optimal FIR differentiators were early investigated in a classical paper by Rabiner and Schafer (1974) and those results are now standard text in most textbooks (e.g., Oppenheim *et al.* 1999; Proakis and Manolakis 2006). We will, however, look at two later developments of FIR differentiators; the maximum flat FIR differentiators (Carlsson 1991; Kumar and Roy 1988; Lebihan 1995), and the methods available through the so-called Parks–McClellan optimization method, sometimes referred to as the Remez method (Parks and McClellan 1972). Both methods, maximum flat filters and Parks–McClellan/Remez optimization, lead to linear-phase FIR filters, ideal for data analysis, and we will compare the two methods next.

The maximum flat FIR differentiators are based on finding the best digital filter that approximates the ideal differentiator with frequency response $H(\mathrm{j}\omega) = \mathrm{j}\omega$ under the constraint of having as many derivatives at DC as possible being zero. This is the 'maximum flat' behavior. To obtain filters with an integer sample delay, we let the length of the filter be $2N + 1$. The FIR filter coefficients $b_k$ have been shown in (Lebihan 1995) to be computable using a recursive formula by first computing coefficients, $c_n$, as

$$c_1 = -\frac{N}{N+1}$$
$$c_n = (-1)c_{n-1}\frac{(n-1)(N-n+1)}{n(N+n)} \tag{3.26}$$

for $n = 1, 2, \ldots N$. These coefficients $c_n$ are then used to build the FIR filter by using the coefficients in the rightmost part of the FIR filter (with MATLAB/Octave definitions), and flipping the coefficients and changing the sign in the leftmost part of the FIR filter vector, and in between we put a zero. For $N = 2$ this leads to a filter of length 5 with coefficients $b_k = -1/12, 8/12, 0, -8/12, 1/12$. The relative error, as defined in Section 3.4.2, denoted $\delta_2$ of this filter is plotted in Figure 3.13 together with the error for $N = 4$, denoted $\delta_3$. As shown in the figure, the filters perform reasonably well for oversampling ratios above 10 times, with a maximum error of $-46$ dB and $-90$ dB, respectively, at $f = 0.1 f_s$. Already at $N = 6$ the relative error is below $-130$ dB for oversampling ratios larger than 10 times.

By using the optimization method of Parks–McClellan, very accurate and yet relatively short filters can be designed for example for differentiation. The essential difference compared to the maximum flat filters is that the Parks–McClellan method leads to filters that behave better close to the Nyquist frequency, at the expense of some extra FIR filter coefficients. In Figure 3.13 the relative error, denoted $\delta_4$, is plotted for an optimized FIR differentiation filter with 41 FIR coefficients (including a zero coefficient in the middle). As can be seen in the plot, the error is below $-120$ dB up to $0.4 f_s$, corresponding to an oversampling ratio of 2.5, which is the standard ratio used in many data acquisition systems. This means that the differentiation is accurate without the need to resample the measurement data.

**Example 3.4.1** *We will illustrate the procedure of differentiating a signal with a simple example. Using MATLAB/Octave, design a maximum flat FIR differentiating filter using Equation (3.26), with $N = 2$ and produce a plot of the relative error, $\delta(f)$.*

*Equation (3.26) gives us $c_1 = -2/3$ and $c_2 = (-1) \cdot (-2/3) \cdot 1/(1 \cdot 1 \cdot 2 \cdot 4) = 2/3 \cdot 1/8 = 1/12$. We then construct the MATLAB/Octave denominator coefficient vector*

*B = [−1/12 2/3 0 −2/3 1/12]*

*where we use the vector with $c_1$ and $c_2$ on the right-hand side of the zero, and the flipped coefficients with changed sign on the left-hand side of the zero.*

*For all FIR filters, the numerator vector is a simple scalar,*

*A = 1.*

*The next thing we need to consider is the delay of the filter. All linear-phase FIR filters with length of $2N + 1$ have a delay of N samples with the notation we use here. (MATLAB uses a different nomenclature, so in MATLAB language a FIR filter of order N is actually $N + 1$ long, so watch out with this.) The delay of N samples corresponds to a phase shift of (using r for relative frequency, i.e., $r = f/f_s$)*

$$\phi = -2\pi N r \tag{3.27}$$

*so in order to get the undelayed filter response, we need to compensate the frequency response by multiplying by $e^{j2\pi N r}$.*

*We are now ready to write the code. We use a sampling frequency of 1 Hz, which means that the frequency axis will 'automatically' be scaled in relative frequency. The code thus becomes*

```
A=1;
B=[-1 8 0 -8 1]/12;
[H,r]=freqz(B,A,1000,1);          % 1000 freq. values, 1 Hz fs
```

```
Hc=H.*exp(j*2*pi*2*r);          % Compensate for delay
Ht=j*2*pi*r;                    % True diff. response
delta=abs((Ht-Hc)./Ht);         % Relative error
figure
semilogx(r,20*log10(delta));    % Error in dB
xlabel('Relative frequency, f/f_s')
```

*This example produces a plot like that for $\delta_4$ in Figure 3.13.*
*End of example.*

It should be mentioned that the paper by Pintelon and Schoukens (1990) that we mentioned in Section 3.4.2, also includes a high-performing IIR filter for differentiation with integer sample delay. This filter is also an acceptable candidate for differentiation along with the methods presented in this section. However, the Parks–McClellan optimized FIR filters perform better closer to the Nyquist frequency, so for data analysis, where computational expense is usually not vital, the increased filter size for the FIR filters is not that crucial. In real-time applications, it is quite a different matter.

### 3.4.4 FFT-based Processing

In this chapter we have presented some digital signal processing concepts and methods to process digital data. It should be mentioned that an alternative method which is common in audio processing, for example, is to use FFT (see Section 9.3) to transform overlapped blocks of time data, then process data in the frequency domain, and finally inverse transform back to time domain. This method, known as *overlap-add* or *overlap-save*, depending on details in the implementation, is described in most standard textbooks on signal processing, for example (Oppenheim *et al.* 1999; Proakis and Manolakis 2006). The method is efficient and performs very well, for example for integration and differentiation. However, as it requires an in-depth understanding of the discrete Fourier transform, among other things, it does not fit very well into this chapter.

## 3.5 Chapter Summary

In this chapter we have presented some important theory and applications of time data processing of measured signals. We started by the sampling theorem, a fundamental theorem for all digital data analysis. We noted that in order to sample a signal, we first need to make sure that the bandwidth of the signal is below $f_s/2$, the Nyquist frequency. Under the assumption that this is fulfilled, then the sampled signal will contain all information in the analog signal, which is represented in the interpolation formula, which states that

$$x(t) = \sum_{n=-\infty}^{\infty} x(n)\text{sinc}[f_s(t - n\Delta t)] \tag{3.28}$$

where

$$\text{sinc}(x) = \frac{\sin(\pi x)}{\pi x}. \tag{3.29}$$

In other words, if we fulfill the sampling theorem, then any value of the analog signal at any time, $t$, can be calculated using the samples of the signal. Strictly, this is true only for infinite signals, so in practice we will get some error due to the finite data length, but for practical purposes this works well. It is, for this reason, always good to keep measured signals as long as possible, or more practically said; always record a portion before and after your important event takes place.

Next, we explained some of the nature of resampling signals, and mentioned that the MATLAB/Octave command **resample** is a favorable way to change the bandwidth of a measured signal. Instead of designing a lowpass filter to reduce the bandwidth of the signal, for the inexperienced user it is safer to resample down to 2 times the lower frequency you want in the signal, and then upsample back to the original sampling frequency. And actually, even for the experienced user, the method proposed provides better accuracy than most other techniques, so it is still highly recommended.

The procedure of downsampling followed by upsampling just mentioned, will produce a higher oversampling ratio, which is often necessary for digital filters to perform as their analog counterparts, and is also necessary for example for min-max-analysis.

We summarized some important notes on the behavior of digital filters that are designed to perform like analog filters. Regardless of which transform (bilinear, impulse invariant, etc.) has been used to produce the digital filter from an analog filter prototype, there are some important points regarding the digital filter behavior;

- the digital filter performs poorer the closer to the Nyquist frequency one gets. Most digital filters that approximate analog filters perform well if the oversampling ratio is kept above a factor of 10 times,
- digital filters usually have some time delay which can be important to understand. For FIR filters with linear-phase characteristics, and of length $2N + 1$, the time delay is always $N$,
- phase distortion, or phase linearity, is another important consideration when time domain analysis of, for example, transients are of interest. The easiest way to produce a linear-phase filter, is by designing a FIR filter with symmetric filter coefficients,
- using the MATLAB/Octave command **filtfilt**, any filter can be used to produce linear-phase characteristics.

We introduced some digital filter procedures for designing digital filters that approximate an analog filter, and used this to illustrate how fractional octave filters can be designed easily in MATLAB/Octave. We noted particularly that the digital filters behave as the analog filters at low frequencies, but that the performance deteriorates more the closer to the Nyquist frequency we get. Also, we observed there is a difference between the **bilinear** commands in MATLAB and Octave, respectively. How high the digital filters perform well depends on the design, and examples were shown for filters for differentiation, for example, that perform well up to 0.4 times the sampling frequency, whereas many other filters perform well only up to approximately 0.1 times the sampling frequency.

In the last section of the chapter we showed some examples of good filters for integration and differentiation of measured signals. It was shown that the simplest and most intuitive filters for both integration and differentiation, which are very commonly used, should be avoided in favor of only slightly more computationally costly filters. A few examples of the application of those filters can be found in the problem section following this.

## 3.6 Problems

Many of the problems following are supported by the accompanying ABRAVIBE toolbox for MATLAB/Octave and further examples which can be downloaded with the toolbox. If you have not already done so, please read Section 1.6, and follow the instructions to download this toolbox together with example files.

**Problem 3.1** *To illustrate the effect of aliasing, produce a time axis corresponding to a sampling frequency of 2 kHz, 0.05 s long, using MATLAB/Octave,. Then calculate a cosine with frequency 800 Hz ($0.8 f_s/2$) and a cosine of 1200 Hz ($1.2 f_s/2$). Verify that the samples are identical (within the computation accuracy) for both signals.*

**Problem 3.2** *Assume that we have a time signal containing two frequencies, that is*

$$x(t) = X_1 e^{j2\pi f_1} + X_2 e^{j2\pi f_2} \tag{3.30}$$

*where $X_1$ and $X_2$ are complex constants and thus include the initial phase relationships of the two frequency components. Show that passing this signal through a filter with frequency response $H(j\omega)$ with linear phase, that is $\angle H = A\omega$ will result in the same relative phase relationship between the two phases, $\angle X_1$ and $\angle X_2$ as they had before the filter.*

**Problem 3.3** *Calculate and plot the frequency response of a simple integrator according to Equation (3.23), using the filter coefficients given in the text near the equation. Overlay with the true frequency response of integration, $H = 1/j\omega$.*

**Problem 3.4** *Calculate the frequency response of a differentiator using the MATLAB/Octave* ***diff****, by using the filter coefficients mentioned in Section 3.4.3. Then calculate the true differentiator frequency response, and use both FRFs to compute and plot the relative error similar to Figure 3.13. Use a relative frequency axis from 1e-3 to 0.5.*

**Problem 3.5** *Design a maximum flat FIR differentiator for $N = 6$ using Equation (3.26) and plot the relative error similar to Example 3.4.1. Make sure the error at $r = 0.1$ is approximately $-130$ dB as the text says.*

**Problem 3.6** *Create a sine signal in a vector x, with a frequency of 230 Hz, using an oversampling ratio of 10 times, and 10 seconds long in MATLAB/Octave, and create the true derivative of the same signal in another vector xp (x prime). Use the accompanying toolbox command* ***timediff*** *to calculate the derivative of the signal in x, in vector y, and compare it with the time vector xp. How much do they differ? Explain (reading the text inside* ***timediff****) why the vector y is shorter than xp? Try the different types that* ***timediff*** *has as option. Do they make a difference? Which one performs best?*

**Problem 3.7** *Repeat Problem 3.6 but instead using the command* ***timeint****, and compare it with the true integral of the sine. Answer the same questions as in Problem 3.6 relating to integration instead of differentiation.*

## References

ANSI S1.11 2004 *Specification for Octave-Band and Fractional-Octave-Band Analog and Digital Filters*. American National Standards Institute.

Carlsson B 1991 Maximum flat digital differentiator. *Electronics Letters* **27**(8), 675–677.

IEC 61260 1995 *Electroacoustics – Octave-Band and Fractional-Octave-Band Filters*. International Electrotechnical Commission.

IEC 61672-1 2005 *Electroacoustics - Sound level meters – Part 1: Specifications*. International Electrotechnical Commission.

ISO 2631-1: 1997 *Mechanical vibration and shock – Evaluation of human exposure to whole-body vibration – Part 1: General requirements*. International Organization for Standardization, Geneva, Switzerland.

ISO 2631-5: 2004 *Mechanical vibration and shock – Evaluation of human exposure to whole-body vibration – Part 5: Method for evaluation of vibration containing multiple shocks*. International Organization for Standardization, Geneva, Switzerland.

ISO 8041: 2005 *Human response to vibration – Measuring instrumentation*. International Organization for Standardization, Geneva, Switzerland.

Kumar B and Roy SCD 1988 Coefficients of maximally linear, FIR digital differentiators for low-frequencies. *Electronics Letters* **24**(9), 563–565.

Lebihan J 1995 Maximally linear FIR digital differentiators. *Circuits Systems and Signal Processing* **14**(5), 633–637.

Mansfield NJ 2005 *Human Response to Vibration*. CRC Press.

Nyquist H 2002 Certain topics in telegraph transmission theory (reprinted from *Transactions of the AIEE*, February 1928, pp 617–644). *Proceedings of the IEEE* **90**(2), 280–305.

Oppenheim AV, Schafer RW and Buck JR 1999 *Discrete-Time Signal Processing*. Pearson Education.

Parks TW and McClellan J 1972 Chebyshev approximation for nonrecursive digital filters with linear phase. *IEEE Transactions on Circuit Theory* **CT19**(2), 189–194.

Pintelon R and Schoukens J 1990 Real-time integration and differentiation of analog-signals by means of digital filtering. *IEEE Transactions on Instrumentation and Measurement* **39**(6), 923–927.

Proakis JG and Manolakis DG 2006 *Digital Signal Processing: Principles, Algorithms, and Applications* 4th edn. Prentice Hall.

Rabiner LR and Schafer RW 1974 On the behavior of minimax relative error FIR digital differentiators. *Bell System Technical Journal* **53**(2), 333–361.

Rimell AN and Mansfield NJ 2007 Design of digital filters for frequency weightings required for risk assessments of workers exposed to vibration. *Industrial Health* **45**(4), 512–519.

Shannon CE 1998 Communication in the presence of noise (reprinted from *Proceedings of the IRE* **37**, 10–21 (1949)). *Proceedings of the IEEE* **86**(2), 447–457.

Wise J 1983 The effects of digitizing rate and phase distortion errors on the shock response spectrum *Proc. Institute of Environmental Sciences, Annual Technical Meeting*, 29th, April 19–21, Los Angeles, CA.

# 4

# Statistics and Random Processes

Noise and vibrations are often produced by sources with random behavior, for example vibrations resulting from a road surface interacting with the tires of a car, or vibrations caused by turbulence around an airplane wing. To understand random vibrations and their analysis it is important to understand applied statistics. In this chapter we will review some fundamental parts of probability theory, especially of the theory of stochastic processes, or random functions, and the way these methods are used in noise and vibration analysis.

Statistical properties are used in many ways in this field. In data quality assessment, covered in Section 4.4, many statistical properties can be used to assess the quality of a set of acquired data. Also in the classification of signals, for example in order to find the damaging effect of signals, the statistical properties of signals are important tools. The treatment here will be practical and focused on statistical analysis methods commonly applied to measured signals.

The theoretical background of fundamental statistical theory is assumed to be familiar to the reader and is therefore only briefly recapitulated here. For a deeper understanding you are referred to standard textbooks, either mathematical, (Papoulis 2002), or engineering oriented, for example (Bendat and Piersol 2010; Newland 2005; Wirsching *et al.* 1995).

## 4.1 Introduction to the Use of Statistics

Before going on we should discuss two different forms of random signals that we are interested in. There are, as mentioned in the introduction above, dynamic forces that are naturally behaving as random signals. In this case we are interested in describing the random signal (process) as accurately as possible. This is usually done by describing the signal by its spectral characteristics (spectral density, see Section 8.3.1), or correlation function (see Section 4.2.12) and its amplitude characteristics, for example the probability density.

There is, however, also another form of random signals, namely measurement noise coming from various sources, i.e., unwanted 'disturbance' added to our measurement signals. First of all, there is electrical noise inherent in all electronics, measurement sensors as well as data recording hardware. Then there is random contributions to essentially deterministic processes, which can be thought of as random signals. An example of the latter is the vibration from a reciprocating engine, which is essentially a periodic signal. But, due to the inexact amount of fuel injected in the cylinder during each combustion, and the uncertainty in the exact time of each combustion, in reality the periodic signal will not be perfectly periodic. It can then be thought of as a periodic signal plus a random contribution, where the random contribution is essentially an 'error contribution' to the periodic signal. By modeling the

*Noise and Vibration Analysis: Signal Analysis and Experimental Procedures,* First Edition. Anders Brandt.

total measurement signal in this way, we often apply some averaging procedure which has the goal of eliminating, or minimizing, the random part of the signal. In Section 2.4, the sound of gunshots was taken as an example where, ideally, each gunshot should sound the same, which means the measured sound pressure should be a deterministic signal. But due to the uncertainty in the exact amount of gunpowder in each shot, there will of course be a random contribution to this deterministic sound.

### *4.1.1 Ensemble and Time Averages*

There is an abstraction in the mathematical way of describing statistics and random processes that I frequently find my students seem to have missed in the maths (or probability) class. I therefore want to point out the concept of realizations, and time dependent statistical measures a bit. A random process as described in the theory of stochastic processes, is an abstract entity which we cannot study practically. An example of such a process is the electrical signal in a certain type of electrical resistor coming from thermal noise. If we call this process $x(t)$, this means it is some theoretical signal which has certain properties. Let us now look at the mean, or average, or *expected value* of this process, which we denote $\mu_x(t)$. What you should observe particularly is that this average is a *time-dependent variable*. How does that come to be? It comes from the concept of *ensemble averaging* from the theory of statistics. This average is namely considered as *the average of infinitely many such resistors*, on each of which we measure the time dependent voltage coming from the resistor. Each such voltage, which we can call $x_i(t)$ is a *realization* of the process $x(t)$ (you can compare this with a stochastic variable, for example the throw of a dice, where each throw of the dice is a realization).

If this was the only type of average available to us, we would be in trouble, as the cost of measurements would be very high. Instead of measuring the noise from one prototype car (or a few) as we do today, we would have to measure very many (20 to 100, perhaps) for everything we would want to measure. But, luckily, there is the concept of *ergodicity*. But first we need to understand stationarity.

### *4.1.2 Stationarity and Ergodicity*

A random process is defined as (strongly or strictly) *stationary* if all statistical properties of the process are independent of time. This means that for example the mean value

$$E[x(t_1)] = E[x(t_2)] \tag{4.1}$$

for any arbitrary times $t_1$ and $t_2$. The symbol E used here is the expected value operator, which will be introduced in Section 4.2.1, but I assume you are already slightly familiar with the concepts here. It should be noted, however, that in terms of an experimental situation, to estimate this time-dependent mean value, we need to measure many so-called ensembles (or realizations) of the process and calculate the mean at each time instant $t$ through the ensembles. That means that, if we would measure for example the road noise in a car, we would have to find many cars and measure the noise in all those cars at the same time (which would cause a problem, because they should also be in the same location at every instant in time, have the same speed, etc., which would lead to a very crowded test track).

Obviously this is not a very practical measure. Therefore, we need to consider *ergodic random processes*. A random process is referred to as ergodic, if it is stationary, and if ensemble properties and time properties are equal, that is, if the mean can be calculated from a single ensemble (realization) as

$$\mu_x = E[x(t)] = \lim_{T\to\infty} \frac{1}{2T} \int_{-T}^{T} x(t)dt \tag{4.2}$$

and this mean is equal to the mean in Equation (4.1). In reality, a measured signal is usually ergodic if it is stationary, so it is enough to analyze one realization of it. However, you should note that this does *not* imply that most measured signals are stationary! Indeed, stationary conditions are often difficult to obtain. This will be addressed in Section 4.3.3 below.

Taking the example of road noise in a car, stationarity, and in this case ergodicity, implies that the noise in a sense 'sounds constant', i.e., we will only have stationary sound if we run the car with constant speed, on the same type of asphalt, etc. If the nature of the road changes (as it often does in practice, more often than we would want it to from a measurement point-of-view) the noise will not be stationary.

## 4.2 Random Theory

When we have noise and vibration signals of random nature, the theory of stochastic processes is used to describe the signals. Some of the most basic concepts of statistics and stochastic process theory will therefore be briefly recapitulated here with some comments on the meaning of each measure. In practice, instead of the theoretical term stochastic process, we often use one of the synonyms *random signal* or *random process*. In this section we will also, unless otherwise noted, assume that the random signal is stationary and ergodic.

### 4.2.1 Expected Value

We will start our discussion with the definition of *expected value* of a random variable $x$, because it is needed to understand the errors introduced next. The expected value can be interpreted as the mean, or average, value of the function, when an infinite number of values are used. Thus the expected value of a random variable (process) $x$ is obtained by

$$\mathrm{E}[x(n)] = \lim_{N \to \infty} \frac{1}{2N+1} \sum_{n=-N}^{N} x. \tag{4.3}$$

The expected value will be further discussed in Section 4.2.7.

### 4.2.2 Errors in Estimates

When calculating statistical estimates, we differentiate between two types of errors; the *bias error*, and the *random error*. We assume that we shall estimate (i.e., measure and calculate) a parameter, $\phi$, which can be for example the standard deviation $\sigma_x(t)$. In the theory of statistical variables, the 'hat' symbol, ˆ, is usually used for a variable estimate, so we denote our estimate $\hat{\phi}$ (in practice for example $\hat{\sigma}_x$). We now define the bias error, $b_{\hat{\phi}}$, as

$$b_{\hat{\phi}} = E[\hat{\phi}] - \phi \tag{4.4}$$

i.e., the bias error is the difference between the expected value of our estimate and the 'true' value $\phi$. In practice we generally divide this error by the true value to obtain the *normalized bias error*, $\varepsilon_b$, as

$$\varepsilon_b = \frac{b_{\hat{\phi}}}{\phi} \tag{4.5}$$

If we have a nonzero bias error, there will thus be a difference between the estimate $\hat{\phi}$ and the true value $\phi$, even if we make a very large number of averages to estimate $\hat{\phi}$. This is of course unwanted, and such an estimate is called a *biased* estimate. The opposite, an estimate that has no bias, is consequently called an *unbiased* estimate, and is usually preferred. In many cases, however, we cannot find an unbiased estimator, as we will see is the case in spectral estimation, e.g., in Chapter 10.

The random error, on the other hand, is defined as the standard deviation, $\sigma_{\hat{\phi}}$, of the difference between our estimated variable and the true variable, that is

$$\sigma_{\hat{\phi}} = \lim_{N \to \infty} \sqrt{\frac{1}{N-1} \sum_{k=1}^{N} \left\{ \hat{\phi}_k - E\left[\hat{\phi}\right] \right\}^2} \tag{4.6}$$

and, as with the bias error, we define the *normalized random error*, $\varepsilon_r$, as

$$\varepsilon_r = \frac{\sigma_{\hat{\phi}}}{\phi}. \tag{4.7}$$

A good estimator should have a random error that approaches zero as the number of averages increases. Such an estimator is referred to as a *consistent* estimator.

In most cases we do not know the true parameter $\phi$, and therefore we generally have to use our estimated parameter, $\hat{\phi}$, in its place if we wish to calculate the normalized bias and random errors.

When we have estimated the normalized random error, $\varepsilon_r$, we can use it, for small errors ($\varepsilon_r < 0.1$), to calculate a 95% *confidence interval*

$$\hat{\phi}(1 - 2\varepsilon_r) \leq \phi \leq \hat{\phi}(1 + 2\varepsilon_r). \tag{4.8}$$

We assume confidence intervals are known from a basic course in statistics. It follows quite straightforwardly from the nature of the Gaussian distribution and the central limit theorem, which we will discuss in Section 4.2.13.

### 4.2.3 Probability Distribution

The *probability distribution*, $P(x)$ of a random (stochastic) signal $x(t)$, is defined as

$$P(x) = \text{Prob}[x(t) \leq x] \tag{4.9}$$

where 'Prob' denotes *probability*. Typical distribution functions for random variables are, for example, the well known Gaussian distribution (see Section 4.2.13) and the theoretically popular, but in real life not so common, uniform distribution.

### 4.2.4 Probability Density

The derivative of the probability distribution function is called the *probability density function*, PDF, which gives the relative occurrence of amplitudes in $x(t)$. This is the derivative of $P(x)$,

$$p_x(x) = \frac{\mathrm{d}}{\mathrm{d}x}[P(x)]. \tag{4.10}$$

Since $P(x)$ is clearly a continuously growing (or constant) function for increasing $x$, because the probability in Equation (4.9) must increase with increasing amplitudes (until it reaches the maximum amplitude), then it follows that

$$p_x(x) \geq 0 \tag{4.11}$$

for all $x$. Furthermore, it follows directly from Equation (4.10), that

$$P(x) = \int_{-\infty}^{x} p_x(x)\mathrm{d}x \tag{4.12}$$

and

$$\int_{-\infty}^{\infty} p_x(x)\mathrm{d}x = 1. \tag{4.13}$$

The interpretation of the probability density is thus, that the *area under the curve* in a certain amplitude range, equals the probability that the random signal is within that range, i.e.,

$$\mathrm{Prob}\,[x_1 \leq x(t) \leq x_2] = \int_{x_1}^{x_2} p_x(x)\mathrm{d}x. \tag{4.14}$$

## 4.2.5 Histogram

The probability distribution and density functions above are theoretical functions that we cannot estimate from real-life signals. Instead, in practice we estimate the *histogram*, or more specifically the amplitude histogram of the signal $x(n)$ (as a histogram can be made from any measure). The (amplitude) histogram consists of a discrete number of values, where each value is the number of samples of the signal in a certain amplitude range. The procedure of creating a histogram is illustrated in Figure 4.1. In principle, first we need to choose the minimum and maximum amplitude we wish to include, which is frequently the ADC range, or alternatively, the minimum and maximum amplitude values of the signal. As most vibration data have zero mean, it may be more practical to use the largest of the minimum or maximum (maximum of the absolute values of all samples), $M_x$, and use a symmetric amplitude interval $[-M_x, M_x]$

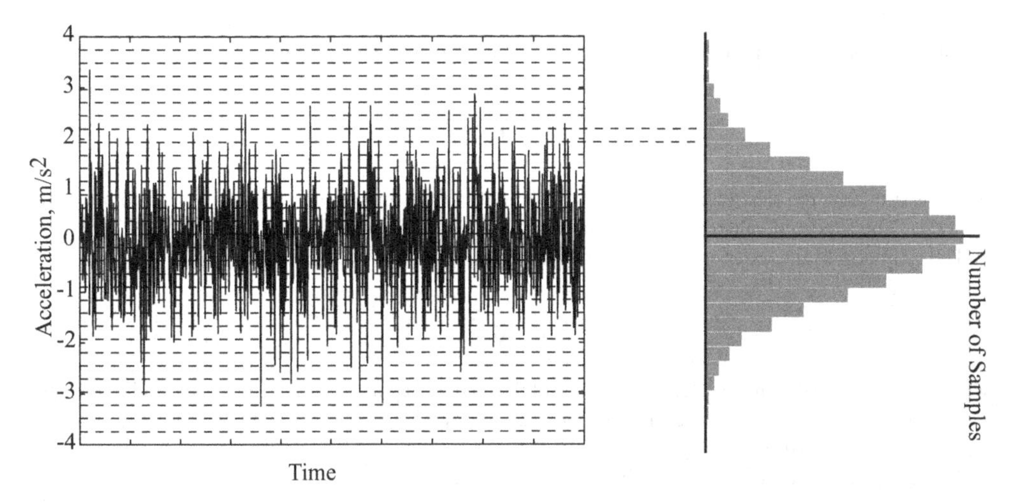


**Figure 4.1** Illustration of the principle of the calculation of a histogram. It is formed by calculating the number of samples in the analyzed signal, that fall within each of a number of amplitude ranges, called *bins*

for the histogram calculation. This amplitude range is then divided into a number of uniformly spaced intervals, with width $\Delta x$. Finally we produce the histogram by counting the number of samples, $N_i$, in our data, $x(n)$, that fall within each bin.

### 4.2.6 Sample Probability Density Estimate

The histogram as we have defined it is not directly comparable to the probability density. By normalizing it by the total number of samples $N$ and the amplitude width $\Delta x$ we obtain the *sample probability density*, $\hat{p}_i$. Thus we have

$$\hat{p}_i = \left(\frac{N_i}{N}\right)\frac{1}{\Delta x}. \tag{4.15}$$

This estimate of the (continuous) probability density function has a bias error due to the limited amplitude ($x$-axis) resolution. It also has a random error due to the limited number of samples used in the calculation. In order to keep these errors small, it is recommended to select the width of each bin so that $\Delta x \leq \sigma_x/5$ and use $N \geq 10{,}000$, where $\sigma_x$ is the standard deviation of the signal, see Section 4.2.7. For data with Gaussian distribution and with zero mean, most of the data is within 4 to 5 times $\sigma_x$, the standard deviation of the signal (see Section 4.2.13). Thus 40 or 50 bins are normally suitable for calculating the sample probability density when the minimum and maximum amplitudes for the histogram calculation are chosen as $\pm 4\sigma_x$ or $\pm 5\sigma_x$, respectively. The bias and random errors for the sample probability density function are approximate and can be found in (Bendat and Piersol 2010). An example of a sample probability density function is found in Figure 4.5.

### 4.2.7 Average Value and Variance

For a function $g(x)$ of a random signal (process) $x(t)$ with probability density $p_x(x)$, the expected value, $\mathrm{E}[g(x)]$, is defined by

$$\mathrm{E}[g(x)] = \int_{-\infty}^{\infty} g(x)p_x(x)\mathrm{d}x. \tag{4.16}$$

It follows directly from Equation (4.16) that the expected value of $x(t)$, is given by

$$\mu_x = \mathrm{E}[x] = \int_{-\infty}^{\infty} x \cdot p_x(x)\mathrm{d}x \tag{4.17}$$

which is similar to the calculation of the center of gravity in one dimension, if you feel more comfortable with that. The expected value, or average value, is essentially the 'amplitude center of gravity'.

The estimator of the mean value based on $N$ samples of $x(n)$, $n = 0, 1, \ldots, N-1$, is

$$\bar{x} = \hat{\mu}_x = \frac{1}{N}\sum_{n=0}^{N-1} x(n) \tag{4.18}$$

which is a consistent estimator, i.e., the variance of the estimate approaches zero as $N$ approaches infinity, see Section 4.2.2. Note that we either use the 'bar', $\bar{x}$, symbol or the hat symbol combined with the symbol of the true estimate in $\hat{\mu}_x$.

The *variance*, $\sigma_x^2$, of a random variable, $x(t)$, is defined by

$$\sigma_x^2 = \mathrm{E}\left[(x-\bar{x})^2\right] = \mathrm{E}[x^2] - \bar{x}^2 \tag{4.19}$$

The square root of the variance, i.e., $\sigma_x$, is called the *standard deviation* of $x$. In order to obtain an unbiased estimator for the variance and standard deviation, the so-called sample variance defined by Equation (4.20) should be used, that is

$$\hat{\sigma}_x^2 = \frac{1}{N-1}\sum_{n=0}^{N-1}(x-\bar{x})^2 \tag{4.20}$$

where the name sample variance is used when we use $N-1$ in the denominator instead of N, to make the estimator unbiased. In practice this makes little difference compared with using $N$ in the denominator, as we typically use many thousand values to compute the variance.

You should note that the standard deviation is similar to the root mean square, RMS, as defined in Equation (2.17). For $N$ samples of a dynamic signal $x(n)$, the RMS level is defined by

$$x_{\mathrm{RMS}} = \sqrt{\frac{1}{N}\sum_{n=0}^{N-1}x^2(n)}. \tag{4.21}$$

The RMS level differs from the standard deviation in that the RMS level includes the mean value (static, or DC value) of the signal $x$. Thus, when the DC value of a signal is zero, the RMS level is essentially equal to the standard deviation (except for the difference of $N$ and $N-1$ in the denominator which can be neglected for large $N$). When a measured signal contains a DC value, it is recommended that this value be removed from the signal and treated separately, anyway. So for noise and vibration signals we regularly use RMS and standard deviation synonymously.

The mean estimator in Equation (4.18) is an unbiased estimator. The random error, $\varepsilon_r[\bar{x}]$ is more complicated because the random signal is time varying. It turns out (Bendat and Piersol 2010), that the random error of the average of a random signal is

$$\sigma[\bar{x}] = \frac{\sigma_x}{\sqrt{2BT}} \tag{4.22}$$

where $B$ is the bandwidth and $T$ the observation time of $x(t)$ (the time it takes to sample the $N$ samples in the averaging process) and you should note that the error *is not* normalized. Because the mean is usually zero, normalizing it would mean dividing by zero.

The estimator for the variance in Equation (4.20) is a consistent and unbiased estimator. The random error of the variance $\hat{\sigma}_x^2$, if $x(t)$ has a bandwidth $B$ and the observation time is $T$, is

$$\varepsilon_r[\hat{\sigma}_x^2] \approx \frac{1}{\sqrt{BT}}. \tag{4.23}$$

It can be shown that the normalized random error of $\hat{\sigma}_x$, under the same assumptions as for the variance error above, is

$$\varepsilon_r[\hat{\sigma}_x] \approx \frac{1}{2\sqrt{BT}}. \tag{4.24}$$

The product $BT$, found in the equations for the random error of the mean, variance, and standard deviation above is called the *bandwidth–time product*, and is central in signal analysis and spectrum estimation (see Section 9.1.1). Clearly it specifies that, in order to obtain a certain random error when calculating an estimate of any of these measures, we must measure during a specific minimum amount of time, and further, the lower the bandwidth of the signal is, the longer time we must measure. This follows naturally from the fact that, loosely speaking, *frequency is change per time unit.*

### 4.2.8 Central Moments

The $i$th *central moment*, $M_i$, of a signal $x(t)$ is defined by

$$M_i = \mathrm{E}\left[(x - \bar{x})^i\right] \tag{4.25}$$

where $i$ is an integer number and $x(t)$ is a time varying signal. From a comparison of Equation (4.19) and Equation (4.25), it follows that

$$\begin{aligned} M_1 &= 0 \\ M_2 &= \sigma_x^2 \end{aligned} \tag{4.26}$$

The statistical moments are used to calculate many higher-order statistical measures. In the next two subsections we will present two commonly used functions in noise and vibration analysis based on higher-order moments, which are called skewness and kurtosis.

### 4.2.9 Skewness

Skewness is a commonly used parameter when analyzing dynamic signals. If the measured signal is denoted $x$, the *skewness*, denoted $S_x$, is defined by

$$S_x = \frac{M_3}{\sigma_x^3}. \tag{4.27}$$

Apparently the skewness is a dimensionless measure, and it measures to what degree the signal is non-symmetric around its mean. If the signal is symmetric, the skewness is zero. For many random vibration signals, the probability distribution is symmetric around the mean (e.g., the normal distribution). Thus skewness differing from zero in many cases indicates that something is wrong.

The similarity and difference between the skewness and the mean value should be considered. The third power in the third moment of the estimate of skewness does not change the sign of the negative part of $x$ (we assume $x$ is zero mean for simplicity). Thus skewness is similar to the mean. The difference between the mean and the skewness is that the values of $x$ are raised to the third power in the skewness estimate. This exaggerates high values and suppresses low values of $x$ compared with the mean. The skewness value is thus more sensitive to an asymmetry in the large values in $x$ than the mean.

### 4.2.10 Kurtosis

Another commonly used parameter is the *kurtosis*, $K_x$ which is also a dimensionless parameter, defined by

$$K_x = \frac{M_4}{\sigma_x^4}. \tag{4.28}$$

Kurtosis resembles the variance, except the values are raised to the fourth power instead of the second. Both these powers are even and thus make all values in the summation positive. The higher power in the kurtosis compared with the variance estimate, will emphasize large values in the signal and suppress small values in the kurtosis, compared with the variance.

The kurtosis can for this reason also be regarded as comparing the tails of the probability density function with that of the normal distribution. For a normally distributed variable, the kurtosis is exactly 3 (see Problem 4.3). If the kurtosis is larger, the distribution has 'higher tails' than the normal distribution, and vice versa if the kurtosis is smaller than 3. For a sine wave, the kurtosis is 1.5. For many other signals, it may be necessary to investigate empirically what kurtosis values are to be regarded as 'normal', see Section 4.4.

An alternative kurtosis value known as the *excess kurtosis* is common in software for noise and vibration analysis. The excess kurtosis is simply the difference of the kurtosis defined above and 3, i.e.,

$$Ke_x = K_x - 3 \tag{4.29}$$

and the meaning of this is of course to obtain values that are 0 if data are Gaussian, instead of the somewhat odd number 3. It is a matter of taste which kurtosis definition one wants to use, but unfortunately, due to the two similar definitions, there is a cause of confusion when discussing kurtosis values, so great care should be exercised to make sure to mention which definition you use.

### 4.2.11 Crest Factor

The *crest factor* is a rather common statistical measure in signal analysis. It is defined as the ratio of the maximum absolute value to the RMS value of the signal. If we thus have a signal, $x(n)$, with zero mean, the crest factor, $c_x$ is thus defined as

$$c_x = \frac{\max(|x(n)|)}{\sigma_x}. \tag{4.30}$$

The crest factor is often an important property, as it tells how 'peaky' data are. A high crest factor implies there is at least one large (positive or negative) peak in the signal. For a Gaussian random signal the crest factor is usually of the order of 4–5, see Section 4.2.13.

### 4.2.12 Correlation Functions

The autocorrelation function, $R_{xx}(\tau)$, for a stochastic, ergodic time signal $x(t)$ is defined as

$$R_{xx}(\tau) = \mathrm{E}[x(t)x(t-\tau)] \tag{4.31}$$

and can be interpreted as a measure of the similarity a signal has with a time-shifted version of itself (shifted by time $\tau$).

Similarly, the cross-correlation between two different stochastic, ergodic functions, $x(t)$ and $y(t)$, where $x(t)$ is seen as the input and $y(t)$ the output of some system, is defined as

$$R_{yx}(\tau) = \mathrm{E}[y(t)x(t-\tau)]. \tag{4.32}$$

The autocorrelation function can be seen as a special case of the cross-correlation, for the case where the two signals are equal.

It is easy to see that the autocorrelation at $\tau = 0$ equals the variance of the signal $x(t)$, since

$$R_{xx}(0) = \mathrm{E}[x(t)x(t)] = \sigma_x^2. \tag{4.33}$$

The definitions of autocorrelation and cross-correlation are sometimes formulated such that, e.g., the cross-correlation is expressed as $R_{yx} = \mathrm{E}[x(t)y(t+\tau)]$, which is easily seen to be equivalent with our definition by substituting $u$ for $t+\tau$ which leads to $t = u - \tau$ and the definition turns into Equation (4.32).

Furthermore, with the definition we have used, if we assume the output signal $y(t)$ is the input signal delayed by $\tau_1$ seconds, i.e., $y(t) = x(t - \tau_1)$, the cross-correlation $R_{yx}$ becomes

$$R_{yx}(\tau) = \mathrm{E}[y(t)x(t-\tau)] = \mathrm{E}\left[x(t-\tau_1)x(t-\tau)\right] \tag{4.34}$$

which by making the variable substitution $u = t - \tau_1$, and thus $t - \tau = u + \tau_1 - \tau$, leads to

$$R_{yx}(\tau) = \mathrm{E}\left[x(u)x(u - (\tau - \tau_1))\right] = R_{xx}(\tau - \tau_1) \tag{4.35}$$

which means the cross-correlation equals the autocorrelation with the maximum shifted to $\tau = \tau_1$. This principle is commonly used to find time delays by use of the cross-correlation.

The principle of the cross-correlation definition is worth a comment. You should note that the definition of cross-correlation in Equation (4.32) is the *mean of the cross product of the signals*. If we assume zero mean signals, this cross product will reveal if there is a dependence between the signal $y(t)$ and the time-shifted signal $x(t - \tau)$. If these two signals are uncorrelated (there is no linear relationship between the two signals) the cross product will have a zero mean, but if there is a dependence between the two signals, both will more often go positive or negative simultaneously, which will lead to a nonzero mean. This is the basic principle of finding correlation between different signals. If you have not yet contemplated this fact, I recommend you take a moment to think about it.

For correlation functions, the following relationships hold for real signals $x(t)$ and $y(t)$.

$$R_{xx}(-\tau) = R_{xx}(\tau) \text{ even function} \tag{4.36}$$

and

$$R_{xy}(-\tau) = R_{yx}(\tau). \tag{4.37}$$

Correlation functions should be estimated in practice using spectra, as a direct implementation of the equations given here is very time consuming. Procedures for estimation of correlation functions will therefore be discussed in Chapter 10.

### 4.2.13 The Gaussian Probability Distribution

The *Gaussian*, or *normal* distribution is the most common probability distribution. We denote a random signal with Gaussian distribution, and with mean $\mu_x$ and standard deviation $\sigma_x$, by $N(\mu_x, \sigma_x)$. The probability density function for this $N(\mu_x, \sigma_x)$ distributed signal is

$$p_x(x) = \frac{1}{\sigma_x\sqrt{2\pi}} \mathrm{e}^{-\frac{1}{2}\left(\frac{x-\mu_x}{\sigma_x}\right)^2}. \tag{4.38}$$

Rather than using this general form, a *standardized* or *normalized* variable, $z$ is usually formed by taking

$$z = \frac{x - \mu_x}{\sigma_x} \tag{4.39}$$

for which $\mu_z = 0$ and $\sigma_z = 1$. For this standardized variable, the Gaussian distribution is simplified to

$$p_z(z) = \frac{1}{\sqrt{2\pi}} \mathrm{e}^{-\frac{z^2}{2}} \tag{4.40}$$

which is apparent from the equations above. The importance of this standardization or normalization is very important. By subtracting the mean from any signal, we get a new signal which has zero mean. By dividing any signal by its standard deviation, we get a new signal with unity standard deviation. This is easily verified, see Problem 4.1.

The probability density function, PDF, of the standardized variable is plotted in Figure 4.2. The interpretation of the standardized variable is that any number $z_0$ along the $x$-axis, corresponds to the value $x - \mu_x = \sigma_x z_0$ for the original variable $x$. As we usually have zero mean vibration signals, we can further simplify this so that $x = \sigma_x z_0$. If you further interpret the standard deviation as the RMS value, which is more practical, the $x$-axis in Figure 4.2 should be interpreted in values times the RMS value of $x$.

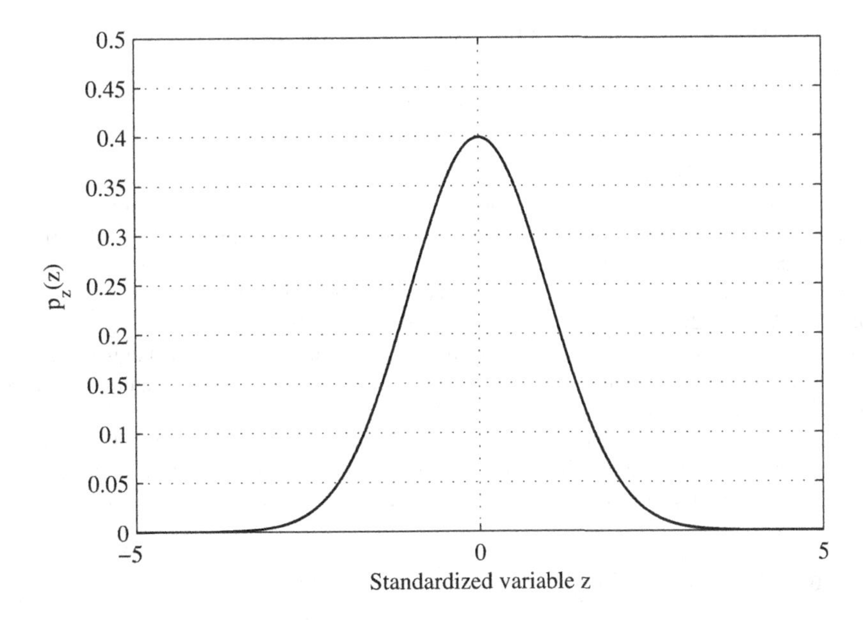


**Figure 4.2** Probability density function, PDF, of the standardized variable, $z$, for normal distribution

We recall it is the area under the PDF which is the probability that the signal is within the limits between which the area is calculated. Thus, it is good to recall (from your statistics class) that for $z$ we have that

$$
\begin{aligned}
\int_{-1}^{1} p_z(z)\mathrm{d}z &= 0.68 \\
\int_{-2}^{2} p_z(z)\mathrm{d}z &= 0.95 \\
\int_{-3}^{3} p_z(z)\mathrm{d}z &= 0.997
\end{aligned}
\tag{4.41}
$$

which are useful to remember in order to get a feeling for signals with Gaussian distribution. The numbers in Equation (4.41) say that the signal $x$ (the non-normalized signal) is within $\pm\sigma_x$, or 1 times its RMS value, 68% of the time, within 2 times its RMS value 95% of the time, and within 3 times its RMS value 99.7% of the time. This says something about the normal distribution.

Since this is a book strongly recommending using software such as MATLAB/Octave, it is worth noting that the values in Equation (4.41), and similar values for other integral limits, can be calculated using the *error function*. This function is available as the command **erf** in MATLAB/Octave, and is defined by

$$
\mathrm{erf}(x) = \int_{0}^{x} \frac{2}{\sqrt{pi}} \mathrm{e}^{-t^2} \mathrm{d}t \tag{4.42}
$$

from which, together with the symmetry of $p_z(z)$, it is evident that

$$\int_{-z_0}^{z_0} p_z(z)\mathrm{d}z = \operatorname{erf}\left(\frac{z_0}{\sqrt{2}}\right). \tag{4.43}$$

The proof of this is left for Problem 4.2.

In conjunction with the error function, it should also be mentioned that the inverse function **erfinv** is available in MATLAB/Octave. This function is very useful when computing limits for hypothesis tests as we will discuss in Section 4.3.1.

The importance of the Gaussian distribution is strongly related to the *central limit theorem* (Bendat and Piersol 2010; Papoulis 2002), which says that any random signal which is produced by a sum of many different contributions, is normally distributed *regardless* of the distribution of each of the contributions. This is usually the case for random sources in nature and many signals that occur naturally are therefore normally distributed.

If a variable is normally distributed, *confidence levels* can be calculated for the likelihood that the variable is inside a certain region. The most common confidence level is the 95% level, which from Equation (4.41) is found to be the probability that $|z| \leq 2$, which is equivalent to $|x - \bar{x}| \leq 2\sigma_x$.

## 4.3 Statistical Methods

When analyzing noise and vibration signals in practice, there is a need for some statistical tools for investigating properties as for example whether data are Gaussian or not, or whether data are stationary or not. For this purpose, a statistical tool called *hypothesis test*, is used. In this section we will therefore introduce some basic concepts of hypothesis testing, and thereafter we will discuss the issues of testing normality and stationarity of data.

### *4.3.1 Hypothesis Tests*

In some of the following subsections we will use *hypothesis tests*. Such tests are common in statistics and we often meet the results of them in everyday life, for example when media report that smoking increases the risk of lung cancer, or that high fat consumption can lead to an increased risk of heart disease, etc. It is important to note that hypothesis tests do not prove anything, but they *test* whether a particular set of observations (data) either agree or disagree with a statement – the hypothesis (Brownlee 1984; Sheskin 2004). As hypothesis tests are not part of most engineering curricula, we will present the principle of hypothesis testing in some detail.

We illustrate the principle of a hypothesis test by an example. Assume we have a random variable, $x$, which has a mean value which we denote $\phi$ (because we want the notation to be general, regardless of which actual statistical measure we would like to test). We now believe that the value of this mean is $\phi = \phi_0$, for a specific value $\phi_0$, and we want to test if the mean value of our observed data is actually $\phi_0$. Note that we are talking about the *theoretical* mean here, which is untouchable for us. What we can obtain by a measurement is an *estimate* of the variable $\phi$, which we denote $\hat{\phi}$.

To define a hypothesis test, we have to know the probability distribution of the tested variable $\phi$, although not necessarily the distribution of the original random variable $x$. The way a hypothesis test is designed is that we set up a *null hypothesis* that $\phi = \phi_0$, which we denote $H_0$. The *alternative hypothesis*, $\phi \neq \phi_0$, we denote $H_1$. The test we perform in this example, is to reject the null hypothesis if our *observed* mean value, $\hat{\phi}$ is outside a certain interval around $\phi_0$, i.e., if

$$\left|\hat{\phi} - \phi_0\right| \geq \delta \tag{4.44}$$

and we want to be able to calculate $\delta$.

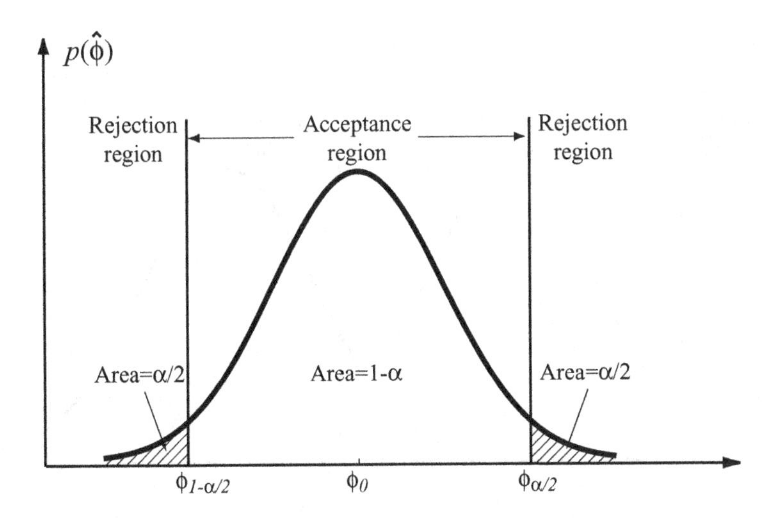


**Figure 4.3** Illustration of acceptance and rejection regions for hypothesis tests, for our example of testing the mean of a random variable. The figure shows the Gaussian probability density function of the estimated mean values, i.e., $p(\hat{\phi})$, which has a mean value of $\phi_0$ if the null hypothesis $H_0$ is correct. The standard deviation depends on the number of values used for the mean calculation. The area $\alpha$ is the significance level of the test, which is the probability that the null hypothesis is rejected when it is actually true

If the observations reject the null hypothesis then our data supports the alternative hypothesis. (More generally there can actually be several alternative hypotheses, but we limit the discussion to two hypotheses here as it serves all our purposes in this book.) The estimate $\hat{\phi}$ of the statistical property under test is based on a sum of independent observations of $x$. According to the central limit theorem, regardless of the probability distribution of $x$, the estimate $\hat{\phi}$ will thus (at least approximately) have a Gaussian probability distribution as in Figure 4.3. It is important to understand that this probability density is the assumed *true* probability density, if our null hypothesis is true.

To test the null hypothesis, $H_0$, we select a *significance level*, $\alpha$ as indicated in Figure 4.3. This means that we will accept the null hypothesis if our estimated mean, $\hat{\phi}$ is inside the acceptance region indicated in the figure. If $\hat{\phi}$ happens to come out as a value in either of the rejection regions, we will reject the null hypothesis. The significance level is thus the probability that we erroneously *reject* the null hypothesis when it is actually true. As can be seen in the figure, this probability is usually chosen to be small, usually in the order of 0.1–5%.

Two errors can occur in a hypothesis test. In the first case, just mentioned, the null hypothesis is true, but we reject it anyway because our tested variable happens to be outside our limits. This is called the *type I error*. The probability of this, as we mentioned above, equals the significance level, which we select to be small when we design the test. This means that if we reject the null hypothesis, then the probability that we did so erroneously is the significance level, $\alpha$, which will make our case strong if we decide on a small $\alpha$.

The other error that can occur is that the null hypothesis is not true, but that we accept it anyway. In other words, our test variable (in our example the mean, $\phi$) is not equal to $\phi_0$, as indicated in Figure 4.4. In this case the probability density function of the variable $\phi$, marked by a solid line in the figure, is different than the probability density function we assume under the null hypothesis. There is then a certain probability, $\beta$, that the estimate $\hat{\phi}$ falls inside the acceptance region, as indicated in Figure 4.4.

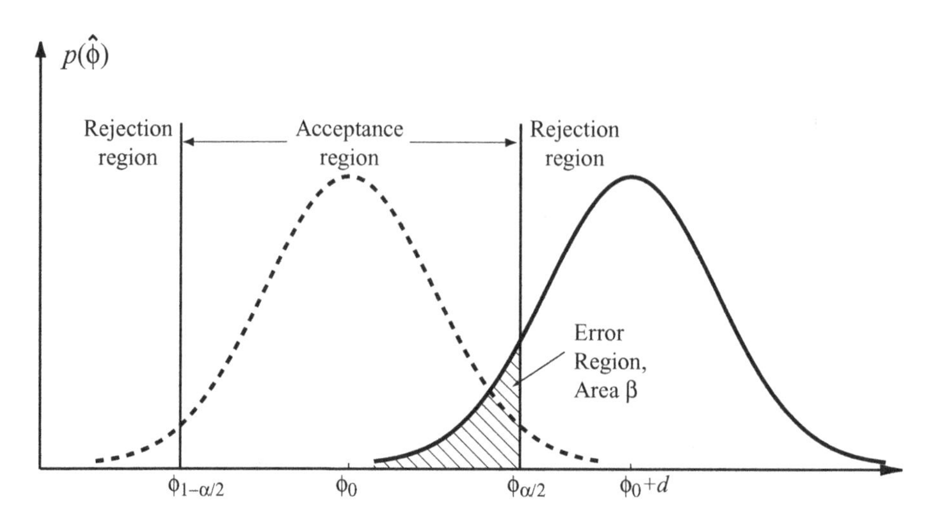


**Figure 4.4** Illustration of the type II error in hypothesis testing, for our example with testing the mean value of a random variable. The type II error occurs when the null hypothesis is not true, but still accepted, which occurs inside the dashed area in the figure. In the figure the probability density of the mean of our estimated variable, (solid) is centered at $\phi_0 + d$ whereas the assumed density function used for the acceptance and rejectance regions (dashed) is centered at $\phi_0$. The 'opposite' of the area $\beta$ indicated in the figure (i.e., $1 - \beta$) is called the *strength* of the test. See text for details

This is called a *type II error*. The error region $\beta$ in Figure 4.4 is the probability of making an error of type II. The 'opposite' of this probability, $1 - \beta$, is called the *strength* of the test, and gives the probability of not accepting the null hypothesis 'by mistake'. Increasing the strength can be accomplished by increasing the significance level, $\alpha$, which is of course generally not a good choice as it increases the risk that we reject the null hypothesis despite the fact that it is true. Alternatively, the strength can be increased by adding more data in the calculation of the variable under test, and thus reducing the variance of $\hat{\phi}$. The latter should be preferred if at all possible.

In our example we have used a simple test variable, the mean of the random variable $x$. In a hypothesis test we can more generally use any variable $\phi$, for example an RMS value, the variance, the kurtosis, etc. The only restriction is that we have to know the probability distribution of the test variable under the null hypothesis.

From the discussion above, it should be clear that the best philosophy when designing hypothesis tests is to use the argument of the 'opponent' as the null hypothesis. Doing this, if the test turns out to reject the null hypothesis, we have a strong case arguing that our own opinion is correct. Then the opponent has a weak case because the probability that the opponent is correct, but the test did not come out in his favor, is a mere $\alpha$. In order to select such a null hypothesis, however, we need to know the distribution of the opponent's case. As we will see in the next two sections, this is often not the case. Instead the test has to be designed so that the null hypothesis is what we want to prove. Then the strength of the test should be made as high as possible, i.e., $\beta$ should be small, since $\beta$ is then the critical probability in the case the test turns out in our favor.

Hypothesis tests somewhat resemble confidence levels, and in some cases either can be used. It is generally considered that, whenever possible, confidence levels should be preferred over hypothesis tests.

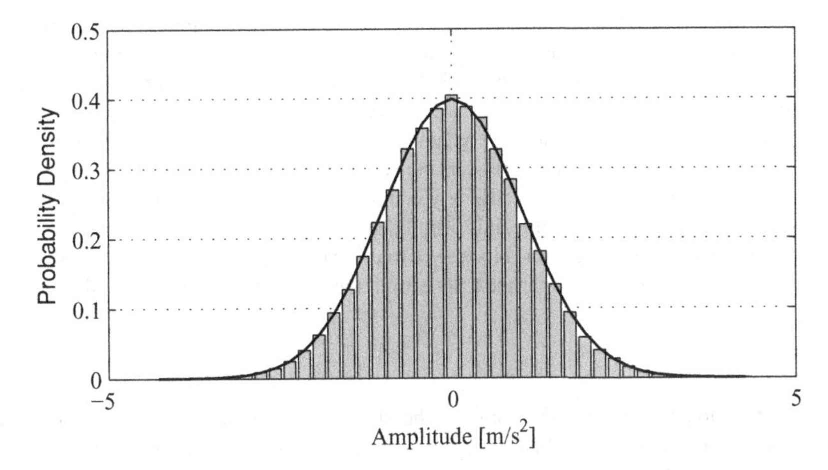


**Figure 4.5** A plot of the sample probability density function of a Gaussian signal with zero mean and unity standard deviation as bar chart, and the theoretical normal probability density as solid line. A total number of 50 000 samples where used and 40 bins within $\pm 4\sigma_x$

## 4.3.2 *Test of Normality*

The normal, or Gaussian, distribution is by far the most common probability distribution encountered in practice when studying noise and vibration signals of random character. This follows from the *central limit theorem*, as was mentioned previously. When Gaussian noise is presumed in an analysis situation, data should be tested for normality.

It is not easy in practice to investigate thoroughly, that is with high statistical certainty, whether measured data are normally distributed or not. For many practical purposes, it is sufficient to calculate the sample probability density defined in Equation (4.15), and compare it with the theoretical curve for the normal distribution, based on the calculated mean and standard deviation of the measured data, i.e.,

$$p_x(x) = \frac{1}{\hat{\sigma}_x \sqrt{2\pi}} e^{-\frac{(x-\hat{\mu}_x)^2}{2\hat{\sigma}_x^2}}. \tag{4.45}$$

An example of this comparison plot is shown in Figure 4.5.

When investigating whether a measured signal is Gaussian or not, it is often the tails of the PDF which are most important. In such cases a linear $y$-axis as in Figure 4.5 is not very convenient as the tails are low. Using a logarithmic $y$-axis is therefore often used, where the bar chart of the histogram is normally replaced by a trace plot, see Problem 4.4.

For cases where an absolute statistical measure of normality is of importance, the chi-square goodness-of-fit test can be used. This is an example of a hypothesis test and is found in many standard textbooks on random data analysis (e.g., Bendat and Piersol 2010; Brownlee 1984; Sheskin 2004).

## 4.3.3 *Test of Stationarity*

When the spectral density of a random signal is to be estimated (see Chapter 10), an underlying assumption is that the signal is stationary. Before such estimates are made, the signal should be tested to verify that it is stationary.

Before proceeding, it may be fruitful to consider for a moment some practical aspects of the concept of stationarity. In practice, few signals can be considered stationary for any longer period of time. Measuring the acoustic pressure from a microphone during a rocket launch, for example, if stationary at all, the signal will definitely only be stationary during some limited time during which the thrust, speed, etc., can be considered constant. The same can be said about a vibration signal measuring road vibrations on a car. This signal will only be stationary for constant speed, type of road surface, etc.

Another aspect of stationarity is that for a given signal, the time frame during which we study the signal is of importance. Of course, any time-varying signal is not stationary if the observation time is too short, and the question if a signal is stationary or not, often turns into a philosophical question. In order for spectral analysis to be applicable, it is usually enough to ensure that some lower-order moment (for example the RMS value) of each time block used in the averaging process is approximately constant (this assumes Welch's method is used for the PSD estimation, but can be extended to other estimators.

### Frame statistics

In order to test stationarity in practice, some of the statistical properties need to be investigated as functions of time. For example, the RMS value can be tested, or the skewness or kurtosis. The simplest way to do this is to divide the measured signal into a number of frames, and then for each frame, the selected properties are calculated and plotted as a function of time, as illustrated in Figure 4.6. If the value plotted does not seem to vary with time, the signal can be considered stationary. Some engineering judgment of what variations are acceptable will be needed, of course.

It is important to select a proper frame time for this type of test. The frame time should be large enough so that the variance is small, which is dependent also on the signal bandwidth (see Section 4.2.7). At the same time it must be sufficiently short for any temporary changes to show. Thus, some compromise is always necessary. If frequency analysis is going to be used, using the same frame size for the stationarity plots as for subsequent FFT analysis is usually a good practice.

### The reverse arrangements test

For a more rigid test of stationarity, this can be done using various statistical methods (Bendat and Piersol 2010; Brownlee 1984; Sheskin 2004). The *reverse arrangements test*, which is usually recommended,

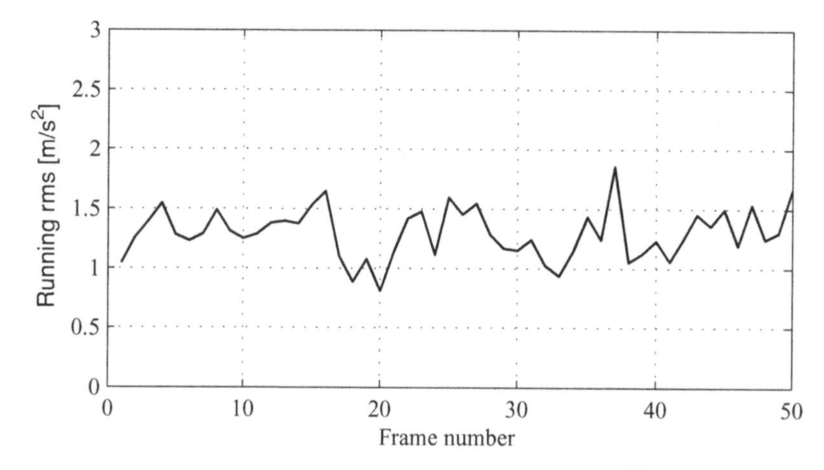


**Figure 4.6** Plot of RMS (root mean square) value as a function of frame number. The data used were tested for stationarity by the reverse arrangements test and found to be stationary with a significance level of 0.02, see Section 4.3.3. Some random variation can thus be tolerated in a plot like this

is a so-called nonparametric method that does not require any *a priori* assumption or knowledge of the statistical distribution of the signal, and is therefore often preferred in practice.

The reverse arrangements test is a hypothesis test that detects if there is a trend in the measured parameter. It is based on a calculation of frame statistics as we discussed in Section 4.3.3. If we do not have any trend in the data (parameter), then each value of our parameter should be greater than, loosely speaking, approximately half the other values, and smaller than half of the other values, on average. The test procedure is as follows.

First, assume we have a sampled time sequence of a signal $x(n)$ of length $M \times N$ samples, $M$ and $N$ being integer numbers. We divide this sequence into $N$ segments, for which we calculate the parameter we want to use to test the stationarity, which we call $y_i$. The RMS value is often chosen, but the skewness or kurtosis may also be chosen, as well as any other statistical parameter that can be calculated from each frame.

Next, we test the sequence of numbers $y_i$ for variations outside what is expected due to sample variations. This is done by calculating a new function $h_{ij}$, as

$$h_{ij} = \begin{cases} 1 \ (y_i > y_j) \\ 0 \text{ otherwise} \end{cases} \tag{4.46}$$

where $i = 1, 2, \ldots, N-1$, and $j = i+1, i+2, \ldots, N$. We now calculate the variables

$$A_i = \sum_{j=i+1}^{N} h_{ij} \tag{4.47}$$

and

$$A = \sum_{i=1}^{N-1} A_i \tag{4.48}$$

where the variable $A$ is called the *number of reverse arrangements*.

Based on $N$ independent observations of a stationary random variable, the mean and variance are given by Equation (4.49) and Equation (4.50), respectively,

$$\mu_A = \frac{N(N-1)}{4} \tag{4.49}$$

$$\sigma_A^2 = \frac{N(2N+5)(N-1)}{72}. \tag{4.50}$$

The null hypothesis of the reverse arrangements test is now defined as: the signal $x$ is stationary if the variable $A$ falls within the acceptance region given by Equation (4.51), defined by

$$A_{N;1-\alpha/2} < A \leq A_{N,\alpha/2}. \tag{4.51}$$

The limits in Equation (4.51) to accept stationarity can easily be calculated from the definition of the error function, erf, in Equation (4.42), which yields that

$$A_{N;1-\alpha/2} = \mu_A - \sigma_A \operatorname{erf}^{-1}(1-\alpha) \tag{4.52}$$

and

$$A_{N;alpha/2} = \mu_A + \sigma_A \operatorname{erf}^{-1}(1-\alpha) \tag{4.53}$$

where $\operatorname{erf}^{-1}$ is the inverse error function, which in MATLAB/Octave is available as the command **erfinv**.

From the discussion in Section 4.3.1 it follows that when the reverse arrangements test fails (the null hypothesis is rejected), the case is strong that the data are not stationary. The probability then that the data are stationary but erroneously tested as nonstationary, is the same as the significance level. However, if

**Table 4.1** Rms value for each frame for reverse arrangements test in Example 4.3.1. The values should be read row by row

| | | | | |
|---|---|---|---|---|
| 2.95 | 2.82 | 3.26 | 3.51 | 3.21 |
| 2.96 | 2.87 | 3.28 | 3.41 | 3.31 |
| 2.99 | 2.94 | 3.16 | 3.42 | 3.27 |
| 3.05 | 2.88 | 3.11 | 3.32 | 3.44 |

the test shows that the data are stationary, the case is somewhat weaker and depends on the strength of the test. The strength of this test can not be defined rigidly, however, since when the signal is not stationary, we do not have a general expression for the probability distribution of it because the nonstationarity can take on many forms. This is a weakness in many hypothesis tests like the present one.

Based on an analysis of the strength of the test versus significance level, when using the reverse arrangements procedure, some guidelines are recommended in (Himmelblau *et al.* 1993). They are as follows:

- $N$ should always be larger than 10, and preferably more than 20. If less than 20, use $\alpha = 0.10$.
- If the measurement time is between 20 and 100 s, use segments of 1 s each, and use $\alpha = 0.05$.
- If the measurement time is larger than 100 s, use $N = 100$ and $\alpha = 0.02$.

**Example 4.3.1** *We illustrate the process of the inverse arrangements test with an example on a small number of data to keep the example simple. Assume we have measured 20 s of an acceleration signal with a sampling frequency of 2000 Hz. Make a reverse arrangements test, based on the RMS of the signal, to see if the data are stationary.*

*The guidelines above lead to a choice of* $N = 20$ *and* $\alpha = 0.05$. *We calculate the RMS values of each of the 20 segments and obtain the values in Table 4.1.*

*We now make an upper-diagonal matrix* $h_{ij}$ *and for each pair* $i, j$, *where* $j > i$ *we set* $h_{ij}$ *to one if* $y_i > y_j$. *For space reasons we omit that result here. We then sum each row to produce* $A_i$ *for* $i = 2, 3, \dots N$ *which becomes*

$$A_i = [4, 0, 9, 16, 8, 3, 0, 7, 9, 7, 2, 1, 3, 5, 3, 1, 0, 0, 0] \tag{4.54}$$

*and finally we sum the values* $A_i$ *and get the total* $A = 78$.

*We have to calculate the limits for the null hypothesis (that our data are stationary), for* $N = 20$ *and* $\alpha = 0.05$, *using Equations (4.52) and (4.53), respectively, which yields the limits*

$$\begin{aligned} A_{N;1-\alpha/2} &= 64 \\ A_{N;\alpha/2} &= 125 \end{aligned} \tag{4.55}$$

*where you should note that we take the integer part of the calculated* $A_{N;1-\alpha/2}$ *and* $A_{N;\alpha/2}$ *since* $A$ *is an integer.*

*Since our calculated* $A = 78$ *is within the limits, we accept the null hypothesis; our data are stationary at the significance level 0.05.*

*End of example.*

### The runs test

The reverse arrangements test in the previous section is not particularly sensitive to periodicities in the signal, since it is based on rearranging the segments (see Problems 4.5 and 4.6). To discover, e.g., periodic fluctuations in the data that make the signal nonstationary, a test based on the so-called runs test, or Wald–Wolfowitz test, is more suitable, (Brownlee 1984; Sheskin 2004). This is a nonparametric test that tests the randomness of a variable, and the idea for using it to test for stationarity is based on testing whether the difference of the measured parameter of each frame (e.g., the RMS value) is a random variable. If there is a slowly varying periodicity in the data, with the RMS value periodically increasing and decreasing, the sequence of RMS values will, of course, not be random. We will illustrate this method by an example.

First, the variable that is going to be tested is calculated for frames just as with the reverse arrangements test. We denote these values by $y_i$ as before. Next the mean, $\mu_y$, of the obtained frame values is calculated. Then, each value $y_i$ is compared with the mean, and assigned a '+' if it is larger than or equal to the mean value, and a '−' if it is less than the mean. Assume we have divided our data into 20 segments, and for each segment we have calculated the RMS value and determined the mean of these RMS values. The result of the comparison with the mean could then look as (the results are taken from the same sequence as in Example 4.3.1)

$$- - + + + - - + + + - - - + + - - - - ++$$

A *run* is defined as a sequence with the same sign ('+' or '−'). Thus in this example we have 8 runs out of the 20 segments. Now, denote the number of frames used by $N$. If the underlying data are stationary, based on $N$ independent observations of these data, it turns out that for large $N$, the variable $r$, the number of runs in our test, is an approximately normally distributed random variable, with mean and variance as follows:

$$\mu_r = \frac{2N_1N_2}{N} + 1 \tag{4.56}$$

and

$$\sigma_r^2 = \frac{2N_1N_2(2N_1N_2 - N)}{N^2(N-1)}. \tag{4.57}$$

The next step is to find the limits for which the runs test indicates a random behavior on the runs (i.e., indicates the data to be stationary). Thus the null hypothesis is formulated: 'the signal $x$ is stationary if the variable $r$ falls within the acceptance region given by $r_{low} \leq r \leq r_{high}$' where the limits are the same as those for the reverse arrangements test in Equation (4.52) and Equation (4.53), with the mean and standard deviation from Equation (4.56) and Equation (4.57), respectively. As for any hypothesis test, a level of significance, $\alpha$, has to be chosen.

In our example, for a level of significance of $\alpha = 0.05$, we obtain the upper and lower numbers 6 and 15, respectively. Our number of runs, 8, falls within this range, and thus the run test indicates that our data are stationary with the level of significance 0.05.

Finally, we should note that a runs test should always be used with more than 20 segments as the test gets weak for smaller numbers. If a check for spectrum estimation is the goal for the runs test (with Welch's method, see Section 10.3), it is recommended to use the same segment (frame) size as the blocksize for the spectrum averaging.

## 4.4 Quality Assessment of Measured Signals

In this section we will look at some examples of how the statistical tools, etc., discussed previously in this chapter can be used to assess the quality of measured data in the field of noise and vibration analysis. In modern data analysis, increasing amounts of data are recorded in field tests as well as lab tests, for later analysis. Vibration measurements are certainly challenging, with many potential causes of measurement data being erroneous, for example due to bad sensors, cable problems, etc. There is therefore a need to assess the quality of measured data and find possible errors.

A means for data quality assessment is to investigate some of the statistical properties of the measured signals. This can be done regardless of whether the data are actually of random character or not, as also deterministic signals from this point of view can be investigated using the same methods. It should be stressed, however, that due to the very different nature of various vibration signals, it is difficult to make a standard data quality test that will work in every case. This is particularly true if one wants automatic detection of different errors. I therefore suggest an approach to this application that uses some statistical measures as indicators of potential (but not necessarily actual) errors, followed by a manual inspection where the engineering experience can be used to interpret the suspicious statistical measures.

It should be stressed that data quality assessment has to be done on time signals. This is one of several arguments for why time data should (almost) always be recorded for noise and vibration signals. The tradition in many areas of this engineering field has been to reduce data to spectra immediately during the data acquisition process, which was a technique introduced in the 1970s when the first analyzers came on the market. This way of working should be abandoned except in monitoring applications, as will be further discussed in Chapter 11.

In general, unless stationarity for some reason can be definitely assumed *a priori* (due to known test conditions, for example), data should be tested for stationarity, as this is usually assumed for the other measures. The reverse arrangements test and the run test described in Sections 4.3.3 and 4.3.3, respectively, should be used to test stationarity in cases where a statistically reliable method is requested. In many other cases, it may be sufficient to use the procedure with frame statistics described in Section 4.3.3. For stationarity, it is often sufficient to investigate the standard deviation (or variance). However, some errors are more easily revealed when studying the skewness or kurtosis as a function of frame number (or time). For example, kurtosis can be used as a 'spike detector'.

Often when data are assumed to be normally distributed *a priori*, this should be investigated using one of the procedures mentioned above. Calculating a sample probability density function is also motivated by the fact that many errors in instrumentation, such as drop-outs or spikes, can be revealed in a histogram by showing an increased occurrence around zero or the amplitude of the spikes, respectively, as shown in Figure 4.7.

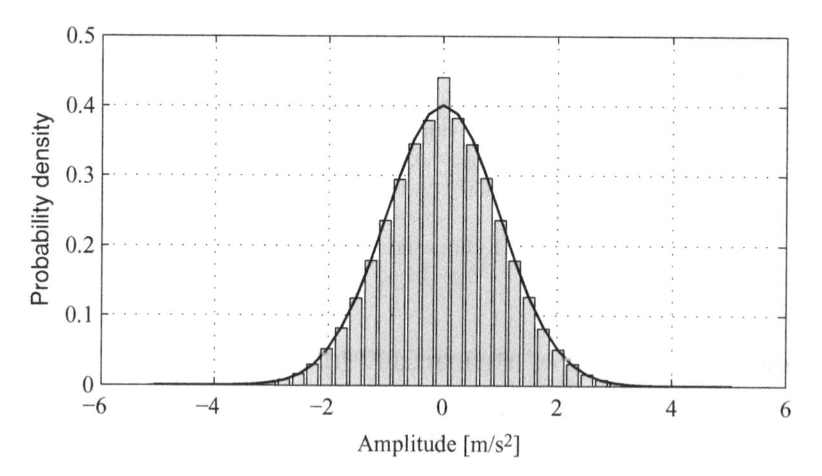


**Figure 4.7** Plot of the sample probability density of a signal with drop-outs, overlaid by the Gaussian PDF using the mean and standard deviation of the signal. As can be seen in the figure, the histogram reveals the increased occurrence of values around zero, which indicates there is something wrong in the signal

When data are verified to be stationary, and if necessary, Gaussian, next some standard statistical values should be calculated using the complete signal. Such standard statistics should include minimum, maximum and mean value, standard deviation or RMS value, skewness and kurtosis. By comparing these values with known values, either from *a priori* theoretical assumptions of the data, or by empirical knowledge of the type of data, many errors can be detected.

When recording many channels, or many measurements, a useful trick is to normalize each statistical measure to the measure of one measurement (channel) which you assume (or perhaps make sure through some manual analysis) is without quality errors. A plot of all those normalized metrics will clearly show a channel with a potential problem.

It should also be pointed out that producing frame statistics of each of the metrics for smaller frames of data has the potential to reveal problems occurring only temporarily in data. For example, spikes due to a loose cable may occur only during extra large vibration levels such as when a car hits a pot hole or passes over a railroad crossing. Calculating, in this case, kurtosis over small, say 1 s frames, is likely to show these spikes, whereas the effect of a few single spikes may not come through in a calculation over the entire data set.

We will now show some results from a data quality analysis on an eight-channel measurement on a truck running on a rough road.

**Example 4.4.1** *We illustrate data quality analysis with an example based on acceleration data that were recorded on a truck driving with constant speed on a stretch of rough road. The dataset used here consists of eight channels which were selected from a larger set from some faulty measurements, to illustrate the procedure.*

*After importing to MATLAB we calculate some basic statistics of each channel based on all data. The results are found in Table 4.2. The statistics in the table reveal that channel 1 has a relatively high mean value, and a kurtosis much higher than remaining channels. This is an indication of something that may be wrong, and we put the channel up for a closer look. Next, the kurtosis of channel 2 and 4 seem to be higher than remaining channels. Now, this could be quite in order if there is a good explanation for it. It could be due to different vibration character in different directions, or it could be that those points with higher kurtosis are in locations near some impulse force due to, e.g., rattling.*

*To find if there are temporary errors in the data such as spikes, for example, frame statistics can be calculated and plotted. For space reasons not all such plots can be shown here. As an example, frame statistics for skewness and kurtosis of channel 4 are plotted in Figure 4.8. We see that the kurtosis is relatively high at some instances, for example in frame 14. This is an indication that could mean there is a spike or similar in the data, and it should be investigated manually. In this case, it turned out to be the result of potholes causing extra high shocks.*

*End of Example.*

**Table 4.2** Statistics results for Example 4.4.1.

| Channel # | Mean | RMS | Skewness | Kurtosis |
|---|---|---|---|---|
| 1 | 1.2 | 3.8 | 0.16 | 11.0 |
| 2 | −0.032 | 4.1 | 0.19 | 8.8 |
| 3 | −0.012 | 1.3 | 0.25 | 4.9 |
| 4 | −0.0035 | 0.4 | 0.04 | 6.8 |
| 5 | 0.2 | 1.8 | 0.04 | 4.3 |
| 6 | −0.014 | 0.5 | −0.02 | 5.0 |
| 7 | 0.047 | 3.3 | 0.19 | 3.0 |
| 8 | 0.027 | 1.3 | 0.7 | 6.1 |

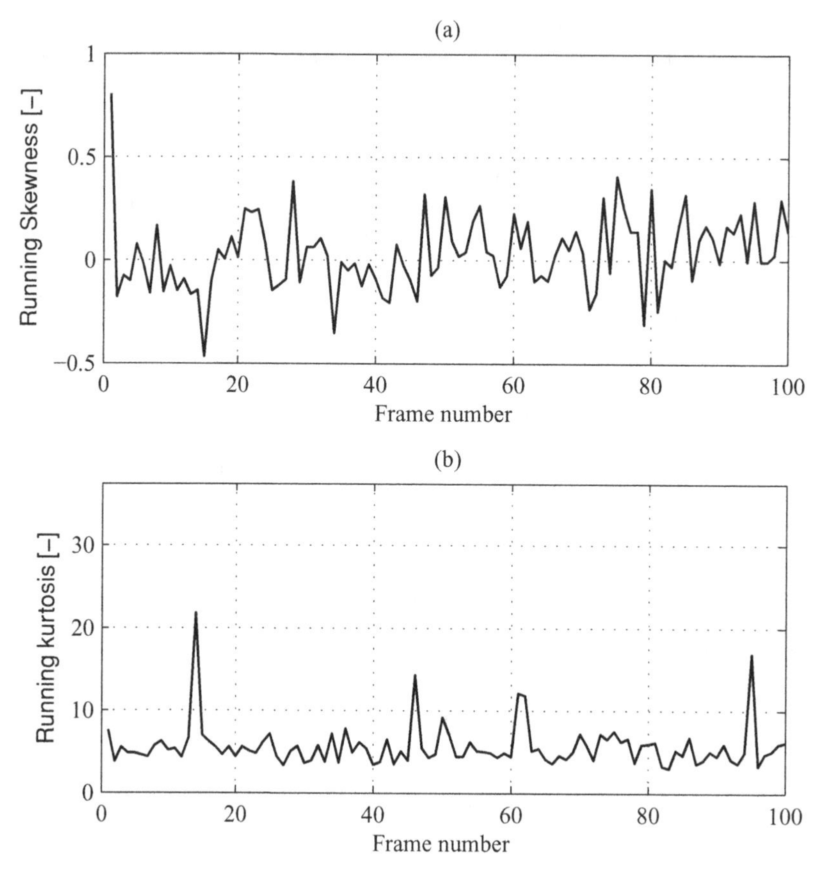


**Figure 4.8** Plots for Example 4.4.1. In (a) frame statistics based on skewness is plotted versus frame number. In (b) a similar plot with frame statistics of kurtosis is shown. The latter shows a few frames with extra high kurtosis, that could be caused by some unnatural spikes, for example. See text for a discussion

## 4.5 Chapter Summary

In this chapter we have presented some basic statistical properties and showed how to estimate them based on experimental data. These are found in Section 4.2 and will not be repeated here.

Next we introduced statistical hypothesis testing and presented two methods to use such tests to verify if a signal is stationary, the *reverse arrangements test* and the *runs test*. The former test reveals trends in the data that violate stationarity, and the latter reveals periodicities in the data. We also noted that in many cases it is sufficient to use frame statistics to evaluate the stationarity of data. In such an analysis, a statistical measure (e.g., the standard deviation, or kurtosis) is calculated for frames of the data, and plotted against frame number. If data are stationary, there should be little variation in the statistical parameter.

An important application of statistics, namely quality assessment of data, was then discussed. The recommended procedure to find anomalies in data is to calculate a number of statistics in two different ways;

- *Overall statistics.* By calculating mean, RMS, skewness, kurtosis, etc. over the entire data record, the nature of the measured data is found. If one or more channels stand out from the others, in terms of one or more statistical properties, then it is an indication that something may be wrong in that channel, and it should be manually inspected.
- *Frame statistics.* The overall statistics can, of course, miss errors that occur only intermittently and are therefore lost in the average over the entire data. The same statistics should therefore preferably be computed also for frames of, e.g., 1 s duration. Either those results can be plotted for every channel, or for large channel counts it might be enough to list the maximum value, as this alone will be an indication of an error in data.

## 4.6 Problems

Many of the problems following are supported by the accompanying ABRAVIBE toolbox for MATLAB/Octave and further examples which can be downloaded with the toolbox. If you have not already done so, please read Section 1.6, and follow the instructions to download this toolbox together with example files.

**Problem 4.1** *Assume a stationary random signal $x(t)$ has a mean $\bar{x}$ and standard deviation $\sigma_x$. Prove that the normalized variable $z = (x - \bar{x})/\sigma_x$ has zero mean, and unity standard deviation.*

**Problem 4.2** *Show that the relation in Equation (4.43), between the integral of the normal distribution probability density and* error function, *is true.*

**Problem 4.3** *There is a theorem in random theory that states that the expected value of a product of four random processes, $x_1$, $x_2$, $x_3$, and, $x_4$, each with normal distribution, is*

$$\mathrm{E}[x_1x_2x_3x_4] = \mathrm{E}[x_1x_2]\mathrm{E}[x_3x_4] + \mathrm{E}[x_1x_3]\mathrm{E}[x_2x_4] + \mathrm{E}[x_1x_4]\mathrm{E}[x_2x_3]. \tag{4.58}$$

*Show that this leads to the fact that kurtosis of a random signal with normal distribution is exactly 3.*

**Problem 4.4** *Create a Gaussian time signal, x with 100 000 samples in MATLAB/Octave. Then create a new non-Gaussian variable, y, by the equation $y = x + 0.1x|x|$. This new variable will, of course, not be Gaussian. Calculate and plot the sample probability density function (PDF) of x overlaid with the Gaussian PDF using the mean and standard deviation of x, with linear and logarithmic y-axis, respectively. Then repeat the plots for y. Compare the plots and watch the differences between linear and logarithmic y-axes.*

*Hint: You can use the* ***apdf*** *command from the accompanying toolbox to compute the PDF.*

**Problem 4.5** *Create a time signal in MATLAB/Octave with 100 000 samples, which is the product of a Gaussian random signal, and a half sine, so that the RMS level of the signal increases over the first half of the data, and then decrease over the last half. Run the reverse arrangements test on the data with a significance level of $\alpha = 0.02$ and $N = 100$ segments. Is the data stationary based on this test? Answer why?*

*(Comment: The data are clearly not stationary.)*

**Problem 4.6** *Perform a runs test on the same data as in Problem 4.5. Does the data pass the test as stationary? Explain the difference between the result of this test and the reverse arrangement test result.*

## References

Bendat J and Piersol AG 2010 *Random Data: Analysis and Measurement Procedures* 4th edn. Wiley Interscience.
Brownlee K 1984 *Statistical Theory and Methodology*. Krieger Publishing Company.
Himmelblau H, Piersol AG, Wise JH and Grundvig MR 1993 *Handbook for Dynamic Data Acquisition and Analysis*. Institute of Environmental Sciences and Technology, Mount Prospect, Illinois.
Newland DE 2005 *An Introduction to Random Vibrations, Spectral, and Wavelet Analysis* 3rd edn. Dover Publications Inc.
Papoulis A 2002 *Probability, Random Variables, and Stochastic Processes* 4th edn. McGraw-Hill.
Sheskin D 2004 *Handbook of Parametric and Nonparametric Statistical Procedures* 3rd edn. Chapman & Hall.
Wirsching PH, Paez TL and Ortiz H 1995 *Random Vibrations: Theory and Practice*. Wiley Interscience.

# 5

# Fundamental Mechanics

It could be argued that a single mass connected to a spring and a damper is a superficial and limited example of a mechanical system, which indeed it is. However, as we will see in later chapters, the *single degree-of-freedom* system, or SDOF system as it is often called, is most essential to understand when it comes to understanding mechanical dynamics, because real structures in some respects behave as if they were constructed of several SDOF systems. It is also common to make approximations in mechanical dynamics based on SDOF assumptions. A thorough understanding of this, the simplest of mechanical systems, is thus absolutely essential in order to understand the dynamics of mechanical systems.

In this chapter we introduce the SDOF system, starting from its basic equations, and deduce important results necessary for understanding the dynamics of it. If you have already studied mechanical dynamics, it is advised that you still read through this chapter at least briefly, as many of the results we emphasize here are often omitted in other texts, as the discussion here is particularly concerned with the experimental results we can obtain from measurements on a mechanical system.

## 5.1 Newton's Laws

The fundamental mechanics theory most commonly used to describe vibrations was established by Isaac Newton (1642–1727). In 1687 he published his three well-known laws in his famous *Principia* (Newton 1687):

- The Law of Inertia
  An object in motion tends to stay in motion, and an object at rest tends to stay at rest, unless the object is acted upon by an outside force.
- The Law of Acceleration
  The rate of change (time derivative) of a body's momentum (= mass times velocity) is proportional to the sum of all forces acting on the body.
- The Law of Action and Reaction
  If an object exerts a force on a second object, the second object exerts an equal but opposite force on the first.

If several forces act on a body, they are summed vectorially. The law that we will use in this chapter is the second law, the Law of Acceleration. It is so interesting, in fact, that we shall study it in great detail.

---

*Noise and Vibration Analysis: Signal Analysis and Experimental Procedures*, First Edition. Anders Brandt.

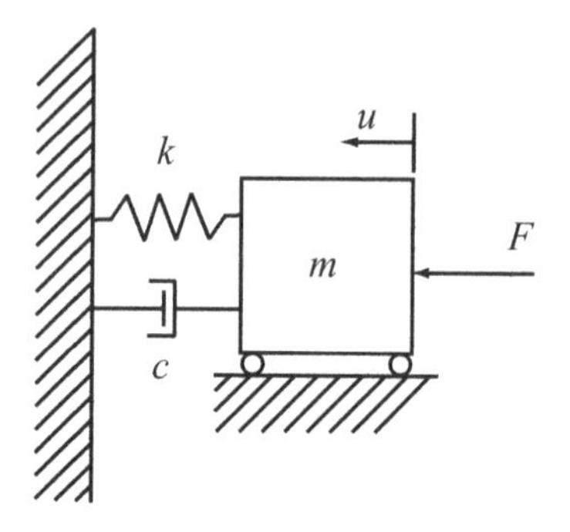


**Figure 5.1** Mechanical system with one degree-of-freedom, i.e., a mass, $m$, moving in one direction. Connected to the mass are a spring with stiffness $k$ and a viscous damper with damping $c$

## 5.2 The Single Degree-of-freedom System (SDOF)

We first assume that we have a mass connected to a spring and a so-called viscous damper, as shown in Figure 5.1. This system is called a *single degree-of-freedom* system, where the term degree-of-freedom refers to the motion along (in this case) one translational axis. In Section 6.2.1 we will discuss the concept of degrees-of-freedom in mechanical systems in more detail. The mass is driven by a dynamic force, $F(t)$. To make the example as simple as possible we choose a horizontal force so that we do not have to consider any effects of gravity and initial spring compression, which might divert attention from the important issues at stake here. We further assume that the mass can move without friction (except for the effect of the viscous damper) around its equilibrium position $u_0$. The spring and the viscous damper are assumed to be massless.

A description of the spring and the viscous damper is appropriate here. A spring which is subjected to a force, $F$, gives a counter force which is proportional to the displacement from its equilibrium position. The constant of proportionality, $k$, is called the *stiffness* of the spring with units of [N/m]. The viscous damper similarly provides a counter force which is proportional to the velocity, that is, to the time derivative of displacement. The constant of proportionality for the damper is given by the symbol $c$ which has units of [Ns/m].

The balance of forces according to Newton's second law gives

$$m\ddot{u} = -c\dot{u} - ku + F(t) \tag{5.1}$$

where the minus sign indicates that the damper and spring provide counter forces (in opposite direction to the force $F$). Equation (5.1) is usually rewritten as

$$m\ddot{u} + c\dot{u} + ku = F(t) \tag{5.2}$$

Time derivatives are indicated by a 'dot' such that velocity is given by $\dot{u} = \mathrm{d}u/\mathrm{d}t = v$ and acceleration by $\ddot{u} = \mathrm{d}^2u/\mathrm{d}t^2 = a$ which are read 'u-dot' and 'u-double-dot,' respectively. These notations are common in mechanics where we often deal with time derivatives of displacement.

### 5.2.1 *The Transfer Function*

We shall now examine the solution to Equation (5.2), which we solve by Laplace transformation. If you are not familiar with the Laplace transform, read Section 2.6.1 carefully before continuing with the present chapter. Taking the Laplace transform of both sides of Equation (5.2) gives

$$(ms^2 + cs + k)U(s) = F(s) \tag{5.3}$$

This expression leads to the *transfer function*, $H(s)$, between $F(s)$ and $U(s)$ given by

$$H(s) = \frac{U(s)}{F(s)} = \frac{1/m}{s^2 + sc/m + k/m} \quad (5.4)$$

where we have divided numerator and denominator by $m$ to render the $s^2$ term alone in the denominator. For second-order systems it is common to write the denominator in the 'standard form'

$$H(s) = \frac{U(s)}{F(s)} = \frac{1/m}{s^2 + s2\zeta\omega_n + \omega_n^2} \quad (5.5)$$

which means that in our case

$$\omega_n = \sqrt{\frac{k}{m}} \quad (5.6)$$

and

$$\zeta = \frac{c}{2\sqrt{mk}} \quad (5.7)$$

where $\omega_n$ is the *undamped natural (angular) frequency*, sometimes called undamped *resonance* frequency, in [rad/s] and $\zeta$ (Greek 'zeta') is the *relative damping*, which is dimensionless. The standard form in Equation (5.5) is often used, because much can be understood about the system from just knowing the parameters $\omega_n$ and $\zeta$. In the following text we do not assume any prior experience, but if you are familiar with second order systems you will recognize much from before.

**Example 5.2.1** *Assume an SDOF system with $m = 1$ [kg], $c = 100$ [Ns/m], and $k = 10^6$ [N/m]. Using Equation (5.6) we find that the undamped natural frequency will be*

$$f_n = \frac{\omega_n}{2\pi} = \frac{1}{2\pi}\sqrt{\frac{k}{m}} = \frac{1}{2\pi}\sqrt{\frac{10^6}{1}} = \frac{1000}{2\pi} \approx 159.2 \ Hz \quad (5.8)$$

*and from Equation (5.7) the relative damping will be*

$$\zeta = \frac{c}{2\sqrt{km}} = \frac{100}{2\sqrt{10^6}} = 0.05 = 5\%. \quad (5.9)$$

*End of example.*

## 5.2.2 The Impulse Response

From Section 2.6.1 we know that the time domain equivalent of the transfer function in Equation (5.5) is the *impulse response*. In order to find the impulse response, we first calculate the *poles* of Equation (5.4), which are

$$s_{1,2} = -\zeta\omega_n \pm j\omega_n\sqrt{1-\zeta^2} = -\zeta\omega_n \pm j\omega_d \quad (5.10)$$

where $\omega_d = \omega_n\sqrt{1-\zeta^2}$ is often referred to as the *damped natural frequency*.

The poles correspond, as we mentioned earlier, to the homogeneous solution to Equation (5.3), which describe the free oscillations, where the displacement of the mass, $u(t)$ is nonzero while the force $F(t)$ equals zero.

Using the poles in Equation (5.10), the transfer function in Equation (5.4) can be rewritten as

$$H(s) = \frac{1/m}{(s-s_1)(s-s_2)}. \quad (5.11)$$

**Table 5.1** Values of the factors $\sqrt{1-\zeta^2}$ and $\sqrt{1-2\zeta^2}$ in Equation (5.15) and Equation (5.21), respectively

| $\zeta$ | $\sqrt{1-\zeta^2}$ | $\sqrt{1-2\zeta^2}$ |
|---|---|---|
| 0.01 | 1.000 | 1.000 |
| 0.05 | 0.999 | 0.997 |
| 0.1 | 0.995 | 0.990 |
| 0.2 | 0.980 | 0.969 |
| 0.3 | 0.954 | 0.906 |

We can now use partial fraction expansion (see Section 2.6.1) to rewrite Equation (5.11) into

$$H(s) = \frac{C_1}{s - s_1} + \frac{C_2}{s - s_2} \tag{5.12}$$

where the constants $C_1$ and $C_2$ can be identified by residue calculus as mentioned in Section 2.6.1. This is referred to the problem section at the end of this chapter.

Applying Laplace transform pair 4 from Table (2.1) on Equation (5.12) results in the impulse response of the SDOF system

$$h(t) = C_1 e^{s_1 t} + C_2 e^{s_2 t}. \tag{5.13}$$

Using the expression for the poles from Equation (5.10) we can rewrite Equation (5.13) as

$$h(t) = C_1 e^{-\zeta\omega_n t} e^{j\omega_d t} + C_2 e^{-\zeta\omega_n t} e^{-j\omega_d t}. \tag{5.14}$$

From the first exponential factor in each term of Equation (5.14) it is clear that the product of the relative damping $\zeta$ and the natural frequency $\omega_n$ determines how quickly the motion of the mass is damped out after excitation. From the second, imaginary exponential factor in each term in Equation (5.14) we will get oscillations with a frequency, $\omega_d$ which is lower than the undamped natural frequency. The solution in Equation (5.14) is only practical in the case where the damping is *under critical*, i.e., $0 \leq \zeta \leq 1$, where the solutions will oscillate, i.e., we will have vibration. The upper limit, $\zeta = 1$, indicates the case where an impulse-shaped excitation results in a similar impulse-shaped response, after which all motion ceases. See Section 5.7 for a general discussion about damping.

It should be noted that the left-hand side of Equation (5.14) is in general real because the force and displacement signals are of course real. The right-hand side of the equation, however, seems to be complex. Since there is an equality sign between the two, naturally the imaginary parts on the right-hand side must equal zero. If we make use of the fact that the two poles are complex conjugate pairs, $s_2 = s_1^*$ and that $h(0) = 0$, then it can be shown after some steps that Equation (5.14) can be simplified to

$$h(t) = A e^{-\zeta\omega_n t} \sin(\omega_d t) \tag{5.15}$$

where $A$ is a real constant. So indeed, the impulse response turns out to be real.

The impulse response of the SDOF system thus consists of an exponentially damped sine oscillating with the frequency $f_d = f_n\sqrt{1-\zeta^2}$, usually called the *damped natural frequency*. From Equation (5.15) it is clear that the frequency of oscillation decreases with increased damping, although slightly, see Table 5.1. Figure 5.2 shows the impulse responses, $h(t)$, for two typical values of relative damping, $\zeta = 0.01$ and $\zeta = 0.05$ for an SDOF system with undamped natural frequency of 159.2 Hz.

**Example 5.2.2** *Taking the SDOF system from Example 5.2.1 we find that the impulse response in this case with relative damping of $\zeta = 5\%$ will oscillate with a frequency of*

$$f_d = f_n\sqrt{1-\zeta^2} = 159.2 \times 0.999 = 159.0 \text{ [Hz]}. \tag{5.16}$$

*End of example.*

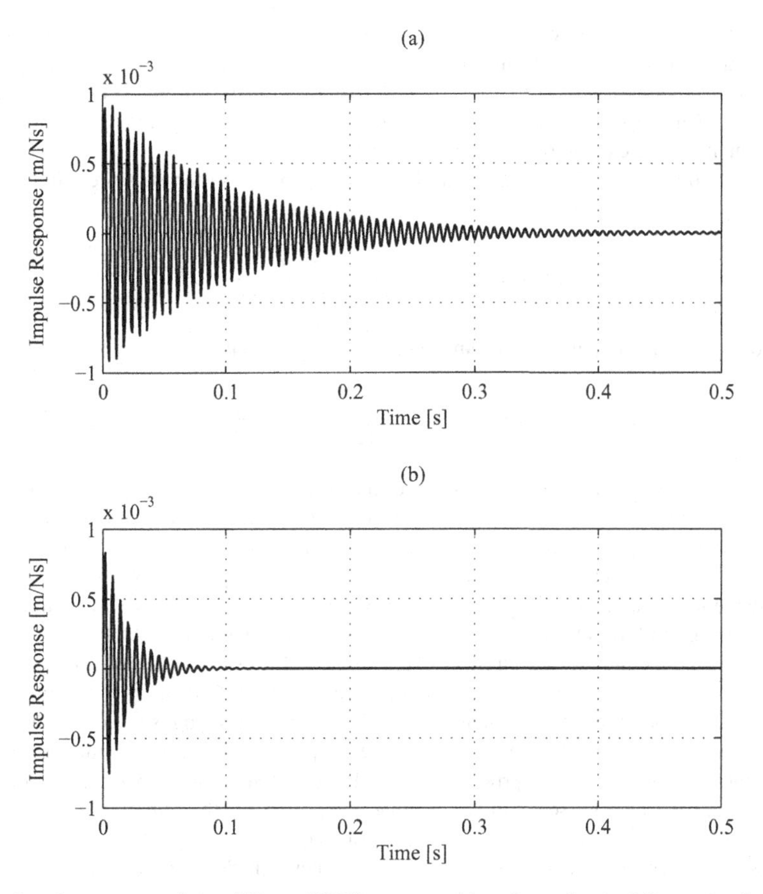


**Figure 5.2** Impulse response of two different SDOF systems with undamped natural frequency $f_n = 159.2$ Hz and relative damping of a) $\zeta = 0.01$, and b) $\zeta = 0.05$

It is well worth recapitulating the meaning of the impulse response here. As we saw in Section 2.6.3, the effect of the impulse response in the convolution process to obtain the output from the input, is to weight input values 'backwards' in time. The longer the impulse response is, the longer is the part of the input (force) which is weighted together to form the output (displacement) at any instance in time. If we consider an input signal which is more or less constant in level, such as stationary noise or a periodic signal, we can therefore predict that the more input values that are weighted together, the higher the output signal will be. Less damping thus leads to a longer impulse response which leads to higher vibrations. If the input is a transient shock, then the amount of damping will determine how long the response continues after the input shock has excited the system.

### *5.2.3 The Frequency Response*

When we measure a dynamic system in order to identify it, we cannot measure the transfer function as in Equation (5.4), as it is a nonphysical entity, as explained in Section 2.6.1. Instead, we usually measure the frequency response, which as we know from Section 2.7.2 is defined as the ratio of the spectrum of the output (response) and the spectrum of the input (force). The most intuitive method to determine the frequency response at a certain frequency is to let the excitation be a sinusoid with the desired frequency and calculate the frequency response magnitude as the ratio of the response and force amplitudes. The phase angle can be determined, if desired, by measuring the phase difference between the response and the force. In Chapters 13 and 14 more refined methods for measuring frequency responses will be described, which are more commonly used in practice.

As we know from Section 2.7.3, we obtain the frequency response for the system in Equation (5.4) by letting $s = j\omega = j2\pi f$. We first get

$$H(f) = \frac{U(f)}{F(f)} = \frac{1/m}{-\omega^2 + j2\zeta\omega_n\omega + \omega_n^2}. \tag{5.17}$$

After a bit of algebraic manipulation we can simplify this equation into

$$H(f) = \frac{U(f)}{F(f)} = \frac{1/k}{1 - (f/f_n)^2 + j2\zeta\,(f/f_n)} \tag{5.18}$$

where we have made the denominator a function of the *relative frequency* $f/f_n$ which is very practical. The shape of the frequency response is evidently independent of the actual natural frequency, and only dependent on the ratio $f/f_n$. You should also note that the numerator in Equation (5.18) has been replaced by $1/k$. This reflects the physical fact that at very low frequencies, where the denominator is approximately equal to unity, the frequency response is approximately $1/k$, which means that we only 'feel' the spring, as the forces from the damper and mass are small at low frequencies.

In Figure 5.3, the magnitude and phase of $H(f)$ is plotted for three different values of the relative damping, $\zeta$. It is clear that for low frequencies (relative to the natural frequency), the response (displacement) magnitude is approximately constant and the phase angle between response and force is close to zero. These relationships indicate that the mass moves in phase with the force which is a result of the fact that at low frequencies only the spring is in effect. If we now suppose that we increase the frequency and keep the force constant, we see in Figure 5.3 that as we approach the natural frequency the response increases greatly and has a maximum near the natural frequency. At higher frequencies the response decreases, inversely proportional to the square of the frequency (which is apparent from Equation (5.18)).

At and around the natural frequency something important occurs in the phase plot. From having been in phase at low frequencies, the response begins to lag behind, and eventually becomes out of phase with the force above the natural frequency. This phase relationship means that we have maximum displacement in one direction (remember we are considering sinusoidal force at each frequency!) while we have maximum force directed the opposite way. The reason is that the reactive force of the mass is proportional to its acceleration while the force of the spring is proportional to its position (displacement from equilibrium).

Exactly at the undamped natural frequency where $f = f_n$, Equation (5.18) shows that the frequency response is purely imaginary. Since the imaginary number is in the denominator, this means that the phase is exactly $\angle H(f_n) = -\pi/2$ or $-90°$ at this frequency. This can also be seen in the phase plots in Figure 5.3.

The damping affects how high the resonance peak is; the lower the damping, the greater the motion around the resonance frequency, for a given force. In Figure 5.3 you should also observe that higher damping leads to a larger frequency range over which the phase curve switches from 0 to $-180°$.

We shall now examine the frequency response in more detail. Writing the expressions for amplitude and phase of $H(f)$ we get

$$|H(f)| = \frac{1/k}{\sqrt{\left(1 - \left(\frac{f}{f_n}\right)^2\right)^2 + \left(2\zeta\frac{f}{f_n}\right)^2}} \tag{5.19}$$

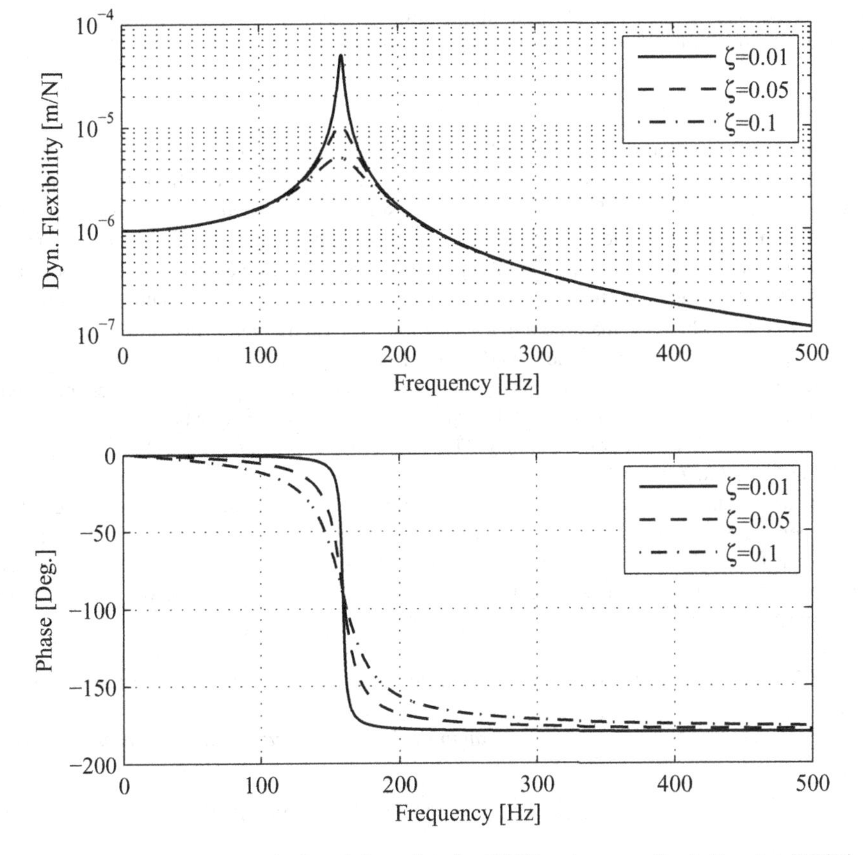


**Figure 5.3** Frequency response magnitude and phase plot of an SDOF system according to Equation (5.18) for three values of the relative damping, $\zeta = 0.01$, $\zeta = 0.05$ and $\zeta = 0.1$. The undamped natural frequency, $f_n$, is 159.2 Hz

and for the phase

$$\angle H(f) = -\arctan\left(\frac{2\zeta\frac{f}{f_n}}{1-\left(\frac{f}{f_n}\right)^2}\right) \tag{5.20}$$

where $\angle$ indicates phase angle of $H(f)$. From Equation (5.19) we can now calculate where the amplitude of the frequency response $H(f)$ has its maximum, by differentiating and finding the zero of the derivative. We find that this frequency, sometimes called the *damped resonance frequency*, $f_{max}$, is given by

$$f_{max} = f_n\sqrt{1-2\zeta^2} \tag{5.21}$$

which is valid for damping values $\zeta \leq 1/\sqrt{(2)}$ and where the peak value of $|H(f)|$ is

$$|H(f_{max})| = \frac{1/k}{2\zeta\sqrt{1-\zeta^2}} \tag{5.22}$$

which is also valid for $\zeta \leq 1/\sqrt{2}$. This maximum value can also be expressed using the $Q$-factor from Equation (5.27), in which case the maximum is approximately

$$|H(f_{max})| \approx Q \cdot \frac{1}{k} \tag{5.23}$$

which means that the $Q$-factor tells how much the resonance amplification is.

It should be noted that the damped resonance frequency $f_{max}$ is different from the damped natural frequency $f_d$ defined in Section 5.2.2. In fact, there is some disagreement between different books on what to call both these frequencies. This is a good reason to always use the undamped natural frequency when communicating results from structural dynamics. It should, however, be understood that there is one frequency by which the impulse response oscillates, $f_d = f_n\sqrt{1-\zeta^2}$, and another frequency, $f_{max} = f_n\sqrt{1-2\zeta^2}$, where the maximum in the magnitude of the frequency response is found, and both these frequencies are lower than the undamped natural frequency. We thus have maximum displacement, for constant force at all frequencies, at a lower frequency than the undamped natural frequency, and $f_{max}$ becomes lower with increasing damping. All of this is valid, however, only when $\zeta \leq 1/\sqrt{2}$, otherwise we have no peak at all at the natural frequency. The values of $\sqrt{1-\zeta^2}$ and $\sqrt{1-2\zeta^2}$ for some different values of $\zeta$ are given in Table 5.1.

It is difficult to state typical values of relative damping, as it can differ much between different structures, depending on design and material choice. In steel structures, for example, the damping is often as low as 0.001 (e.g., airplane wings) up to possibly 0.1 (structures with screw joints, etc.). The value of $\zeta = 0.05$ is often used as a typical value of damping in steel structures. We shall discuss more about damping in Section 5.7.

**Example 5.2.3** *Taking again the SDOF system from Example 5.2.1, calculate the damped resonance frequency,* $f_{\max}$.

*From Equation (5.16) and the results from Example 5.2.1, we find that the damped resonance frequency will be*

$$f_{\max} = f_n\sqrt{1-2\zeta^2} = 159 \times 0.997 = 158.8 \text{ [Hz]}. \tag{5.24}$$

*End of example.*

### 5.2.4 The Q-factor

The *half-power bandwidth* or *3 dB bandwidth* or *resonance bandwidth* of a resonance peak is defined by

$$B_r = f_u - f_l \tag{5.25}$$

where the lower and upper frequencies, $f_l$ and $f_u$ are defined by

$$|H(f_l)|^2 = |H(f_u)|^2 = \frac{1}{2}|H(f_{max})|^2 \tag{5.26}$$

that is, the upper and lower frequencies are defined so that the power (amplitude squared) of $|H(f)|$ has been halved in relation to the peak value of $|H(f_{max})|$. In electrical circuit theory, a common concept for resonant circuits is the *quality factor*, or $Q$-factor, which can be calculated as the ratio between the center frequency and the half-power bandwidth, i.e.,

$$Q = \frac{f_n}{B_r}. \tag{5.27}$$

(It should perhaps be particularly noted here that the word 'quality' relates to radio receiver applications, where a good-quality resonance is one with small bandwidth which picks up the radio signal, but

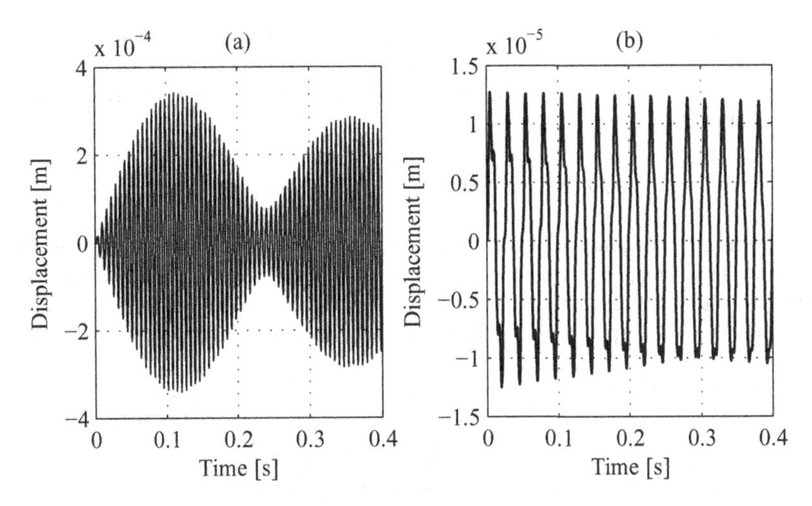


**Figure 5.4** Illustration of transient forced response of SDOF system, causing in (a) strong beating, and in (b) some distortion. The SDOF system has a undamped natural frequency of 159.2 Hz and relative damping of 0.2%. The exciting force is sinusoidal with 10 N amplitude and, in (a), 155 Hz, and, in (b), 40 Hz frequency. The beating phenomenon is most prominent for systems with low damping and is more severe the closer to the natural frequency the excitation frequency is, until the beating disappears when the excitation frequency coincides with the natural frequency (although there is still a rather long transient part until the correct level is obtained in the latter case)

not surrounding frequencies. In mechanical applications a high $Q$ value is usually rather 'bad quality' as it causes large vibrations.) The $Q$-factor is common in vibration fatigue and shock applications, see Section 16.1. The relative damping, $\zeta$, is related to the $Q$-factor, as will be discussed in Section 5.5.3.

### 5.2.5 *SDOF Forced Response*

The response of the SDOF system to a force input is, like the response of any linear system, composed of a transient and a steady-state response, as we discussed in Section 2.7.4. This can sometimes result in a forced response that exhibits a phenomenon called *beating* which is illustrated in Figure 5.4 (a). Beating is most prominent if the damping is low and the excitation frequency is close to the natural frequency of the SDOF system. For the example in Figure 5.4 we use an SDOF system with the same undamped natural frequency of 159.2 Hz as in the previous examples, but with relative damping of 0.2%. The plot in Figure 5.4 (a) illustrates the beating that results if we apply a sinusoidal excitation force of 10 N amplitude and a frequency of 155 Hz to the system. In (b) the result of a force with 10 N amplitude and 40 Hz frequency is shown. In the latter case the result is a clearly distorted sine. In both cases, after this transient behavior has died out, the response is a steady-state sine with 155 and 40 Hz, respectively, but with the low damping of this system it will take a long time before this is achieved, see also Problem 6.8.

## 5.3 Alternative Quantities for Describing Motion

When we experimentally measure frequency response, as for example in *experimental modal analysis*, we normally measure responses in the form of accelerations, not displacements, because acceleration is easier to measure, as we will see in Chapter 7. The frequency response according to Equation (5.18), consisting of the ratio of displacement with force, is often called *dynamic flexibility* or *receptance* and has the units [m/N = $s^2$/kg]. As mentioned above, we can instead consider that we, in place of displacement, $u(t)$, measure velocity, $v(t) = du/dt$. Through differentiation of Equation (5.18), which in the frequency domain corresponds to a multiplication by $j2\pi f (= j\omega)$, we obtain the *mobility*, $H_v(f)$, with units of [m/Ns = s/kg], that is

$$H_v(f) = \frac{V(f)}{F(f)} = j2\pi f \frac{1/k}{1 - (f/f_n)^2 + j2\zeta\,(f/f_n)}. \tag{5.28}$$

It is more common to measure acceleration, rather than velocity, although velocity sensors do exist, and for example laser Doppler vibrometers typically measure velocity. The most common frequency response is thus that of acceleration with force which is called *accelerance* or sometimes *inertance* and has the units [m/Ns$^2$ = 1/kg]. Through differentiation of Equation (5.28) we obtain the accelerance as

$$H_a(f) = \frac{A(f)}{F(f)} = -(2\pi f)^2 \frac{1/k}{1 - (f/f_n)^2 + j2\zeta\,(f/f_n)} \tag{5.29}$$

where the minus sign comes from the square of the imaginary number.

By analogy with what we discussed for dynamic flexibility above, we could take the absolute values of $H_v$ and $H_a$ in Equations (5.28) and (5.29), respectively, differentiate and set the derivatives to zero to determine at which frequency the peak in the absolute value occurs. It can be shown that the magnitude of mobility has a maximum at exactly $f = f_n$ while the magnitude of accelerance has a maximum at a somewhat higher frequency. In Figure 5.5 the three different types of frequency response are plotted using log–log plot format to obtain straight lines for the asymptotes (see Appendix B).

Instead of calculating response divided by force we could calculate the inverted expression, i.e., force divided by displacement, velocity, or acceleration. There is, however, a very good reason to measure

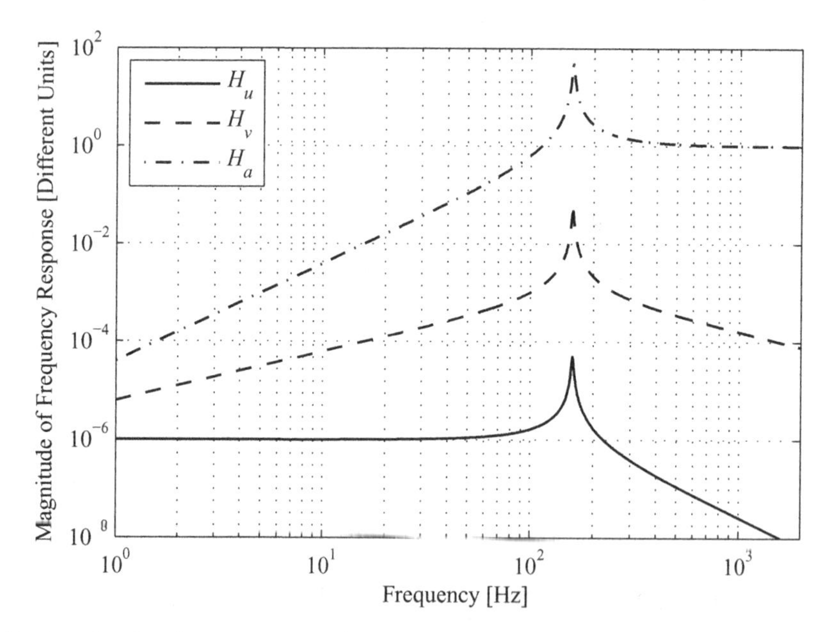


**Figure 5.5** Log–log plot of the three common forms of frequency response, dynamic flexibility, $H_u$, mobility, $H_v$, and accelerance, $H_a$. Note especially the asymptotic slopes for low and high frequencies indicated in the figure. Also note that the three forms of frequency response have different units and are therefore not comparable in the same plot. They are plotted in the same plot here for shape comparison only

**Table 5.2** Names of frequency responses between different response signals, $R$, and force, $F$. The names in boldface are those recommended by the international standard (ISO 2641: 1990)

| Response quantity, $R$ | $R/F$ | | $F/R$ | |
|---|---|---|---|---|
| Displacement | $u/F$ | **Dynamic flexibility**<br>Receptance<br>Compliance | $F/u$ | **Dynamic stiffness** |
| Velocity | $v/F = j2\pi f \cdot u/F$ | **Mobility**<br>Admittance | $F/v = 1/(j2\pi f) \cdot F/u$ | **Mechanical impedance** |
| Acceleration | $a/F = -(2\pi f)^2 \cdot u/F$ | **Accelerance**<br>Inertance | $F/a = -1/(2\pi f)^2 \cdot (F/u)$ | **Apparent mass** |

functions in the middle column in Table 5.2, i.e., 'response over force' type of frequency response, as will be further discussed in Section 6.4.1. Although the 'force over response' type of frequency response is thus rarely used, it is still good to know the names as they do occasionally occur in texts on vibrations. All six possible kinds of frequency response and their respective names are therefore given in Table 5.2.

## 5.4 Frequency Response Plot Formats

As our aim is to understand the frequency response of the SDOF system from an experimental point of view, we will now turn to different ways of presenting the frequency response in any of the 'response over force' types of dynamic flexibility, mobility, and accelerance. As we will see, there are several common ways of plotting the frequency response, each of which emphasizes different aspects of it. Experimentally, the usual response units are either acceleration or velocity, but using modern software for analysis, it is easy to convert between any type of response units. In this section we will therefore discuss all three forms of response units.

### *5.4.1 Magnitude and Phase*

The first and most common plot format is the magnitude-phase plot that we have been using already. This plot format can in turn be plotted either using linear or logarithmic scales for the $x$- and $y$-axes. It is most common to use logarithmic scale for the magnitude $y$-axis, as otherwise a lot of detail is lost, as is shown in Appendix B. For the $x$-axis both formats are common and each format has its own merits. The main reasons to use a logarithmic scale for the $x$-axis are twofold. First, it produces straight-line asymptotes as we mentioned in conjunction with Figure 5.5. Second, as indicated by Equation (5.25), the relative bandwidth, i.e., the ratio of the resonance bandwidth and the resonance frequency, is proportional to the relative damping. If the damping is the same for all resonances on real structures with several resonances, the width of each resonance will therefore look equal when using a logarithmic frequency axis.

In Figure 5.6 the magnitude and phase of the three frequency response types are plotted for the same SDOF system as we used in the examples previously in this chapter, with an undamped resonance frequency of 159.2 Hz, and a relative damping of 5%.

It is probably equally popular among engineers working with structural dynamics to plot frequency responses with linear frequency axis. This plot format has the advantage of not extending the

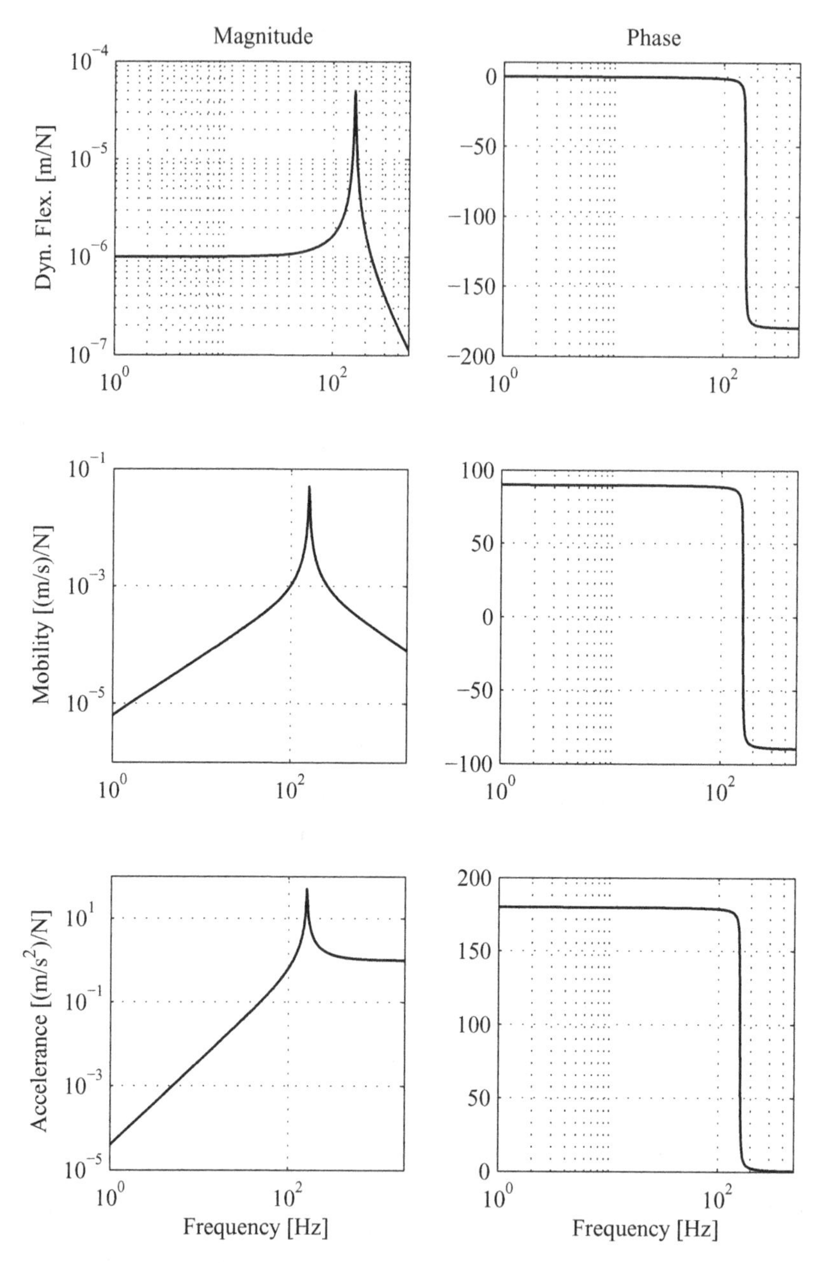


**Figure 5.6** Log-frequency plot of frequency responses with magnitude and phase. The magnitude is plotted with a logarithmic $y$-axis whereas the phase is plotted with a linear $y$-axis

low-frequency part and compressing the upper-frequency part, and many engineers find this format better for many purposes, see also Appendix B. A plot in magnitude/phase format and with linear frequency axes, of all three frequency response types, is shown in Figure 5.7. As before, the $y$-axis of the magnitude plot is logarithmic and the $y$-axis of the phase plot is linear.

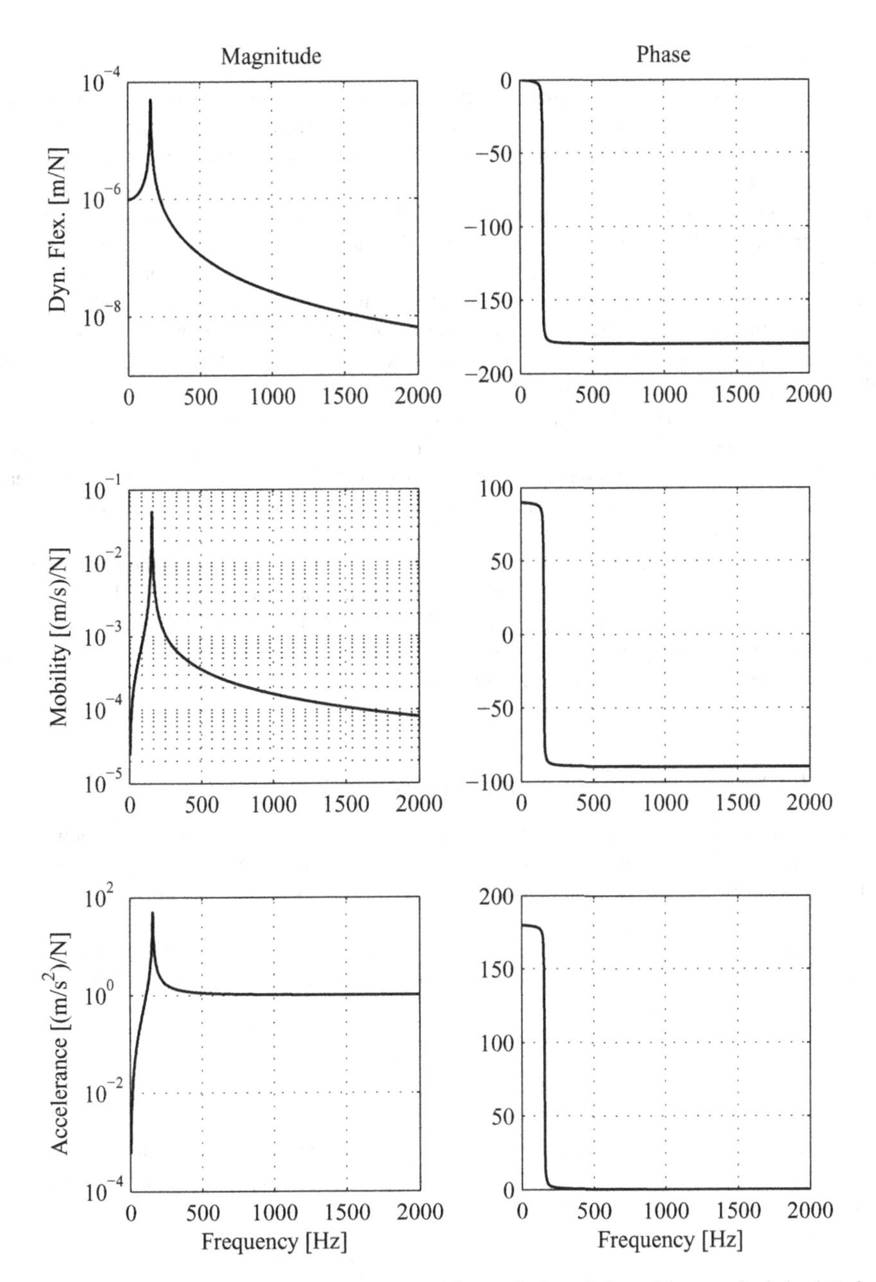


**Figure 5.7** Lin-frequency plot of frequency responses with magnitude and phase. The magnitude is plotted with a logarithmic $y$-axis whereas the phase is plotted with a linear $y$-axis

### 5.4.2 Real and Imaginary Parts

A careful examination of the denominator of the frequency response of dynamic flexibility type in Equation (5.18) reveals that where $f \ll f_n$ the unity term dominates the denominator, and thus the frequency response is (approximately) real. On the other side of the resonance where $f \gg f_n$ the term $-(f/f_n)^2$ dominates, and the frequency response is again real. It is only at frequencies around $f = f_n$ that the frequency response becomes significantly complex, and exactly at $f = f_n$ it is purely imaginary. The exact appearance of the real and imaginary parts of dynamic flexibility are shown in the upper plots of Figure 5.8. As can be seen in the figure, the real part is positive for low frequencies, and then makes a characteristic 'bend' around the resonance frequency. At higher frequencies the frequency response is negative, due to the 180° phase shift that occurs at the resonance. The imaginary part, on the other hand, is zero at most frequencies, and exhibits a dip at the resonance, as a result of the fact that the phase is exactly −90° when $f = f_n$, see Equation (5.20).

As the conversion of dynamic flexibility into mobility and accelerance is accomplished by multiplying once and twice, respectively, with the factor $j\omega$ which includes the imaginary number, the real and imaginary parts will swap when converting between dynamic flexibility and mobility so that the real part in one of the functions becomes the imaginary part in the next format, and vice versa. This is shown in the subsequent plots in Figure 5.8.

Of the real and imaginary parts, it is most common to plot the part that exhibits the peak (or dip), that is the imaginary part of the dynamic flexibility or accelerance, and the real part of the mobility. This plot is sometimes used for estimating the resonance frequency, as is described in Section 5.5.2. Another common use is for examining the quality of frequency responses with force and response measured in the same point, so-called *driving point* FRFs, see Section 13.12.2.

### 5.4.3 The Nyquist Plot – Imaginary vs. Real Part

The last plot format we shall discuss was first described by (Kennedy and Pancu 1947) in a well-known paper. They showed that for mobility, a perfect circle is formed in the Nyquist diagram, i.e., in a plot with the imaginary part versus the real part of $H_v(f)$. Nyquist plots of dynamic flexibility, mobility and accelerance are shown in Figure 5.9.

If we begin with the transfer function for dynamic flexibility from Equation (5.4) above, convert to mobility, and evaluate on the frequency axis $s = j\omega$, we find that the mobility can be written as

$$H_v(\omega) = \frac{j\omega}{k - \omega^2 m + j\omega c} = \frac{\omega^2 c + j\omega(k - \omega^2 m)}{\left(k - \omega^2 m\right)^2 + (\omega c)^2} \tag{5.30}$$

where we have multiplied the second expression in the numerator and denominator by the complex conjugate of the denominator of the second expression to obtain the last expression. From this equation we split the real and imaginary parts and obtain

$$H_{vR} = \mathrm{Re}\{H_v\} = \frac{\omega^2 c}{\left(k - \omega^2 m\right)^2 + (\omega c)^2} \tag{5.31}$$

and

$$H_{vI} = \mathrm{Im}\{H_v\} = \frac{\omega\left(k - \omega^2 m\right)}{\left(k - \omega^2 m\right)^2 + (\omega c)^2}. \tag{5.32}$$

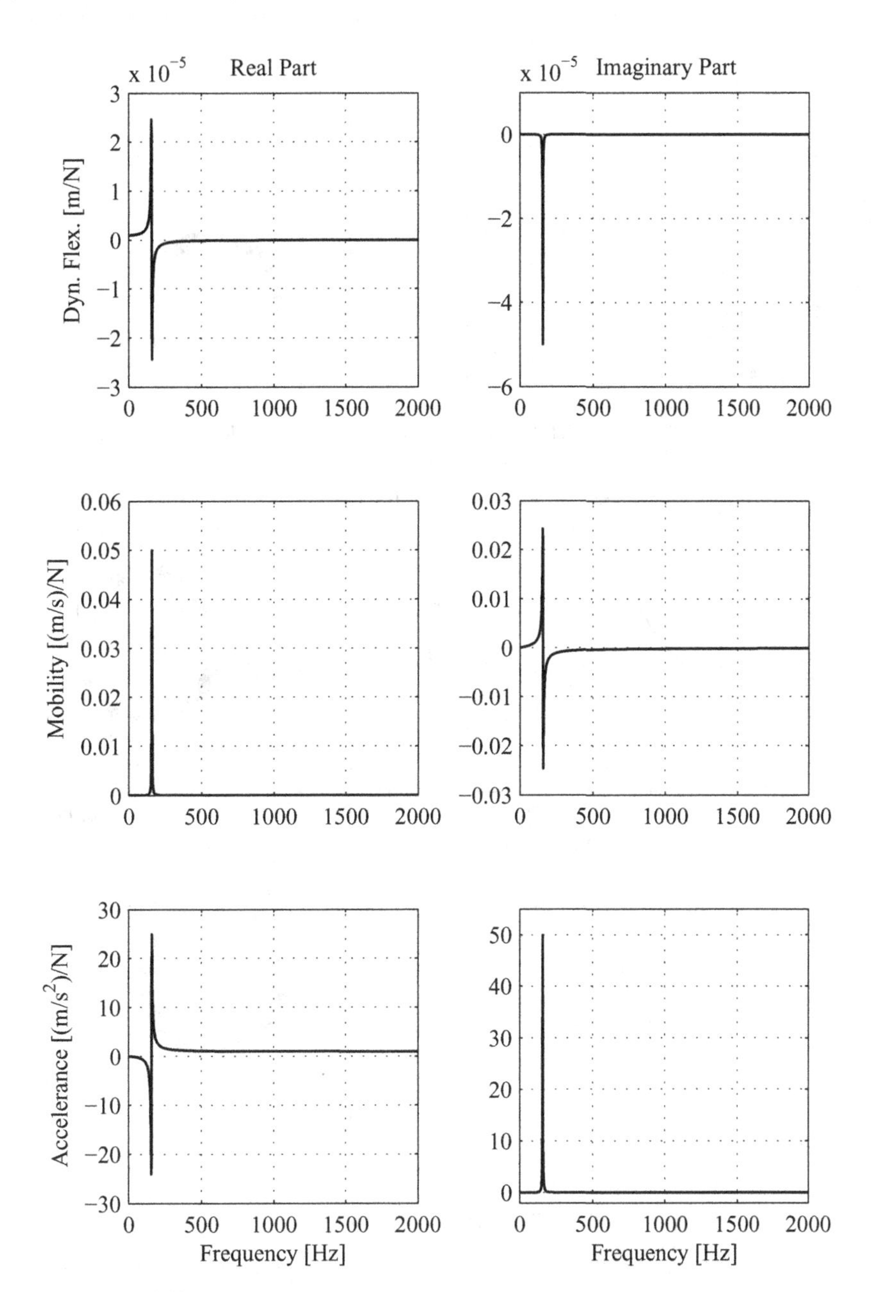


**Figure 5.8** Real part and imaginary part for dynamic flexibility, mobility and accelerance

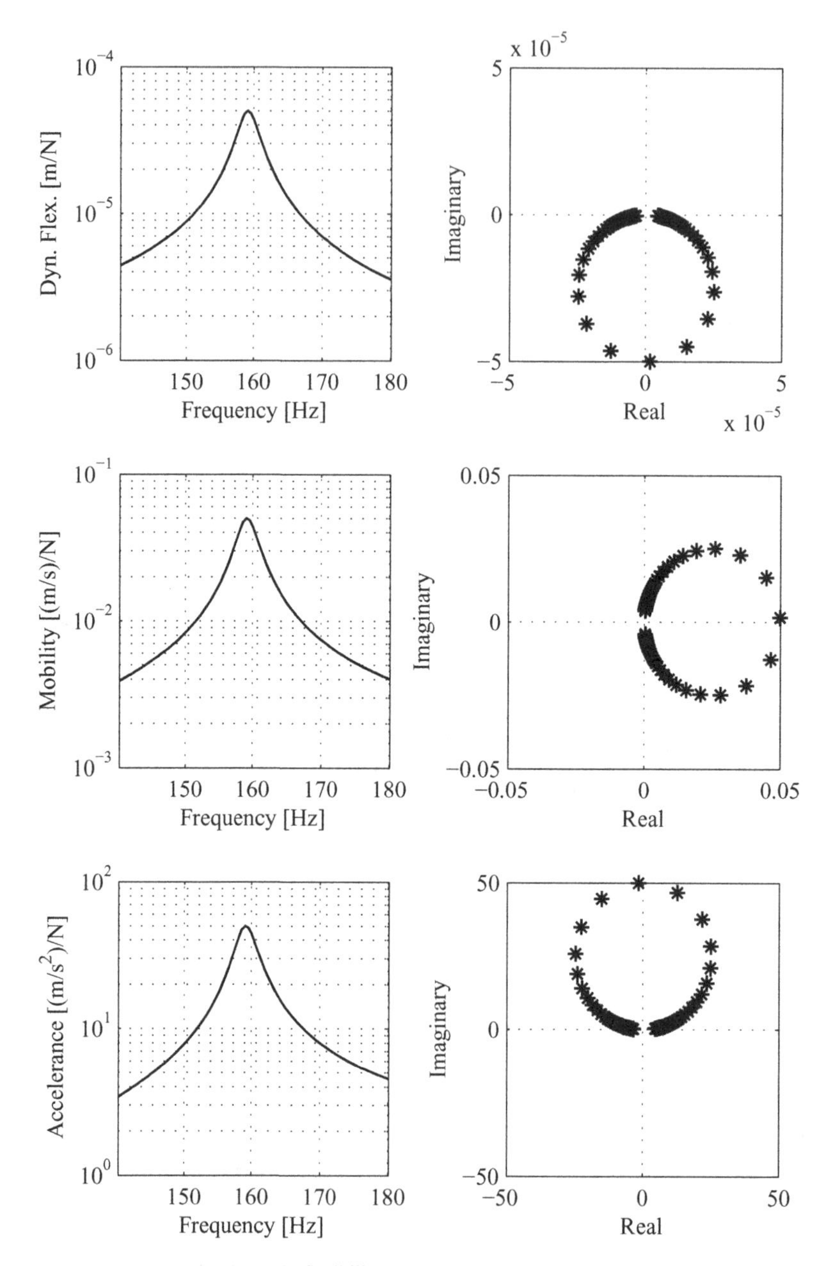


**Figure 5.9** Nyquist diagram for dynamic flexibility, mobility and accelerance for an SDOF system with viscous damping. The Nyquist curve for mobility displays a perfect circle

Now comes the trick. We let

$$X = H_{vR} - \frac{1}{2c} \tag{5.33}$$

and

$$Y = H_{vI}. \tag{5.34}$$

With these definitions it can be shown that

$$X^2 + Y^2 = \frac{\left[(\omega c)^2 + (k - \omega^2 m)^2\right]^2}{4c^2\left[(\omega c)^2 + (k - \omega^2 m)^2\right]^2} = \left(\frac{1}{2c}\right)^2 \tag{5.35}$$

which is thus the same as

$$\left(H_{vR} - \frac{1}{2c}\right)^2 + (H_{vI})^2 = \left(\frac{1}{2c}\right)^2. \tag{5.36}$$

Equation (5.36) is the equation of a circle with radius $1/2c$ and center at ($H_{vR} = 1/2c$, $H_{vI} = 0$). This fact has been used in experimental modal analysis as a specific method for curve fitting, called the *circle fit method*, see Section 5.5.4.

## 5.5 Determining Natural Frequency and Damping

When characterizing mechanical systems it is often desired to determine $f_n$ and $\zeta$. We shall therefore discuss a number of techniques for accomplishing this task. In the present section we limit our analysis to some easy-to-use methods that can be used for a first rough estimate. In Section 16.7 we will take this one step further and use a more accurate mathematical curve fitting technique to estimate the parameters.

Depending on which type of frequency response we look at (dynamic flexibility, mobility, or accelerance), there will be (small) differences in peak magnitude location, etc., as we have mentioned before. If the damping is low, say less than 0.1, then those differences are small. The purpose of the discussion here is to provide some approximate means of roughly estimating $f_n$ and $\zeta$. We will therefore not present all details and exact formulas, but limit the discussion to practical formulas to be used. Keep in mind, however, that these are approximate! More details can be found in most textbooks on vibration analysis, e.g., (Den Hartog 1985; Ewins 2000; Inman 2007).

### *5.5.1 Peak in the Magnitude of FRF*

The simplest method follows directly from the discussion above. For $\zeta \ll 1$ we have $f_n \approx f_d$ and we can define $f_n$ as the frequency where the magnitude of the frequency response, $|H(f)|$, has its maximum. This method is approximately valid for dynamic flexibility or accelerance, whereas for mobility it is exact, that is, the peak in the magnitude of $H_v(f)$ is located exactly at $f_n$.

### *5.5.2 Peak in the Imaginary Part of FRF*

An alternative method of finding $f_n$ is to look at the location of the peak or dip that occurs in the imaginary parts of dynamic flexibility or accelerance FRFs, or in the real part in mobility FRFs. I have chosen to call this method 'peak in the imaginary part' since it is most common to measure accelerance. In many cases the peak of the imaginary part of (for example) an accelerance FRF is more pronounced than the peak in the magnitude plot, particularly if two resonances are closely spaced, or highly damped.

Although the peak or dip in the real part of mobility and the imaginary part of accelerance do not match $f_n$ exactly, for low values of the relative damping, $\zeta$, the peak is located near $f_n$.

### 5.5.3 Resonance Bandwidth (3 dB Bandwidth)

The resonance bandwidth, $B_r$, from Equation (5.25) is often used to determine the critical damping ratio $\zeta$. Many textbooks, for example (Ewins 2000), include a proof that for any relative damping factor $\zeta$

$$\zeta = \frac{f_u^2 - f_l^2}{2f_d^2} \tag{5.37}$$

where $f_l$ and $f_u$ are the half-power lower and upper frequencies from Equation (5.26). For small damping ratios, say $\zeta < 0.1$, Equation (5.37) can be approximated by

$$\zeta \approx \frac{f_u - f_l}{2f_d}. \tag{5.38}$$

Simplifying Equation (5.38) by using the bandwidth $B_r$ from Equation (5.25), we obtain a useful expression for the damping as

$$\zeta \approx \frac{B_r}{2f_d} \tag{5.39}$$

Since the quality factor, $Q$, in Equation (5.27) was defined as the ratio of the resonance frequency and the resonance bandwidth in Equation (5.27), we can combine this definition with Equation (5.39) to obtain the following relationship between the $Q$-factor and the relative damping

$$Q \approx \frac{1}{2\zeta}. \tag{5.40}$$

The derivations behind the expressions above are based on mobility frequency response. However, the expressions are approximately valid also for dynamic flexibility and accelerance, if the damping is low.

### 5.5.4 Circle in the Nyquist Plot

For dynamic flexibility and accelerance, a perfect circle is not obtained, but for low values of the relative damping, $\zeta$, approximate circles are obtained. It can be shown that the undamped natural frequency, $f_n$, lies at the frequency at which the rate of change of angle is largest between two frequency values, as one moves along the circle (or approximate circle if the frequency response is not mobility). If we want to measure the damping by the circle fit method, it is necessary to use mobility. Of course it is not the viscous damping $c$ we would wish to estimate, but rather the relative damping, $\zeta$. The relative damping can be estimated from the rate of angle change, although this is hardly practical without using a computer routine. More information on the circle fit technique can be found in textbooks on modal analysis, for example (Maia and Silva 2003; Ewins 2000). One can thus measure both resonance frequency and damping using the circle method.

## 5.6 Rotating Mass

An interesting case in vibration analysis is a mass which rotates around a center point. This case arises whenever there exists an imbalance, for example in a rotating engine part, and is one of the most common

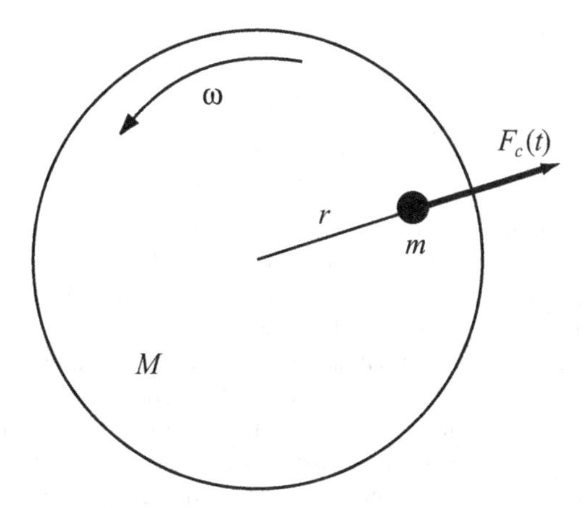


**Figure 5.10** This figure illustrates the centrifugal 'force', $F_c$, arising when a small mass, $m$, is placed on the disk a distance $r$ from the center of rotation. The disk mass, $M$, is assumed to be evenly distributed over its area

causes of vibrations. One case, which can be studied relatively easily, is that of a round, homogeneous disk of mass $M$, see Figure 5.10, which rotates around its center of gravity (center of the disk). If we create an imbalance by attaching a small mass, $m$, at a distance $r$ [m] from the center, and assume that the disk rotates with angular frequency $\omega$ [rad/s], a 'force' is generated, the so-called *centrifugal force*, $F_c(t)$ [N],

$$F_c = mr\omega^2. \tag{5.41}$$

This new 'force' arises because of the necessity to balance the current system, whose center of gravity, because of the small mass, $m$, no longer lies at the point of rotation. A simple way to balance the system is to attach another identical mass, $m$, symmetrically around the point of rotation (across from the first mass), so that the center of gravity for the whole system, of total mass $M + 2m$, now agrees with the point of rotation. This balancing method is called *balancing in one plane*, and is what is used for example when the auto mechanic balances the wheels of your car.

In more complex cases, for example an axle suspended on two bearings, the number of degrees-of-freedom is increased, and the balancing is more complicated. The methods for balancing are well developed, although they are a bit beyond the scope of this book. More details on this topic can be found in, for example, (Norfield 2006).

We shall touch upon one more phenomenon regarding rotating objects. If we imagine that we have an axle suspended on a bearing, which in turn is clamped onto a flexible mount, we can model the system as a single degree-of-freedom system (SDOF) in the tangential direction. If we look at one degree-of-freedom (translational direction), the centrifugal 'force' from the imbalance mass, $m$, which now corresponds to the net result of the total mass distribution around the center of rotation, will then be compensated for by the SDOF system, that is, the whole system can be modeled as

$$M\ddot{u} + c\dot{u} + ku = mr\omega^2 \sin(\omega t) \tag{5.42}$$

where $M$, $c$, and $k$ come from the bearing and its mounting. This equation is well known, and its solution will also be a sinusoid which, if we assume a complex solution, $u$, can be written as

$$u = u_0 e^{j\omega t} \tag{5.43}$$

for which the solution can be written as

$$u_0 = \frac{\frac{m}{M} r \left(\frac{f}{f_n}\right)^2}{1 - \left(\frac{f}{f_n}\right)^2 + j2\zeta\left(\frac{f}{f_n}\right)}. \tag{5.44}$$

The implication of the solution in Equation (5.44) is very interesting. Because we apparently have a resonance at the (undamped) frequency $f_n$, the vibration levels will be much higher at and around this rotation speed than at other rotation speeds. In order to get small vibration levels at the operating rotation speed we should obviously operate this system away from the resonance. If we operate far below the resonance, we get relatively low vibrations; this is called running the machine *sub-critically*. Alternatively, letting the rotation speed increase above the resonance is sometimes a better option, if the resonance frequency is too low to allow operating at a considerably lower frequency. When the operating rotation speed is higher than the first resonance frequency of the machine, it is said to run *super-critically*. A disadvantage with the super-critical operating speed case is naturally that we must pass the resonance when starting the machine. Ensuring the machine is passing its critical speed reasonably fast, however, is usually sufficient to avoid that problem. Super-critical operation is the most common case.

## 5.7 Some Comments on Damping

So far in this chapter we have discussed different SDOF models considering viscous damping. As mentioned a few times already, damping is a difficult issue in vibration engineering. There are many models for different forms of damping, but there is limited knowledge on how to calculate the total effects of different forms of damping in an actual structure. Many of the known forms of damping, for example Coulomb friction, are nonlinear. In many cases, particularly with low damping, however, it works relatively well to approximate the effect of the various forms of damping by a linear model. Therefore the usual situation is that, regardless of the actual damping, it is approximated by viscous damping. This often works well, but there is one other common form of damping that we will briefly mention, namely hysteretic damping, also known as structural damping. More thorough treatment of different damping forms is found in most books on mechanical vibrations, for example (Craig and Kurdila 2006; Inman 2007; Rao 2003).

### 5.7.1 *Hysteretic Damping*

For many real-life structures, experimental results show that the model with viscous damping which we have used so far, does not completely agree with the frequency responses obtained experimentally. An alternative to the viscous model is therefore sometimes favored by replacing the viscous damping, $c$, by a complex spring constant. This form of damping, called *structural*, or *hysteretic*, damping is achieved by replacing the frequency dependence on the damping term by $\eta = c/\omega$, where $\eta$ is called the *loss factor*. With this damping model, Newton's equation can be written as

$$m\ddot{u} + k(1 + j\eta)u = F(t). \tag{5.45}$$

The mathematical background to Equation (5.45) is not as rigid as the equation for viscous damping which we have studied earlier in this chapter, because it is not an ordinary differential equation with real coefficients. Therefore we cannot use the Laplace transform, nor can we define free oscillations for this model. However, the frequency response corresponding to Equation (5.45) can be solved, and resembles the one for viscous damping. The dynamic flexibility frequency response for an SDOF system with hysteretic damping is

$$H_u(f) = \frac{U(f)}{F(f)} = \frac{1/k}{1-(f/f_n)^2 + j\eta}. \tag{5.46}$$

The important difference between the viscous and the structural damping models lies in the frequency dependence of the damping.

We can observe from the similarity between Equations (5.18) and (5.46) that exactly at the natural frequency, $f_n$, we have that

$$\eta = 2\zeta. \tag{5.47}$$

Thus we can use Equation (5.47) in Equation (5.39) to calculate $\eta$ from the 3 dB bandwidth, $B_r$, and we find that

$$\eta \approx \frac{B_r}{f_r}. \tag{5.48}$$

It can further be shown (Kennedy and Pancu 1947) that, for structural damping, a circle may be obtained in the Nyquist plot, just as for viscous damping. For hysteretic damping, however, the circle is formed when we plot dynamic flexibility, and the circle has its center at $1/(2\eta)$ and a radius of $1/(2\eta)$. Hysteretic damping is sometimes available as an option in software for experimental modal analysis.

## 5.8 Models Based on SDOF Approximations

After this introduction of the SDOF system, we will now discuss a few cases where SDOF models are commonly used with great success. The first application, presented in Section 5.8.1, is vibration isolation, which is very common. Vibration isolators are found in a large variety of products such as cars, washing machines and electronic devices such as computers. In Section 5.8.2 we will deduce two very useful relationships between static deflection and resonance frequencies and point to some applications of those relationships. A third application where the SDOF system is used is the shock response spectrum, SRS, which is presented in Section 16.1.

### *5.8.1 Vibration Isolation*

The super-critical operation mentioned in Section 5.6 leads us to the application of vibration isolation. Vibration isolation design is often based on the single-degree-of-freedom (SDOF) model. There are two different reasons to use vibration isolation which we will discuss separately:

1. We wish to protect a sensitive device from a vibrating environment, for example the control electronics box for engine control mounted on top of an engine with high vibration levels.
2. We wish to protect the environment around a vibrating device from the vibration force from the device, for example an engine in a vehicle, whose vibrations we do not wish to propagate through the vehicle body.

These two cases are illustrated in Figure 5.11. In the first case, in Figure 5.11(a), we can define the amount of vibration isolation as the ratio of the vibration of the device and the vibration of the base. This is independent of whether we choose displacement, velocity or acceleration as the vibration parameter, as the ratio (for harmonic excitation) will remain the same. We start by setting up Newton's equation for the mass of the device

$$m\ddot{y} = -c(\dot{y} - \dot{x}) - k(y - x). \tag{5.49}$$

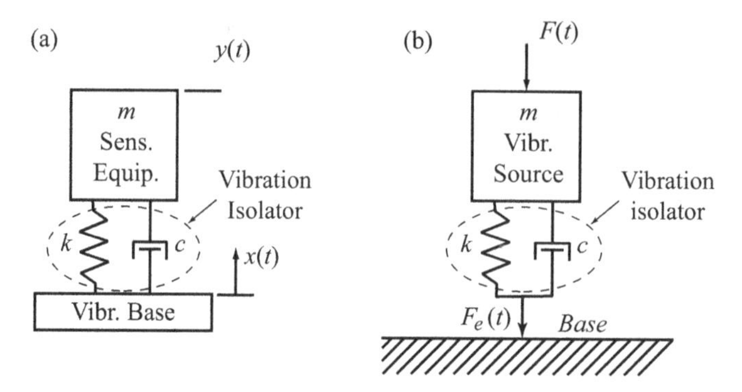


**Figure 5.11** Two cases of vibration isolation. (a) Illustrates the case of some sensitive equipment to be protected from environment vibrations, whereas (b) illustrates the case of a vibrating source, for example an engine, from which the environment is to be protected

We solve this equation by Laplace transforming it and regrouping the terms, which results in

$$\left[ms^2 + cs + k\right] Y(s) = (cs + k)\, X(s). \tag{5.50}$$

From Equation (5.50) it follows that the transfer function of the vibration of the sensitive device, with the vibration of the base, is

$$\frac{Y(s)}{X(s)} = \frac{(cs + k)/m}{s^2 + s \cdot c/m + k/m}. \tag{5.51}$$

The denominator of Equation (5.51) is the same as the one we obtained for an SDOF system in Equation (5.18), and thus we have that

$$\omega_n = 2\pi f_n = \sqrt{\frac{k}{m}} \tag{5.52}$$

and

$$\zeta = \frac{c}{2\sqrt{mk}}. \tag{5.53}$$

Substituting Equations (5.52) and (5.53) into Equation (5.51) and simultaneously setting $s = j\omega$ leads (after a few steps) to the frequency response

$$\frac{Y(f)}{X(f)} = \frac{1 + j2\zeta\left(f/f_n\right)}{1 - \left(f/f_n\right)^2 + j2\zeta\left(f/f_n\right)}. \tag{5.54}$$

The frequency response in Equation (5.54) is plotted in Figure 5.12 as a function of the relative frequency $f/f_n$. It is clear from this figure that, for frequencies larger than $\sqrt{2}f_n$, the vibration level of the sensitive device is lower than that of the base and thus we have vibration isolation. A drawback, however, is that at frequencies lower than $\sqrt{2}f_n$, the sensitive device will actually have higher vibration levels than the base. Therefore, vibration isolation works best when the frequency content in operation is limited to frequencies above a certain frequency, so that the isolator can be designed to give isolation for all operating conditions. This is usually the case for most engines, and pumps, etc., except during startup of the device. In most cases, where the startup is sufficiently fast, this does not pose a problem.

As seen in Figure 5.12, the isolation effect increases with decreasing damping. Using an insufficient damping level can, however, result in large displacements, especially when transient vibration occur,

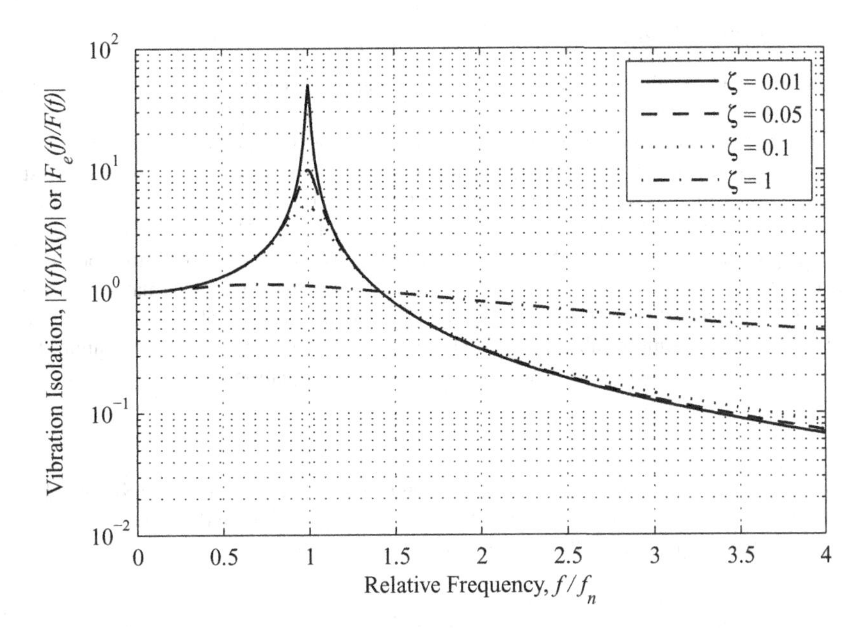


**Figure 5.12** Vibration isolation defined as the frequency response, for different relative damping ratios, $\zeta$, of both vibration isolation cases, (a) and (b), as illustrated in Figure 5.11. In case (a) the curve in the present figure shows the frequency response of the vibration level of the sensitive device, $Y(f)$, with the level of the base, $X(f)$ as given by Equation (5.54). In case (b) the curve in the present figure shows the frequency response of the force level of the base (environment) with the force from the device, as given by Equation (5.55). As shown by the figure, actual vibration isolation is obtained for relative frequencies above $\sqrt{2}$

such as a vehicle passing a bump, causing transient vibration (shock). In such cases, the vibration isolator has to be designed with enough damping so that the displacement does not become larger than can be handled by the spring, as every real spring has a limit for how much it can be compressed. If the spring is compressed to its limit, the resulting acceleration shock level can become very high, causing the device to break.

For the second case, in Figure 5.11(b), the vibration isolation is defined as the ratio of the force on the base and the force on the side of the device producing the vibrations, $F_e/F$, as shown in the figure. An analysis of this case, which is left for Problem 5.7, yields a frequency response identical to that of Equation (5.54). Note, however, that in this case the frequency response is the ratio of two forces,

$$\frac{F_e(f)}{F(f)} = \frac{1 + j2\zeta\left(f/f_n\right)}{1 - \left(f/f_n\right)^2 + j2\zeta\left(f/f_n\right)}. \tag{5.55}$$

The above equations are naturally approximations to any real case, as any structure we mount on vibration isolators will of course not be moving in just a single direction. And, in practice, many vibration isolator designs will not even be linear, for example because they contain rubber parts to add damping. Rubber is a highly nonlinear material with temperature sensitivity and many other properties that may make the isolator nonlinear. The equations deduced above are still very useful from a design standpoint as a first approximation. Usually the manufacturers of isolators can aid in the selection and application of a particular isolator design.

### 5.8.2 *Resonance Frequency and Stiffness Approximations*

Another case where an SDOF model can be used for a first approximation, is to determine the resonance frequency that will be added to a structure when mounting a new component to an existing dynamic structure. In Chapter 6 we will show that continuous structures (if damping is low enough) exhibit resonant behavior. If we add another dynamic structure to the first structure, the new combined structure will have dynamic properties that are rather complicated to predict. This field is called *substructuring* and more details can be found in for example (Craig and Kurdila 2006; Ewins 2000; Inman 2007).

As a first approximation, however, there is a simple relationship that can sometimes be successfully used, when the second structure which is added to the first structure can be approximated as a rigid mass. If this mass, $m$, is added to the structure, and we can calculate the static displacement, $d$, caused by the mass (which can often be simply calculated if we have a finite element model of the dynamic structure), we can calculate an approximate point stiffness, $k$, by the simple equation,

$$k = \frac{mg}{d} \tag{5.56}$$

where $g$ is the gravitational constant, $g = 9.806$ m/s$^2$ and all units are assumed to be SI units. If we now use the stiffness of Equation (5.56) to define an SDOF system together with the mass, we also have that the resonance frequency, $f_r$, of this SDOF system will be

$$f_r = \frac{1}{2\pi}\sqrt{\frac{k}{m}}. \tag{5.57}$$

Substituting $k$ from Equation (5.56) into Equation (5.57), by a simple calculation we get that

$$f_r = \frac{\sqrt{g}}{2\pi\sqrt{d}} \approx \frac{0.5}{\sqrt{d}}. \tag{5.58}$$

From this equation we see that the resonance frequency is only dependent on the static deflection, $d$, caused by the mass.

Equation (5.58) can be rewritten in the form

$$d = \frac{g}{(2\pi f_r)^2} \approx \frac{0.25}{f_r^2}. \tag{5.59}$$

The result in Equation (5.59) is useful to consider in experimental modal analysis, when trying to achieve free–free boundary conditions by supporting a test structure with a soft spring. From the equation we see that obtaining a particular resonance frequency when hanging a structure in a soft spring, will result in a particular displacement, or extension of the spring from its unloaded condition. If, for example, we want the resonance frequency to be 1 Hz, Equation (5.59) yields that the extension (or compression) of the spring will be 25 cm.

## 5.9 The Two-degree-of-freedom System (2DOF)

A mechanical system consisting of two masses is of special interest in some applications. We will therefore now study such a system in some detail. In Chapter 6 we will introduce the more general MDOF system with an arbitrary number of degrees-of-freedom. To simplify the equations, we will restrict the treatment in this section to an undamped 2DOF system. A general illustration of a 2DOF system is shown in Figure 5.13. Newton's equations for each of the two masses give an equation system:

$$\begin{cases} m_1\ddot{u}_1 = F_1 - k_1u_1 - k_2(u_1 - u_2) \\ m_2\ddot{u}_2 = F_2 - k_2(u_2 - u_1) - k_3u_2. \end{cases} \tag{5.60}$$

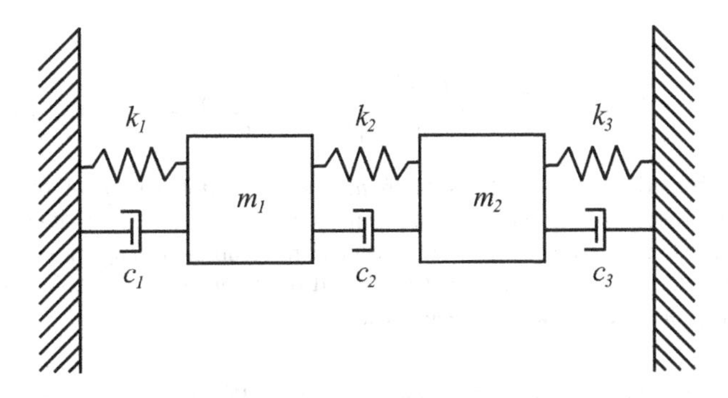


**Figure 5.13** Mechanical system with two degrees-of-freedom, the 2DOF system

We are now interested in finding the free vibrations, the (undamped) natural frequencies, which we obtain when the forces are zero. If we rewrite the equations somewhat, we obtain

$$\begin{cases} m_1\ddot{u}_1 + k_1u_1 + k_2(u_1 - u_2) = 0 \\ m_2\ddot{u}_2 + k_2(u_2 - u_1) + k_3u_2 = 0. \end{cases} \tag{5.61}$$

The solutions to Equation (5.61) can be found by assuming a trial solution. This is an alternative technique to the Laplace transform, which we use to illustrate a common approach found in many textbooks on vibration. In Chapter 6 we will use a more general way to solve this equation. We can surely assume (referring to the results for the SDOF system), that if there are any solutions to Equation (5.61), they will be harmonic (oscillating). Thus we assume a solution of the form

$$\begin{cases} u_1(t) = U_1 \sin(\omega t) \\ u_2(t) = U_2 \sin(\omega t). \end{cases} \tag{5.62}$$

This gives us the second derivatives

$$\begin{cases} \ddot{u}_1(t) = -\omega^2 u_1(t) \\ \ddot{u}_2(t) = -\omega^2 u_2(t). \end{cases} \tag{5.63}$$

Substituting the two last sets of equations into Equation (5.61) gives

$$\begin{aligned} \left[\left(-m_1\omega^2 + k_1 + k_2\right) U_1 - k_2U_2\right] \sin(\omega t) = 0 \\ \left[-k_1U_1 + \left(-m_2\omega^2 + k_2 + k_3\right) U_2\right] \sin(\omega t) = 0. \end{aligned} \tag{5.64}$$

Solutions to Equation (5.64) will now be those combinations of $U_1$, $U_2$ and $\omega$ that satisfy the equations. Generally, there will not be any unique such solutions, but if we set the ratio $U_1/U_2 = U$ we obtain, from the first equation in Equation (5.64) that

$$U = \frac{-k_2}{m_1\omega^2 - k_1 - k_2} \tag{5.65}$$

and from the second equation

$$U = \frac{m_2\omega^2 - k_2 - k_3}{-k_2}. \tag{5.66}$$

The two last equations must simultaneously apply, and thus

$$\frac{-k_2}{m_1\omega^2 - k_1 - k_2} = \frac{m_2\omega^2 - k_2 - k_3}{-k_1}. \tag{5.67}$$

This equation leads to a fourth-order polynomial in $\omega$

$$\omega^4 - \omega^2 \left[ \frac{k_1 + k_2}{m_1} + \frac{k_2 + k_3}{m_2} \right] + \frac{k_1 k_2 + k_1 k_3 + k_2 k_3}{m_1 m_2} = 0. \tag{5.68}$$

Equation (5.68), sometimes referred to as the *frequency equation*, has four solutions in $\omega$. As for the SDOF system, these roots will come in complex conjugate pairs. This means that our system with two degrees-of-freedom has two natural frequencies, which are solutions to Equation (5.68). For each of those two solutions, there will be a unique ratio of the displacements of the masses, $U = U_1/U_2$, i.e., the two masses move in a specific way relative to each other.

**Example 5.9.1** *Let us illustrate the above discussion of a 2DOF system with an example. Assume both masses are equal, as well as all the springs, i.e.,*

$$\begin{cases} m_1 = m_2 = m \\ k_1 = k_2 = k_3 = k. \end{cases} \tag{5.69}$$

*Substituting this into Equation (5.68), we obtain*

$$\omega^4 - \omega^2 \frac{4k}{m} + \frac{3k^2}{m^2} = 0 \tag{5.70}$$

*with the solutions*

$$\omega_{1,2}^2 = \frac{2k}{m} \pm \sqrt{\frac{4k^2}{m^2} - \frac{3k^2}{m^2}} = \frac{(2 \pm 1)\, k}{m} = \begin{cases} k/m \\ 3k/m. \end{cases} \tag{5.71}$$

*Inserting the first of these frequencies into either Equation (5.65) or Equation (5.66) gives*

$$\frac{U_1}{U_2} = +1. \tag{5.72}$$

*The second frequency in Equation (5.71) in the same way gives*

$$\frac{U_1}{U_2} = -1. \tag{5.73}$$

*End of example.*

The solutions in Equations (5.72) and (5.73) are very important. They implicate that for the system with two degrees-of-freedom, there are two *natural frequencies*, *eigenfrequencies*, or *resonance frequencies*, by which the system can oscillate by itself, without any applied force, just as was the case for one frequency in an SDOF system. For each of these frequencies, there is a unique relation between the displacement of each of the two masses. We call this relative motion a *mode shape*, which is related to the fact that a resonance is also called a *mode* (which we will discuss further in Section 6.1). An undamped 2DOF system will always have one mode where the two masses move in phase, and one mode where they move out of phase. The size of the relative displacements depend on the sizes of the masses and springs, so the reason they were equal in our example above, Equations (5.72) and (5.73), were special cases, since we chose the masses and springs equal to each other. For physical systems, which will of course have damping, the problem becomes somewhat more difficult. We will discuss this in more detail in Chapter 6.

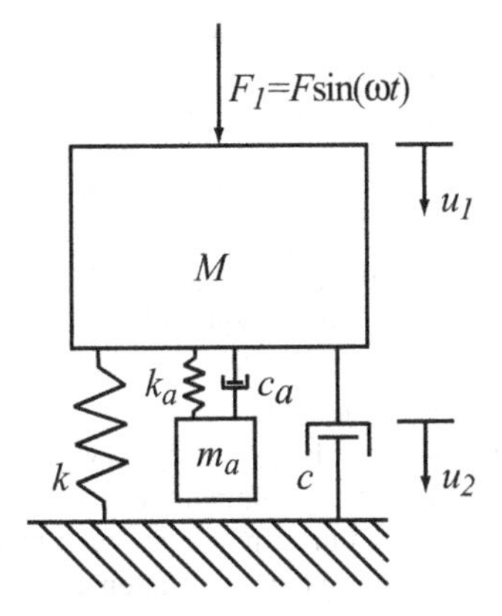


**Figure 5.14** Illustration of the tuned damper. The damper consists of $m_a$, $c_a$, $k_a$, which are attached to the SDOF system consisting of $M$, $c$, $k$

## 5.10 The Tuned Damper

We will conclude this chapter by studying a particular application of the 2DOF system that can be used to reduce vibrations. Assume that we have an SDOF system with a mass $M$, a spring with stiffness $k$, and damping $c$. We then add a second SDOF system with $m_a$, $k_a$ and $c_a$, respectively, as illustrated in Figure 5.14. The first SDOF system could, for example, be a (simplified) model of a machine on its foundation, modeled as a single mass moving translationally, with stiffness and damping from the machine mounts. The second SDOF system is called a *tuned damper* or a *tuned absorber*.

Assume we force the system in Figure 5.14 with a harmonic force, corresponding to the operating speed of the machine. Since the system is linear, if we force it with a harmonic force, the resulting vibration, $u_1$, will be harmonic.

To find the displacement of the SDOF mass, $M$, for a particular force, we formulate Newton's equations and get

$$\begin{aligned}\left[\left(-M\omega^2 + k + k_a\right)U_1 - k_a U_2\right]\sin(\omega t) &= F\sin(\omega t)\\ \left[-kU_1 + \left(-m_a\omega^2 + k_a\right)U_2\right]\sin(\omega t) &= 0.\end{aligned} \tag{5.74}$$

The second equation in Equation (5.74) gives

$$U_2 = \frac{k}{k_a - m_a\omega^2}U_1 \tag{5.75}$$

which, substituted into Equation (5.74), gives that

$$U_1 = \frac{k_a - m_a\omega^2}{\left(k_a - m_a\omega^2\right)\left(k + k_a - M\omega^2\right) - kk_a}F. \tag{5.76}$$

Equation (5.76) may seem complicated at first, but the interesting part for us at the moment is the numerator. It follows from the numerator that there is an angular frequency $\omega_a$ for which the mass $M$ will not move at all. This is an *antiresonance*, a phenomenon that will be discussed more in detail in Chapter 6. By choosing the mass $m_a$ and the spring $k_a$ appropriately such that $\omega_a$ corresponds to the natural frequency of the SDOF system $M, c, k$, we can apparently entirely remove the vibrations at that frequency. Thus we choose $m_a$ and $k_a$ so that

$$\omega_a = \sqrt{\frac{k_a}{m_a}} = \omega_r = \sqrt{\frac{k}{M}}. \tag{5.77}$$

Note especially in Equation (5.77) that the resonance of the tuned damper, considered as a separate SDOF system equals the frequency at which the vibration becomes zero. This principal applies also to the case where a tuned damper is applied to a continuous structure. Tuned dampers can be purchased 'off the shelf' from vibration isolator manufacturers.

When we add damping to the tuned damper, Equation (5.77) will turn into an expression similar to the poles of the SDOF system, see for example the denominator of Equation (5.18). Thus, the denominator in the new expression will not equal zero at the tuned frequency, but rather be some low number, depending on the damping. In practice, we can choose how much attenuation we want to achieve at the tuned frequency by changing the damping of the tuned damper.

The vibration absorption does of course not come without cost. While we can reduce the vibration at a particular frequency, $\omega_a$, the complete 2DOF system will have two resonance frequencies. In Figure 5.15 an example of a frequency response of the displacement, $u_1$, with the force, $F_1$, of the system in Figure 5.14 with an added tuned damper is plotted. Example values of $M$, $c$, and $k$ have been used, and the tuned damper values were chosen so that the damping of the tuned damper was 10 times that of the SDOF system and the frequency of the tuned damper was chosen equal to the resonance frequency of the SDOF system. It should be especially noted in Figure 5.15, that an effect of the higher damping in the tuned damper is that both resonances of the new system, including the tuned damper, obtain higher damping. This is a general property of MDOF systems, as energy dissipation anywhere on the structure naturally helps to attenuate all vibrations in the structure.

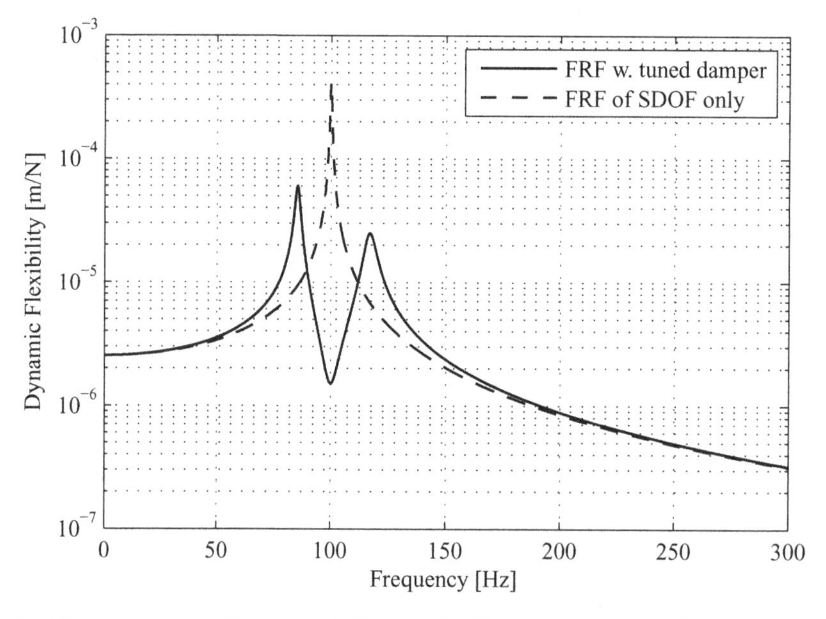


**Figure 5.15** Frequency response (dynamic flexibility) $U_1/F$ in Figure 5.14, before and after attachment of a tuned damper. As can be seen in the figure, after attaching the tuned damper, an antiresonance occurs at the tuned frequency. In the figure it has been assumed that the damping of the tuned damper was 10 times the damping of the SDOF system

## 5.11 Chapter Summary

In this chapter we have studied a mechanical system with one degree-of-freedom, the SDOF system. If we have a system with mass, damping and stiffness $m$, $c$, and $k$, respectively, we found that such a system will have an undamped natural frequency $f_n$ in [Hz] of

$$f_n = \frac{\omega_n}{2\pi} = \frac{1}{2\pi}\sqrt{\frac{k}{m}} \tag{5.78}$$

where $\omega_n$ is the natural angular frequency in [rad/s] and $\omega_n = 2\pi f_n$. The SDOF system will have a relative damping, $\zeta$, of

$$\zeta = \frac{c}{2\sqrt{mk}}. \tag{5.79}$$

The impulse response $h(t)$ of the SDOF system consists of an exponentially decaying sine wave described by

$$h(t) = Ae^{-\zeta\omega_n t}\sin(2\pi f_d t) \tag{5.80}$$

where the damped natural frequency, $f_d$ is defined by

$$f_d = f_n\sqrt{1-\zeta^2} \tag{5.81}$$

and where $A$ is a constant left for Problem 5.2. Example impulse responses were plotted in Figure 5.2.

The frequency response function $H(f)$ in the form of dynamic flexibility, is

$$H_u(f) = \frac{U(f)}{F(f)} = \frac{1/k}{1-(f/f_n)^2 + \mathrm{j}2\zeta(f/f_n)}. \tag{5.82}$$

The magnitude of this frequency response function, FRF, (plotted in Figure 5.3) will have a maximum at the damped resonance frequency

$$f_{max} = f_n\sqrt{1-2\zeta^2} \tag{5.83}$$

which is slightly different from the damped natural frequency $f_d = f_n\sqrt{1-\zeta^2}$ by which the impulse response oscillates. Experimentally it is more common to obtain the mobility FRF,

$$H_v(f) = \frac{V(f)}{F(f)} = \frac{\mathrm{j}(2\pi f)/k}{1-(f/f_n)^2 + \mathrm{j}2\zeta(f/f_n)} \tag{5.84}$$

or the accelerance,

$$H_a(f) = \frac{A(f)}{F(f)} = \frac{-(2\pi f)^2/k}{1-(f/f_n)^2 + \mathrm{j}2\zeta(f/f_n)}. \tag{5.85}$$

## 5.12 Problems

Many of the problems following are supported by the accompanying ABRAVIBE toolbox for MATLAB/Octave and further examples which can be downloaded with the toolbox. If you have not already done so, please read Section 1.6, and follow the instructions to download this toolbox together with example files.

**Problem 5.1** *Calculate the constants $C_1$ and $C_2$ in the partial fraction expansion of $H(s)$ in Equation (5.12) by using residue calculus (the hand-over method).*

**Problem 5.2** *Calculate the constant A for the impulse response in Equation (5.15) using the results from Problem 5.1.*

**Problem 5.3** *Using the results from Problem 5.2 above, prove that the impulse response in Equation (5.14) can be written as in Equation (5.15).*

**Problem 5.4** *Write a MATLAB/Octave function which calculates the impulse response $h(t)$ given the undamped resonance frequency $f_n = \omega_n/2\pi$ and the relative damping $\zeta$. The function could be defined as for example*

```
function [h, t] = fz2impresp(fn,z)
```

*Run the function for different values of $f_n$ and $\zeta$ and plot the results. Then observe, for a given $f_n$, how many periods of oscillation you see for different values of the damping $\zeta$.*

**Problem 5.5** *Write a MATLAB/Octave function with a definition*

```
function [Hv, f] = sdofmob(fn, z)
```

*which calculates the mobility frequency response $H_v(f)$ in Equation (5.28), given the undamped resonance frequency $f_n = \omega_n/2\pi$ and the relative damping $\zeta$. Use the function to plot the mobility for $f_n = 1$ Hz, and $\zeta = 0.01, 0.05,$ and $0.1$. Plot the three frequency responses overlaid, with linear scales as well as logarithmic y-scale (using the MATLAB/Octave* **semilogy** *command), and log–log scale (***loglog** *command). Observe that with a nominal resonance frequency of 1 Hz the frequency axis will be equal to the normalized frequency $r = f/f_n$.*

*Then write a new MATLAB/Octave function* **sdofacc** *similar to the previous function, but which calculates the accelerance frequency response. Plot the accelerances for the same values of frequency and damping as for mobility. Compare the results with those for mobility.*

**Problem 5.6** *Calculate the undamped resonance frequency $f_n$ and relative damping $\zeta$ of a mechanical single-degree-of-freedom system with the following parameters:*

*$m = 4$ kg, $k = 10^6$ m/N, and $c = 80$ m/Ns.*

*Use the* **sdofacc** *function from Problem 5.5 to compute the accelerance and plot it. Then use a suitable method from Section 5.5 to find the mass, stiffness and damping values. How correct can you get it? (The accuracy of particularly the relative damping estimate will be poor from a visual inspection, so do not expect a very high accuracy for c. m and k should be much easier to obtain good results for.)*

*Hint: Use the asymptotic behavior at higher frequencies to obtain the mass, then the peak in the imaginary part to find the resonance frequency, from which the stiffness can be calculated using the mass estimate. Finally find the damping $\zeta$ using the 3-dB bandwidth and use the appropriate equation from Section 5.5 to find the viscous damping.*

**Problem 5.7** *Prove that the vibration isolation case B leads to Equation (5.55).*

**Problem 5.8** *We assume that you are adding a mass to a beam. The beam gets a static deflection of 1 mm due to the added mass. Use Equation (5.58) to obtain an approximate estimate of what the resulting resonance frequency will be.*

## References

Craig RR and Kurdila AJ 2006 *Fundamentals of Structural Dynamics*. John Wiley & Sons, Ltd.
Den Hartog JP 1985 *Mechanical Vibrations*. Dover Publications Inc.

Ewins DJ 2000 *Modal Testing: Theory, Practice and Application* 2nd edn. Research Studies Press, Baldock, Hertfordshire, England.

Inman D 2007 *Engineering Vibration* 3rd edn. Prentice Hall.

ISO 2641: 1990 *Vibration and shock – Vocabulary*. International Organization for Standardization, Geneva, Switzerland.

Kennedy C and Pancu C 1947 Use of vectors in vibration measurement and analysis. *J. of the Aeronautical Sciences* **14**(11), 603–625.

Maia N and Silva J (eds) 2003 *Theoretical and Experimental Modal Analysis*. Research Studies Press, Baldock, Hertforsdhire, England.

Newton I 1687 *Philosophi Naturalis Principia Mathematica*. London.

Norfield D 2006 *Practical Balancing of Rotating Machinery*. Elsevier Science.

Rao S 2003 *Mechanical Vibrations* 4th edn. Pearson Education.

# 6

# Modal Analysis Theory

Modal analysis, which is a part of the wider subject of structural dynamics, is the theory dealing with the dynamics of mechanical systems described by *modes*. In this chapter we will show a general approach related to vibrations in mechanical systems with more than one (or two) degrees-of-freedom. We will show that these systems can be condensed down to their poles and mode shapes, and we will show how this condensation is done. The theory in this chapter is essential to understand vibrations, as we will get answers to such questions as: what will the vibration level be in (e.g.,) DOF 63 in Z direction, if we apply a harmonic force of 10 N in DOF 12 in the X direction (see Section 6.2.1 for a discussion of degrees-of-freedom, DOF)? We will see that the answer to such questions lies in the mode shapes and poles of the system (structure).

Modal analysis is a comprehensive subject and this chapter will by necessity be limited. It includes, however, all the essential information for the beginner to understand the concepts of modal analysis from an experimental perspective. In addition, engineers working with analytical modal analysis may also find it useful as an overview, helping them to understand experimental results. This chapter does not include all the necessary information to fully understand mechanical system simulation such as used in for example the Finite Element Method, FEM. The reader interested in more in-depth material can find that in dedicated books on modal analysis (Maia and Silva 2003; Ewins 2000; Heylen *et al.* 1997) and structural dynamics (Craig and Kurdila 2006).

The outline of this chapter is different from many textbooks on mechanical vibration because its focus is on what we can obtain experimentally. This means that synthesis of frequency response functions and mode shape scaling form a central part of the chapter. In addition, in Section 6.5 we present a powerful method to simulate forced response solutions for time domain force inputs. This is a very important section, as it gives us a tool to use for the remainder of the book to produce simulated measurement data which we can use for testing signal analysis techniques, etc.

## 6.1 Waves on a String

Most of this chapter will deal with so-called *lumped parameter systems*, i.e., systems with discrete masses, dampers, and springs. Modal analysis is, however, also strongly related to *wave theory*, the study of waves in continuous structures. Indeed, modes in continuous structures are identical to the modes we will obtain later in this chapter, at the points where we define our mass locations (assuming we select these carefully, at least; I am referring to a principal equality). I therefore think it is appropriate to include a short introduction to the concept of modes for continuous systems to remind the reader that modes and

*Noise and Vibration Analysis: Signal Analysis and Experimental Procedures,* First Edition. Anders Brandt.

waves are dual theories, which is to say they answer the same question: what vibrations do we get for particular force inputs?

I will assume you are at least briefly acquainted with the modes of an ideal string, fixed at both ends. The modes on a fixed string are shaped as sines, and have the spatial shape

$$v(x) = sin\left(\frac{n\pi x}{L}\right) \tag{6.1}$$

where $x$ is the coordinate along the string, from 0 to $L$, the length of the string. Each mode is a *standing wave*, where all points move either in phase, or out of phase, and for each mode, $n$, there is an associated *natural frequency* $f_n$ [Hz], by which the mode $n$ is oscillating, so that an individual point along the mode at, say, point $x_0$ has a sinusoidal motion

$$u(x_0, t) = v(x_0)\sin(\omega_n t) \tag{6.2}$$

where the angular natural frequency $\omega_n = 2\pi f_n$ as usual.

The important thing now, is to realize how the modes are used to describe a particular vibration pattern. Let us say that at $t = 0$ we have a particular deformation of the string, and we then release it. To describe the vibration pattern that results, the modes are then 'tuned' so that the sum of all the modes at $t = 0$ equal the initial deformation, and after the release of the string each mode starts to oscillate at its own natural frequency. The 'tuning' consists of setting a particular amplitude and phase of each mode, so that the sum of all the modes at $t = 0$ results in the initial deformation shape. (Note that we know from the theory of Fourier series that there is indeed such a solution for any deformation pattern, since all the modes are sinusoids. We can always take the deformation pattern between $x = 0$ and $x = L$, make it repeat periodically, outside this region, and split it into a Fourier Series. The result of the Fourier series is the amplitudes and phases of all sines.) The result is two 'waves', or deformation patterns, each half as large as the original pattern (since the total pattern is the sum of the two) – one moving in each direction on the string, as you probably have seen many times. The waves hit the boundaries at $x = 0$ and $x = L$ where they are reflected (change sign, so if the wave shape was pointing upwards, for example, it is pointing downwards after the reflection) and move back and forth for all eternity. That is, if there is no damping.

An alternative way to describe the deformation patterns moving back and forth, is by using *wave equations* of the form

$$u(x, t) = U_+(x)e^{j(\omega t - kx)} + U_-(x)e^{j(\omega t - kx)} \tag{6.3}$$

for example. In order to use wave equations, however, there must be a known closed- form solution; we need to have the equation. The versatility of modes, lies in the fact that the modes give exactly the same answers (the modes can be calculated from the wave equations), but the modes can be calculated or experimentally determined, in many cases for which we do not know the closed-form solutions to the wave equations. Furthermore, even rather basic mechanical geometries such as plates and beams, have several wave types which sum up to the total solution. In most cases these are very difficult (read: impossible) to obtain on real-life structures such as bridges, buildings, airplanes and refrigerators. However, the modes can be obtained in several ways; most commonly from an FE model, or from experimental modal analysis. Once the modes are known, simulations can be made as we will see later in this chapter.

## 6.2 Matrix Formulations

We will develop a general theory for multiple degrees-of-freedom systems by looking at the 2DOF system from Section 5.9, illustrated in Figure 5.13. To make the matrix notation easy to read we introduce the following symbols for matrix formulations. A rectangular matrix will be denoted by brackets, e.g., $[M]$, a diagonal matrix will be denoted by e.g., $\lceil M_r \rfloor$, column vectors will be denoted by curly brackets, e.g., $\{u\}$ and row vectors will be denoted by lower brackets, e.g., $\lfloor H_p \rfloor$, see also Appendix D.

For the 2DOF system in Figure 5.13, Newton's second law gives the following equation system,

$$\begin{cases} m_1\ddot{u}_1 = F_1 - c_1\dot{u}_1 - k_1u_1 - c_2(\dot{u}_1 - \dot{u}_2) - k_2(u_1 - u_2) \\ m_2\ddot{u}_2 = F_2 - c_2(\dot{u}_2 - \dot{u}_1) - k_2(u_2 - u_1) - c_3\dot{u}_2 - k_3u_2 \end{cases} \tag{6.4}$$

which we can write in matrix form if we define the following matrices. We define the *mass matrix*, $[M]$, as

$$[M] = \begin{bmatrix} m_1 & 0 \\ 0 & m_2 \end{bmatrix} \tag{6.5}$$

the *damping matrix*, $[C]$, as

$$[C] = \begin{bmatrix} c_1 + c_2 & -c_2 \\ -c_2 & c_2 + c_3 \end{bmatrix} \tag{6.6}$$

and the *stiffness matrix*, $[K]$, as

$$[K] = \begin{bmatrix} k_1 + k_2 & -k_2 \\ -k_2 & k_2 + k_3 \end{bmatrix}. \tag{6.7}$$

Furthermore we define the *displacement vector*, $\{u\}$, as a column vector

$$\{u\} = \begin{Bmatrix} u_1 \\ u_2 \end{Bmatrix} \tag{6.8}$$

and the *force vector*, $\{F\}$, as a column vector

$$\{F\} = \begin{Bmatrix} F_1 \\ F_2 \end{Bmatrix}. \tag{6.9}$$

With these definitions Equation (6.4) can be written more compactly as

$$[M]\{\ddot{u}\} + [C]\{\dot{u}\} + [K]\{u\} = \{F(t)\}. \tag{6.10}$$

The formulation in Equation (6.10) is of course not limited to our 2DOF system, but general to all MDOF systems with proper formulations of the mass, damping, and stiffness matrices, and the displacement and force vectors.

### *6.2.1 Degree-of-freedom*

The concept of degree-of-freedom, DOF, is important for the discussion in this chapter. In this chapter we discuss MDOF systems built up by *lumped parameters*, i.e., discrete masses, dampers, and springs. For such a system, the motion of each mass is a DOF, i.e., we always have as many DOFs as we have masses. Conversely, we can define the number of degrees-of-freedom necessary to describe a system (structure) as the number necessary to specify the instantaneous position of all points on the structure at a given time. Thus for a three-dimensional structure, we will typically have six DOFs for each point on the structure we wish to describe the motion of: three translational and three rotational.

The most common situation is that the lumped parameter model represents some continuous structure, in which case it is, of course, a matter of approximating the infinite number of DOFs on the continuous structure by a discrete set of lumped masses. Producing these lumped masses, dampers and springs to correctly represent a continuous structure is an important part of structural dynamics, see for example (Craig and Kurdila 2006). In this chapter we will not discuss producing the matrices, etc., as we are focused on the principal results of them in order to understand what experimental data will look like, and why. It is, however, worth mentioning that in FE models, currently millions of DOFs can be used to produce accurate models of structures, whereas experimentally, few of us can afford to use more than a few hundred, and very often just a few tens. Perhaps it should be mentioned for clarity that what a FEM software does, is essentially building mass and stiffness matrices.

It should be noted that what we mean by a degree-of-freedom in experimental mechanics (see for example Sections 16.6 and 16.7), is a particular point and a certain direction where we have one or more transducers. An 'experimental DOF', can for example be point 22 in the -Y direction, usually written as '22Y-' or similar. Thus, we can have up to three translational DOFs in each point. In addition to that we can have three rotational DOFs in each point, although transducers for rotational DOFs are not readily available, so it is rare to obtain rotational DOFs experimentally.

## 6.3 Eigenvalues and Eigenvectors

To find general solutions for our 2DOF (and generally, MDOF) system, we will look at three cases separately;

- an *undamped system*, for which $[C] = 0$,
- a proportionally damped system, for which $[C] = a[M] + b[K]$ for real constants $a$ and $b$, and
- a generally damped system where $[C]$ is any damping matrix.

### 6.3.1 Undamped System

We will start by looking at the *free vibrations*, i.e., solutions to Equation (6.10) when $\{F\} = \{0\}$. For the undamped system and for free vibrations, we have a special case of Equation (6.10)

$$[M]\{\ddot{u}\} + [K]\{u\} = \{0\} \tag{6.11}$$

where we use the zero column vector $\{0\}$ to emphasize that it is a matter of a vector on the right-hand side. We start by Laplace transforming Equation (6.11) which yields

$$\left[s^2[M] + [K]\right]\{U(s)\} = \{0\}. \tag{6.12}$$

This equation can be reformulated, by multiplying by the inverse of $[M]$ and rearranging the two terms in the equation, into

$$\left[[M]^{-1}[K] + s^2[I]\right]\{U(s)\} = \{0\}. \tag{6.13}$$

If we compare this equation with the 'standard form' of an eigenvalue problem with eigenvalues $\lambda$

$$[[A] - \lambda[I]]\{x\} = \{0\} \tag{6.14}$$

we see that Equation (6.13) is an eigenvalue problem with

$$[A] = [M]^{-1}[K] \tag{6.15}$$

and the eigenvalues

$$\lambda = -s^2 \tag{6.16}$$

and the *undamped natural frequencies* are given by $s = j\omega = \pm\sqrt{-\lambda}$.

You should remember that eigenvalue problems have solutions in terms of eigenvalues and eigenvectors, see also Appendix E. The eigenvalues, $\lambda_r$, are the values of $\lambda$ that satisfy the eigenvalue equation, and for each eigenvalue there is a particular vector, the eigenvector, $\{\psi\}_r$, that satisfies the equation

$$[[A] - \lambda_r[I]]\{\psi\}_r = \{0\}. \tag{6.17}$$

(You should not be surprised that we find that the solution to Newton's equations turn out to be an eigenvalue problem. Eigenvalue problems are, among other things, giving the solutions to differential equations.)

The non-trivial solutions to the standard eigenvalue problem in Equation (6.14) are obtained by setting the determinant equal to 0, i.e.,

$$\det\left[[A] - \lambda\,[I]\right] = 0 \tag{6.18}$$

which in our case is equivalent to the equation

$$\det\left[[M]^{-1}\,[K] - \lambda\,[I]\right] = 0 \tag{6.19}$$

which leads to a polynomial in $\lambda$, the *characteristic equation*. The roots of this polynomial are known from linear algebra as the *eigenvalues*, denoted $\lambda_1, \lambda_2, \ldots \lambda_N$ where $N$ is the number of dimensions in the matrix equation, that is, the number of masses (DOFs) in the system.

The matrices $[M]$ and $[K]$ are in most cases *positive definite*, which means that the eigenvalues are strictly larger than zero. In cases where there are so-called *rigid body modes*, the matrices $[M]$ and $[K]$ are *positive semidefinite*, which means that the eigenvalues are larger than or equal to zero. Rigid body modes are modes where all masses move without any relative motion in between each other, that is, the system has no vibrations, but is translating along, or rotating around, any of the axes.

Since the eigenvalue problem in Equation (6.13) is formulated such that $\lambda = -s^2$, the *poles* of the system, which will give us the frequencies of free vibrations, are

$$s_r = \pm\sqrt{-\lambda_r} = \pm j\sqrt{\lambda_r} \tag{6.20}$$

where we use the fact that we know that $\lambda_r \geq 0$. We thus find that the poles always come in complex conjugate pairs, and lie on the imaginary axis in the Laplace domain (the latter a direct result of the fact that we have no damping – the real part of $s$ corresponds to damping as we know from Section 2.6.1). The vector $\{u\} = \{\psi\}_r$ which satisfies Equation (6.17) for a particular eigenvalue $\lambda_r$, is the *eigenvector* of the system for that eigenvalue. Within structural dynamics the eigenvectors, $\{\psi\}_r$, obtained for an undamped system are called *normal mode shapes*, or simply *normal modes* and are unique for each given structure and boundary conditions. Because eigenvectors can be arbitrarily scaled, the mode shapes are only determined in shape, that is, only the relative motion of the points is unique. For example, the two mode shapes

$$\{\psi\}_r = \begin{Bmatrix} 1 \\ -1 \\ 1 \end{Bmatrix}_r \tag{6.21}$$

and

$$\{\psi\}_r = \begin{Bmatrix} 10 \\ -10 \\ 10 \end{Bmatrix}_r \tag{6.22}$$

are both the same eigenvector but with different scaling. Note that we use a subscript $\{\}_r$ to denote mode number $r$.

Naturally, as the alert reader has already observed, there is another quite similar way to formulate an eigenvalue problem for mechanical systems as in Equation (6.12). We could have multiplied this equation by the inverse of the stiffness matrix instead of the inverse of the mass matrix, and then divided by $s^2$ and obtained an equation similar to Equation (6.13) namely

$$\left[[K]^{-1}[M] + \frac{1}{s^2}[I]\right]\{U(s)\} = \{0\}. \tag{6.23}$$

This equation is an eigenvalue problem with eigenvalues $\lambda_r = -1/s^2$ from which the poles can be solved similarly to Equation (6.16). Solving Equation (6.23) will (of course) yield the reciprocal eigenvalues to the ones we find solving Equation (6.13), which means that the poles will be the same in both cases. Verifying this is left as Problem 6.2 at the end of this chapter. There are in fact several other eigenvalue problem formulations, some of which are sometimes preferred for better stability and computational efficiency. Those methods can be found in for example Craig and Kurdila (2006) and Inman (2007).

**Example 6.3.1** *Determine the eigenvalues, poles and eigenvectors of the 2DOF system in Figure 5.13 if* $m_1 = m_2 = 1$ *[kg],* $k_1 = k_3 = 100$ *[N/m],* $k_2 = 150$ *[N/m], and dampers* $c_1 = c_2 = c_3 = 0$.

*We start by writing the mass and stiffness matrices, which according to Equation (6.5) and Equation (6.7), respectively become*

$$[M] = \begin{bmatrix} 1 & 0 \\ 0 & 1 \end{bmatrix} \tag{6.24}$$

*and*

$$[K] = \begin{bmatrix} 250 & -150 \\ -150 & 250 \end{bmatrix}. \tag{6.25}$$

*The next step in the solution is to calculate* $[A]$ *as*

$$[A] = [M]^{-1}[K] = [K] \tag{6.26}$$

*since the mass matrix is the identity matrix. We now formulate the determinant of* $A - \lambda I$ *and set it to zero, in order to find the non-trivial solutions to Equation (6.14). We thus get*

$$\begin{vmatrix} 250 - \lambda & -150 \\ -150 & 250 - \lambda \end{vmatrix} = (250 - \lambda)^2 - 150^2 = 0. \tag{6.27}$$

*This equation can easily be solved (for example by using the MATLAB/Octave* **roots** *command, see Problem 6.2) and we find the solutions*

$$\begin{aligned} \lambda_1 &= 100 \\ \lambda_2 &= 400 \end{aligned} \tag{6.28}$$

*from which we obtain the poles of the system*

$$\begin{aligned} s_1 &= \pm j\sqrt{100} = \pm j10 \\ s_2 &= \pm j\sqrt{400} = \pm j20 \end{aligned} \tag{6.29}$$

*rad/s. We thus have two undamped natural frequencies of this system:* $f_1 = 10/2\pi \approx 1.6$ *Hz, and* $f_2 = 20/2\pi \approx 3.2$ *Hz.*

*Having found the eigenvalues the next step is to find the corresponding eigenvectors. These we obtain by substituting one of the eigenvalues into Equation (6.14) and find the vector coefficients in* $\{\psi\}_r$ *which satisfy the equation. In our example we first get*

$$\begin{bmatrix} 250 - 100 & -150 \\ -150 & 250 - 100 \end{bmatrix} \begin{Bmatrix} \psi_1 \\ \psi_2 \end{Bmatrix}_1 = \begin{Bmatrix} 0 \\ 0 \end{Bmatrix} \tag{6.30}$$

*where the subscript index 1 after the column vector* $\{\psi\}_1$ *indicates it is the first eigenvector (mode shape) corresponding to the first eigenvalue. Equation (6.30) contains two equal equations. Either one can be used to find the eigenvector, and we get*

$$150\psi_1 - 150\psi_2 = 0 \tag{6.31}$$

*which means* $\psi_1 = \psi_2$ *and thus the eigenvector is*

$$\{\psi\}_1 = \begin{Bmatrix} 1/\sqrt{2} \\ 1/\sqrt{2} \end{Bmatrix}_1 \tag{6.32}$$

*if we scale it to unity length.*

*Similarly for the second eigenvalue we get*

$$\begin{bmatrix} 250-400 & -150 \\ -150 & 250-400 \end{bmatrix} \begin{Bmatrix} \psi_1 \\ \psi_2 \end{Bmatrix}_2 = \begin{Bmatrix} 0 \\ 0 \end{Bmatrix} \tag{6.33}$$

*from which we take any row and get the equation*

$$-150\psi_1 - 150\psi_2 = 0 \tag{6.34}$$

*which yields the eigenvector*

$$\{\psi\}_2 = \begin{Bmatrix} 1/\sqrt{2} \\ -1/\sqrt{2} \end{Bmatrix}_2. \tag{6.35}$$

*End of example.*

## 6.3.2 Mode Shape Orthogonality

As we have seen in Section 6.3.1 the mode shape vectors can be arbitrarily scaled. Furthermore, the mode shapes are generally not independent (although they were in the above example), as we will see in this section, but instead they have a *weighted orthogonality* property which we will now address. From Equations (6.12) and (6.16), for a particular eigenvalue, $\lambda_r$, and its associated eigenvector, $\{\psi\}_r$, we have

$$-\lambda_r[M]\{\psi\}_r + [K]\{\psi\}_r = \{0\}. \tag{6.36}$$

If we premultiply Equation (6.36) by another eigenvector transposed, $\{\psi\}_s^T$, we get

$$-\lambda_r\{\psi\}_s^T[M]\{\psi\}_r + \{\psi\}_s^T[K]\{\psi\}_r = 0. \tag{6.37}$$

Now remember that for vectors and matrices, transposing a product results in reversing the order of the factors in the product, and transposing each factor separately, i.e., $(ab)^T = b^T a^T$. Furthermore, since our matrices $[M]$ and $[K]$ are symmetric, we have that $[M]^T = [M]$ and $[K]^T = [K]$. The transpose of Equation (6.37) thus yields

$$-\lambda_r\{\psi\}_r^T[M]\{\psi\}_s + \{\psi\}_r^T[K]\{\psi\}_s = 0. \tag{6.38}$$

If we formulate Equation (6.36) again, for another eigenvalue, $\lambda_s$, with its eigenvector, $\{\psi\}_s$, and premultiply with $\{\psi\}_r^T$, we similarly get

$$-\lambda_s\{\psi\}_r^T[M]\{\psi\}_s + \{\psi\}_r^T[K]\{\psi\}_s = 0. \tag{6.39}$$

We now take Equation (6.39) minus Equation (6.38), which yields

$$(\lambda_r - \lambda_s)\{\psi\}_r^T[M]\{\psi\}_s = 0 \tag{6.40}$$

for any two eigenvalues $\lambda_r$ and $\lambda_s$. Thus, if $r \neq s$ we must have

$$\{\psi\}_r^T[M]\{\psi\}_s = 0. \tag{6.41}$$

If we now use Equation (6.41) in either Equation (6.37) or Equation (6.38), then

$$\{\psi\}_r^T[K]\{\psi\}_s = 0 \tag{6.42}$$

must also hold for any $r \neq s$. Equation (6.41) and Equation (6.42) are the equations for the *weighted orthogonality* properties of modal vectors.

For the case where $r = s$ and the two modal vectors premultiplying and postmultiplying the mass or stiffness matrix are the same vectors, we define the *modal mass* of mode $r$, $m_r$, by

$$\{\psi\}_r^T [M]\{\psi\}_r = m_r \tag{6.43}$$

and similarly the *modal stiffness* of mode $r$, $k_r$, by

$$\{\psi\}_r^T [K]\{\psi\}_r = k_r. \tag{6.44}$$

By replacing $-\lambda_r$ by the undamped natural frequency $(j\omega_r)^2 = -\omega_r^2$ in Equation (6.38), it follows that

$$k_r = \omega_r^2 m_r. \tag{6.45}$$

Because the mode shapes $\{\psi\}_r$ have an arbitrary scaling, the modal mass and stiffness are clearly not well defined numbers for a particular system. They can rather be any number, depending on the scaling of the mode shapes. But the important use of the modal mass (particularly, but in principle also the modal stiffness) is in its use for scaling mode shapes. Thus we can scale the mode shapes *so that the modal mass becomes*, for example, unity, i.e., $m_r = 1$. This is a very common way of scaling mode shapes. Other ways of scaling mode shapes are for example to unit length, so that $||\{\psi\}_r||_2 = 1$, or so that the largest coefficient in the mode shape is unity. The concepts of modal mass and stiffness are very important also because they are used for several purposes when we deduce other fundamental properties such as for example *modal coordinates* as we will see in Section 6.3.3, and for frequency responses of MDOF systems, as we will see in Section 6.4.

**Example 6.3.2** *Calculate the modal mass and stiffness of each mode shape in Example 6.3.1. Rescale the mode shapes (just as an example) to unity modal stiffness.*

*With the scaling of unity length we calculate the modal mass of mode 1 as*

$$m_1 = \{\psi\}_1^T [M]\{\psi\}_1 = (1/\sqrt{2})^2 + (1/\sqrt{2})^2 = 1 \tag{6.46}$$

*Similarly, the modal mass of the second mode is $m_2 = 1$ because the numbers are the same (verify this if you are not sure!). The modal stiffness of the first mode becomes*

$$k_1 = \{\psi\}_1^T [K]\{\psi\}_1 = \ldots = 100 \tag{6.47}$$

*and for the second mode*

$$k_2 = \{\psi\}_2^T [K]\{\psi\}_2 = \ldots = 400. \tag{6.48}$$

*Here we alternatively could have used the relation in Equation (6.45) to find the modal stiffness from the modal masses and natural frequencies.*

*Since the equations for mode shape orthogonality are square-form matrices (the mode shapes multiply twice) it is obvious that we have to divide by the square root of the 'current' modal stiffness to obtain a square product of unity. Therefore the new, scaled mode shapes for unity modal stiffness become*

$$\{\psi\}_1 = \left\{ \begin{array}{c} 1/(10\sqrt{2}) \\ 1/(10\sqrt{2}) \end{array} \right\}_1 \tag{6.49}$$

*and the second mode shape scaled for unity modal stiffness becomes*

$$\{\psi\}_2 = \left\{ \begin{array}{c} 1/(20\sqrt{2}) \\ -1/(20\sqrt{2}) \end{array} \right\}_2 . \tag{6.50}$$

*End of example.*

### 6.3.3 Modal Coordinates

The concept of *modal coordinates* or *principal coordinates* is very important in modal analysis. It follows directly from linear algebra theory of eigenvalues and eigenvectors, that eigenvectors diagonalize matrices, as we saw in the orthogonality criteria in Section 6.3.1 (see also Appendix E). Nevertheless, we will formulate the proper coordinate transformation here and see what it leads to. We do this by defining the coordinate transformation

$$\{u(t)\} = [\Psi]q(t) \tag{6.51}$$

and we call these new coordinates $q$ the modal coordinates. The transformation matrix $[\Psi]$ is the *mode shape matrix* which is a matrix with the mode shapes in columns, i.e., the $r$th column in $[\Psi]$ is $\{\psi\}_r$. Newton's equation can now be written in these new coordinates as

$$[M][\Psi]\{\ddot{q}\} + [K][\Psi]\{q\} = \{F(t)\}. \tag{6.52}$$

We premultiply this equation by $[\Psi]^T$ which gives the equation

$$[\Psi]^T[M][\Psi]\{\ddot{q}\} + [\Psi]^T[K][\Psi]\{q\} = [\Psi]^T\{F(t)\}. \tag{6.53}$$

From Equations (6.43) and (6.44) it follows that, replacing the vectors $\{\psi\}_r$ by the mode shape matrix $[\Psi]$, will result in diagonal modal mass and modal stiffness matrices

$$[\Psi]^T[M][\Psi] = \lceil M_r \rfloor \tag{6.54}$$

and

$$[\Psi]^T[K][\Psi] = \lceil K_r \rfloor \tag{6.55}$$

where the matrix $\lceil M_r \rfloor$ have the modal masses, $m_r$ on its diagonal, and $\lceil K_r \rfloor$ has the modal stiffnesses, $k_r$ on its diagonal.

Using these relations in Equation (6.53) gives us the equation

$$\lceil M_r \rfloor \{\ddot{q}\} + \lceil K_r \rfloor \{q\} = \{F'(t)\} \tag{6.56}$$

where we have renamed the forces in the new coordinate system $\{q\}$ as

$$\{F'(t)\} = [\Psi]^T\{F(t)\}. \tag{6.57}$$

Equation (6.56) is a very important property. It shows that in the modal coordinates, each mode is uncoupled from all the other modes, and each row in Equation (6.56) corresponds to the equation of an uncoupled SDOF system

$$m_r\ddot{q}_r + k_r q_r = F_r'. \tag{6.58}$$

One important implication of the modal coordinates, and the use of these coordinates, will become clear in the next section where we introduce a special form of damping, proportional damping. But first we will illustrate the concept of modal coordinates with an example.

**Example 6.3.3** *Set up the system of uncoupled forced response equations for the system in Example 6.3.1 (with numbers).*

*Using the modal masses and stiffnesses already calculated in Example 6.3.2 we get the uncoupled equations*

$$\begin{bmatrix} 1 & 0 \\ 0 & 1 \end{bmatrix} \begin{Bmatrix} \ddot{q}_1 \\ \ddot{q}_2 \end{Bmatrix} + \begin{bmatrix} 100 & 0 \\ 0 & 400 \end{bmatrix} \begin{Bmatrix} q_1 \\ q_2 \end{Bmatrix} = \begin{bmatrix} 1/\sqrt{2} & 1/\sqrt{2} \\ 1/\sqrt{2} & -1/\sqrt{2} \end{bmatrix} \begin{Bmatrix} F_1 \\ F_2 \end{Bmatrix}. \tag{6.59}$$

*You should note especially that the mode vectors multiplying the force vector are horizontal, since the mode shape matrix is transposed.*

*End of example.*

### 6.3.4 Proportional Damping

The concept of proportional damping is defined as the case where we have a damping matrix which can be written as a linear combination of the mass and stiffness matrices, i.e.,

$$[C] = a[M] + b[K] \tag{6.60}$$

where $a$ and $b$ are real constants. This is often called *Rayleigh damping*. From the orthogonality criteria in Equations (6.54) and (6.55) it follows that the damping matrix will also be a diagonal matrix in the modal coordinates, since

$$[\Psi]^T[C][\Psi] = a\lceil M_r \rfloor + b\lceil K_r \rfloor = \lceil C_r \rfloor. \tag{6.61}$$

This means that with this special form of damping, the MDOF system is decoupled in modal coordinates and we get a set of uncoupled equations

$$\lceil M_r \rfloor\{\ddot{q}\} + \lceil C_r \rfloor\{\dot{q}\} + \lceil K \rfloor[\Psi]\{q\} = \{F'(t)\} \tag{6.62}$$

which is a sufficient condition for the damped system to have the same mode shapes as for the undamped system. Strictly speaking these are not eigenvectors of the damped system, as there is no eigenvalue problem defined for this case, however, the mode shapes are usually still referred to as eigenvectors because they are eigenvectors of the undamped system. In each row of the equation system in Equation (6.62) we have an equation

$$m_r\ddot{q} + c_r\dot{q} + k_r q = F_r'(t) \tag{6.63}$$

which is Newton's equation of an SDOF system with the modal mass, damping and stiffness as parameters. From Section 5.2.1 we then know that each mode (SDOF system) will have an undamped natural frequency

$$\omega_r = \sqrt{\frac{k_r}{m_r}} \tag{6.64}$$

and relative damping

$$\zeta_r = \frac{c_r}{2\sqrt{m_r k_r}}. \tag{6.65}$$

Using these equations, we can calculate the poles for mode $r$ as

$$s_r = -\zeta_r\omega_r \pm j\omega_r\sqrt{1-\zeta_r^2} = \sigma_r + j\omega_{dr}. \tag{6.66}$$

Using the relation for the real part of the poles that

$$\sigma_r = -\zeta_r\omega_r = -\zeta_r\sqrt{\frac{k_r}{m_r}} \tag{6.67}$$

with Equation (6.65) we find that

$$c_r = 2\zeta_r\sqrt{m_r k_r} = \frac{-2\sigma_r\sqrt{m_r k_r}}{\sqrt{k_r/m_r}} = -2\sigma_r m_r. \tag{6.68}$$

**Example 6.3.4** *We continue with the same SDOF system as in the previous examples, but we now add proportional damping defined by Equation (6.60) with* $a = 2/15 \approx 0.1333$ *and* $b = 1/1500$. *Calculate the poles of the system.*

*The damping matrix is*

$$[C] = \begin{bmatrix} 2/15 & 0 \\ 0 & 2/15 \end{bmatrix} + \begin{bmatrix} 250/1500 & -150/1500 \\ -150/1500 & 250/1500 \end{bmatrix} = \begin{bmatrix} 0.3 & -0.1 \\ -0.1 & 0.3 \end{bmatrix}. \quad (6.69)$$

*The next step to solve the damped case would be to find the eigenvectors (mode shapes) of the undamped system, which we already have from Example 6.3.1. We can therefore now compute the diagonal modal damping matrix by computing, for example by using MATLAB/Octave,*

$$[\Psi]^T[C][\Psi] = \begin{bmatrix} 0.2 & 0 \\ 0 & 0.4 \end{bmatrix}. \quad (6.70)$$

*We have the modal masses and stiffnesses from Example 6.3.2, so we can now calculate the relative damping factors*

$$\zeta_1 = \frac{c_1}{2\sqrt{m_1 k_1}} = \frac{0.2}{2\sqrt{1 \cdot 100}} = 0.01 \quad (6.71)$$

*and*

$$\zeta_2 = \frac{c_2}{2\sqrt{m_2 k_2}} = \frac{0.4}{2\sqrt{1 \cdot 400}} = 0.01. \quad (6.72)$$

*It is common in practice to express this as 1% relative damping for both modes. The poles, finally, are $s_1$ and $s_1^*$, where using the undamped natural frequencies from Example 6.3.1 we get*

$$s_1 = -\zeta_1\omega_1 + j\omega_1\sqrt{1-\zeta_1^2} = -0.1 + j10\sqrt{1-0.01^2} \quad (6.73)$$

*for the first mode, and for the second mode we have the poles $s_2$ and $s_2^*$, where*

$$s_1 = -\zeta_1\omega_1 + j\omega_2\sqrt{1-\zeta_2^2} = -0.2 + j20\sqrt{1-0.01^2}. \quad (6.74)$$

*End of example.*

The definition of proportional damping given in Equation (6.60), where the two parameters $a$ and $b$ controls the damping matrix, is not the most general definition leading to an uncoupled damping matrix as in Equation (6.62). It can be shown that the mode shapes of a damped system are equal to the modes of the undamped system if the equation

$$\left([M]^{-1}[C]\right)\left([M]^{-1}[K]\right) = \left([M]^{-1}[K]\right)\left([M]^{-1}[C]\right) \quad (6.75)$$

is satisfied, although this will not be proven here, see for example Craig and Kurdila (2006) or Ewins (2000). The most common proportional damping used in simulation of mechanical systems, is referred to as *modal damping* and is obtained by adding an individual damping factor, $\zeta_r$, to each undamped natural frequency to produce the poles, and use the mode shapes of the undamped system, see Section 6.4.3.

Although the assumption of proportional damping as defined either by Equation (6.60) or Equation (6.75) is certainly not always valid, in many cases there seem to be good reasons to assume this form of damping. First it can be argued that many types of damping are related to the stiffness elements, e.g., internal material damping, or to the mass elements, e.g., for friction damping. A stronger argument for the validity of proportional damping as a total approximation of damping, however, is perhaps the empirical evidence. Practical experience from experimental modal analysis has shown that mode shapes are often indeed real or near-real. In many cases, therefore, proportional damping seems to be a valid assumption.

### 6.3.5 General Damping

In the case of general, or *nonproportional*, damping, the eigenvalue problem we used above cannot be used, because the normal modes do not decouple the damping matrix. An alternative solution can, however, be found by reformulating the second-order system into a so-called *state-space* formulation, a common technique developed in the field of control engineering. We thus define a new vector with $2N$ elements, $\{z(t)\}$,

$$\{z(t)\} = \begin{Bmatrix} \{u(t)\} \\ \{\dot{u}(t)\} \end{Bmatrix} \tag{6.76}$$

whereby the first derivative $\dot{z}(t)$ is

$$\{\dot{z}(t)\} = \begin{Bmatrix} \{\dot{u}(t)\} \\ \{\ddot{u}(t)\} \end{Bmatrix}. \tag{6.77}$$

Newton's equation for our MDOF system is now extended by adding $N$ extra lines

$$[M]\{\dot{u}\} - [M]\{\dot{u}\} = \{0\} \tag{6.78}$$

by introducing two new matrices

$$[A] = \begin{bmatrix} C & M \\ M & 0 \end{bmatrix} \tag{6.79}$$

and

$$[B] = \begin{bmatrix} K & 0 \\ 0 & -M \end{bmatrix} \tag{6.80}$$

and finally the force vector is appended by $N$ zeros so that

$$F' = \begin{Bmatrix} F(t) \\ 0 \end{Bmatrix}. \tag{6.81}$$

With these definitions, we set up the $2N$-by-$2N$ equation system

$$\begin{bmatrix} C & M \\ M & 0 \end{bmatrix} \begin{Bmatrix} \dot{u} \\ \ddot{u} \end{Bmatrix} + \begin{bmatrix} K & 0 \\ 0 & -M \end{bmatrix} \begin{Bmatrix} u \\ \dot{u} \end{Bmatrix} = \begin{Bmatrix} F \\ 0 \end{Bmatrix}. \tag{6.82}$$

or more compactly

$$[A]\{\dot{z}\} + [B]\{z\} = F' \tag{6.83}$$

which is a linear first-order differential equation in $z(t)$. The solutions to Equation (6.83) are of the form

$$\{z(t)\} = \{\Phi\}e^{\lambda t} = \begin{Bmatrix} \{\psi\} \\ \lambda\{\psi\} \end{Bmatrix} e^{\lambda t} \tag{6.84}$$

where $\{\Phi\}$ is an eigenvector of length $2N$, and the lower half includes a multiplication by the eigenvalue because the lower half of $\{z\}$ is the derivative of the first half, and $\lambda$ is the inner derivative from the $e^{\lambda t}$ factor.

To find the free vibrations of the system described by Equation (6.83), we first Laplace transform Equation (6.83), which results in

$$[s[A] + [B]]\{Z(s)\} = F'(s). \tag{6.85}$$

With the same procedure as for the undamped system, we premultiply Equation (6.85) by $[A]^{-1}$ and rearrange, and set the force to zero to find the free vibrations. We get the equation

$$\left[[A]^{-1}[B] - \lambda[I]\right]\{\Phi\} = \{0\} \tag{6.86}$$

which is a standard eigenvalue problem with eigenvalues equal to minus the poles, $\lambda_r = -s_r$. Solving the eigenvalues and eigenvectors gives $2N$ eigenvalues and $2N$ corresponding eigenvectors of length $2N$. Since the coefficient matrices $[A]$ and $[B]$ are real, the eigenvalues must be real or come in complex conjugate pairs. If they are real the system is overdamped, so we concentrate on the case where they are complex, and the system exhibits free vibrations. In that case the system will have $N$ complex conjugate pairs of eigenvalues, and $N$ complex conjugate pairs of eigenvalues.

An important point to note is that because the eigenvalue problem in the case of nonproportional damping is of first order, the eigenvalues are directly related to the Laplace operator, i.e., $\lambda = -s$, and thus the poles of the system, not to the square of the poles as for the undamped case. As the poles come in complex conjugate pairs, we take every second pole from the eigenvalues, and we can therefore write the poles with positive imaginary part as

$$s_r = -\zeta_r w_r + j w_r \sqrt{1 - \zeta_r^2} \tag{6.87}$$

for $r = 1, 2, 3 \ldots, N$. The complex conjugate poles $s_r^*$ are of course also poles of the system.

The angular frequencies $w_r$ for nonproportionally damped systems are called *natural frequencies* and are not strictly speaking the same as the *undamped* natural frequencies. The main difference between the mode shapes of a nonproportionally damped system and those of the proportionally damped system is that the mode shapes of the nonproportionally damped system are complex. This means that each point on the structure has its own phase angle relative to the other points, which in turn means that each point on the structure reaches its maximum deflection at different time instances. The result, if the complex mode has phase angles that differ substantially from 0 and 180° relative to each other (a large *mode complexity*), is that the mode shape is not a standing wave as for the normal modes, but rather consists of a 'traveling wave' whose maximum moves around over the structure.

In experimental modal analysis it is very easy to obtain complex modes due to errors in the parameter extraction, because of the rapid phase shift in the FRF around the natural frequency. This rapid phase shift can make a small error in estimated frequency yield a large phase error in the mode shape. It is therefore important to understand mode complexity in order to understand if the obtained complex modes are 'true' or a result of erroneous data and/or curve fitting. On most structures with light damping, the mode shape complexity is not particularly large, as most mode shapes have phase angles of close to 0 or 180°, even if the damping is strongly nonproportional. It turns out (Ewins 2000) that in order to get highly complex modes on ordinary structures (not rotating, for example, for which modes are normally highly complex), in addition to having a nonproportional damping matrix, it is also necessary that at least two modes are very close in frequency. A result of this is that, if experimental modal analysis results in highly complex modes, it is good practice to treat the results with some suspicion. The most likely cause of highly complex modes is poor curve fitting.

It can be shown that there are orthogonality criteria similar to those for the undamped system in Equation (6.54) and Equation (6.55) such that

$$[\Phi]^H [A][\Phi] = \lceil M_A \rfloor \tag{6.88}$$

and

$$[\Phi]^H [B][\Phi] = \lceil M_B \rfloor \tag{6.89}$$

where coefficients in the diagonal matrices are called *modal A* and *modal B*, respectively. These coefficients, $m_a$ and $m_b$, respectively, can be used for mode shape scaling, just like the modal mass and stiffness numbers for the proportionally damped system, and as we will see in Section 6.4.4, there is a good reason for scaling mode shapes to unity modal A.

**Example 6.3.5** *To illustrate the concept of nonproportional damping, we change the damping matrix from Example 6.3.4 by adding 0.5 to element (1,1), which produces the nonproportional damping matrix*

$$[C] = \begin{bmatrix} 0.8 & -0.1 \\ -0.1 & 0.3 \end{bmatrix}. \tag{6.90}$$

*Find the poles and mode shapes of the system.*

*The generalized eigenvalue problem can be solved in MATLAB/Octave by building the matrices* $[A]$ *and* $[B]$ *etc. There are some steps necessary that we will not discuss in great length here, but instead summarize by the following lines of MATLAB/Octave code that, together with the comment code, should be sufficient. We assume that the mass, damping and stiffness matrices are already defined as in previous examples. You should note that we use eig*($-A \setminus B$)*, with a minus sign, which gives correct poles directly, since we have that* $\lambda = -s$*. This is not necessary but simplifies the sorting process of the poles a little if we want the poles with positive imaginary part to come first. We also assume that the absolute values of the eigenvalues are larger than unity.*

```
A=[C    M;   M   0*M];
B=[K  0*M; 0*M    -M];
[V,D]=eig(-A\B)
% Sort in descending order
[Dum,I]=sort(diag(abs(imag(D))));
p=diag(D(I,I));
V=V(:,I);
% Scale to unity Modal A
Ma=V.'*A*V;
for col = 1:length(V(1,:))
    V(:,col)=V(:,col)/sqrt(Ma(col,col));
end
```

*The results of this code are the column vector p with the poles of the system, and the matrix V with the mode shapes in columns. After sorting and scaling for unity modal A, these variables contain*

$$p = \begin{Bmatrix} -0.225 + 9.999i \\ -0.225 - 9.999i \\ -0.325 + 19.995i \\ -0.325 - 19.995i \end{Bmatrix} \tag{6.91}$$

*and the eigenvalues of the first mode are*

$$\{\phi\}_1 = \begin{Bmatrix} -0.11 + 0.11i \\ -0.11 + 0.11i \\ -1.10 - 1.13i \\ -1.10 - 1.15i \end{Bmatrix}_1 \tag{6.92}$$

*and the complex conjugate,* $\{\phi\}_1^*$*, whereas for the second mode we have the eigenvector*

$$\{\phi\}_2 - \begin{Bmatrix} 0.08 - 0.08i \\ -0.08 + 0.08i \\ 1.53 + 1.63i \\ -1.58 - 1.58i \end{Bmatrix}_2 \tag{6.93}$$

*and the complex conjugate,* $\{\phi\}_2^*$.

*The natural frequencies are* $f_1 = 1.59$ *Hz and* $f_2 = 3.18$ *Hz, and the relative damping coefficients* $\zeta_1 = 0.0225$ *and* $\zeta_2 = 0.0162$*. The mode shapes, finally,* $\{\psi\}_1$ *and* $\{\psi\}_2$ *are the upper halves of* $\{\phi\}_1$ *and* $\{\phi\}_2$*, respectively.*

*End of example.*

The important parts of the mode shape vectors are obviously the upper halves as the lower halves can be reconstructed by multiplying the upper half by the corresponding eigenvalue (pole) if needed, for example, to calculate modal A or modal B.

## 6.4 Frequency Response of MDOF Systems

We have now come to a point where we can formulate relations for the frequency responses (FRF) of MDOF systems. As we will see there are two ways to synthesize frequency responses; either directly from Newton's equation, or by using the modal parameters, i.e., poles and mode shapes. It is worth pointing out the great potential offered by FRFs, as these functions can be used to calculate the steady-state response in a particular point (degree-of-freedom), for a particular force input in a point (the same as the response point or another point). This section will thus answer the important question we raised in the introduction to the present chapter: how much vibration do we get in a point (DOF) on our structure for a particular force input in a particular point? We will comment more below as we come to the results.

### 6.4.1 *Frequency Response from* $[M]$, $[C]$, $[K]$

We first look at how frequency responses can be calculated from known mass, damping and stiffness matrices. The most intuitive frequency responses would be those obtained from the Laplace transform of Newton's equation, i.e.,

$$\left[s^2[M] + s[C] + [K]\right]\{U(s)\} = \{F(s)\} \tag{6.94}$$

which can be rewritten as

$$[Z(s)]\{U(s)\} = \{F(s)\} \tag{6.95}$$

where the matrix $[Z(s)]$ is called the *system impedance matrix*. The frequency responses from this equation would be obtained by evaluating the equation along the imaginary axis in the $s$-plane, i.e., $Z(\mathrm{j}\omega) = Z(s)|_{s=\mathrm{j}\omega}$. There is an important implication, however, of this matrix, which makes it rather unsuitable for the purpose of general description of mechanical systems. If we look at the formulation in Equation (6.95), an individual element $z_{pq}(\mathrm{j}\omega)$ implies

$$z_{pq}(\mathrm{j}\omega) = \left.\frac{F_p}{U_q}\right|_{U_k=0,k\neq q} \tag{6.96}$$

that is, in order to experimentally measure an individual element $z_{pq}$ in $[Z(s)]$, we would have to ground every point except point $q$, to ensure the displacement of all other points would be equal to zero. Of course this is impossible in most cases.

We must therefore reformulate Equation (6.95) by inverting the impedance matrix, and we obtain an alternative, useful formulation by introducing $[H(s)] = [Z(s)]^{-1}$, whereby we get the equation

$$[H(s)]\{F(s)\} = \{U(s)\}. \tag{6.97}$$

To get the frequency responses we evaluate Equation (6.97) on $s = \mathrm{j}\omega$ and get the *receptance* frequency responses (or *dynamic flexibility*) defined by the matrix equation

$$[H(\mathrm{j}\omega)]\{F(\mathrm{j}\omega)\} = \{U(\mathrm{j}\omega)\}. \tag{6.98}$$

Measuring an individual element $H_{pq}(\mathrm{j}\omega)$ in $[H(\mathrm{j}\omega)]$ implies measuring

$$H_{pq}(\mathrm{j}\omega) = \left.\frac{U_p}{F_q}\right|_{F_k=0,k\neq q} \tag{6.99}$$

which is usually very much easier than keeping the displacements to zero. It is simply a matter of 'not touching' the points except where we are inputting a force. If we wish we can input several forces, so-called *multiple-input* excitation, which is no problem considering we measure all forces exciting the system. This will be described in detail in Chapter 14.

The frequency response formulation in Equation (6.98) is 'physical' in the sense that it represents the displacement at each point as a superposition of the contribution of each nonzero force, which is exactly what the physics of mechanical systems imply.

To compute the frequency responses in Equation (6.98), the procedure is very simple:

- Compute the system impedance $[Z(j\omega)]$ by Equation (6.95).
- Compute the receptance matrix by inverting the impedance matrix at each frequency $\omega$, i.e., $[H(j\omega)] = [B(j\omega)]^{-1}$ at each frequency.

When working with experimentally obtained frequency responses, it is most common to use the frequency in [Hz] as the variable. We will therefore later in this book refer to, for example, the receptance matrix $[H(f)]$.

### 6.4.2 Frequency Response from Modal Parameters

The development of frequency responses from mass, damping and stiffness matrices in Section 6.4.1 required the entire matrices to be known, and included a matrix inversion of the entire system impedance matrix at each frequency. For large systems this is computationally inefficient. Also, it is rare to know the damping matrix, so the equations developed in Section 6.4.1 are often not practically useful. In Section 6.4.3 we will see an alternative way of synthesizing FRFs which is more practical, and which is based on the results we obtain in the present section. In this section we will also show that the modal parameters provide a much more computationally efficient way to compute the frequency responses. Furthermore, the development in this section is the key to *experimental modal analysis*, which will be described briefly in Section 16.7, as we will now show the relation between measured frequency responses and modal parameters (natural frequencies, damping coefficients, and mode shapes).

We will develop a general form of an expression for the receptance frequency response as defined in Equation (6.98) for the case of proportional damping as described in Section 6.3.4, because this is somewhat easier to follow than the general case for nonproportional damping. It should perhaps be mentioned that for the undamped case we cannot define any frequency responses, as they would go to infinity at $f = f_r$; frequency responses require damping.

We start by noting that the frequency response matrix we want is the inverse of the system impedance matrix in the Laplace domain (eventually setting $s = j\omega$, but we wait until Equation (6.110) at the end of this argument), so we have that

$$\left[s^2[M] + s[C] + [K]\right] = [H]^{-1}. \tag{6.100}$$

We now premultiply both sides with the mode shape matrix transposed, $[\Psi]^T$ and postmultiply with $[\Psi]$ which yields

$$[\Psi]^T \left[s^2[M] + s[C] + [K]\right][\Psi] = [\Psi]^T[H]^{-1}[\Psi]. \tag{6.101}$$

Next, we make use of the fact that we have proportional damping and the orthogonality criterion therefore makes the matrix on the left-hand side diagonal, that is we have

$$\left[s^2 \lceil M_r \rfloor + s \lceil C_r \rfloor + \lceil K_r \rfloor\right] = [\Psi]^T[H]^{-1}[\Psi]. \tag{6.102}$$

We then note that the inverse of a product $[ABC]^{-1} = C^{-1}B^{-1}A^{-1}$ and take the inverse of both sides of the equation, which results in

$$\left[s^2 \lceil M_r \rfloor + s \lceil C_r \rfloor + \lceil K_r \rfloor\right]^{-1} = [\Psi]^{-1}[H]\left([\Psi]^T\right)^{-1} \tag{6.103}$$

and then we premultiply this equation by $[\Psi]$ and postmultiply by $[\Psi]^T$ and reverse the equation to get

$$[H] = [\Psi]\left[s^2 \lceil M_r \rfloor + s^2 \lceil C_r \rfloor + \lceil K_r \rfloor\right]^{-1}[\Psi]^T. \tag{6.104}$$

Now we note that the inverse of a diagonal matrix is nothing but the reciprocal of each value on the diagonal. Therefore we define a new matrix, the *inverse pole matrix*, $[S^{-1}]$ which is a diagonal matrix where each element on the diagonal, $s_{rr}$, is

$$s_{rr} = \frac{1}{s^2 m_r + s c_r + k_r} = \frac{1/m_r}{(s - s_r)(s - s_r^*)} \tag{6.105}$$

for mode $r$. Using this matrix we can simplify the result in Equation (6.104) to

$$[H] = [\Psi]\left\lceil S^{-1} \right\rfloor [\Psi]^T. \tag{6.106}$$

It is particularly important to look at what Equation (6.106) means for a particular function $H_{pq}(s) = X_p/F_q$. A careful study of the equation reveals that the frequency response $H_{pq}(s)$ can be written as

$$H_{pq}(s) = \sum_{r=1}^{N} \frac{\psi_{pr}\psi_{qr}}{m_r(s - s_r)(s - s_r^*)} \tag{6.107}$$

where $\psi_{pr}$ is mode shape coefficient in point $p$ for mode $r$, etc.

To find a more general description of the transfer function in Equation (6.107) we can apply a partial fraction expansion (see Section 2.6.1) of each term in the sum to split it into a sum of the *residues*, $A_{pqr}$ divided by $(s - s_r)$. Since the numerator coefficient in Equation (6.107) is real, and the poles are complex conjugate pairs, it is relatively easy (see Problem 6.1) to show that the frequency response can be written as

$$H_{pq}(s) = \sum_{r=1}^{N} \frac{A_{pqr}}{s - s_r} + \frac{A_{pqr}^*}{s - s_r^*} \tag{6.108}$$

where the residues, $A_{pqr}$ are composed of a *modal scaling constant*, $Q_r$ for each mode $r$, and the mode shape coefficients in the two points $p$ and $q$,

$$A_{pqr} = Q_r \psi_{pr} \psi_{qr} = \frac{1}{j2\omega_{dr} m_r} \psi_{pr} \psi_{qr}. \tag{6.109}$$

Equation (6.108) is a general form of the partial fraction expansion of the transfer function $H_{pq}(s)$, which, together with the expression for the residues in Equation (6.109), forms the standard description of transfer functions from modal parameters. The equation is valid for all mode shapes, also complex mode shapes in the case of nonproportional damping.

We will now replace the transfer functions used for the development here by setting $s = j\omega$, which gives us the following general expression for MDOF frequency response functions,

$$H_{pq}(j\omega) = \sum_{r=1}^{N} \frac{A_{pqr}}{j\omega - s_r} + \frac{A_{pqr}^*}{j\omega - s_r^*} \tag{6.110}$$

where the residues $A_{pqr}$ are defined in Equation (6.109).

The result in Equation (6.110), which is called the *modal superposition equation* is very important because it is the key to the topic of experimental modal analysis, as it relates frequency responses that can be experimentally estimated, with the modal parameters; the poles and mode shapes. This equation also shows why scaling mode shapes to unity modal mass is so convenient; the factor $m_r$ in the denominator can then be neglected which makes the calculations a little easier.

It is clear, comparing Equation (6.107) with the expression of the transfer function of an SDOF system in Equation (5.4), that Equation (6.107) (and, of course, also Equation 6.110) is a sum of SDOF transfer functions or frequency response functions, respectively. This is the reason for the great interest we took in the SDOF system; MDOF systems have frequency response functions that consist of sums of SDOF frequency response functions, where each mode corresponds to one SDOF system. In the case of MDOF systems, this does not necessarily mean that every mode produces a clear peak in the frequency response, like an SDOF system, because two or more natural frequencies can coincide or be very close, which will result in just one peak in the frequency response. However, if modes are well separated, each mode will show a peak very similar to the peak of an SDOF system (but not identical, because surrounding modes will interfere at least a little, and sometimes a lot).

An important implication of Equation (6.110) and Equation (6.109) is that the frequency response matrix $[H]$ is obviously symmetric, so that

$$H_{pq}(\mathrm{j}\omega) = H_{qp}(\mathrm{j}\omega). \tag{6.111}$$

This rather interesting equation is called *Maxwell's reciprocity relation* and shows that the frequency response between two points is the same if we excite in point $q$ and measure the response in point $p$, as if we reverse the force and response points.

The residues can easily be formulated in matrix notation, whereby the *residue matrix*, $[A]_r$ becomes

$$[A]_r = Q_r\{\psi\}_r\{\psi\}_r^T = Q_r \begin{bmatrix} \psi_{1r}\psi_{1r} & \psi_{1r}\psi_{2r} & \psi_{1r}\psi_{3r} & \dots \\ \psi_{2r}\psi_{1r} & \psi_{2r}\psi_{2r} & \psi_{2r}\psi_{3r} & \dots \\ \psi_{3r}\psi_{1r} & \psi_{3r}\psi_{2r} & \psi_{3r}\psi_{3r} & \dots \\ \dots & \dots & \dots & \dots \end{bmatrix} \tag{6.112}$$

which is a matrix of rank one (since it is composed only of linear combinations of one vector, $\{\psi\}_r$. It should be noted that each column in the residue matrix is the mode shape $\{\psi\}_r$, scaled by the modal constant $Q_r$ and the mode shape coefficient corresponding to the column, that is

$$[A]_r = Q_r \left[ \psi_{1r}\{\psi\}_r \;\; \psi_{2r}\{\psi\}_r \;\; \psi_{3r}\{\psi\}_r \;\; \dots \right]. \tag{6.113}$$

With this expression for the residue matrix, the entire frequency response matrix $[H(\mathrm{j}\omega)]$ can be written compactly as

$$[H(\mathrm{j}\omega)] = \sum_{r=1}^{N} \frac{[A]_r}{\mathrm{j}\omega - s_r} + \frac{[A^*]_r}{\mathrm{j}\omega - s_r^*} \tag{6.114}$$

which is the equation for synthesizing frequency responses from modal parameters, and the basic expression used for modal parameter extraction in the frequency domain, see Section 16.7. It should be noted here that although Equation (6.114) was developed here for a system with proportional damping, it is also valid for systems with general damping, i.e., with complex mode shapes.

A very important implication of Equation (6.112) for experimental modal analysis, is found by observing that any row or column in the residue matrix, $[A]_r$ contains the mode shape $\{\psi\}_r$. This assumes that there are not two poles which coincide, however. Nevertheless we can conclude from this, that *the minimum amount of data necessary to be able to extract a mode shape from a measurement of a frequency response matrix, is one row or column.* In the special case of two coinciding modes, there are multi-reference techniques which can separate the two if two rows or columns of $[H]$ are measured.

In experimental modal analysis we often write Equation (6.114) using a pole matrix, $[\Lambda^{-1}]$, similar to the inverse pole matrix $[S^{-1}]$ in Equation (6.106) but expanding the matrix to a size of $2N$, formulating it in the frequency domain instead of in Laplace domain, and renumbering the poles to $s_r, r = 1, 2, 3, ..., 2N$.

The inverse pole matrix $[\Lambda^{-1}]$ is thus

$$[\Lambda^{-1}(j\omega)] = \begin{bmatrix} \frac{1}{j\omega - s_1} & 0 & 0 & \cdots \\ 0 & \frac{1}{j\omega - s_2} & 0 & \cdots \\ \cdots & \cdots & \cdots & \cdots \\ \cdots & \cdots & \cdots & \frac{1}{j\omega - s_{2N}} \end{bmatrix} \tag{6.115}$$

and if we further redefine the $2N$-by-$2N$ mode shape matrix $[\Psi']$, including the complex conjugate mode shapes in columns, Equation (6.114) can be written compactly as

$$[H(j\omega)] = [\Psi'][\Lambda^{-1}][\Psi']^T. \tag{6.116}$$

Another important implication of Equation (6.110) is the great information saving it offers. Essentially, knowing the poles (frequencies and relative damping coefficients) and the mode shapes of all modes, the frequency response between any two points on the structure can be synthesized. Furthermore, if a limited frequency range is of interest, as it always is in practice, we only need the first $N_M$ modes which can offer a great saving. The number of coefficients necessary for this are $N_M$ complex poles ($2N_M$ real numbers) plus $N_M \times N$ mode shape coefficients. This should be compared with storing all frequency responses with, say, $N$ frequency values each, which would correspond to $N^3$ complex numbers. For $N = 1000$ DOFs and $N_M = 25$ modes, for example, this offers a saving of a factor of approximately 80,000 (approximately 25000 values instead of $2 \times 10^9$).

**Example 6.4.1** *Calculate and plot the frequency responses $H_{11}(j\omega)$ and $H_{12}(j\omega)$ for the 2DOF system from Example 6.3.4*

*For the first FRF, $H_{11}$ we have the first residue, according to Equation (6.109)*

$$A_{111} = Q_r \psi_{11} \psi_{11} \tag{6.117}$$

*with the modal scaling constant*

$$Q_1 = \frac{1}{j2\omega_{d1} m_1} \approx \frac{1}{j2\omega_n m_1} = \frac{1}{j20} \tag{6.118}$$

*which results in the residue*

$$A_{111} = \frac{1/2}{j20} = \frac{-j}{40} = -0.025j \tag{6.119}$$

*For the second mode, similarly we have*

$$A_{112} = \frac{1/2}{j40} = \frac{-j}{40} = -0.0125j \tag{6.120}$$

*Thus, the frequency response becomes*

$$H_{11}(j\omega) = \frac{-0.025j}{j\omega - s_1} + \frac{0.025j}{j\omega - s_1^*} + \frac{-0.0125j}{j\omega - s_2} + \frac{0.0125j}{j\omega - s_2^*} \tag{6.121}$$

*For the second mode, similarly we get*

$$H_{11}(j\omega) = \frac{-0.025j}{j\omega - s_1} + \frac{0.025j}{j\omega - s_1^*} + \frac{0.0125j}{j\omega - s_2} + \frac{-0.0125j}{j\omega - s_2^*} \tag{6.122}$$

*The two frequency responses are plotted in Figure 6.1.*
*End of example.*

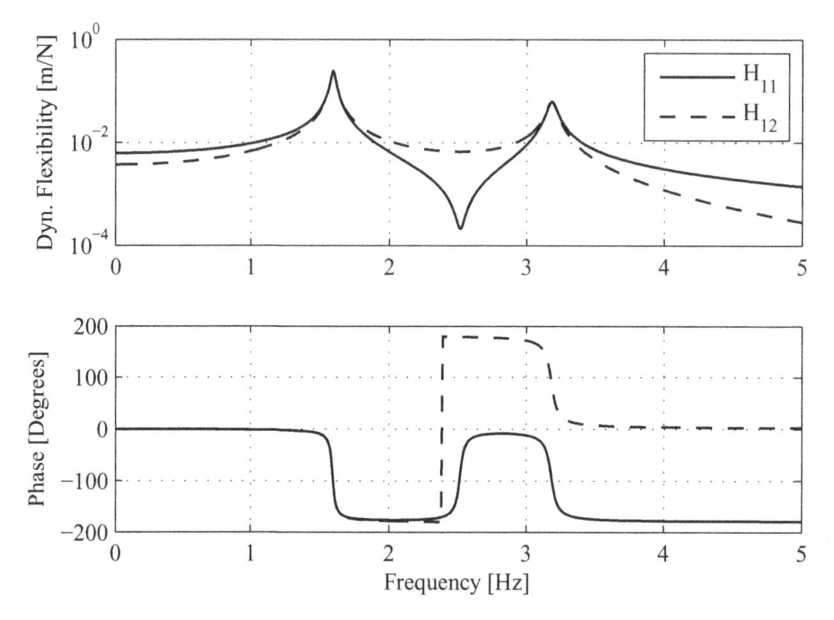


**Figure 6.1** Plots of 2DOF frequency responses for Example 6.4.1, magnitude (upper plot) and phase (lower plot)

### *6.4.3 Frequency Response from [M], [K], and $\zeta$ – Modal Damping*

As was noted earlier, the equations in Section 6.4.1 are of somewhat limited practical use since we usually do not know the damping matrix $[C]$. In most cases where frequency responses are wanted for simulation purposes, we are therefore forced to use some other means of synthesizing frequency responses. One such way, using the results from Section 6.4.2, is by using the normal modes and undamped natural frequencies, $[\Psi]$ and $\omega_r$ from a solution of the undamped system, and adding a relative damping factor, $\zeta_r$ to each mode, to form a complex pole, and then using Equation (6.110) to synthesize frequency responses. This is the most common method used in simulation of mechanical systems, and the damping is then usually referred to as *modal damping*.

### *6.4.4 Mode Shape Scaling*

Mode shapes are, as we have noted several times, arbitrarily scaled. The relation necessary to synthesize correctly scaled frequency responses from modal parameters is the modal scaling constant, $Q_r$, in, for example, Equation (6.109). As we have mentioned, mode shapes can be scaled many different ways; for largest coefficient of 1, unity length, etc. The two most common and most important scaling conventions are, however, unity modal mass, and unity modal A. The convenience of the former was pointed out in conjunction with Equation (6.107). The convenience of the latter will be apparent if we look at what it implies for a proportionally damped system (remember, for nonproportional damping, unity modal A is the usual scaling).

We start with the definition of the matrix $A$ in the state-space formulation from Equation (6.79), repeated here for convenience

$$[A] = \begin{bmatrix} C & M \\ M & 0 \end{bmatrix} \tag{6.123}$$

and we remember that the state-space eigenvectors are of the form

$$\{\phi\}_r = \left\{ \begin{array}{c} \psi_r \\ \lambda_r \psi_r \end{array} \right\}. \tag{6.124}$$

If we use the above equations to calculate modal A for a proportionally damped system, for one mode $r$, we get

$$\begin{aligned} m_a &= \{\phi\}_r^T [A] \{\phi\}_r = \\ &= \lfloor \{\psi\}_r^T \quad \lambda_r \{\psi\}_r^T \rfloor \begin{bmatrix} C & M \\ M & 0 \end{bmatrix} \left\{ \begin{array}{c} \{\psi\}_r \\ \lambda_r \{\psi\}_r \end{array} \right\} \end{aligned} \tag{6.125}$$

which results in

$$m_a = \{\psi\}_r^T [C] \{\psi\}_r + \lambda_r \{\psi\}_r^T [M] \{\psi\}_r + \{\psi\}_r^T [M] \lambda_r \{\psi\}_r \tag{6.126}$$

and we obtain

$$m_a = c_r + 2\lambda_r m_r. \tag{6.127}$$

From Equation (6.68) we have a relation $c_r = -2\sigma_r m_r$, which leads to

$$m_a = -2\sigma_r m_r + 2(\sigma_r + \mathrm{j}\omega_{dr}) m_r = \mathrm{j}2\omega_{dr} m_r \tag{6.128}$$

which is our final relation between modal A and modal mass. Equation (6.128) can be used to rescale mode shapes for proportionally damped systems so that we get, e.g., unity modal A. The motivation for scaling modes to unity modal A is found by comparing Equation (6.128) with Equation (6.109), which shows that if the modes are scaled so that $m_a = 1$, then the residue $A_{pqr}$ is simply

$$A_{pqr} = \psi_{pr} \psi_{qr}. \tag{6.129}$$

For convenience, we also not that for unity modal mass scaling, the mode shapes are scaled so that

$$A_{pqr} = \frac{1}{\mathrm{j}2\omega_{dr}} \psi_{pr} \psi_{qr} \tag{6.130}$$

which follows straight from Equation (6.109).

In other words, if we scale mode shapes to unity modal A, the modal constant is unity and the FRF synthesis according to Equation (6.110) becomes simple. Therefore this mode shape scaling is often preferred in experimental modal analysis. It should be noted, however, that in analytical modal analysis, where the undamped system is usually solved, it is more common to use unity modal mass. It should also be recalled from Example 6.3.5 that scaling mode shapes to unity modal A means that real mode shapes, such as those from proportionally damped systems, become complex, although all mode shape coefficients are, of course, either in phase or out of phase.

### 6.4.5 The Effect of Node Lines on FRFs

An effect of the expression of the residue in Equation (6.109) is of particular importance. What happens if one of the mode coefficients is zero? Such a point, which is a point with no motion in the mode, is often called a *node* or a *nodal point* (not to be confused by a 'node' in the sense of a point on an element in a finite element model). On continuous structures there are lines along which all points have zero motion for a particular mode, called *node lines*. Apparently, from Equation (6.109), the residue becomes zero for this mode, and the consequence is that the mode does not appear in the FRF, as is seen in Figure 6.2.

An implication of this is that it is very important in experimental modal analysis to select measurement points carefully so that reference points are not located on node lines. If a reference point with one or

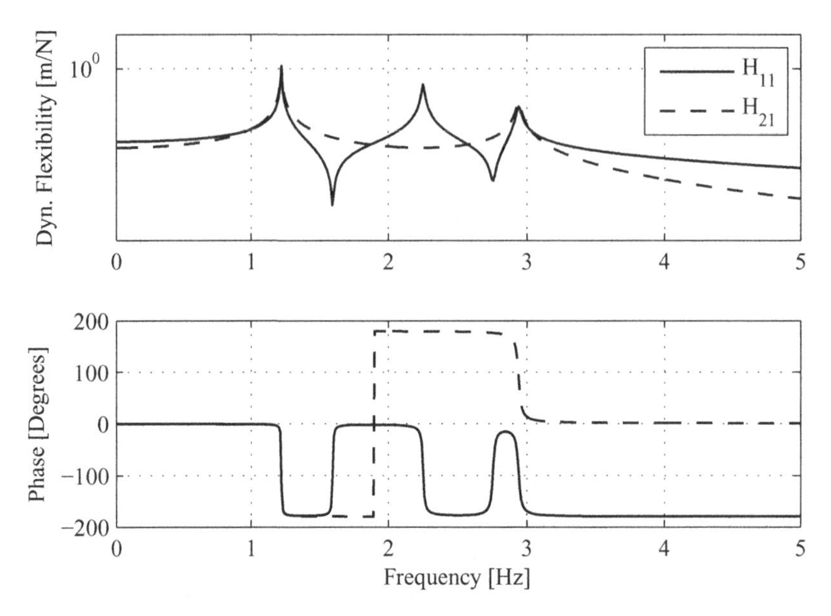


**Figure 6.2** Plot of two frequency responses from a 3DOF system; $H_{11}$ which in this case shows all three modes (solid), and a cross-FRF, $H_{21}$, where mode two obviously has a nodal line, since this mode is invisible in $H_{21}$ (dashed). As the figure shows, there is no peak at all in $H_{21}$ at the natural frequency of the second mode, where the residue is zero

more zeros in the mode shape vectors is chosen, the entire mode (or modes) will be impossible to detect. This is an important reason why multiple references are usually preferable, as will be discussed in more detail in Section 16.7.5.

### 6.4.6 *Antiresonance*

The results of Example 1, plotted in Figure 6.1, revealed a phenomenon known as an *antiresonance* at approximately 2.5 Hz, for the cross-frequency response $H_{11}(f)$. This phenomenon is due to a simple relation between phases, and although not due to any global property of the structure, it is still interesting to understand why there sometimes occurs an antiresonance between two modes, and sometimes not.

The answer to this question lies in the phase of the frequency responses and the expression of the residues according to Equation (6.109). A natural frequency (resonance) produces a phase change in the FRF of $-180°$. Thus, if there is no antiresonance between two resonances, the phase of the FRF at the two natural frequencies will necessarily have opposite sign, as is seen in the phase of $H_{12}$ in Figure 6.1. The phase at the first natural frequency is $-90°$, and at the second natural frequency it is $+90°$ (or $-270°$, but arctan() gives phases only between $\pm 180°$). Thus, since the sign change has to come from the residue in the numerator, one of the two mode shape coefficients must have changed sign between the two modes. If, as we know in this case, the two points (masses) are in phase in the first mode, the two points must be out of phase in the second mode, as we know indeed they are. If you instead look at the phase of $H_{11}$ in Figure 6.1, the antiresonance lifts the phase by $+180°$, so that the phase relationship at the second natural frequency is the same as the phase at the first natural frequency. In this FRF the force and displacement are in the same point, and the mode shape coefficient therefore cannot, of course, change sign since it is actually the mode shape coefficient squared that makes the residue. Consequently this FRF shows an antiresonance between the two modes. This is a necessary requirement for all *driving point* FRFs, where force and response are in the same point, and is used as a quality check when FRFs are measured experimentally using a shaker, see Section 16.7.2.

### 6.4.7 *Impulse Response of MDOF Systems*

Equation (6.113) can be inverse transformed into a corresponding equation for the time domain. The impulse response matrix $[h(t)]$ is thus

$$[h(t)] = \sum_{r=1}^{N} [A]_r e^{s_1 t} + [A^*]_r e^{s_1^* t} \tag{6.131}$$

which is the time domain formulation used for synthesizing impulse responses from modal parameters, or for formulating parameter extraction methods in the time domain.

## 6.5 Time Domain Simulation of Forced Response

The problem of computing the motion (displacement, velocity, or acceleration) solution for a particular force vector input to a mechanical system is called *forced response* computation, or forced response simulation. There is a wide range of methods available for solving this for arbitrary forces. If it is sufficient to calculate a steady-state solution, the simplest way is often to use the FFT to solve the solution in the frequency domain. This method can also be used to solve the steady-state time response, provided the time record is short enough so that the IFFT can be calculated. This used to be a major drawback, but due to the increase in computer power and memory size, today FFT can be calculated on data with several million samples and the major drawback with the FFT method is that it handles the steady-state response only, and does not include the transient part of the solution.

In order to include the transient part of the solution, a time domain method has to be used. There are many methods of solving the forced response in time domain, the most common class of methods being the so-called ODE (Ordinary Differential Equations) solvers, for example using the Runge–Kutta method. These methods are attractive because of their generality for solving any type of differential equations, linear or nonlinear. However, there are some drawbacks with these methods, the main ones being risk of instability, difficulties choosing proper time step size, and computational inefficiency. Many techniques are presented in standard textbooks such as (Craig and Kurdila 2006; Inman 2007; Rao 2003).

Using digital filter theory, Ahlin *et al.* (2006) have described a very accurate time domain method using simple digital filters. The method is not unique; equivalent models can be formulated using state-space techniques or so-called ARMA models, see for example Kozin and Natke (1986). The formulation of the method by Ahlin is, however, attractive in its simplicity and transparency, and the method has proven to be fast and to have superior dynamic range and speed compared to, e.g., ODE-based methods (Brandt and Ahlin 2003). It will therefore be presented here as an example of an accurate and fast method to produce forced response time data. Such data are essential in many method development cases where data from known mechanical systems are needed to verify, for example, a signal processing method or an experimental modal analysis parameter extraction method. The example of beating in the forced response of an SDOF system in Figure 5.4, for example, was computed by this method. Many of the examples later in this book have also been produced using data simulated with the method described here. To keep the presentation here simple, we will restrict the analysis to linear mechanical systems, although Ahlin *et al.* (2006) have also shown how the method can be extended to nonlinear systems. That is, however, beyond the scope of this book.

The method is based on modal superposition which we discussed in Section 6.4.2, i.e., the equation for the frequency response between points $p$ and $q$, expressed as

$$H_{pq}(\mathrm{j}\omega) = \sum_{r=1}^{N} \frac{A_{pqr}}{\mathrm{j}\omega - s_r} + \frac{A_{pqr}^*}{\mathrm{j}\omega - s_r^*} \tag{6.132}$$

where the residues, $A_{pqr}$, if we assume scaling to unity modal A for each mode $r$, only depends on the mode shape coefficients in the two points $p$ and $q$,

$$A_{pqr} = \psi_{pr}\psi_{qr} \tag{6.133}$$

which should be familiar by now. Using modal superposition for the simulation has some obvious advantages that we can conclude from Chapter 6, namely that:

- we can use known mass, damping and stiffness matrices to obtain the poles and residues,
- we can use known mass and stiffness matrices to obtain undamped poles and mode shapes, and then add modal (viscous) damping, to obtain the complex poles,
- we can use the modal model in Equation (6.132) directly, whether it comes from a FE model, an experimental modal analysis, etc., and
- we can easily exclude certain modes, if we wish to simulate what would happen if we remove a mode from a particular frequency range (by using a modal model and exclude whichever modes we want to).

To solve the forced response problem, we now formulate digital filters for each mode in the summation in Equation (6.132). This is the key to the method, and Ahlin *et al.* (2006) shows that the best method to transform the analog filters in the modal superposition, is in most cases a so-called *ramp invariant transform*. Using this transform, the difference equation (see Section 3.3.2) for the positive pole (the left term in Equation 6.132 for one mode) is

$$\begin{aligned} y(n) = \mathrm{e}^{-s_r\Delta t}y(n-1) + \frac{s_r\Delta t + \mathrm{e}^{-s_r\Delta t}}{s_r^2\Delta t}A_{pqr}x(n) + \\ + \frac{1 - s_r\Delta t\mathrm{e}^{-s_r\Delta t} - \mathrm{e}^{-s_r\Delta t}}{s_r^2\Delta t}A_{pqr}x(n-1) \end{aligned} \tag{6.134}$$

where $s_r$ is mode number $r$ and $\Delta t = 1/f_s$ is the sampling increment. From Equation (6.134) we can derive the numerator digital filter polynomial

$$n_{pqr} = \frac{A_{pqr}}{s_r^2\Delta t}[-s_r\Delta t - 1 + \mathrm{e}^{s_r\Delta_t}, 1 + s_r\Delta t\mathrm{e}^{s_r\Delta t} - \mathrm{e}^{s_r\Delta t}] \tag{6.135}$$

and the denominator digital filter polynomial

$$d_{pqr} = [1, -\mathrm{e}^{s_r\Delta t}] \tag{6.136}$$

which are defined using the MATLAB/Octave nomenclature here.

The total digital filter coefficients, including the complex conjugate term in Equation (6.132), can be conveniently computed from the numerator and denominator polynomials $n_{pqr}$ and $d_{pqr}$ using polynomial multiplication, which as we know is equivalent to convolution. The total numerator polynomial $A(n)$ is thus

$$A = \mathrm{Re}[d_{pqr} * d_{pqr}^*] \tag{6.137}$$

and

$$B = 2\mathrm{Re}[n_{pqr}] * \mathrm{Re}[d_{pqr}] + 2\mathrm{Im}[n_{pqr}] * \mathrm{Im}[d_{pqr}] \tag{6.138}$$

where in both equations $*$ stands for convolution and the superscript $^*$ stands for complex conjugate. The similarity between the symbols should cause no problem. The details of these calculations are found in Ahlin *et al.* (2006).

Although the formulation presented here has been formulated for displacement output, it can also be extended to velocity output. The modal superposition solution does not exist for acceleration output, because the degree of the numerator polynomial in this case is equal to the degree of the denominator polynomial. The displacement or velocity output can, however, conveniently be differentiated, e.g., by the methods presented in Section 3.4.3 to produce velocity or acceleration output with very high accuracy.

The method presented here is, of course, not free of error. The error of most concern with the filters presented here, is a bias error. The error is, however, relatively easy to calculate, as the filter coefficients, once computed, can be used to synthesize the frequency response using the MATLAB/Octave command **freqz** as was explained in Section 3.3.2 This frequency response can then be compared with the true frequency response of the system, based on the synthesis formulas from Chapter 6. The bias error is only dependent on the sampling frequency, and is usually negligible if at least 10 times oversampling is used for the filters.

For an example of the filters used here, see e.g., Section 14.5.1, where the presented method is used to calculate input/output data of a known 2DOF system. The method will be used extensively later in this book.

## 6.6 Chapter Summary

In this chapter we have seen how vibration problems in discrete mechanical systems with masses, viscous dampers, and springs can be solved. Newton's equation can be formulated in matrix form for general MDOF systems using mass, damping and stiffness matrices, and a displacement vector $\{u\}$, as

$$[M]\{\ddot{u}\} + [C]\{\dot{u}\} + [K]\{u\} = \{F(t)\}. \tag{6.139}$$

It was shown that the undamped system has eigenvectors, called *normal modes*, where all points move in phase or out of phase, which means that the mode shapes are real (can be described by real numbers with + or - sign). The *undamped natural frequency*, $\omega_r$, of each mode is found by taking the square root of the eigenvalues. For a system with $N$ degrees-of-freedom ($N$ masses) it was shown that we get $N$ normal modes.

Normal modes were then shown to diagonalize the mass and stiffness matrices into *modal mass* and *modal stiffness* by the weighted orthogonality criteria

$$[\Psi]^T[M][\Psi] = \lceil M_r \rfloor \tag{6.140}$$

where we use the mode shape matrix $[\Psi]$ with each mode shape as a column, and the diagonal matrix $\lceil M_r \rfloor$ on the diagonal has the modal mass $m_r$ of mode $r$. Similarly we showed that the diagonal modal stiffness matrix $\lceil K_r \rfloor$ is obtained by

$$[\Psi]^T[K][\Psi] = \lceil K_r \rfloor \tag{6.141}$$

and further $\omega_r^2 \lceil M_r \rfloor = \lceil K_r \rfloor$.

The *modal coordinates* or *principal coordinates*, $\{q\}$, defined by $\{u\} = [\Psi]\{q\}$ were used to decouple the coupled equations into uncoupled SDOF systems with the modal mass and stiffness values.

For proportionally damped systems where the viscous damping matrix is a linear combination of the mass and stiffness matrices, i.e., $[C] = a[M] + b[K]$, it was then shown that the mode shapes are the same as the normal modes, and complex poles can be obtained in the modal coordinates, by first calculating the diagonal damping matrix

$$[\Psi]^T[C][\Psi] = \lceil C_r \rfloor \quad (6.142)$$

and then obtaining the poles by

$$s_r = \sigma_r \pm j\omega_{dr} = -\zeta_r\omega_r \pm j\omega_r\sqrt{1-\zeta^2} \quad (6.143)$$

where

$$\omega_r^2 = \frac{k_r}{m_r} \quad (6.144)$$

is the *undamped natural frequency* of mode $r$, and the relative damping is

$$\zeta_r = \frac{c_r}{2\sqrt{m_r k_r}} \quad (6.145)$$

i.e., the poles are calculated for an SDOF system with the modal mass, damping, and stiffnesses.

For the general damping case, we then showed that a state-space formulation can be used which leads to similar complex conjugate pairs of poles, and complex mode shapes, i.e., all points no longer necessarily move exactly in and out of phase; we may have traveling waves.

For any type of damping we showed that a frequency response matrix can be formulated as

$$[H(j\omega)]\{F(t)\} = \{u(t)\} \quad (6.146)$$

and we showed that this frequency response matrix can be decomposed by using the residue matrix and poles by the *modal superposition* equation, where each frequency response $H_{pq} = X_p/F_q$ is

$$H_{pq}(j\omega) = \sum_{r=1}^{N} \frac{A_{pqr}}{j\omega - s_r} + \frac{A^*_{pqr}}{j\omega - s^*_r} \quad (6.147)$$

where the residues, $A_{pqr}$ are composed of a *modal scaling constant*, $Q_r$ for each mode $r$, and the mode shape coefficients in the two points $p$ and $q$,

$$A_{pqr} = Q_r\psi_{pr}\psi_{qr} = \frac{1}{j2\omega_{dr}m_r}\psi_{pr}\psi_{qr}. \quad (6.148)$$

We discussed that a common mode shape scaling method, to unity modal A, is equal to setting the modal scale constant to $Q_r = 1$, which obviously simplifies the frequency response synthesis as no scaling needs to be done in the modal superposition.

Finally we presented an accurate method of calculating forced response in the time domain, using a method which defines digital filters for each SDOF system in a modal superposition formulation. The method is very convenient to produce accurate time data for use in simulation examples of noise and vibration signals.

## 6.7 Problems

Many of the problems following are supported by the accompanying ABRAVIBE toolbox for MATLAB/Octave and further examples which can be downloaded with the toolbox. If you have not already done so, please read Section 1.6, and follow the instructions to download this toolbox together with example files.

**Problem 6.1** *Show that the residues $A_{pqr}$ for each mode can be split as given in Equation (6.108), by applying the hand-over method from Section 2.6.1 to Equation (6.107) and using the expression for the poles*

$$s_r = \sigma_r + j\omega_{dr}. \tag{6.149}$$

**Problem 6.2** *Find the eigenvalues, poles, natural frequencies (in Hz) and mode shape vectors for a 2DOF system like the one in Figure 5.13 for which $m_1 = 2$ and $m_2 = 5$ kg, $k_1 = k_2 = k_3 = 10^4$ N/m, and proportional damping with $[C] = 5[M] + 0.0001[K]$.*

*Use manual calculations to find the characteristic equation and use the MATLAB/Octave* **roots** *command to find the roots of this equation. Use both the formulation according to Equation (6.13) and the formulation according to Equation (6.23) and verify that you get exactly the same results. This exercise is valuable to learn where the two eigenvalue formulations diverge and where they come together again.*

**Problem 6.3** *Use the MATLAB/Octave command* **eig** *to solve Problem 6.2.*

**Problem 6.4** *Synthesize frequency response functions between all degrees-of-freedom for the system in Problem 6.2 using the formulation in Equation (6.104). Write a MATLAB/Octave script to calculate the matrix inverse at every frequency, and plot the frequency responses in mobility form overlaid in one plot. How many FRFs do you see? Explain why!*

**Problem 6.5** *Use MATLAB/Octave to formulate the solution to Problem 6.2 using the state-space form and verify that you get the same results as in Problem 6.2.*

**Problem 6.6** *Synthesize frequency response functions in mobility form between all degrees-of-freedom for the system in Problem 6.2 using the formulation in Equation (6.110) and the results from Problem 6.5. Write the equations calculating each residue, and then write a MATLAB/Octave script to plot the frequency responses. Compare with the results from Problem 6.4.*

**Problem 6.7** *Add 5 to element (1,1) of [C] in Problem 6.2 and solve the poles and mode shapes of the nonproportionally damped system. Check how the poles are affected and how complex the new mode shapes become.*

**Problem 6.8** *Use the forced response method of Section 6.5 to create the beating and distorted signals mentioned in Section 5.2.5. Check how long it takes for the transient to die out. Then reduce the damping of the system to 5% and rerun the simulation, and see how long it takes for the transient to disappear then.*

*(Hint: You can use the accompanying toolbox command* **timefresp** *to create the displacement output of the SDOF system.)*

## References


Ahlin K, Magnevall M and Josefsson A 2006 Simulation of forced response in linear and nonlinear mechanical systems using digital filters *Proc. ISMA2006, International Conference on Noise and Vibration Engineering*, Catholic University, Leuven, Belgium.

Brandt A and Ahlin K 2003 A digital filter method for forced response computation *Proc. 21st International Modal Analysis Conference*, Kissimmee, FL.

Craig RR and Kurdila AJ 2006 *Fundamentals of Structural Dynamics*. John Wiley & Sons, Ltd.

Ewins DJ 2000 *Modal Testing: Theory, Practice and Application* 2nd edn. Research Studies Press, Baldock, Hertfordshire, England.

Heylen W, Lammens S and Sas P 1997 *Modal Analysis Theory and Testing* 2nd edn. Catholic University Leuven, Leuven, Belgium.

Inman D 2007 *Engineering Vibration* 3rd edn. Prentice Hall.

Kozin F and Natke HG 1986 System-identification techniques. *Structural Safety* **3**(3-4), 269–316.

Maia N and Silva J (eds) 2003 *Theoretical and Experimental Modal Analysis*. Research Studies Press, Baldock, Hertforsdhire, England.

Rao S 2003 *Mechanical Vibrations* 4th edn. Pearson Education.

# 7

# Transducers for Noise and Vibration Analysis

A large variety of sensors and instrumentation for measurements are in use in the various fields of noise and vibration measurement and analysis. A book such as this can by no means include any comprehensive description of all these sensor types. However, it is still reasonable to give a short description of the most common types of sensors, to serve as an introduction to the newcomer to this field. In this chapter we will therefore present some of the most commonly used transducers for vibration measurements, the piezoelectric transducer, and the most common type of microphone used for precision acoustic measurements, the condenser microphone. At the end of the chapter there is a section also on electromagnetic shakers used for producing the excitation force for measurements of frequency response. Excellent information on sensor technology can be obtained from the manufacturers, who are the real experts on their own sensors, of course. There is a lot of information available on Internet at the many manufacturers' web sites.

To measure particularly vibration accurately, but sometimes also noise, is often somewhat of a challenge and care must be taken to assure good quality measurements. Becoming a good experimentalist is certainly very difficult through reading a book. If you are new to this field and are going to make vibration measurements, I encourage you to start by making a number of (seemingly) simple measurements on setups where you know the correct answer. In this chapter I have devoted Section 7.12 to some comments on good measurement procedures that can help you obtaining good measurement practice.

A word about the nomenclature used in this chapter is perhaps necessary. Transducers and sensors are both common names for devices that transform some physical entity into an electrical voltage that can readily be measured by an instrument. Although the two names are sometimes used with slightly different meanings, I will use the two as synonyms, as is quite common among measurement engineers.

## 7.1 The Piezoelectric Effect

The most common transducers used in vibration analysis are based on the piezoelectric effect. This effect is common, occurring naturally in, for example, quartz crystals, where a charge, $q$, is produced across the crystal when a force is applied to it. The charge is proportional to the force, $F$, such that

$$q = S \cdot F \tag{7.1}$$

*Noise and Vibration Analysis: Signal Analysis and Experimental Procedures,* First Edition. Anders Brandt.

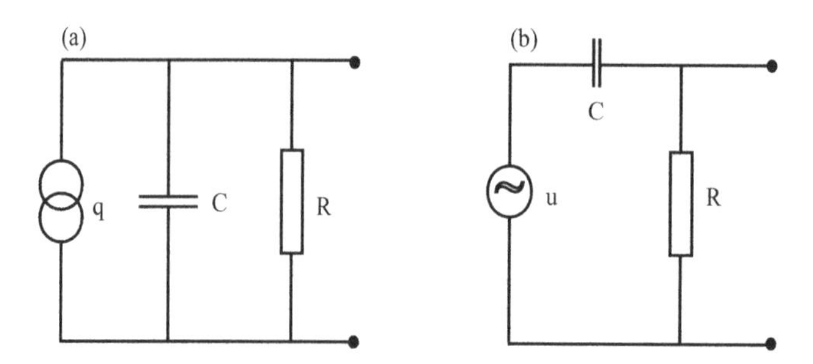


**Figure 7.1** Charge and voltage models for piezoelectric transducers

where $S$ is called the *sensitivity factor*. The direction-dependent piezoelectric effect is usually explained through the existence of electric dipoles in the material. Under certain conditions these dipoles can be forced into a common direction, which then becomes the direction of the material's piezoelectric effect.

There are two main types of materials used in piezoelectric transducers for vibration measurements. Traditionally, ceramic materials artificially polarized to obtain piezoelectric characteristics were used. This type of crystal is still used in accelerometers of charge mode type. The most common piezoelectric material type today in IEPE transducers is quartz crystals. The quartz crystals have relatively low piezoelectric (charge) sensitivity, which in the past made them unsuitable, but today this is overcome by adding signal conditioning inside the sensor, see Section 7.3.

An electrical model of piezoelectric transducers can be formulated in two ways. First, the transducer can be described by a charge model, where the transducer signal is seen as a charge generator coupled in parallel with a resistance and a capacitor. Alternatively, the transducer can be described by a voltage model, where the transducer signal is an electric voltage and the resistor and capacitor are instead in series with the voltage generator, see Figure 7.1. Both of these models are frequently used.

## 7.2 The Charge Amplifier

Since the traditional so-called *charge mode* piezoelectric transducer produces charge, and not voltage, the measured signal must be converted to a voltage before we can measure it with normal measuring instruments. The conversion is accomplished using a *charge amplifier*, see Figure 7.2. In the figure, $C_c$

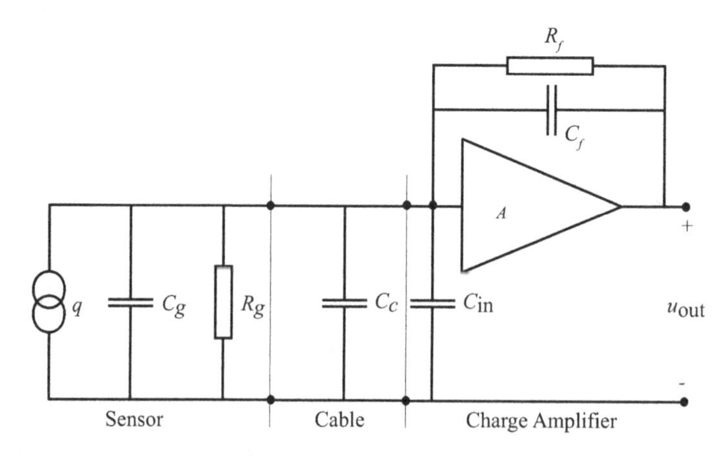


**Figure 7.2** Circuit diagram of the charge amplifier

symbolizes the cable capacitance, while $R_f$ and $C_f$ are feedback resistance and capacitance, respectively. $R_g$ and $C_g$ are the transducer's inner resistance and capacitance, respectively, and $C_{in}$ is the charge amplifier input capacitance. If we calculate the amplification, it can be shown (Serridge and Licht 1986) that

$$\frac{u_{out}}{q} = -\frac{1}{\left(1+\frac{1}{A}\right)C_f + \frac{1}{A}C_s} \tag{7.2}$$

where $C_s$ is the net input capacitance, that is, $C_s = C_g + C_c + C_{in}$. For large values of the amplification, $A$, the total amplification consequently can be written approximately as

$$\frac{u_{out}}{q} \approx -\frac{1}{C_f}. \tag{7.3}$$

What is particularly interesting in Equation (7.3) is that the output signal from the charge amplifier is independent of the capacitances of the transducer and cable, and only dependent upon the characteristics of the amplifier. However, the signal from a charge mode transducer is very low, and therefore susceptible to noise. Using such a sensor, cables therefore need to be kept as short as possible.

Charge amplifiers are relatively expensive to manufacture and the signal from the transducer is sensitive to disturbances. Another disadvantage with this type of transducer is the so-called *triboelectric effect*. This effect is due to changes in cable capacitance resulting from bending the cable, giving rise to a noise signal. Special cables must therefore be used to reduce this effect. Even then, great caution must be observed so that the cable does not move during the measurement, usually by taping down the cable.

Among the advantages of charge mode transducers, compared with the more common IEPE sensors we are going to introduce in the next section, are their ability to handle higher operating temperatures, and that they can be used over a large dynamic measurement range. The charge amplifier can easily be reconfigured for different measurement ranges by changing the feedback capacitor, which according to Equation (7.3) controls the amplification. Charge amplifiers therefore often have an amplification range of many decades.

## 7.3 Transducers with Built-In Impedance Converters, 'IEPE'

To reduce the cost and increase signal quality when measuring with piezoelectric transducers, the US transducer manufacturer Kistler in the 1960s developed the built-in impedance converter which led to a breakthrough in transducer technology in the 1980s. This method includes a built-in, semiconductor-based impedance converter as in Figure 7.3. The upper transistor is a CMOS-FET, a field-effect transistor with extremely high input impedance. Together with the bipolar transistor it forms an impedance

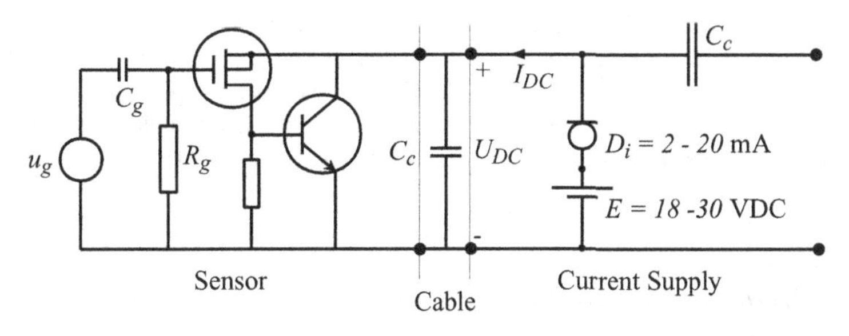


**Figure 7.3** Circuit diagram of the IEPE principle and the current supply providing signal conditioning for an IEPE sensor

**Table 7.1** Common names for the technique of built-in impedance conversion in piezoelectric transducers, for a few of the largest manufacturers. Other manufacturers use other names. ICP is currently the most common name used by analyzer manufacturers, although IEPE is starting to take over

| Manufacturer | Brand name |
|---|---|
| Brüel & Kjær | Deltatron™ |
| Dytran Instruments | LIVM™ |
| Endevco | Isotron™ |
| KistlerInstrument Corp. | Piezotron® |
| PCB Piezotronics Inc. | ICP® |

converter, which converts the transducer's output voltage, $u_g$, with a high impedance, to a voltage with low impedance, as described below.

As mentioned earlier, this type of transducer is usually based on quartz crystals which are less expensive than the previously used ceramic crystals. The lower sensitivity of the quartz crystal material can be compensated for by the amplification in the built-in impedance converter. Depending upon manufacturer's brand names, this technique is known by a number of different names, see Table 7.1. Recently a new name, IEPE (Integrated Electronics Piezo Electric) has been introduced as a 'manufacturer neutral' name for the technology.

The IEPE impedance converter, which has become a *de facto* standard, is based on the converter being fed with a constant current of between 2 and 20 mA. The input current to the amplifier turns it on and produces a DC 'working point' voltage, $U_{DC}$, of typically 9–12 volts. The measurement signal from the transducer is superimposed as a variation on this DC voltage. The obtained alternating current (AC) voltage from the transducer, and with that the calibration factor, are independent of the measurement current. That said, it is important that the voltage $E$ is sufficiently large so that it can generate the constant current, to ensure the measured signal is not clipped. Also it should be mentioned that a weak point with the IEPE 'standard' is that unfortunately the current needed for proper operation of different IEPE sensors is not firmly standardized. Some IEPE sensors perform badly for low currents, which causes problems when using some measurement systems with built-in current supply, because instrument manufacturers sometimes design the built-in current supply with rather low current to keep power consumption and heat down. Therefore, one must carefully check the manufacturer's specifications and verify that a particular sensor can work properly with a particular instrument.

The circuit feeding the transducer with built-in impedance conversion, is quite simple, see Figure 7.3. A battery or stabilized voltage, $E$, is connected to a current diode, $D_i$, which ensures that a constant DC current flows through it. A so-called coupling capacitor, $C_c$, usually of tantalum type, ensures that the direct current flows to the transducer and not to the measuring instrument. An indicator device is often found on the current supply unit. The indicator measures the DC voltage $U_{DC}$. A voltage of approximately 9–12 volts indicates that the unit has contact with the impedance converter. A $U_{DC}$ voltage of 0 or $E$ volts indicates, on the other hand, either a short-circuit in the cable or an open connection (e.g., broken cable), respectively. These indications can be of considerable help when trouble-shooting measurement connections, and some measurement systems with built-in supply for this type of transducer also have an indicator in the software.

## 7.3.1 Low-frequency Characteristics

Piezoelectric transducers can never measure frequencies down to DC because of the piezoelectric sensitivity to charge (at DC there is no change in charge, i.e., electrical current). This is on the other hand rarely

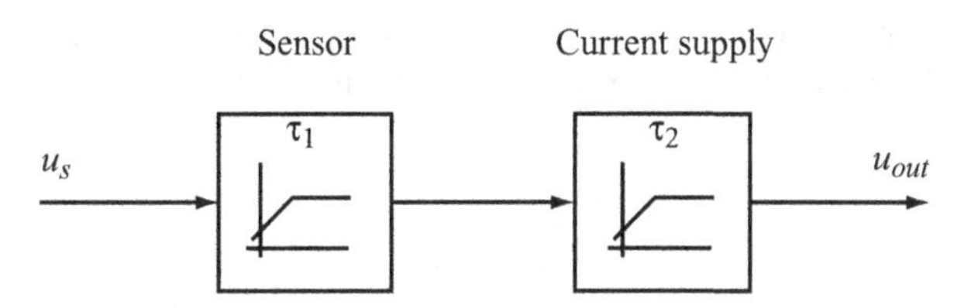


**Figure 7.4** Equivalent model for time constant for a piezoelectric transducer. The 'true' transducer signal $u_s$ passes first the time constant $\tau_1$, which is the transducer's inner time constant, and then the time constant $\tau_2$, which originates in the current supply unit or charge amplifier, depending on the sensor type

necessary for vibration analysis. When it is necessary, however, there are other types of transducers that can be used, for example piezoresistive or capacitive accelerometers. With well-chosen time constants, relatively low frequencies can be measured using piezoelectric transducers, down to approximately 0.1–0.5 Hz with maintained accuracy. In such cases it is essential to understand the time constants involved. An equivalent model for this purpose is shown in Figure 7.4, where $\tau_1 = R_1 C_1$ is the transducer's built-in time constant (dependent upon its inner capacitance and resistance), and $\tau_2 = R_2 C_2$ is the time constant of the current supply (or charge amplifier, if such a device is used). It consists of the coupling capacitor and the shunt resistance (actually in parallel with the input impedance of the measuring instrument, but the latter is normally much higher than the shunt resistance).

The total time constant resulting from the cascade coupling of circuits with two highpass filter time constants can be approximated by

$$\frac{1}{\tau_{tot}} \approx \frac{1}{\tau_1} + \frac{1}{\tau_2} \tag{7.4}$$

which can also be written as

$$\tau_{tot} \approx \frac{\tau_1 \tau_2}{\tau_1 + \tau_2} \tag{7.5}$$

which we recognize as a formula similar to resistors in parallel coupling, or springs in serial coupling, whichever is most familiar.

In the frequency domain, for a simple high-pass filter with an RC link, the lower frequency limit is given by

$$f_c = \frac{1}{2\pi\tau} \approx \frac{0.16}{\tau} \tag{7.6}$$

where $f_c$ is the cutoff frequency ($-3$ dB frequency) in Hz and $\tau$ is the time constant in seconds. In most cases, one of the time constants is much smaller than the other, and the smaller of the two hence determines the total time constant. This is particularly the case when using measurement hardware with built-in IEPE supply. The built-in coupling capacitor is generally smaller than that used in most external current supply units, which gives a higher frequency limit when using a built-in supply. In cases where the time constant needs to be long, for example when measuring slow pulses (pulse times of 50 ms or more), then an external current supply should be considered.

### *7.3.2 High-frequency Characteristics*

The high-frequency limit of most vibration measurements is due to sensor or mounting limitations such as accelerometer resonance frequency or stiffness in the attachment point. However, when using IEPE sensors with very long cable lengths, the cable capacitance and/or the power supply drive current can introduce a frequency limit. According to accelerometer manufacturers, the high-frequency limit when using IEPE sensors is affected by the power supply drive current, $I$ in [A], the cable capacitance per meter in [C/m], $C$, the cable length in [m], $L$, and the peak voltage of the sensor, $V_p$ in [V], by the equation

$$f_{\max} = \frac{I}{2\pi V_p C L}. \quad (7.7)$$

This equation shows that if long cables are required, the drive current may have to be increased. Normally, for cables used for IEPE sensors, no special care has to be taken for cable lengths shorter than at least 20 meters (60 feet).

### *7.3.3 Transducer Electronic Data Sheet, TEDS*

A recent standard, (IEEE 1451.4 2004), has brought a new concept to modern instrument hardware. The TEDS standard allows the sensor manufacturer to build a small computer chip into the sensor which stores information about the sensor, such as its sensitivity (calibration factor), type, etc. Many modern data acquisition systems can read this sensor information, which eliminates the possibility of making errors when entering information about the sensor into the measurement systems. Most IEPE sensors can be ordered with TEDS circuits installed.

## 7.4 The Piezoelectric Accelerometer

The piezoelectric accelerometer is fundamentally designed according to the left-hand part of Figure 7.5, which illustrates the pressure mode design. At the bottom the sensor consists of a stiff base, which ideally should follow the motion of the measured object. On the base the silicon crystal and then the seismic mass are glued. The mass is preloaded as indicated in the left-hand illustration in Figure 7.5 so that there is always a positive force onto the crystal within the operating range of the accelerometer. When the base of the accelerometer is subjected to acceleration, the mass generates a reaction force onto the crystal, which is proportional to the acceleration according to Newton's second law, $F = ma$. The sensitivity, $S_a$, expressed in [pC/ms$^{-2}$] for a charge mode accelerometer, or in [mV/ms$^{-2}$] for an IEPE accelerometer, therefore depends both on the crystal's charge sensitivity and on the size of the seismic mass.

The pressure mode sensor suffers from several drawbacks, such as a high base strain sensitivity and relatively high temperature sensitivity, see Section 7.4.4. In order to avoid some of those negative properties, the *shear mode* design has been developed, which is illustrated on the right-hand side of Figure 7.5. The basic principle of the shear mode design is that the piezoelectric crystal is mounted so that it is submitted to a shear force from the seismic mass instead of a pressure force. The seismic mass is

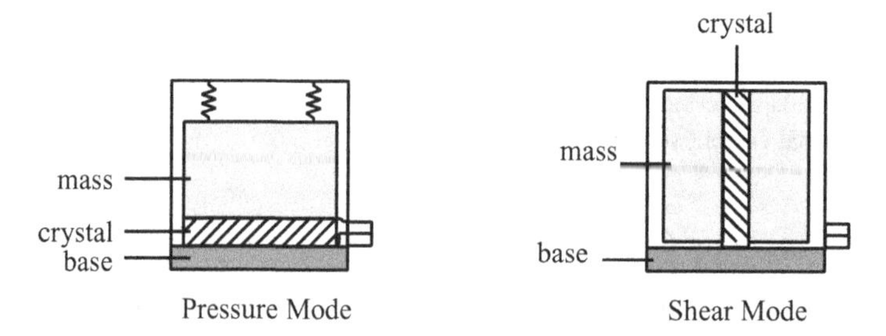


**Figure 7.5** Principle design of pressure mode type (left) and shear mode accelerometers (right). The shear design gives several advantages, such as lower transverse sensitivity and base strain sensitivity, see Section 7.4.4

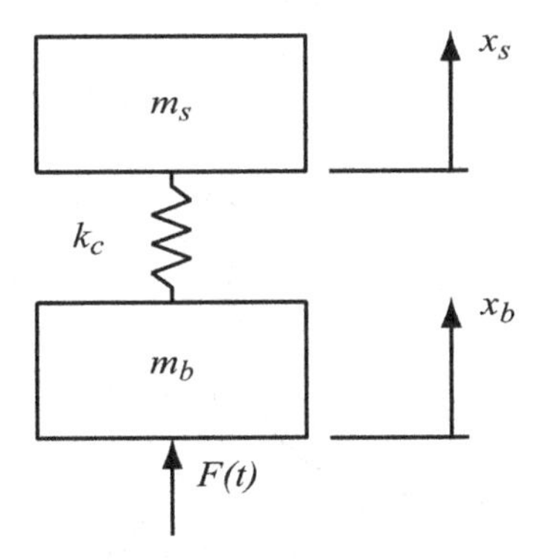


**Figure 7.6** Equivalent model of the piezoelectric accelerometer. $m_b$ is the base mass, $m_s$ the seismic mass, and $k_c$ the stiffness of the silicon crystal

furthermore usually divided into several pieces, which are mounted around the crystal. The advantages with this sensor design, is first of all that piezoelectric crystals are usually more sensitive to shear force than pressure force. Thus a shear mode sensor is usually more sensitive than a pressure mode sensor with equal seismic mass. Furthermore, the design with several seismic masses mounted around the crystal, results in less transverse sensitivity, see Section 7.4.4. Because of the reduced crystal area in contact with the base in the shear design, this sensor type will also exhibit less sensitivity to base strain.

### *7.4.1 Frequency Characteristics*

In order to understand the frequency characteristics of the accelerometer, we shall study the simplified, equivalent model of Figure 7.6. In this model we reduce the accelerometer to the mass of the base, the seismic mass, and the stiffness of the silicon crystal. This model contains no damping, which works quite well since the accelerometer is built with as little damping as possible, to minimize phase error.

The model in Figure 7.6 gives rise to a resonance frequency, which influences the output signal from the transducer. If we set up Newton's equations for the seismic mass we have

$$m_s \ddot{x}_s = -k(x_s - x_b). \tag{7.8}$$

If we assume harmonic motion, i.e., observe each frequency independently, we get

$$\begin{aligned} x_s &= X_s \sin(wt) \Rightarrow \ddot{x}_s = -\omega^2 X_s \sin(wt) \\ x_b &= X_b \sin(wt) \end{aligned} \tag{7.9}$$

Substituting Equation (7.9) into Equation (7.8) yields

$$-\omega^2 m X_s = -k_c (X_s - X_b) \tag{7.10}$$

This equation can be rewritten as

$$\frac{x_s}{x_b} = \frac{\ddot{x}_s}{\ddot{x}_b} = \frac{\frac{k_c}{m_s}}{\frac{k_c}{m_s} - \omega^2} = \frac{1}{1 - (\omega/\omega_m)^2} \tag{7.11}$$

where

$$\omega_m^2 = \frac{k_c}{m_s}. \tag{7.12}$$

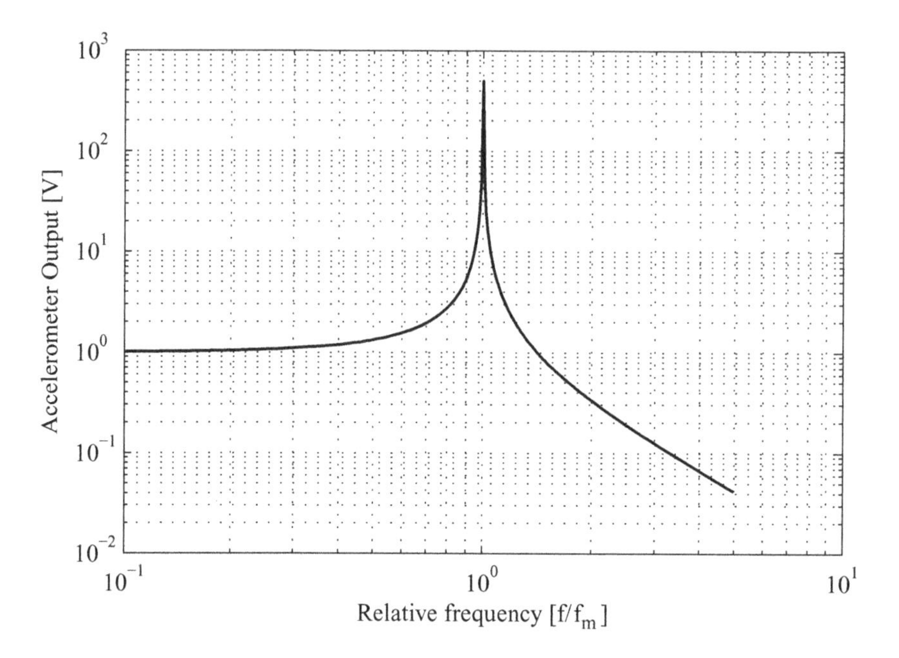


**Figure 7.7** Output signal from an ideally mounted accelerometer as a function of frequency, when a constant sinusoidal acceleration is applied at each frequency, according to the model in Figure 7.6. As seen in the figure the output signal displays a sharp peak at the mounted resonance frequency. In the plot the frequency axis is normalized to the mounted resonance frequency

From Equation (7.11) it follows that when the base of the accelerometer is exposed to a constant acceleration we obtain a resonance at the frequency $f_m$, where

$$f_m = \frac{1}{2\pi}\sqrt{\frac{k_c}{m_s}}. \tag{7.13}$$

which is called the accelerometer's *mounted resonance frequency*.

In order to measure $f_m$ experimentally for a specific accelerometer, it is mounted onto a rigid mass, which is subjected to constant sinusoidal acceleration of increasing frequency. With this kind of measurement, we obtain an output signal from the accelerometer according to Figure 7.7.

As seen in Figure 7.7, the accelerometer signal rises sharply when the frequency approaches the (mounted) resonance frequency, $f_m$. Accelerometers are usually designed with very low damping, in order to reduce the phase error, which could otherwise become significant, see Figure 5.3. Normally, a frequency range up to $f_m/4$ is used for a maximum error of approximately 5%, or $f_m/3$ for a maximum error of approximately 10%. The output signal follows the formula

$$a_{out} = \frac{a_{true}}{1 - (f/f_m)^2} \tag{7.14}$$

which of course is not exactly correct all the way up to the resonance frequency, since Equation (7.14) does not contain any damping, but the formula works well for all usable frequencies.

### 7.4.2 Mounting Accelerometers

Various methods may be used when mounting the accelerometer. The most common are screws, glue, wax, and magnets. Many types of glue are used: rigid glue such as dental cement, and softer glue such as superglue or even glue from a glue gun. When accelerometers are mounted this way, the mount itself will naturally have a certain stiffness. This 'spring', along with the accelerometer's mass, gives rise to a resonance which is often lower than the previously mentioned ideally mounted resonance frequency, thereby limiting the measurement range. It is impossible to predetermine where this resonance will be, since it depends on the measured object's stiffness at the mounting point, which influences the mass $m_b$ according to the above model. A very useful method of verification of the usable frequency range of a particular accelerometer will be discussed in Section 7.12.

When mounting an accelerometer it is very important to ensure that the surface is clean and smooth. The smallest bit of dirt or surface roughness may give rise to elasticity in the mounting, which can seriously influence the measurement. When mounting with screws either the fingers alone should be used, or even better a torque wrench. Suitable torque is normally given in the user's instructions of the transducer. When removing a transducer, it is also very important, especially when using glue, that it is twisted loose, that is, turned in the same plane as that on which it is mounted. All accelerometers designed for glue mounting have a six-sided base (like a nut), so that the transducer may be twisted loose without damaging the casing.

### 7.4.3 Electrical Noise

It should be noted, particularly when using IEPE transducers, that these often have their housing connected to the shield of the cable. Power line 50 Hz (or 60 Hz) noise can easily arise during measurements if there is a difference in electrical potential between the points where the accelerometers are mounted, and if the transducer is mounted on conductive material. As an initial measure, the transducer should be insulated from the measured object using a special insulating plate. There are also some transducers with insulated bases, often consisting of an anodic coating. This coating can sometimes be scratched off, after which it no longer insulates. In environments especially susceptible to noise, special IEPE transducers with insulated housings that are galvanically separated from the transducer itself should be used. These transducers, however, require special cables with two conductors, and are usually heavier industrial transducers made for measuring on foundations and other stiff objects.

To avoid noise when measuring acceleration, particularly with charge mode transducers, another important consideration is that the cables are fixed so as not to hang loose and vibrate. Therefore, cables should be taped down on the structure as tightly as possible. Furthermore, one should always strive for as short cables as possible between the transducer and either the charge amplifier or the current supply unit (or the measurement system if it has a built-in current supply). Another tip when using grounded measurement equipment is to make sure that the same mains outlet supplies all units in the measurement setup. Otherwise potential differences (voltage differences) can exist because different outlets have different ground potentials.

### 7.4.4 Choosing an Accelerometer

In addition to what has already been mentioned about frequency range, etc., of the accelerometer, we shall discuss a few other characteristics of the accelerometer necessary to understand in order to select a proper accelerometer for a particular measurement. An ideal accelerometer should be as light as possible in order not to affect the vibrations the sensor is intended to measure. If the mass of the accelerometer is too high a phenomenon usually referred to as *mass loading* occurs. Mass loading is often misunderstood to be related to the weight of the accelerometer with respect to the weight of the measurement object.

Since mass loading is a dynamic phenomenon, however, this is not correct. Instead, it is the *dynamic stiffness* at the point where the accelerometer is mounted which affects how much mass the accelerometer can have without (seriously) affecting the vibrations. It is often surprising how light an accelerometer has to be in order not to give a mass loading effect. My recommendation is therefore to test experimentally if mass loading is a problem, if there is any doubt, see Section 7.12.

*Temperature sensitivity* manifests itself in an increasing sensitivity with increasing temperature, and is given in [%/°C]. How high the temperature sensitivity is depends much on the accelerometer design, and the shear mode type described in Section 7.4 usually has lower temperature sensitivity than other designs. Great caution must be observed if temperature fluctuations occur while measuring with an accelerometer.

Accelerometers are also sensitive to *base strain*, i.e., to bending of the base. For example, if we mount an accelerometer on a bending beam, then the accelerometer will give a signal due to the deformation of the base making the silicon crystal bend. Also the base strain sensitivity is typically lower for accelerometers based on the shear mode design than for pressure mode accelerometers. The base strain sensitivity is typically given in $\left[(\text{m/s}^2)/\mu\text{Strain}\right]$ ('microstrain') at approximately 250 $\mu$Strain (1 $\mu$Strain is a strain of 1 $\mu$m per m).

When an accelerometer is mounted on a measurement object, there will of course in many cases be vibrations in directions other than the direction intended to measure. These vibrations give rise to an undesired signal from the transducer. The sensitivity to cross-directional vibration is identified by what is called *transverse sensitivity*, which is given as a percentage. If the transverse sensitivity is 1% this indicates that a 100 m/s$^2$ acceleration in the cross-direction will produce an output as large as an acceleration level of 1 m/s$^2$ in the transducer's main direction. The transverse sensitivity depends on the transducer design and is lower for the shear mode type than for other accelerometer designs. The transverse sensitivity can be a significant cause of error in acceleration measurements where the acceleration levels are considerably higher in one direction than in other directions, as the following example shows.

**Example 7.4.1** *An example of a case where there are often considerably higher acceleration levels in one direction than in the other directions is on combustion engines, particularly with straight cylinders. On such objects, great care must be taken when interpreting the results. Consider, for example, an engine with 10 times higher vibration in the vertical direction than in the horizontal directions. Let us assume we use accelerometers with transverse sensitivity of (a common value) 5%. If we denote the low acceleration level in horizontal direction by A, the vertical vibration level will be* $10A$. *Then the output of the vertical accelerometer will be*

$$A_z = 10A + 0.05 \cdot A = 10.05A. \tag{7.15}$$

*In Equation (7.15) the error (the output compared with the true vertical acceleration of* $10A$*) is approx. 0.5%, which is negligible. But what happens with the output of the horizontal transducer? Its output will be*

$$A_x = 0.05 \cdot 10A + A = 1.5A. \tag{7.16}$$

*The error in the horizontal accelerometer is 50%! This example shows that even a seemingly small transverse sensitivity of 5% can easily cause a very large error when the acceleration levels are considerably different in different directions.*

*End of example.*

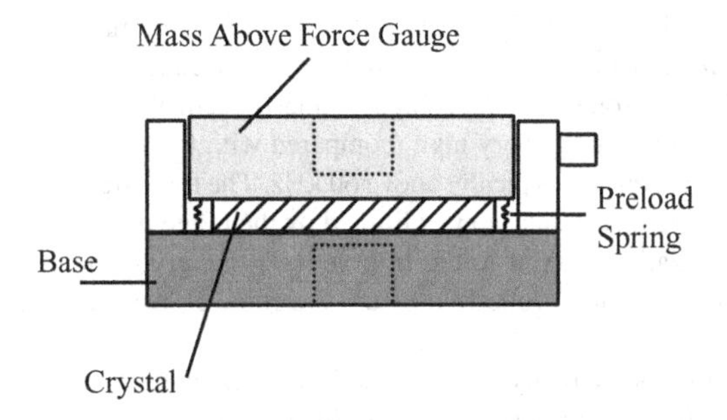


**Figure 7.8** Principle design of the piezoelectric force transducer. A preload is applied over the crystal so the net force is always a pressure over the crystal, illustrated by the two preload springs in the figure. The side marked 'Mass above force gauge' should be attached to the test structure

## 7.5 The Piezoelectric Force Transducer

Besides measuring motion (acceleration) it is often necessary in vibration analysis to measure dynamic force. For that purpose a piezoelectric force transducer is normally used, which has a principal design as illustrated in Figure 7.8. The design of the force transducer is relatively simple with a base and a small mass referred to as the *mass above the force gauge*, which are glued onto the piezoelectric crystal. To allow tensile (pulling) forces being measured by the transducer, the crystal is preloaded so that the force on the crystal is always a compressive force within the operating range of the transducer. The applied force is transferred over the mass above the force gauge directly to the silicon crystal. Because of the limited preload provided by the springs, a piezoelectric force transducer is often specified to allow more compression force than tension force.

Because of the direct contact between the silicon crystal and the base of the transducer, the force transducer is very sensitive to transverse forces. Therefore, a *stinger* (in British English 'rod') should always be used between the transducer and the shaker to eliminate any transverse forces. The stinger should be flexible in the transverse direction, and is often made of piano wire on which small threaded screws are glued or soldered for attachment to the shaker and the force transducer, see Figure 7.9. Stingers with different axial stiffness are required for various measurements, since the stinger itself must be stiff enough to transfer the force within the excitation frequency range, but not stiffer than necessary. Usually the stiffness can be adjusted by using stingers of different length. Practical aspects of the use of stingers will be discussed in Chapter 13.

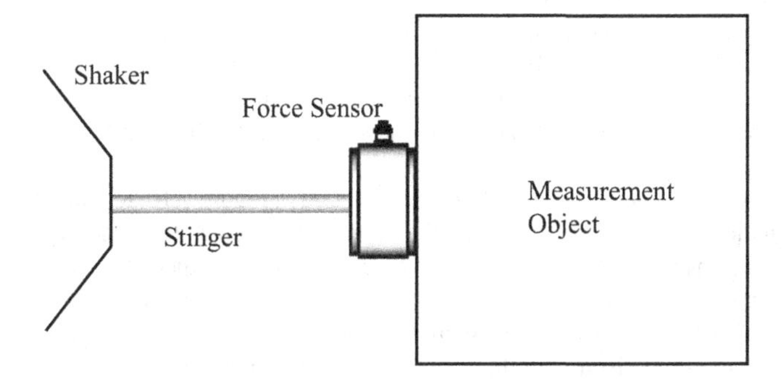


**Figure 7.9** When using a force transducer, transverse forces must be eliminated, which is accomplished by a so-called *stinger*

When excited with no load, the force transducer will display a curve similar to that for the accelerometer in Figure 7.7, with a resonance caused by the mass above the force gauge (see Figure 7.8) and the stiffness of the silicon crystal. This resonance is specified similarly with the accelerometer as the *mounted resonance without load* and is usually very high (compared with the frequency range you can measure accurately with the force transducer), typically above 60 kHz. The force transducer's operating frequency range is rarely limited by this factor; instead it is limited by the range in which the force applied through the force transducer is actually an axial force. In practice it is very difficult (but not impossible!) to excite a structure with a shaker and properly measure the force above, say, 1–2 kHz, see Section 7.12 and Section 13.9.

The force transducer does not actually measure the force applied to the measurement object. Instead, the force measured is the force acting on the crystal; this means that the mass above the force gauge is effectively added to the measurement object when the transducer is attached. It is therefore necessary that this mass is small. For the same reason it is essential to mount the correct side of the force transducer onto the measurement object, as otherwise the much heavier base mass is added to the structure and a large error results. Most transducers have some marking which shows the correct side to be attached to the object. Also see the discussion of mass loading in Section 7.12 for a discussion of the apparent mass of the structure.

Piezoelectric force transducers usually have relatively long time constants compared with accelerometers. This feature is necessary for accurate calibration, as force transducers are generally calibrated semi statically, i.e., by applying a known static force to the transducer. For example, if the transducer is quickly loaded with an accurately known mass, then the resulting, slowly decaying force signal can be measured before the time constant has generated too large an error.

## 7.6 The Impedance Head

The impedance head is a transducer that contains a force transducer and an accelerometer in a single housing. The name comes from the fact that, in the past, the transducer was used to measure mechanical point impedance, i.e., the ratio between (dynamic) force and velocity (the acceleration was integrated). Today, most measurements of frequency response are measured as mobility or accelerance, as we saw in Chapters 5 and 6, but the name of this transducer has remained.

When measuring point accelerance, that is, the ratio between acceleration and force at the same point, it is often impossible to place a force transducer and an accelerometer so that they measure in the same point. Sometimes, for example on thin plates, the transducers can each be placed on opposite sides of the plate. In many cases, however, the force sensor and accelerometer have to be mounted next to each other. This results in an error because of the difference in acceleration between the desired point where the force transducer is placed, and the point at which the accelerometer is possible to mount. This error depends on the difference in mode shape coefficients between the two points, as follows from the discussion in Section 6.4, and thus the error is larger for smaller measurement objects or for higher-order mode shapes where the wavelength of the mode shapes are shorter. The impedance head overcomes this problem.

A schematic illustration of an impedance head can be found in Figure 7.10, together with an equivalent model. As seen in the figure, both the force and the acceleration signals from this type of transducer, as those from separate transducers, have bias errors. One advantage with the impedance head is, however, that when measuring point accelerance, a part of these errors can be compensated for (Håkansson and Carlsson 1987). When measuring flexible structures where the mass above the force gauge, $m_f$ in the figure, is significant in comparison with the structure's apparent mass, the measured point accelerance can rather easily be compensated. The compensation is done by first measuring accelerance with the impedance head without load. The inverse of this measured accelerance, $H_{0a}(f)$, is then subtracted from

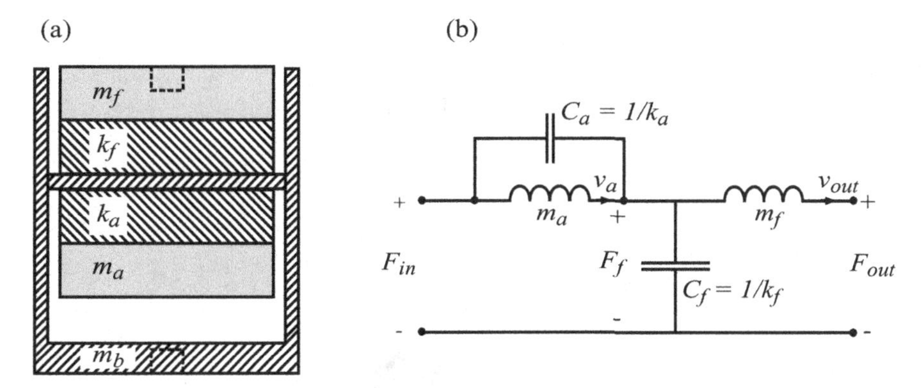


**Figure 7.10** Schematic illustration of the impedance head. (a) mechanical design, and (b) equivalent electric circuit. Force $F_{\text{out}}$ is the force to be measured, whereas the signal given by the transducer is proportional to force $F_f$ in the figure. If $m_f$ is small relative to the measurement object's dynamic mass, then the error is small. Since current in the impedance analog corresponds to mechanical velocity, currents in the figure are given as velocities instead of the acceleration signals actually produced by the transducers. The acceleration signal which the transducer thus gives corresponds to velocity $v_a$, which, as seen in the schematic, does not correspond to the desired acceleration $a_{out}$ (analogous to $v_{out}$). The error in the acceleration signal is analogous to the error of a regular accelerometer, see Section 7.4

the inverse of the measured accelerance of the structure, $H_a(f)$. The corrected accelerance, $H_a'(f)$, is thus calculated by

$$H_a'(f) = \frac{1}{\frac{1}{H_a(f)} - \frac{1}{H_{0a}(f)}}. \tag{7.17}$$

The drawback with impedance heads is that they are considerably taller than separate force transducers. This means that, even using good stingers, the frequency range over which the force will be in line with the impedance head is relatively low. At relatively low frequencies, the impedance head has a tendency to start to 'wobble'. For higher frequencies, where the impedance head would solve the problem of short wavelengths discussed above, one is still often obliged to use a force transducer with a lower form factor.

## 7.7 The Impulse Hammer

An impulse hammer is used to excite vibration, for example to measure frequency response when studying a mechanical system. It consists of a handle, a head, and a force transducer with an interchangeable tip, see Figure 7.11. Impulse hammers come in various sizes from that of a pen, with head mass of a few grams, to sledgehammers with 1.5 m handle and a head mass of 10 kg or more. In using the hammer, you make a distinct (but usually rather soft) impact, and the force transducer will measure the resulting force pulse; the harder the tip the shorter the pulse, and therefore the wider the frequency content. See Section 13.8 for more details on practical aspects of impulse excitation.

## 7.8 Accelerometer Calibration

Several different methods are used to calibrate accelerometers. The method giving highest accuracy, and therefore used by calibration laboratories, is so-called reference calibration, where the accelerometer is subjected to the same acceleration as another, accurately calibrated reference accelerometer. The latter

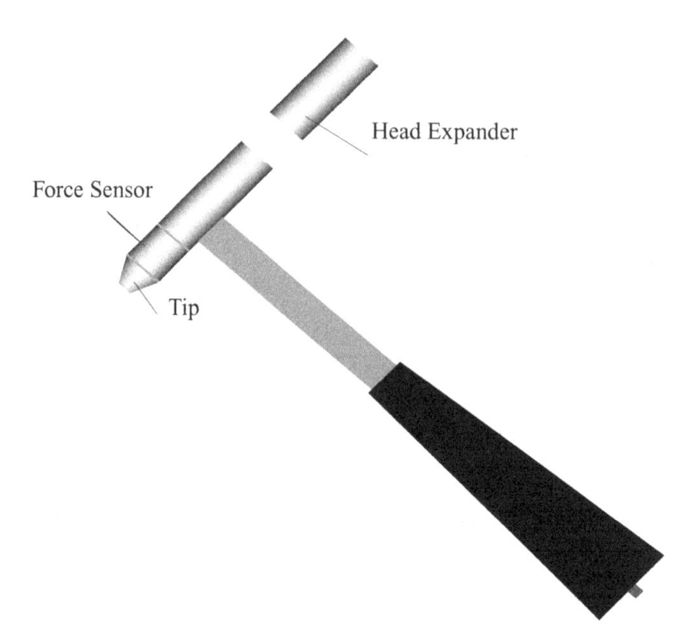


**Figure 7.11** The impulse hammer. The hammer consists of a handle, a head, and a force transducer with an interchangeable tip. The head mass can often be adapted to the measurement object by attaching an extra mass, increasing the head mass and giving more energy in the pulse (longer pulse for approximately same strength strike)

is in certain cases a laser interferometer, which has very high precision. The accelerometer must be calibrated at a specific frequency to provide traceability, and in Europe it is in most cases calibrated at 159.2 Hz, which corresponds to the angular frequency 1000 rad/s. The reason behind using this calibration frequency is that acceleration, velocity and displacement at this frequency are easy to convert, as they are related to each other by angular frequency. In the US, 100 Hz is often used as the calibration frequency, which is considered unsuitable in Europe as it corresponds to the first harmonic of the mains voltage, 50 Hz.

A similar method with somewhat less accuracy uses a calibrator, in this case a shaker with a precisely determined acceleration. This type of calibrator is available as a tabletop model or a practical hand-held model. To use this method of calibration the accelerometer is mounted on the calibrator. Then the level is measured on the same measurement system to be used later for the actual measurement. The measurement system is adjusted so that it gives the correct level. In this way the entire measurement chain is calibrated, which is good from an accuracy point of view. At the same time, all possible sources of error such as an incorrectly configured charge amplifier are eliminated.

A third method of accelerometer calibration can be used in two cases: when one has access to a well-calibrated force transducer, or when one is going to measure the frequency response between force and acceleration (usually with an impulse hammer). In the latter case neither of the transducers' sensitivities need to be known in advance. This method is usually called mass calibration, since it is based on the fact that a reference mass is known with high accuracy. Since Newton's second law states that $F = ma$, the accelerance of a measurement on a solid mass will be $a/F = 1/m$, independent of frequency. This frequency response should thus be a constant straight line, as long as the measurement is correct. The method is good since mass (weight) is relatively easy to measure with high accuracy, and it is applied as shown in Figure 7.12. See also Section 7.12 for a practical discussion of the use of mass calibration.

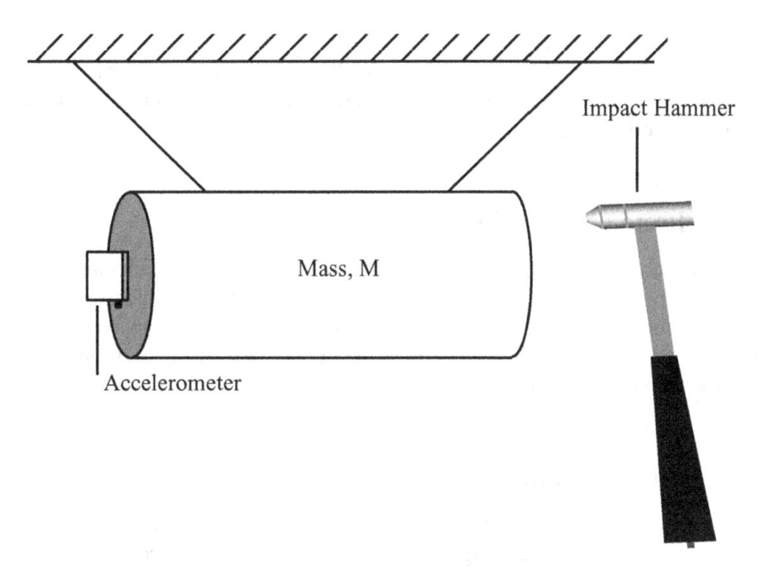


**Figure 7.12** Calibration of force transducer and accelerometer using a calibrated mass. The principle behind this calibration is that the measured frequency response between force and acceleration (acceleration/force) according to Newton's law $F = m \cdot a$ gives a constant value for all frequencies, equal to $1/m$. By adjusting the calibration factor of one of the measurement system channels so that this relationship is fulfilled, then all measurements are correct, with traceability to the mass

## 7.9 Measurement Microphones

Generally when measuring sound pressure, specially developed condenser microphones are used. These microphones exist in two types; externally polarized, or prepolarized. The principle behind the condenser microphone is that there are two plates, as in a capacitor, across which a DC voltage is applied. One plate works as the microphone diaphragm, and when it experiences air fluctuations, and consequently vibrates, the capacitance between the plates varies, as the distance between the plates varies. This variation in capacitance gives rise to a charge change, which is converted to a voltage in the microphone's preamplifier.

The two microphone types differ primarily in that, for the externally polarized microphone, the polarization voltage, usually 200 V DC, must be applied externally by a power supply. In the prepolarized microphone, the plates have already been charged during manufacturing, so that they do not require any power supply. Prepolarized microphones are less sensitive to noise, but have the disadvantage that they can give faulty measurements, for example when the diaphragm is pressed together with the charged plate. Externally polarized microphones are therefore most often preferred, since they either function correctly, or not at all, which will be evident during calibration.

Microphones have a certain *directivity*, that is, they give different measurement values depending on from which direction the sound wave hits the diaphragm. Most microphones used for acoustical measurements in Europe (according to IEC standards), compensate for the influence the microphone itself has on the surrounding sound field, so that the measured sound pressure matches as closely as possible the sound pressure at the measurement position without the microphone present. However, this compensation works only for sound waves directed straight onto the diaphragm. Microphones of this type, called free-field microphones, should be positioned directly toward the sound source. A different type of microphone is usually used in the US (according to ANSI standards), called diffuse-field microphones. This type of microphone is designed to give correct sound pressure when the sound field is diffuse, i.e., the sound comes from all directions.

Depending upon which frequency range shall be measured, and to some degree which sound pressure range, microphones of differing sizes are used; the smaller the microphone the higher the frequencies which can be measured. The most common microphones are 1/2 inch in diameter, and also 1 inch, 1/4 inch and 1/8 inch are common.

## 7.10 Microphone Calibration

Microphones should always be calibrated before (and often after) every measurement. Microphone calibration is done by placing the microphone into a microphone calibrator, which is a tube containing an accurately known sound source. It is important that the ring sealing the microphone to the tube is tight, otherwise the sound pressure inside the calibrator is inaccurate. Most calibrators have various adapters available for use with microphones of different diameters.

## 7.11 Shakers for Structure Excitation

When accurate measurements of frequency responses are required, the best excitation method is often the *electrodynamic shaker*. Shakers of this design come in a variety of sizes from coffee cup size up to several feet in diameter. Even larger electrodynamic shakers are used for vibration testing, and the largest shakers today are up to a few meters in diameter and can give several hundred thousand Newtons output force. Electrodynamic shakers are available for a large variety of output forces and frequency ranges and you are encouraged to check manufacturers' web pages for details. In addition to the electrodynamic shakers, also *hydraulic* shakers are common for some applications where, for example, very low frequencies (even static) are required.

In this section we will limit our discussion to the type of electrodynamic shaker which is used for exciting structures for measurement of frequency responses, for example for experimental modal analysis. The principle of the electrodynamic shaker is that of a moving electric coil in a magnetic field, as illustrated in the principle drawing in Figure 7.13. A cylinder with a coil is suspended in a magnetic field. In the smallest shakers the magnetic field is created by a permanent magnet, whereas in larger

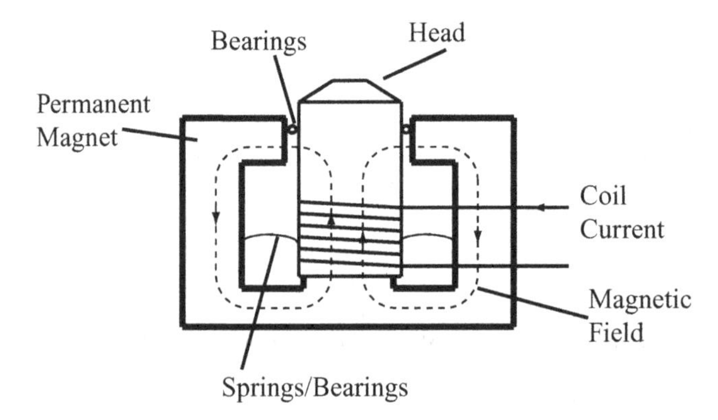


**Figure 7.13** Illustration of the principles of an electrodynamic shaker. A moving coil in a magnetic field created by a permanent magnet is supplied with an AC current, causing the coil to move. Larger electrodynamic shakers can have the permanent magnet replaced by a static coil fed by a DC current to create the magnetic field. Springs are supporting the coil statically, so it stays in an equilibrium position when no current is passing the coil. To ensure movement in only one direction, the cylinder with the coil and the head is often supported by bearings

shakers there are coils fed by DC current to create the magnetic field. The cylinder with the head, where the force output is taken, is supported by springs to hold the head in an equilibrium position, and often supported by bearings to ensure movement of the head in one direction.

The force output, $F(t)$, in [N] from an electrodynamic shaker is dependent on the magnetic flux, $B$, in [Wb/m$^2$], the length of the coil in the magnetic field, $L$, in [m], and the current in the coil, $i(t)$, in [A], through the relationship

$$F(t) = BLi(t) \tag{7.18}$$

where the product $BL$ can be seen as a shaker constant for a particular design. The electrical input impedance of the coil (i.e., of the shaker input) will be dependent on the mechanical load of the shaker. If optimum performance is required from an electrodynamic shaker, it must therefore be fed by a *current controlled* amplifier, which is an amplifier that gives a current output proportional to the voltage input of the amplifier. The current controlled amplifier is particularly important if the shaker is to be driven at very low frequencies.

Due to the physics of the electrodynamic shaker, only dynamic force can be output. If a static force is required, elaborate attachments can be considered, however a hydraulic shaker can be a better choice. The actual performance of a shaker in frequency response measurements is complicated by the *stinger* necessary to avoid transverse vibrations in the force sensor, see Figure 7.9. It is therefore often best to use a trial-and-error approach to find the best attachment of a shaker to a test structure. We will discuss the practical issues with using a shaker in Section 13.12.2.

## 7.12 Some Comments on Measurement Procedures

In the text so far in this chapter some theoretical descriptions of various noise and vibration transducers have been given. It is necessary to discuss briefly how to use these sensors for the best measurement accuracy. Vibration measurements can be very tricky and it is important to be a very critical measurement engineer to ensure the signals you measure are the actual vibration signals, and not some artifacts due to bad sensor installation, bad cables, overload in the measurement equipment, etc., that can so easily ruin the best of measurements. It is therefore essential to obtain experience with the type of errors which often occur, and to know what good results look like.

The most important behavior of a good experimentalist is that he or she *never trusts his/her measurement results*. Another important thing in vibration measurements is to remember that *you have no intuition* of what is going on. Intuition should not be confused by knowledge learned by experience; experience is exactly what you want to acquire, but in order to get experience, you need to question and reevaluate your measurement results over and over again. What if you make a new measurement, do you get the same result? What if you change the accelerometer to another one, do you get the same result? What if you place an extra accelerometer next to the one you are measuring with, do you get the same result? (If not, you have mass loading! See Section 7.4.4.) In addition to all these questions about the sensor and attachment of it, there is also a question of whether the measurement system is correctly set up. This will be discussed further in Chapter 11.

One measurement setup that I advocate you always keep within reach, is the calibration mass mentioned in Section 7.8 and an impact hammer. Using the hardest tip of the impact hammer, you can check that your accelerometer gives a good frequency range for a particular measurement by investigating for how high frequencies the FRF between acceleration and force stays a straight line, independent of frequency. By using this method, you verify that the accelerometer is still working properly throughout its operating frequency range. If an accelerometer is dropped on, e.g., a hard floor, it often does not break entirely, but instead some parts inside it can become loose, which causes 'resonances' inside it at certain frequencies. Calibrating such an accelerometer on a single-frequency calibrator may not reveal this problem, but a mass calibration ensures you find it.

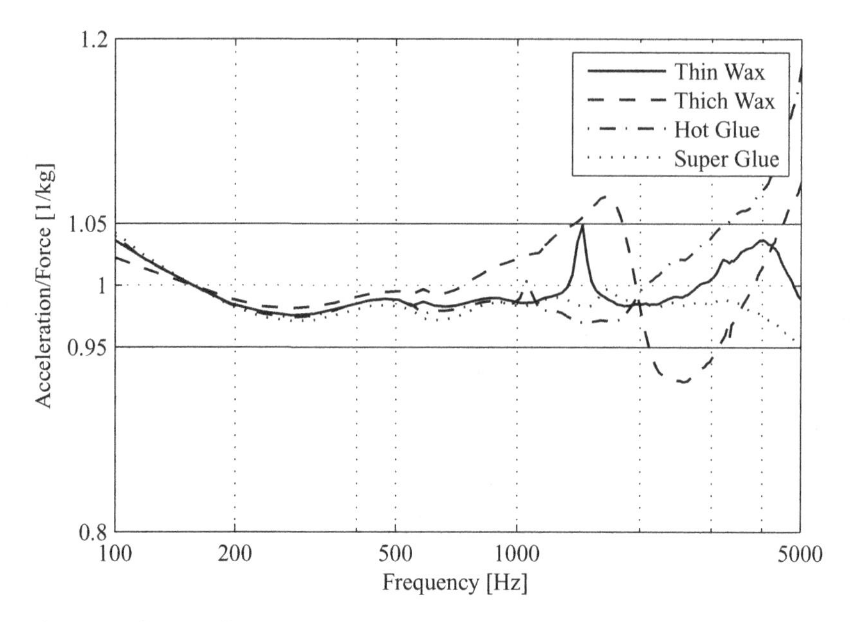


**Figure 7.14** Result of mass calibration with an accelerometer with a mass of approximately 12 g, on a mass of 1 kg. The accelerometer was mounted with three different mounting methods, whereof the first method, wax, was used twice; once with a thin layer (solid plot), and once with a thick layer (dashed plot). In the figure are also shown the results for hot glue (dash dot), and super glue (dotted). The sensitivity of each measurement was adjusted so that the value at 159.2 Hz was correct. The solid lines at 0.95 and 1.05 indicate ± 5% accuracy limits. As can be seen in the plot, the thin layer wax and super glue both give results within the accuracy limits up to 5 kHz, whereas hot glue is slightly worse and the thick layer of wax made the accuracy poor in comparison. The example shows that you need to be careful in order to get good results at higher frequencies

Figure 7.14 shows the result of mass calibration of an accelerometer mounted with three different mounting methods: wax, hot glue, and super glue (cyanoacrylate adhesive). The wax mounting was repeated twice; once with a thin layer covering the entire accelerometer base, and once with a thick layer to 'simulate' a sloppy test engineer. As can be seen in the plot, the thick layer of wax limits the usable frequency range to approximately 1.3 kHz, whereas a thin layer wax and super glue both perform within the specified ± 5% limits up to 5 kHz. Note that the frequencies depend on the mass of the accelerometer, so the results can not be extrapolated to any other sensor. The procedure described here is a good procedure for investigating the frequency range of proper operation for a combination of a particular accelerometer and mounting method. You should, however, remember to include a frequency 'margin', because when the accelerometer is mounted on an actual structure, which is more flexible than the rigid calibration mass, the usable frequency range of the accelerometer is reduced.

I finally want to encourage you to read the many excellent instructions provided by the manufacturers of vibration equipment. They are the experts on their own products and provide many important hints on how to best use their products. But – always remember to be critical about your measurements, no matter what precautions you have taken.

## 7.13 Problems

Many of the problems following are supported by the accompanying ABRAVIBE toolbox for MATLAB/Octave and further examples which can be downloaded with the toolbox. If you have not already done so, please read Section 1.6, and follow the instructions to download this toolbox together with example files.

**Problem 7.1** *A certain transducer manufacturer specifies a time constant of 1000 s for an IEPE accelerometer that you are considering for a measurement. Your measurement system manufacturer specifies a lower cutoff frequency of 2 Hz for the IEPE current supply. What is the total time constant if you are using the specified accelerometer?*

**Problem 7.2** *If you want to be able to measure harmonic vibrations within 5% accuracy, and the accelerometer and measurement errors are assumed to be negligible, what is the lowest frequency you can measure with this accuracy using the data from Problem 7.1?*

**Problem 7.3** *Assume you are planning on using an IEPE accelerometer with a sensitivity of 100 mV/g for measurements of accelerations up to 300 m/s*$^2$*. You are planning to use 50 m cable, and the cable manufacturer specifies a cable capacitance of 94 pF/m and your measurement instrument includes a 4 mA current supply for IEPE sensors. What is the maximum frequency you can measure with this cable length and current supply?*

**Problem 7.4** *This is a recommended exercise rather than a problem, but we place it here for lack of a better place. Using the measurement system of your choice, use an accelerometer and an impact hammer from your supply of sensors (provided you have access to all this, of course). If you do not have a calibration mass, make one from an approx 60 mm long, 40 mm diameter rod of steel or similar material (it should be a hard material so you do not get a deformation when you hit it with the hard tip of your impact hammer, so aluminum is not a good choice, for example). Weigh the mass and write it with a permanent marker on the mass. Then attach your accelerometer using your choice of method; wax, glue, etc. Make an impact measurement (you may need to read appropriate parts of Chapter 13) and make sure you get a coherence of very near unity. Store the frequency response, and thereafter detach the accelerometer, clean it and the calibration mass. Add the accelerometer again, using the same method as the first time. Make a new measurement and compare (overlay) with the previous measurement. Do you get agreement between the measurements? With what accuracy? Remember to regularly try this measurement with your accelerometers and mounting techniques you use, to ensure the accelerometers and your measurement equipment, etc., are operating normally.*

## References

Håkansson B and Carlsson P 1987 Bias errors in mechanical impedance data obtained with impedance heads. *J. Sound & Vibration* **113**(1), 173–183.

IEEE 1451.4 2004 *A Smart Transducer Interface for Sensors and Actuators – Mixed-mode Communication Protocols and Transducer Electronic Data Sheet (TEDS) Formats*. IEEE Standards Association.

Serridge M and Licht T 1986 *Piezoelectric Accelerometer and Vibration Preamplifier Handbook*. Brüel & Kjær, Nærum, Denmark.

# 8

# Frequency Analysis Theory

Frequency analysis is a central part of noise and vibration analysis because of the properties of linear systems which we described in Chapter 2. Frequency analysis is a complicated subject, incorporating Fourier transform theory, statistics, and digital signal processing. In order to get a comprehensive understanding of this important topic, we will dedicate several chapters to it. In this chapter we start with a discussion of frequency analysis by introducing some theoretical aspects of the subject. In Chapter 9, we will then describe the discrete Fourier transform, which is the most important tool for estimating spectra. Then we are ready to spend Chapter 10 on a discussion of how to experimentally estimate the theoretical spectra defined in the present chapter.

When we analyze a signal with respect to its frequency content, it turns out there are three different types (classes) of signals which we must theoretically and practically handle in different ways. These signals are

- periodic signals, e.g., from rotating machines
- random signals, e.g., vibrations in a car caused by the road–tire interaction
- transient signals, e.g., shocks arising when a train passes rail joints.

Each of these three types of signals, or *signal classes*, have spectra of quite different nature, and therefore they must be treated separately. We will start by treating periodic signals, and then move on to random and transient signals.

## 8.1 Periodic Signals – The Fourier Series

Jean Baptiste Joseph Fourier, who was a French scientist around the start of the 19th century, discovered that all *periodic signals* can be split up into a (potentially infinite) sum of sinusoids, where each sinusoid has its individual amplitude and phase, see Figure 8.1. A periodic signal thus has the special property that it contains only (sinusoids with) discrete frequencies. These frequencies are integer harmonics of the fundamental frequency $1/T_p$, i.e., the only frequencies present in the signal are $1/T_p, 2/T_p, 3/T_p$, etc., where $T_p$ is the *period* of the signal.

The mathematical theory of *Fourier series* states that every periodic signal $x_p(t)$ can be written as

$$x_p(t) = \frac{a_0}{2} + \sum_{k=1}^{\infty} a_k \cos\left(\frac{2\pi k}{T_p}t\right) + \sum_{k=1}^{\infty} b_k \sin\left(\frac{2\pi k}{T_p}t\right) \tag{8.1}$$

*Noise and Vibration Analysis: Signal Analysis and Experimental Procedures*, First Edition. Anders Brandt.

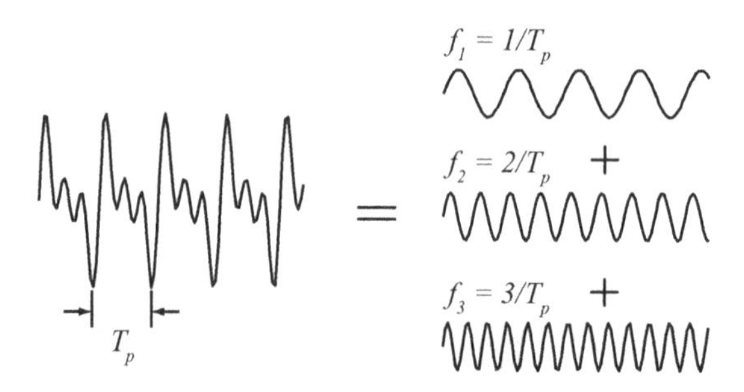


**Figure 8.1** Using the theory of Fourier series, every periodic signal can be split into a (potentially infinite) number of sinusoidal signals, each with individual amplitude and phase. In the figure a periodic signal is shown which consists of the sum of the three frequencies $1/T_p$, $2/T_p$, and $3/T_p$, where $T_p$ is the period of the signal

where the coefficients $a_k$ and $b_k$ can be calculated by

$$
\begin{aligned}
a_k &= \frac{2}{T_p} \int_{t_1}^{t_1+T_p} x_p(t) \cos\left(\frac{2\pi k}{T_p} t\right) dt \text{ for } k = 0, 1, 2, \ldots \\
b_k &= \frac{2}{T_p} \int_{t_1}^{t_1+T_p} x_p(t) \sin\left(\frac{2\pi k}{T_p} t\right) dt \text{ for } k = 1, 2, 3, \ldots
\end{aligned}
\tag{8.2}
$$

where the integration occurs over an arbitrary period of $x_p(t)$.

To make Equation (8.1) easier to interpret physically, by simple trigonometry it can be rewritten as one sinusoid at each frequency, with individual phase angle, $\phi_k$, for each sinusoid. We then obtain the alternative expression

$$
x_p(t) = \frac{a_0}{2} + \sum_{k=1}^{\infty} a'_k \cos\left(\frac{2\pi k}{T_p} t + \phi_k\right) \tag{8.3}
$$

where $a_0$ is the same as in Equation (8.1). Comparing Equations (8.1) and (8.3) we see that the coefficients in Equation (8.3) can be obtained from $a_k$ and $b_k$ in Equation (8.1) by

$$
\begin{aligned}
a'_k &= \sqrt{a_k^2 + b_k^2} \\
\phi_k &= \arctan\left(\frac{b_k}{a_k}\right).
\end{aligned}
\tag{8.4}
$$

By making use of complex coefficients, $c_k$, instead of the real $a_k$ and $b_k$, the Fourier series can alternatively be written as a complex sum as in Equation (8.5).

$$
x_p(t) = \sum_{k=-\infty}^{\infty} c_k e^{\frac{j2\pi k}{T_p} t} \tag{8.5}
$$

where the coefficients $c_k$ are given by

$$
\begin{aligned}
c_0 &= \frac{a_0}{2} \\
c_k &= \frac{1}{2}(a_k - jb_k) = \frac{1}{T_p} \int_{t_1}^{t_1+T_p} x_p(t) e^{-\frac{j2\pi k}{T_p} t} dt, \quad \text{for } k > 0
\end{aligned}
\tag{8.6}
$$

and the integration occurs over an arbitrary period of the signal $x_p(t)$ as before. Note in Equation (8.5) that the summation occurs over both positive and negative frequencies, i.e., $k = 0, \pm 1, \pm 2, \ldots$. Since the left-hand side of the equation is real (we assume that the signal $x_p$ is an ordinary, real signal), the right-hand side must also be real. Since the cosine function is an even function and the sine function is odd, then the coefficients $c_k$ must consequently comply with

$$\begin{aligned} \mathrm{Re}[c_{-k}] &= \mathrm{Re}[c_k] \\ \mathrm{Im}[c_{-k}] &= -\mathrm{Im}[c_k] \\ \because c_{-k} &= c_k^* \end{aligned} \tag{8.7}$$

for all $k \neq 0$ and where $^*$ represents complex conjugation. Hence, the real part of the coefficients $c_k$ is even and the imaginary part is odd. For real-life signals $x_p$, which are necessarily band limited (because of energy limitations), the Fourier series summation can be done over a smaller frequency interval, $k = 0, \pm 1, \pm 2, \ldots, \pm N$, where the coefficients for $k > N$ are negligible when $N$ is sufficiently high.

Note also that each coefficient $c_k$ is half of the signal's amplitude at the frequency $k$, which is evident from Equation (8.6). Thus, the fact that we introduce negative frequencies gives the result that the physical frequency content is split, in a symmetrical way (or anti symmetrical for the imaginary part) so that half the true amplitude of each frequency component is located at its positive frequency, and half at the corresponding (virtual) negative frequency. This is similar to the properties of the continuous Fourier transform, as we saw in Chapter 2.

## 8.2 Spectra of Periodic Signals

To describe a periodic signal, either a *linear spectrum* or a *power spectrum* are used in practice, as we will define in Section 10.2. These two practical spectrum estimators are closely related with the most intuitive spectrum for periodic signals, which is the amplitude spectrum of Figure 8.2, which basically consists of a specification of the coefficients for amplitude and phase angle according to Equation (8.3).

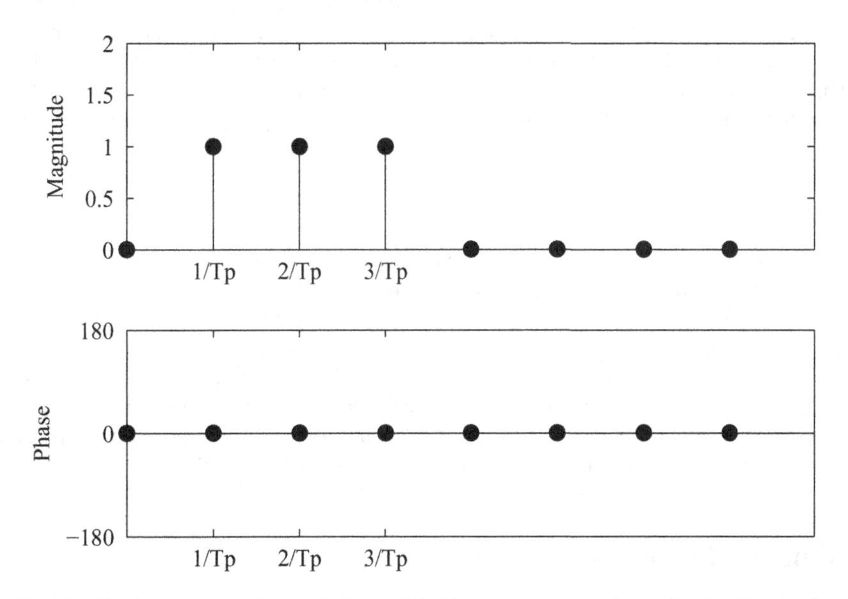


**Figure 8.2** Amplitude spectrum of a periodic signal. The spectrum contains only the discrete frequencies $1/T_p, 2/T_p, 3/T_p$, etc., where $T_p$ is the signal period

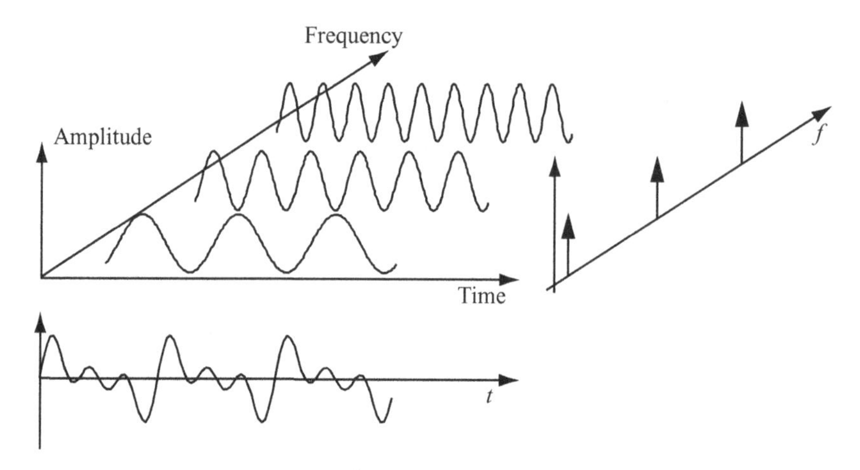


**Figure 8.3** The time and frequency domains can be regarded as the signal appearance from different angles. Both planes contain all the signal information (actually, for this to be completely accurate, the phase spectrum must also be included along with the above amplitude spectrum)

We will later see that when estimating spectra for periodic signals, many times one cannot simply compute this spectrum, since due to superimposed noise, it requires averaging, see Section 10.2.2. Therefore, the so-called *power spectrum* is usually available in noise and vibration analysis software. The power spectrum is not recommended to use, since its name is too easily mistaken for the power spectrum density of random signals, see Section 8.3. The power spectrum is, however, always an intermediate result in the averaging process and (unfortunately) in some systems it is the only available spectrum for periodic signals. This spectrum generally consists of the squared RMS value for each sinusoid in the periodic signal, and is obtained by squaring the coefficients $a_k'$ in Equation (8.4) and dividing by 2. The phase plot in Figure 8.2 is thus missing in the power spectrum and linear spectrum as there is no phase reference available. Estimating phase spectra will be discussed in Section 10.2.3. The linear spectrum, which is the recommended spectrum for periodic signals, usually consists of the RMS levels of the frequency components in the periodic signal, and is thus the square root of the power spectrum.

### *8.2.1 Frequency and Time*

To understand the difference between the time and the frequency domain, i.e., the information that can be retrieved from the different domains, we can study the illustration in Figure 8.3. In the time domain we see the sum of all included sine waves, while in the frequency domain we see each isolated sine wave as a spectral component. Therefore, if we are interested in, for example, the time signal's minimum or maximum value, we must look in the time domain. However, if we want to see which spectral components exist in the signal, we should rather look in the frequency domain. Remember that it is the same signal we see in both cases, that is, all signal information is contained in both domains. Various properties of this information, however, can be more easily identified in one or the other domain.

## 8.3 Random Processes

As discussed in Chapter 4, random processes are signals that vary randomly with time. They do, however, many times have constant average characteristics such as mean value, RMS value, and spectrum, and

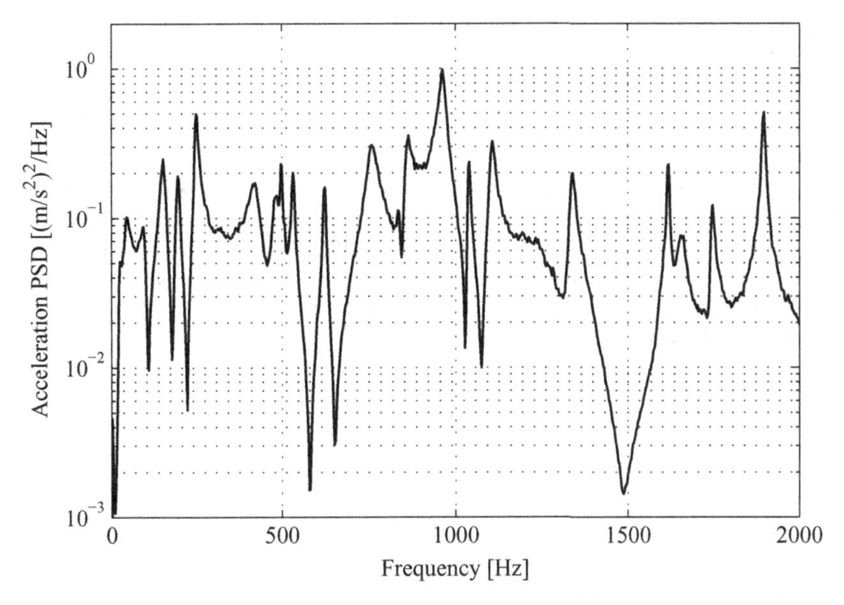


**Figure 8.4** Power spectral density, PSD, of a random acceleration signal. The spectrum is characterized by being continuous in frequency and is a density function, that is, the units are (m/s$^2$)$^2$/Hz if acceleration is measured in m/s$^2$

are then called *stationary signals*. As discussed in Chapter 4, most real-life stationary signals are also *ergodic*, for which ensemble averages can be replaced by time averages. The signals we will study in this section are assumed to be both stationary and ergodic.

## 8.3.1 Spectra of Random Processes

As opposed to periodic signals, random signals have continuous spectra, that is, they contain all frequencies and not only discrete frequencies. Hence we cannot display the amplitude or RMS value of each frequency, but we must instead describe the signal with a density type of spectrum (compare, for example, with discrete and continuous probability functions). The unit for noise spectra is therefore, for example, (m/s$^2$)$^2$/Hz if the signal is an acceleration measured in m/s$^2$. This spectrum is called Power Spectral Density, PSD. An example is shown in Figure 8.4.

The theoretical derivation of the PSD usually involves the autocorrelation function. Alternately, the PSD could be defined as a spectral density with an area under the PSD equal to the mean-square value of the time signal in the corresponding frequency range. In most standard textbooks on random signal analysis (Bendat and Piersol 2010; Newland 2005; Wirsching *et al.* 1995), it is shown that the double-sided (or two-sided) *autospectral density*, or power spectral density, PSD, denoted $S_{xx}(f)$, is the forward Fourier transform of the autocorrelation function, i.e.,

$$S_{xx}(f) = \mathcal{F}[R_{xx}(\tau)] = \int_{-\infty}^{\infty} R_{xx}(\tau) e^{-j2\pi f \tau} d\tau. \tag{8.8}$$

Analogous with the autospectral density, for an input signal $x(t)$, producing an output signal $y(t)$ through some arbitrary system, the double-sided *cross-spectral density*, CSD, $S_{yx}(f)$ is the forward

Fourier transform of the cross-correlation, i.e.,

$$S_{yx}(f) = \mathcal{F}\left[R_{yx}(\tau)\right] = \int_{-\infty}^{\infty} R_{yx}(\tau)\mathrm{e}^{-\mathrm{j}2\pi f\tau}\mathrm{d}\tau. \tag{8.9}$$

The relations in Equations (8.8) and (8.9) are often called the *Wiener–Khinchine relations*, after the mathematicians who (independently of each other) first proved these relationships. The negative frequencies in the Fourier transform has the property that half of the physical frequency content appears at positive frequencies, and the other half at negative frequencies. Therefore, it makes sense to define the physically interpretable *single-sided* (or one-sided) spectral densities, autospectral density, denoted by $G_{xx}(f)$, and cross-spectral density, denoted by $G_{yx}(f)$, as

$$\begin{aligned} G_{xx}(f) &= 2S_{xx}(f) \text{ for } f > 0 \\ G_{xx}(0) &= S_{xx}(0) \end{aligned} \tag{8.10}$$

and

$$\begin{aligned} G_{yx}(f) &= 2S_{yx}(f) \text{ for } f > 0 \\ G_{yx}(0) &= S_{yx}(0) \end{aligned} \tag{8.11}$$

respectively. The spectral densities at zero frequency do not repeat and therefore have to be treated separately.

In Section 4.2.12 we established that the autocorrelation is a real, even function, whereas the cross-correlation is real and has the property that $R_{yx}(\tau) = R_{xy}(-\tau)$. For the double-sided spectral densities, $S_{xx}(f)$ and $S_{yx}(f)$, this property of the cross-correlation and the properties of the Fourier transform from Section 2.7, because the autocorrelation is the cross-correlation of the signal $x$ with itself, lead to the fact that

$$S_{xx}(-f) = S^*_{xx}(f) = S_{xx}(f) \tag{8.12}$$

i.e., the autospectral density is real and even, because the autocorrelation is real and even. It can also be shown that $S_{xx}(f) > 0$ for all frequencies $f$, because the integral between any two frequencies is the mean-square value of the signal between those two frequencies, apparently a positive number, also see Section 8.5. For the double-sided cross-spectral density, it follows that

$$S_{yx}(-f) = S^*_{yx}(f) = S_{xy}(f) \tag{8.13}$$

i.e., the cross-spectral density $S_{yx}$ is a complex function with an even real part, and an odd imaginary part, see Problem 8.3.

For the single-sided autospectral density, we have that $G_{xx}(f)$ is real, and for the single-sided cross-spectral density, we have

$$G_{yx}(f) = G^*_{xy}(f) \tag{8.14}$$

i.e., the cross-spectral density of the reversed signals (we change $y$ to be input and $x$ to be output instead of the usual opposite situation) leads to a complex conjugate. If we consider the phase of $G_{yx}$ and $G_{xy}$ this means that

$$\angle G_{yx} = -\angle G_{xy} \tag{8.15}$$

which is intuitive, because if $x$ leads $y$, it is equivalent to the fact that $y$ lags $x$.

## 8.4 Transient Signals

In addition to the previously mentioned periodic and random signals, we have transient signals. Like the random signals, transient signals have continuous spectra. However, as opposed to random signals, transient signals do not continue indefinitely. It is therefore not possible to scale their spectra by the power in the signal, because power is energy per unit time. Instead, transient signals are generally scaled by their energy, and thus such spectra can have units of for example $(\mathrm{m/s^2})^2\mathrm{s/Hz}$.

The spectrum most commonly used for transient signals is called the Energy Spectral Density, ESD. Since energy is power times time, we obtain the definition of the ESD

$$\mathrm{ESD} = T \cdot \mathrm{PSD} \tag{8.16}$$

where $T$ is the time scale of the PSD, the time it takes to collect one time block if using FFT, see Section 10.3. The ESD is interpreted such that the area under the curve corresponds to the energy in the signal. In Section 10.5 we will discuss more on spectrum estimation of transient signals.

An alternative linear, and therefore perhaps more intuitive, spectrum of a transient signal is obtained by using the continuous Fourier transform without further scaling. The transient spectrum of a signal $x(t)$ is consequently defined as

$$T_x(f) = \mathcal{F}[x(t)] = \int_{-\infty}^{\infty} x(t)\mathrm{e}^{-\mathrm{j}2\pi f t}\mathrm{d}t \tag{8.17}$$

where $\mathcal{F}$ denotes the forward Fourier transform. The transient spectrum $T_x(f)$ in Equation (8.17) is of course a double-sided spectrum and we will return to a discrete approximation of this spectrum in Section 10.5. The reason there is only a single $x$ in the sub index, is that there is no square involved, as there is in, for example, the PSD.

## 8.5 Interpretation of spectra

We now need to discuss what can be interpreted from the different spectra presented in the preceding sections. For a periodic signal, this is relatively simple, as it consists of a sum of individual sinusoids. By knowing what these sinusoids are, i.e., their amplitudes, phase angles and frequencies, we can recreate the measured signal at any specific time, if we so choose. We could, in principle, tabulate the amplitudes (or RMS values), phase angles, and frequencies of each frequency component in the periodic signal.

We may also want to know for example the RMS value of the signal, in order to know how much power the signal generates. This can be done using Parseval's theorem, see Table 2.2. For a periodic signal, which has a discrete spectrum, we obtain its total RMS value by summing the included signals using

$$x_{\mathrm{RMS}} = \sqrt{\sum R_{xk}^2} \tag{8.18}$$

where $R_{xk}$ is the RMS value of each sinusoid for $k = 1, 2, 3, \ldots$. The RMS value of a signal consisting of a number of sinusoids is consequently equal to the square root of the sum of the RMS values. This result could also be explained by noting that sinusoids of different frequencies are *orthogonal*, and can therefore be summed like vectors (using Pythagoras' theorem).

For a random signal we cannot interpret the spectrum in the same way. As we have stated earlier, the PSD of a random signal contains all frequencies in a particular frequency band, which makes it impossible to add the frequencies up. Instead, as the PSD is a density function, the correct interpretation

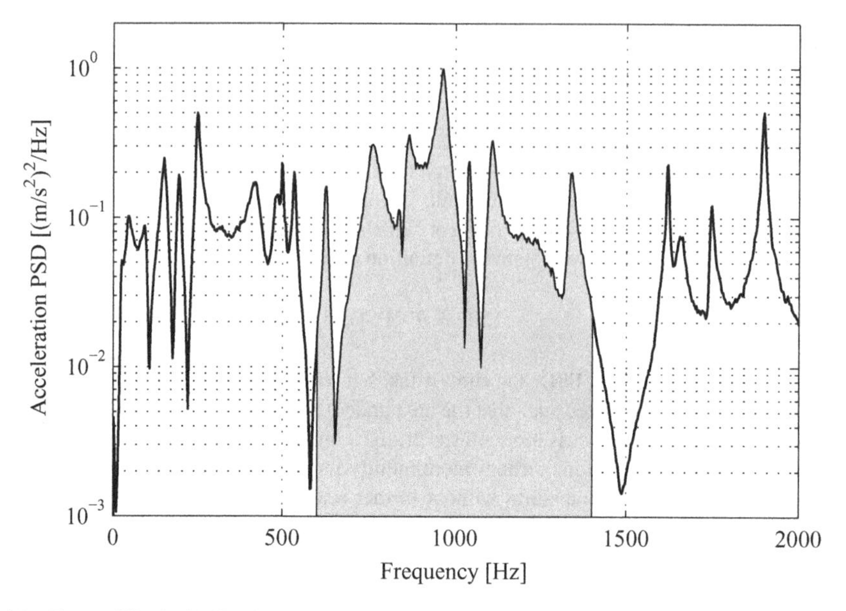


**Figure 8.5** From a PSD the RMS value can be calculated as the square root of the area under the curve. In the figure the area between 600 and 1400 Hz has been highlighted to indicate the area in this frequency range as an example

is to sum the *area* under the PSD in a specific frequency range, which then is the square of the RMS, i.e., the mean-square value of the signal, see Figure 8.5. To calculate the random signal's RMS value from the PSD we use

$$x_{\mathrm{RMS}} = \sqrt{\int G_{xx}(f)df} = \sqrt{\text{area under the curve}}. \tag{8.19}$$

In a similar fashion, we can determine the energy in a transient signal by calculating the square root of the area under the ESD curve.

Spectral density functions and energy density functions are difficult to interpret directly from a plot since it is the area under the curve that is interpreted as power or energy. A suitable function for easier interpretation of spectral density functions is therefore the *cumulated function*. For a PSD, one can thus build the cumulated mean square value, $P_{\mathrm{ms}}$, which is calculated as

$$P_{\mathrm{ms}}(f) = \int_0^f G_{xx}(f)\mathrm{d}f. \tag{8.20}$$

Note the similarity between this function and the statistical distribution function, which is equal to the integral of the probability density function (or the sum if we have a discrete probability distribution). The function $P_{\mathrm{ms}}(f)$ is consequently equal to the mean-square value in a frequency range from the lowest frequency in the spectrum up to the frequency value $f$. In Figure 8.6 a plot of the cumulated PSD of the same acceleration signal used previously in Figure 8.4 is shown. In a plot of the function $P_{\mathrm{ms}}(f)$, one can easily calculate the RMS value in any frequency interval, for example $f \in (f_1, f_2)$ by calculating

$$x_{\mathrm{RMS}}(f_1, f_2) = \sqrt{P_{\mathrm{ms}}(f_2) - P_{\mathrm{ms}}(f_1)}. \tag{8.21}$$

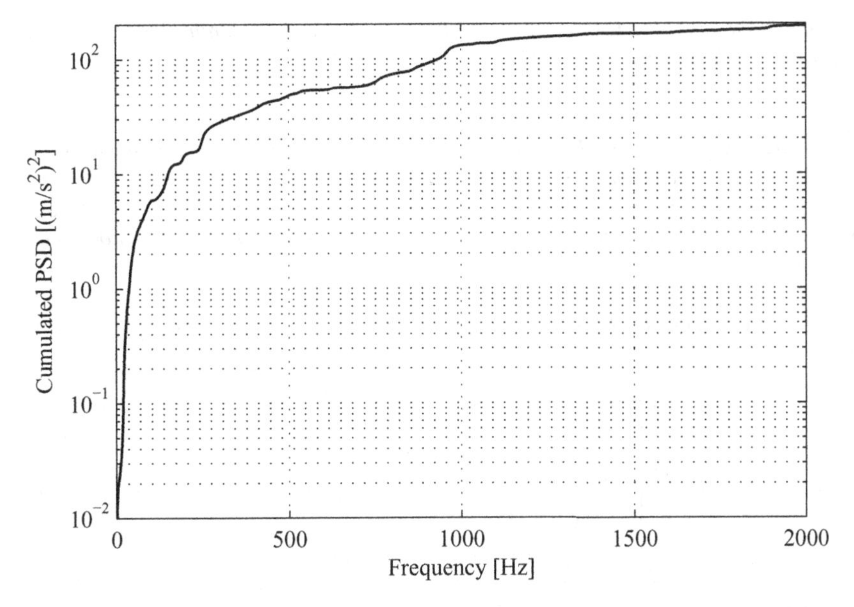


**Figure 8.6** Cumulated mean-square value calculated for the spectral density in Figure 8.4

## 8.6 Chapter Summary

In this chapter we have introduced some theoretical spectra for the three signal classes

- periodic signals,
- random signals, and
- transient signals.

We have noted that whereas periodic signals have discrete spectra, i.e., only certain frequencies exist in the signal, random and transient signals have continuous spectra.

The preferred theoretical spectra we use for the three signal classes are

- linear spectrum, for periodic signals,
- power spectral density, for random signals, and
- transient spectrum, for transient signals.

We have also reviewed some relations between the correlation functions and spectral densities for random signals, by the so-called *Wiener–Khinchine relations*, which state that each spectral density function (auto and cross) is the forward Fourier transform of the respective correlation function.

## 8.7 Problems

Many of the problems following are supported by the accompanying ABRAVIBE toolbox for MATLAB/Octave and further examples which can be downloaded with the toolbox. If you have not already done so, please read Section 1.6, and follow the instructions to download this toolbox together with example files.

**Problem 8.1** *Determine the Fourier series coefficients $a'_k$ and $\phi_k$ according to Equation (8.3) of a square wave signal defined by*

$$x_s(t) = 5\mathrm{sgn}\left[\sin(2\pi f_1 t)\right]$$

*where the frequency $f_1 = 20$ Hz. Plot the signal and an approximation of the signal using the first 5, 10, and 20 Fourier coefficients, respectively. Study what happens by including more harmonics in the Fourier series. (This is the so-called Gibbs phenomenon, which states that the overshoots do not disappear regardless of how many harmonics is included. However, the approximation naturally gets better and better.)*

**Problem 8.2** *Determine the Fourier series coefficients of a triangle wave signal defined by the plot below.*

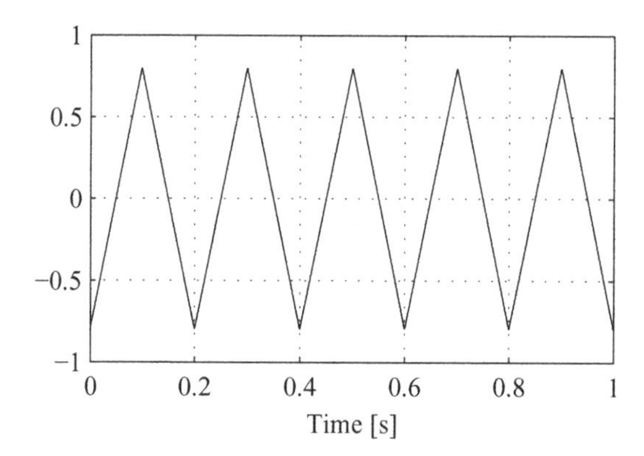


**Problem 8.3** *Prove the properties for autospectral density in Equation (8.12) and for cross-spectral density in Equation (8.13).*

## References

Bendat J and Piersol AG 2010 *Random Data: Analysis and Measurement Procedures* 4th edn. Wiley Interscience.

Newland DE 2005 *An Introduction to Random Vibrations, Spectral, and Wavelet Analysis* 3rd edn. Dover Publications Inc.

Wirsching PH, Paez TL and Ortiz H 1995 *Random Vibrations: Theory and Practice*. Wiley Interscience.

# 9

# Experimental Frequency Analysis

In practice, frequency analysis in the field of noise and vibration analysis is generally based on the discrete Fourier transform. In this chapter we will present the fundamental techniques for estimating a spectrum from measured samples of a signal. In particular we will go into some depth on the discrete Fourier transform, DFT, and its properties. In addition we will discuss some alternative means of estimating the frequency content of signals, e.g., octave filter analysis, which was introduced in Section 3.3.4. The actual estimators for spectrum estimation using DFT, will be discussed in Chapter 10.

## 9.1 Frequency Analysis Principles

When investigating the spectral content of a signal, there are many methods available. First of all, there are two classes of frequency (or spectral) analysis. *Nonparametric* techniques are those techniques that do not require any *a priori* information about the signal. This is probably what you would normally consider as frequency analysis – you want to know the frequency content of a signal, and consequently apply some frequency analysis method to obtain the spectrum. The most common nonparametric methods available are DFT/FFT, octave filters, and the relatively new so-called *wavelet analysis* technique. *Parametric* techniques, on the other hand, are methods that use some *a priori* information about the signal. For example, such information may be that the signal is periodic, and consists of 10 sinusoids. There are many parametric techniques known in signal processing engineering, for example *maximum entropy*, different forms of ARMA-based (Auto Regressive Moving Average) techniques, and the relatively new *MUSIC* (multiple signal classification) method, which can all be found in many standard text books on signal processing, for example in Proakis and Manolakis (2006).

Experience in the noise and vibration field has unfortunately shown that the, in theory, so promising parametric methods usually fail to deliver reliable results. Except in very rare special cases, they should be avoided, as the risk of misinterpreting the results is very high. The reason these methods usually fail when applied to noise and vibration signals, may be related to the fact that, in most cases, vibration data are not produced by simple models, but rather by a variety of different causes, including spurious sources that we do not immediately think about, but which are still existent in the data. An accelerometer mounted on the engine of a car, for example, does not contain vibrations solely produced by the engine, but there will also be road-induced vibrations, etc. The parametric methods in general only work for data where the model is well defined.

Cases where parametric techniques are indeed used successfully in the noise and vibration field are, for example, the 'Vold–Kalman' adaptive filters for order tracking that we will discuss in Chapter 12, and for tracking sine waves during sine sweep testing in vibration control applications. In general, the

*Noise and Vibration Analysis: Signal Analysis and Experimental Procedures,* First Edition. Anders Brandt.

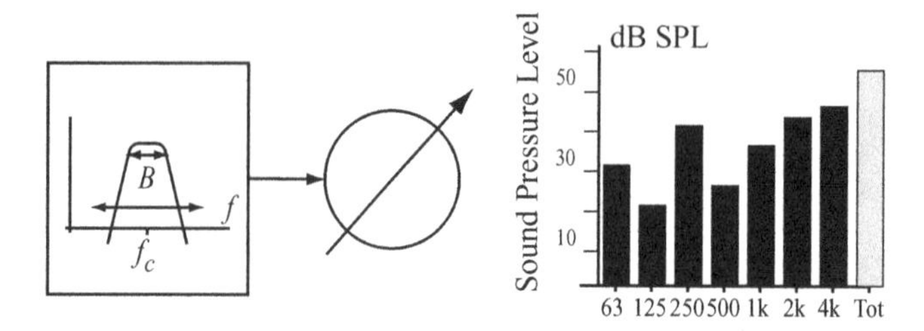


**Figure 9.1** Principle method for measuring spectrum with nonparametric techniques. An adjustable bandpass filter is stepped through a number of center frequencies, and for each frequency band the RMS value is measured with a voltmeter. All (nonparametric) frequency analysis methods are, in principle, using this technique, whether called FFT (DFT), octave analysis, or wavelet analysis, etc.

nonparametric techniques are superior, because of their simplicity and reliability. In the present chapter we will therefore focus our attention on these methods.

### *9.1.1 Nonparametric Frequency Analysis*

The nonparametric methods most common for analysis of noise and vibration signals are DFT/FFT analysis, which we will introduce in this chapter, and *octave band analysis* which we described in Section 3.3.4. In the latter case the frequency content is obtained by passing the time data through a series of parallel filters, after which the RMS value of each frequency band is measured. Those two methods will be discussed further in the remainder of this chapter.

There are other nonparametric methods for spectrum estimation, for example wavelets (Newland 2005). Those methods will not be discussed here. For the particular feature of constant relative bandwidth analysis, which is one of the main benefits of wavelets, octave and fractional octave band analysis are already well established in noise and vibration analysis, and offers the same principle advantages. Also, see Section 10.3.6 for a method using FFT to obtain constant relative bandwidth spectra.

All nonparametric methods have a common principle; they, in effect, calculate the RMS value of the signal in a specific frequency band, during a certain time, the *integration time*. This principle is illustrated in Figure 9.1.

It is important to realize that, given that we do not use any *a priori* information about the spectrum of the signal, this is *the only possible principle*. It does not matter if you find a method given a fancy name (such as wavelet analysis, or Wiegner–Ville etc.) – it is using the principle in Figure 9.1. To see what qualities this type of spectrum estimation has, we can study how an RMS value can be calculated for a bandpass-filtered signal. The RMS value, $x_{\text{RMS}}$, of a signal $x(n)$, is calculated for a sampled signal using the formula

$$x_{\text{RMS}} = \sqrt{\frac{1}{N}\sum_{n=1}^{N} x_n^2} \tag{9.1}$$

We recall from Section 4.2.7 that for a random signal, the RMS value has a normalized random error given by

$$\varepsilon_r = \frac{1}{2\sqrt{BT}} \tag{9.2}$$

where $B$ is the signal's bandwidth and $T$ is the measurement time, i.e., $T = N\Delta t$ if $N$ is the number of samples in the RMS computation. The bandwidth–time product (see also Section 4.2.7) is thus central to spectrum estimates. The finer frequency resolution we require, the longer measurement time we must use to obtain a particular maximum random error. This compromise follows naturally from the fact that, loosely speaking, *frequency is change per time unit.* It can be shown that for all spectrum estimation methods the bandwidth–time product is always smaller than $BT \leq 1/(4\pi)$, see, e.g., Bendat and Piersol (2010). For DFT, the $BT$ product is unity.

## 9.2 Octave and Third-octave Band Spectra

The way of measuring a spectrum prior to our current computer age was to have an adjustable bandpass filter and a voltmeter for AC voltage, as was shown in Figure 9.1. To be able to compare spectra from different measurements, the frequencies and bandwidths used were standardized at an early stage, as we discussed in Section 3.3.4. At that time it was natural to choose a constant relative bandwidth, so that the bandwidth increased proportionally with the center frequency (remember that, for example, the resonance bandwidth of modes on a structure is relative to the resonance frequency, as we saw in Chapter 5). Thus, if we denote the center frequency by $f_m$ and the bandwidth of the filter by $B$, we have

$$\frac{B}{f_m} = \text{constant} \tag{9.3}$$

where the standardized center frequencies and bandwidths were discussed in Section 3.3.4.

### 9.2.1 *Time Constants*

If the signal we measure is non-stationary, the RMS level of the bandpass filtered signal will, of course, vary as a function of time. Due to the bandpass filter's *time constant* there is, however, a limit to how fast the filter output signal can change when the input varies. The filter time constant describes how quickly the signal rises to $(1 - e^{-1})$ or about 63% of the final value, when the level of the input signal is suddenly altered (a step input). For a bandpass filter with bandwidth $B$, the time constant, $\tau$, is approximately

$$\tau \approx \frac{1}{B}. \tag{9.4}$$

For octave and third-octave band measurements with constant relative bandwidth, the different frequency bands consequently have different time constants, with longer time constants for lower frequency bands.

On the right-hand side in Figure 9.1 a typical octave band spectrum for a vibration signal is shown. Note that to the far right the so-called *total signal level*, that is, the signal's RMS value (within the whole frequency range) is shown. This value is usually shown on either side of the octave bands.

### 9.2.2 *Real-time versus Serial Measurements*

To measure an acoustic signal's spectral contents using octave bands, in the simplest case a regular sound level meter with an attached filter bank can be used. With this technique, a set of adjustable filters are used which, often automatically, step through the desired frequency range and stores the result for each frequency band. This type of measurement is called *serial* since the frequency bands are measured one after the other and was common with old sound level meters used for acoustic measurements. Naturally, this method only works when the signal (sound) is stationary during the entire time it takes to step through all frequency bands of interest. In order for the measurement to go faster, or if the signal is not stationary, a *real-time analyzer* can be used, which is designed with all of the third-octave bands in parallel, so that the same time data can be used for all frequency bands simultaneously. Another approach is to record the time signal and do the analysis with digital filters as described in Section 3.3.4.

## 9.3 The Discrete Fourier Transform (DFT)

The Discrete Fourier Transform, DFT, is the method used to transform measured samples of a signal into a spectrum using one of the estimators which will be described in Chapter 10. In order to understand and correctly interpret estimated spectra, it is essential to understand many of the properties of the DFT. We shall therefore study the DFT in some depth in this section.

Let us assume that we have a sampled signal $x(n) = x(n\Delta t)$. We further assume that we have collected $N$ samples, the *blocksize*, of the signal where $N$ is usually an integer power of 2, that is, $N = 2^p$ where $p$ is an integer number, see Section 9.3.1. The (finite) discrete Fourier transform, DFT, $X(k) = X(k\Delta f)$, of the sampled signal $x(n)$ is defined as

$$X(k) = \sum_{n=0}^{N-1} x(n)\mathrm{e}^{-\mathrm{j}2\pi kn/N} \tag{9.5}$$

for $k = 0, 1, \ldots, N-1$. Equation (9.5) is called the *forward DFT* of $x(n)$. To calculate the time signal from the spectrum $X(k)$, we use the *inverse DFT*, or IDFT, defined by

$$x(n) = \frac{1}{N}\sum_{k=0}^{N-1} X(k)\mathrm{e}^{\mathrm{j}2\pi nk/N} \tag{9.6}$$

for $n = 0, 1, \ldots, N-1$.

It should be pointed out immediately that the definition of the DFT presented in Equation (9.5) is by no means the only possible definition. There are several available definitions with different scaling factors in front of the sums. When confronted with new software, one should therefore test a known signal to find out which definition of the DFT is used. A simple way to test this is to create a signal with an integer number of periods and with a suitable blocksize, $N$, say 1024 samples. See Section 9.3.4 on how to create such periodicity. By checking the result of an FFT and comparing with the formulas above, the definition used can be identified. The definitions of the DFT and IDFT in Equation (9.5) and Equation (9.6), respectively, are common, and are the definitions used by MATLAB/Octave and by many noise and vibration software packages.

The number of samples, $N$, is called the *blocksize* (sometimes frame size) of the DFT, and consequently we often refer to the $N$ samples of $x(n)$, $n = 1, 2, \ldots, N-1$ as a *block* of data. Each value, $X(k)$, of the DFT is referred to as a *frequency line* or sometimes a *frequency bin*.

The spectrum, $X(k)$, obtained from the above definition of the DFT is not physically scaled. This is clearly seen by first observing the value for $k = 0$. The discrete frequency $k = 0$ corresponds to the DC component of the signal, that is, the average value of the signal. But, according to Equation (9.5) above, we have

$$X(0) = \sum_{n=0}^{N-1} x(n) = N \cdot \bar{x} \tag{9.7}$$

where $\bar{x}$ denotes the mean value of $x(n)$. Furthermore, if we, as an example, define $x(n)$ as a sine with two periods during the measurement time $n = 0, 1, 2, \ldots, N-1$, where we let $N = 8$, we get

$$x(n) = \cos(4\pi n/N) \tag{9.8}$$

for $n = 0, 1, \ldots, 7$. The DFT of the signal, $x(n)$, will be

$$X(k) = \begin{cases} 0, & \text{for } k \neq 2, 6 \\ 4, & \text{for } k = 2, 6 \end{cases} \tag{9.9}$$

which is apparently not the amplitude of $x(n)$. But if we divide the spectrum, $X(k)$, by $N = 8$, we obtain the value 0.5 at the two frequencies $X(2)$ and $X(6)$, which is half the amplitude of the sine at the two frequencies. We will soon come back to the interpretation of the frequencies of the DFT. We can conclude from this example, however, that by dividing the DFT as we have defined it by $N$, we get physically interpretable results. It could be questioned why we do not define the DFT including the division by $N$. Indeed this would be a more appropriate definition. However, as the definition in Equation (9.5) is the one used by MATLAB/Octave, and many other software packages, it makes sense to use it.

We should also point out some differences between the discrete Fourier transform in Equation (9.5) and the continuous Fourier transform, in Section 2.7 (of course, one is a continuous integral, and the other a discrete sum, which are strictly speaking not comparable at all; I still hope you will make my point).

- The DFT is computed from a finite number of samples, whereas the analog Fourier transform is an integral from minus infinity to infinity.
- The DFT is not scaled in the same units as the analog Fourier transform, since the differentiator $\mathrm{d}t$ is missing. The analog Fourier transform of a signal with units of $\mathrm{m/s^2}$ would have units of m/s, whereas the DFT will have units of $\mathrm{m/s^2}$. In Chapter 10 this will be clear as we present how to compute scaled spectra from the DFT results.
- The DFT is calculated in a nonsymmetrical way, from $n = 0$ to $n = N - 1$, and not symmetrically as the analog Fourier transform.

We will address these issues as we proceed.

### 9.3.1 The Fast Fourier Transform, FFT

Before proceeding with the details of the properties of the DFT, it is appropriate to briefly discuss the FFT, which is an abbreviation of the *Fast Fourier Transform*. The FFT, first published, as we know it, in Cooley and Tukey (1965, 1993). The history of the FFT algorithm is interesting in many ways. It was clear early on that many people had known parts of what was rediscovered (in Cooley's own words) in the 1965 paper, (Cooley *et al.* 1967). Later it was discovered that the famous mathematician Gauss, and many others, apparently knew it more than 150 years earlier, (Heidemann *et al.* 1985). These facts in no way diminishes the revolutionary impact the Cooley and Tukey (1965) paper has had.

The fast Fourier transform is an algorithm (or, as a matter of fact, several), that computes the DFT in Equation (9.5) in a much faster way than the direct implementation of Equation (9.5). The rediscovery of this algorithm revolutionized applications involving spectrum computations, as well as many other signal processing tasks. We will not go into any detail of the FFT algorithm, but only mention a few basic facts about it. The interested reader is referred to standard texts on signal processing (e.g., Oppenheim and Schafer 1975; Proakis and Manolakis 2006) for details on the algorithm.

A careful study of Equation (9.5) reveals that the number of computations to compute $N$ frequency values $X(k)$, is proportional to $N^2$ floating point operations. The FFT algorithm makes use of the symmetry of the exponential terms $\mathrm{e}^{-\mathrm{j}2\pi kn/N}$ in Equation (9.5). It turns out that many of the multiplications in this equation can be saved and the cost of computing the FFT is proportional to $N \log_2(N)$ multiplications instead of $N^2$ for the direct DFT. For large blocksizes, $N$, this is a substantial saving of operations. For example, for $N = 1024$, the direct DFT requires approximately 2 million operations, whereas the FFT requires approximately 30 000 operations ($log_2(1024) = 10$), a saving by a factor of 68. For larger blocksizes the time saving increases very rapidly, and already for $N = 32 \cdot 1024$ the saving is more than 1400 times.

It is often believed that the FFT algorithm requires $N$ to be a power of 2, .i.e., $N = 2^p$ for some integer, $p$, e.g., $N = 512, 1024, 2048$ samples, etc. This is, however, not true. There are FFT algorithms which do not require $N$ to be a power of two, but they are more complicated than those which assume an integer power of 2. Thus the power of two choice of $N$ is common. MATLAB/Octave include several algorithms that can take any length, $N$, and usually optimizes the performance based on the number of samples.

## 9.3.2 *The DFT in Short*

Before proceeding with many of the details of the DFT, to bring some clarity to some of the DFT properties, we will present an overview first published in Thrane (1979), reproduced here by permission of Brüel and Kjær, which elegantly describes the different properties of the DFT. In Figure 9.2 the different steps in the DFT are shown and the following text explains the different steps.

We start with a continuous time signal as in Figure 9.2 (A.1). Figure 9.2 (B.1) shows the Fourier transform of this continuous (infinite) signal, which is of course also continuous, but band-limited so that

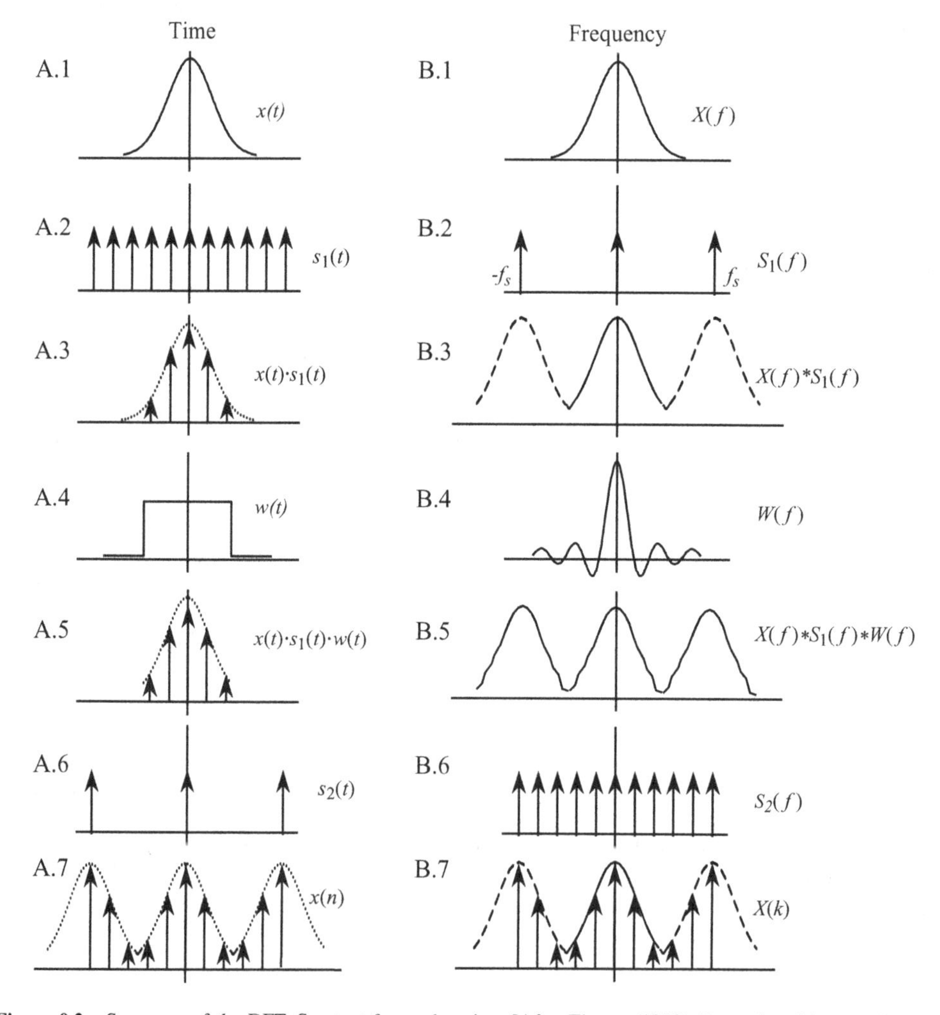


**Figure 9.2** Summary of the DFT. See text for explanation. [After Thrane (1979). Reproduced by permission of Brüel & Kjær]

we fulfill the sampling theorem. For the sake of simplicity we (and Thrane) have used a time function which is a Gaussian function, which has the same shape in time and frequency (see Table 2.2). Of course, a true Gaussian pulse is not band-limited, so the shape should only be viewed as a principal sketch.

The next step in the process is to sample the time signal, which is equivalent to multiplying the signal by an ideal train of pulses with unit value at each sampling instant and zero between, see Figure 9.2 (A.2) and (A.3). In the frequency domain, this operation corresponds to a convolution with the equivalent Fourier transform, which is a train of pulses at multiples of the sampling frequency, $f_s$. We consequently obtain a repetition of the spectrum at each $k \cdot f_s$ .This is actually an illustration of the sampling theorem, because it shows that if the bandwidth of the original spectrum would be wider than $\pm f_s/2$, the periodic repetition of the spectra would overlap, see Figure 9.2 (B.2) and (B.3).

The next step is due to the truncation in time, since we measure only during a finite time. This is equivalent to multiplying the continuous time signal by a rectangular window in the time domain, as illustrated in Figure 9.2 (A.4) and (A.5). In the frequency domain this operation is equivalent to the convolution with a sinc function as in (B.4) and (B.5). The result of this truncation is uncertainty in amplitude in the frequency domain, which can be seen in the ripple of the spectrum in (B.5).

The final step is carried out in the frequency domain, Figure 9.2 (B.6) and (B.7). With the DFT we calculate the spectrum only at discrete frequencies. Because of the symmetry of the Fourier transform, this operation is equivalent with the step in (A.2), i.e., it is equivalent with a multiplication of the spectrum with a train of pulses, only now with frequency increment $\Delta f = 1/T$. In the time domain this step implies a convolution with a train of pulses with separation $T$, as in (A.6), which finally gives the periodicity in the time domain in Figure 9.2 (A.7).

### 9.3.3 *The Basis of the DFT*

We will now take a closer look at the DFT definition in Equation (9.5). First, we should note that the sum is actually two sums, for the real and imaginary parts, respectively. Thus we can rewrite Equation (9.5)

$$X(k) = \sum_{n=0}^{N-1} x(n)\cos(2\pi kn/N) - \mathrm{j}\sum_{n=0}^{N-1} x(n)\sin(2\pi kn/N). \tag{9.10}$$

Each of these sums makes use of the orthogonality properties of sines and cosines, similarly to the continuous Fourier transform, see Section 2.2.4 and Section 2.7. The sum of the product in Equation (9.10) will be nonzero only if there is some frequency content in $x(n)$ at or around the frequency $k\Delta f$, otherwise it will be zero. (If you find it difficult to comprehend the sum here, you could think of the mean value of the product instead, which is, of course, the sum divided by the number of samples. This does not make any significant difference, as our discussion is concerning if the value is zero or nonzero. The actual value, when nonzero, will be clear as we proceed.)

Furthermore, the factors $\cos(2\pi kn/N)$ and $\sin(2\pi kn/N)$, when $k = 1, 2, \ldots, N-1$, is a cosine and a sine with $k$ periods within the blocksize, $N$. Thus, each frequency line, $k$, in the DFT, is the result of a multiplication of the signal $x(n)$ by a cosine or a sine, respectively, with $k$ periods within the blocksize (measurement time). More specifically, the real part of $X(k)$ is the *sum* of the values of the multiplication of the signal $x(n)$ by a cosine and the imaginary part of $X(k)$ is the similar result of a multiplication by a sine.

### 9.3.4 *Periodicity of the DFT*

As evident from Equation (9.5), and the discussion in Section 9.3.2, the discrete Fourier transform $X(k)$ is *periodic* with period $N$, that is

$$X(k + N) = X(k). \tag{9.11}$$

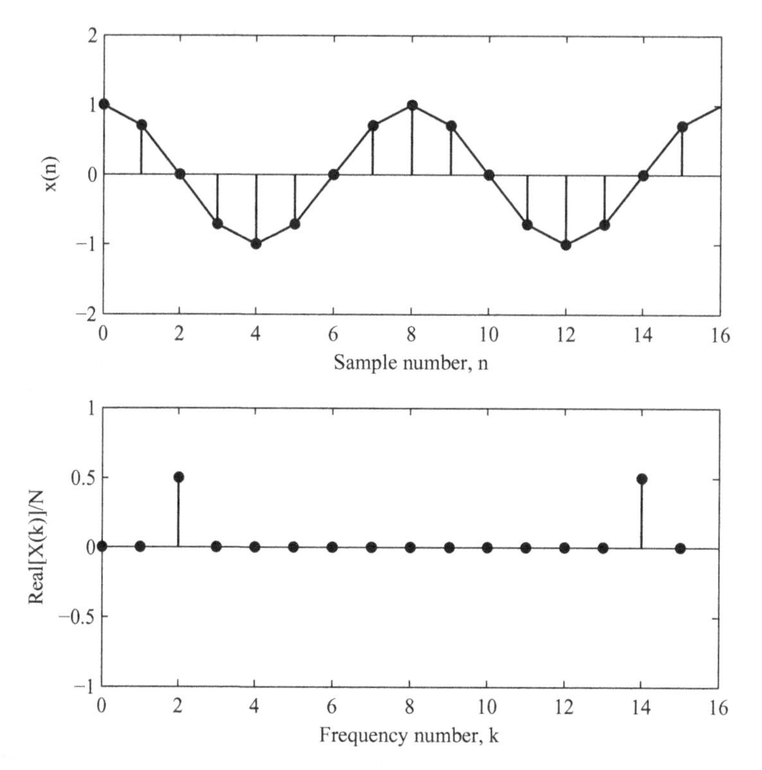


**Figure 9.3** Cosine with $N = 16$ and with two periods in the measurement window (upper), and the real part of the DFT of the same signal (lower) divided by $N$. As evident from the figure, by dividing the FFT by $N$ we obtain half the cosine amplitude at $X(2)$, and half at $X(14)$. The imaginary part is in this case equal to zero because the cosine is an even function. Each frequency line $k$ corresponds to a cosine (or sine for the imaginary part) with $k$ periods in the measurement window. Also note that the measurement window stops at $k = N$ whereas the DFT is only calculated up to $k = N - 1$

Similarly, the time signal $x(n)$ according to Equation (9.6) is periodic with period $N$, so that

$$x(n + N) = x(n). \tag{9.12}$$

In Figure 9.3, a cosine with two periods during the measurement window is plotted together with the real part of the DFT divided by $N$, i.e., $X(k)/N$. (The imaginary part is in this case zero, which will be evident as we proceed. Therefore it is not included in the figure.) The first thing to observe in Figure 9.3 is that the spectrum $X(2)$ contains the first peak. This is a result of the fact that the cosine in $x(n)$ contains two periods in the time window (measurement time), and that the DFT at $k = 2$ is the result of the product of $x(n)$ and a cosine (since we are looking at the real part) with two periods during the time window, and the orthogonality of cosines, see Equation (9.10).

From Figure 9.3 it is clear that the actual 'measurement time', i.e., the period of the cosine, $x(n)$, is one sample more than the samples actually measured. In other words, the signal we have sampled is periodic in the blocksize if the next sample, $x(N)$, in our example $x(16)$, equals the first sample, $x(0)$. Combining this with the fact that the spectrum at each frequency line, $k$, corresponds to a signal with $k$ periods in the blocksize, we can conclude that the frequency increment, $\Delta f$, is

$$\Delta f = \frac{1}{T} = \frac{1}{N\Delta t} = \frac{f_s}{N}. \tag{9.13}$$

The relationships in Equation (9.13) are important to keep in mind when we use frequency analysis software, to keep count of how long the measurement time will be for a certain frequency increment. It should be particularly noted that the frequency increment is the reciprocal of the measurement time, $T$, (which, oddly, is one sample longer than it actually took to gather the data $x(n)$). Thus, if we want a frequency increment of 0.1 Hz, for example, we need to measure a block of data for 10 seconds (or, actually, one time increment less – you start to get the picture).

Equation (9.13) also implies that the Nyquist frequency, $f_s/2$ is found at $k = N/2$. This frequency is of special interest not only because it is the upper limit of the interesting frequencies (i.e., the positive frequencies, we will come to that). If we go back to the definition of the DFT in Equation (9.5), we find that this value is

$$X(N/2) = \sum_{n=0}^{N-1} x(n)e^{-j2\pi(N/2)n/N} = \sum_{n=0}^{N-1} x(n)e^{-j\pi n} = \sum_{n=0}^{N-1} x(n)(-1)^n \tag{9.14}$$

which is a real number. This number is important because, as we will see soon, whereas the remaining values $X(k)$ for $k > N/2$ can be calculated from the first values $X(k)$ for $k = 1, 2, \ldots N/2 - 1$, the value for $k = N/2$ must be stored. To understand this, however, we first need to discuss the symmetry properties of the DFT, see Section 9.3.5.

We now put the focus on the right-hand half of the spectrum in Figure 9.3. The values above $k = N/2$ can easily be seen to be the negative frequencies, if we note the periodicity of $X(k)$. This is illustrated in Figure 9.4, where the spectrum in Figure 9.3 has been repeated once below and once above the earlier values. In the figure it is clearly seen that shifting the upper $N/2$ values of the 'original' spectrum $X(k)$ to the left of the first $N/2$ values (remember the value $k = 0$), we obtain a symmetric spectrum with positive and negative frequencies. (This can be done conveniently in MATLAB/Octave by the **fftshift** command).

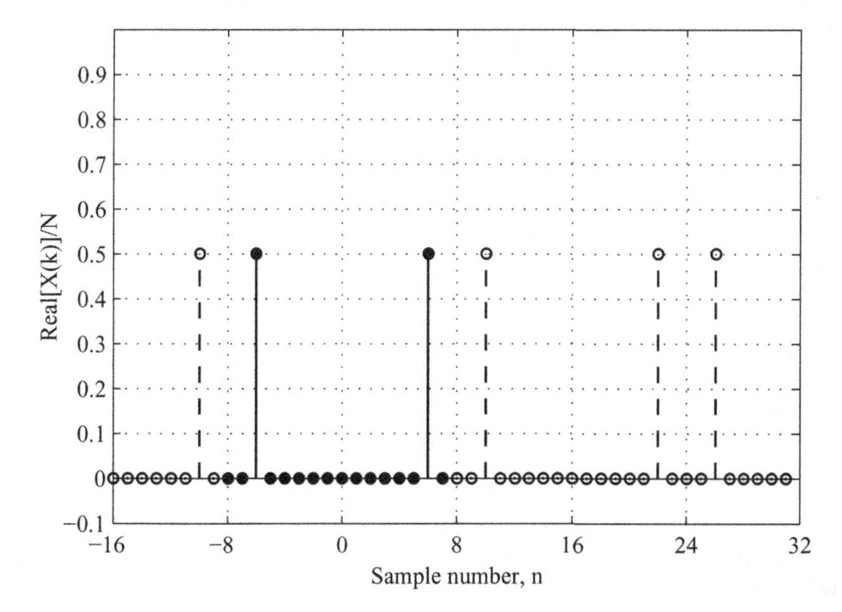


**Figure 9.4** Illustration of the periodicity of the DFT. In the figure, the DFT result $X(k)$ for $k = 1, 2, \ldots, N - 1$ for the signal used in Figure 9.3 has been repeated once to the left, and once to the right, of the original sequence. The frequency lines between $k = -8$ and $k = 7$ are highlighted, indicating a double-sided spectrum. Note that the frequency line corresponding to the Nyquist frequency, $X(8)$ in the original sequence, has been shifted to $X(-8)$

**Table 9.1** Some important transform pairs for the DFT. See Section 9.3.12 about circular convolution as in pairs 5 and 6

| # | Description | $x(n)$ | $X(k)$ |
|---|---|---|---|
| 1 | Periodicity | $x(n+N) = x(n)$ | $X(k+N) = X(k)$ |
| 2 | Constant | 1 | $X(0) = N$ |
| | | | $X(k) = 0, k \neq 0$ |
| 3 | Dirac pulse | $\delta(n)$ | $1, k = 0, 1, \ldots N-1$ |
| 4 | Complex conjugation | $x^*(n)$ | $X^*(N-k)$ |
| 5 | Multiplication | $x(n)y(n)$ | $\frac{1}{N} X(k) \circledast Y(k)$ |
| 6 | Convolution | $x(n) \circledast y(n)$ | $X(k)Y(k)$ |
| 7 | Parseval's theorem | $\sum_{n=0}^{N-1} x(n)y^*(n)$ | $\frac{1}{N} \sum_{n=0}^{N-1} X(k)Y^*(k)$ |

### 9.3.5 *Properties of the DFT*

Most of the properties of the continuous Fourier transform apply in a similar fashion to the DFT. Some important DFT transform pairs are presented in Table 9.1, where the most notable difference compared to the transform pairs for the continuous Fourier transform is the form of Parseval's theorem, which has a scaling factor in the frequency domain for the DFT. Also, as we will show in Section 9.3.12, the multiplication and convolution works differently for DFT than for the continuous Fourier transform.

If $x(n)$ is a real sequence, like our typical measurement signals, then the real part of $X(k)$ comes from the even part of $x(n)$, and the imaginary part of $X(k)$ comes from the odd part of $x(n)$, due to the nature of even and odd signals that we presented in Section 2.7.1. Furthermore, for a real signal, $x(n)$, the real part of its DFT, $X(k)$, is an even function and the imaginary part is an odd function, i.e.,

$$\mathrm{Re}\left[X(-k)\right] = \mathrm{Re}\left[X(k)\right] \tag{9.15}$$

and

$$\mathrm{Im}\left[X(-k)\right] = -\mathrm{Im}\left[X(k)\right]. \tag{9.16}$$

These qualities, called the Fourier transform *symmetry properties*, are valid also for the similarly to the continuous Fourier transform. Thus, according to Equations (9.15) and (9.16), the negative frequencies (the frequency lines $X(k)$ for $k > N/2$) are superfluous. It is therefore customary to discard these data to save (almost) half the storage space; 'almost' because we need to store the real frequency line corresponding to the Nyquist frequency, i.e., $X(N/2)$ as was mentioned above. Thus, for a blocksize of, say, 1024 time samples, we store the first 513 values from the DFT. If needed for performing an inverse DFT, we then simply recreate the upper $N/2 - 1$ values in $X(k)$ using the symmetry properties in Equations (9.15) and (9.16).

### 9.3.6 *Relation between DFT and Continuous Spectrum*

The periodicity of the DFT can be interpreted such that the DFT $X(k)$ of the signal $x(n)$ is the spectrum of the *periodic repetition* of $x(n)$. We can divide this periodic repetition into three different cases, depending on the signal, $x(n)$, if we use an $N$-size DFT, namely

1. the signal $x(n)$ is periodic in $N$, which is unlikely to happen for a measured signal (except for synchronous sampling, see Chapter 12), but is often used when considering the DFT and its effects (as in the present chapter, for example),

2. the signal is transient, of length $L < N$, i.e., it dies out inside $N$, or
3. the signal is continuous (either periodic or random), or transient with a length $L > N$.

In the first case, the DFT, $X(k)$, is an exact representation of the continuous Fourier series of $x(t)$, if we apply appropriate scaling, and there is no frequency content between the discrete frequencies $k\Delta f = k/(N\Delta t)$.

In the second case with a transient signal that dies out inside the blocksize, the analog signal sampled at $x(n\Delta t)$, has a continuous Fourier transform, $X(f)$ which is *sampled* at the discrete frequencies of the DFT, i.e.,

$$X(k) = X(f)|_{f=k\Delta f}\,. \tag{9.17}$$

Furthermore, in the second case, if we wish to calculate $X(f)$ at other frequencies than those sampled by the DFT, we can do so using the samples of $x(n)$ by

$$X(f) = \Delta t \sum_{n=0}^{N-1} x(n)\mathrm{e}^{-\mathrm{j}2\pi fn}. \tag{9.18}$$

For such signals, zero padding is an easy and appropriate method to find values of $X(f)$ by the DFT at arbitrary resolution, see Section 9.3.13. This relation is given here without any proof, as it can be found in any standard textbook on signal processing, (e.g., Oppenheim *et al.* 1999; Proakis and Manolakis 2006).

In the third case, finally, when the signal is truncated before the DFT can be computed, there will be an error, i.e., a difference between the true signal spectrum, and the spectrum of the truncated, periodically repeated, signal. This error is called leakage, and will be dealt with in the next section.

### *9.3.7 Leakage*

What happens if we, for example, compute the DFT with a frequency increment of $\Delta f = 2$ Hz, but the measured signal is a sinusoid of, say, 51 Hz, so that the signal frequency is located right between two spectral lines in the DFT (50 and 52 Hz)? The result is that we get two high DFT values at 50 Hz and at 52 Hz. However, as shown in Figure 9.5, both values are lower than the true value, and there are a number of nonzero values of the DFT around the two peak values. To easily observe this error in the figure, we have scaled the DFT by dividing by $N$ and taken the absolute value of the result, as we discussed in Section 9.3. We have also made the spectrum single-sided, by multiplying all values (except the DC value) by 2, as we discussed in Chapter 8. Thus the correct value should be 1, the amplitude of the sinusoid.

As seen in Figure 9.5 the resulting peak is incorrect, by as much as 36%. Furthermore, it looks as if the frequency content has 'leaked' away on both sides of the true frequency of 51 Hz. This phenomenon is therefore called *leakage*.

One way to explain the leakage effect is by studying what happens in the frequency domain when we limit the measurement time to a finite time, as we do when we acquire only $N$ samples of the continuous signal $x(n)$. This procedure corresponds to multiplying the original, continuous signal by a time window which is zero outside the interval $\Delta t \in (-T/2, T/2)$, and unity within this same interval. A multiplication with this function, $w(t)$, in the time domain, is equivalent to a convolution with the corresponding Fourier transform, $W(f)$ in the frequency domain. We thus obtain the weighted Fourier transform of $x(t) \cdot w(t)$, denoted $X_w(f)$, as

$$X_w(f) = X(f) * W(f) = \int_{-\infty}^{\infty} X(u)W(f-u)\mathrm{d}u \tag{9.19}$$

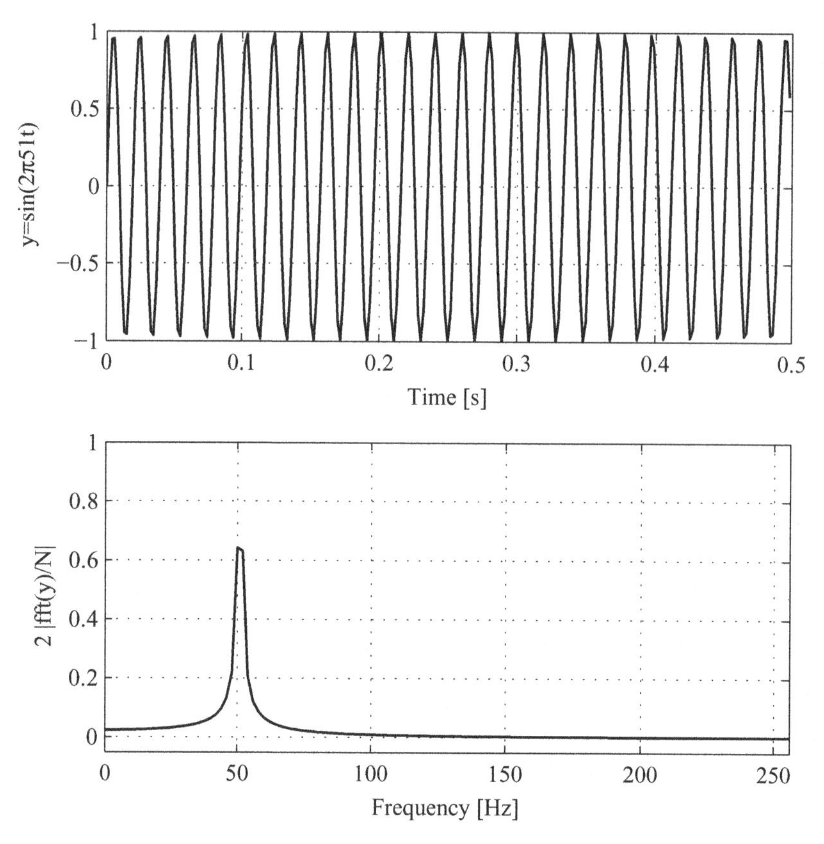


**Figure 9.5** Time block (upper) and scaled DFT (lower) of a 51 Hz sinusoid. 256 time samples have been used, giving 129 spectral lines. The frequency increment is $\Delta f = 2$ Hz, since the measurement time is 0.5 s. Instead of the expected value of 1, that is, the amplitude of the sinusoid, we get a peak much too low (in this case approx. 36% too low). There are also more nonzero frequency values to the left and right of the 50 and 52 Hz values. This phenomenon is called leakage since the frequency content in the signal seems to 'leak' out to surrounding frequencies

where $*$ denotes convolution. $W(f)$ is the transform of a *rectangular time window*, as in the example shown in Figure 9.5. This Fourier transform is

$$W(f) = T\frac{\sin(\pi f T)}{(\pi f T)} = T \operatorname{sinc}(fT) \tag{9.20}$$

which is plotted (in dB magnitude scale) in Figure 9.6 as a dotted line. You should particularly note that the main lobe of the window is exactly zero at all integer $k$, except $k = 0$. The rectangular window is sometimes called a *uniform window* and in MATLAB/Octave it is defined by the **boxcar** command.

We will illustrate the convolution process in Equation (9.19) for two cases; (i) when the frequency of the sine coincides with a frequency line, i.e., the sine is periodic in the time window; and (ii) when the frequency of the sine is located exactly in the middle of two frequency lines. (If you are not already familiar with convolution, you are strongly recommended to study Section 2.6.4 before proceeding.) In both cases the convolution between the Fourier transform of our (continuous) sine wave and that of the time window implies that we allow the former, $X(f)$, to sit at its frequency $f_0$. Then, for each frequency line $k$ where we calculate $X(k)$, we shift the Fourier transform of the window, $W(f)$, $k$ samples (if $k$ is negative it is a shift to the left, if $k$ is positive, we shift to the right), i.e., to $W(f - k\Delta f)$. (Actually,

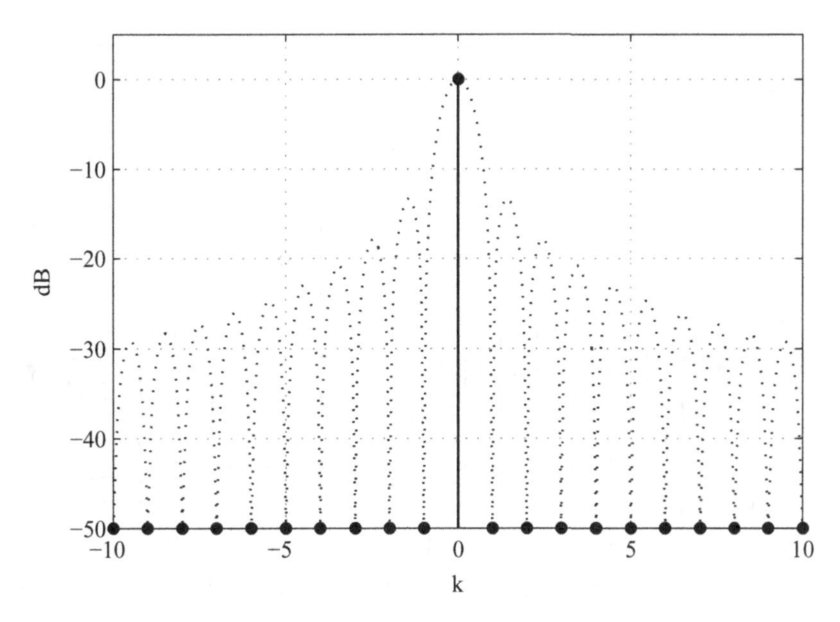


**Figure 9.6** DFT of a sinusoid which coincides with a spectral line ($k = 0$). The convolution between the transform of the (rectangular) time window, $W(f)$, (dotted) and the sinusoid's true spectrum, $\delta(f_0)$, (solid) results in a single spectral line. Note that the scale is in dB, i.e., a logarithmic scale, as opposed to the linear scale in Figure 9.5

we should also reverse the window Fourier transform, but it is symmetric, so nothing really happens in that step.) Finally, we multiply the two and sum all the values (for all frequencies $k$), but since $X(f)$ is a single spectral line there is nothing to sum – it will be a single value for each $k$.

In case (i), when the frequency of the sine coincides with a frequency line, i.e., when $f_0 = k_0 \Delta f$, the result obtained is plotted in Figure 9.6. The reason there is only one nonzero value, is that for all integer numbers $k$ where we place $W(f - k\Delta f)$, except for $k = k_0$, the spectral line of the sinusoid corresponds to a zero in $W(f)$.

In Figure 9.7 the result of the convolution as described above, for case (ii), where the frequency of the sinusoid is located exactly between two spectral lines (we have an integer number plus one half period in the time window) is illustrated. At each shift of $k$ samples during the convolution, the frequency line of the sine will now coincide approximately with the maximum of each side lobe. We see in the picture that the result is that we obtain many nonzero frequency lines which slowly decrease to the left and right, and we get two peaks for $k = 0$ and $k = 1$, which are the same height, although much lower than the sinusoidal amplitude. It can be shown that if the RMS values of all spectral lines are summed as will be discussed in Section 10.6.5 the result is equal to the true RMS value of the sinusoid. (This is a result of the fact that Parseval's theorem always ensures that the power of the signal in the spectrum is equal to the power in the time domain, see Table 9.1.) Hence, the power in the signal seems to 'leak' out to nearby frequencies, giving this phenomenon the name *leakage*.

### *9.3.8 The Picket-fence Effect*

An alternative way to look at the discrete spectrum $X(k)$ from the DFT, is to see each spectral line as the result of a bandpass filtering, followed by an RMS computation (or amplitude detection, we will discuss scaling in Chapter 10) of the signal after the filter. This process is often illustrated as in Figure 9.8 with

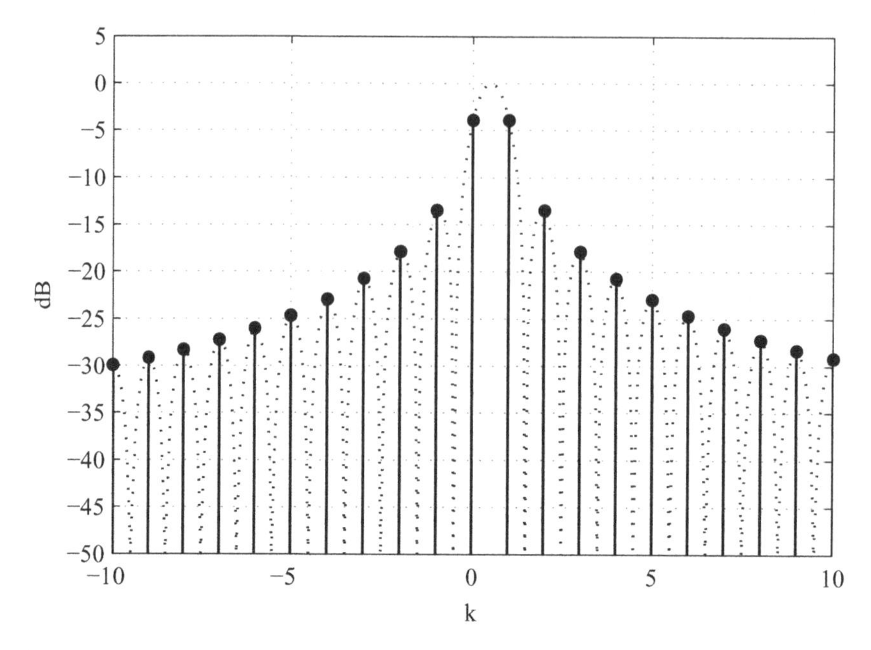


**Figure 9.7** Leakage with rectangular window. The frequency of the sine wave is located at $f_0 = 0.5\Delta f$, exactly mid way between two frequencies, $k = 0$ and $k = 1$, corresponding to an integer number of periods plus one half period in the time window. The DFT result, $X(k)$ is illustrated by black dots and the dotted line is the rectangular window spectrum. When a periodic signal does not have an integer number of periods in the measurement window, then due to the finite measurement time, the convolution results in too low a frequency peak. At the same time the power seems to 'leak' into nearby frequencies although the total power in the spectrum is still the same. Note that the scale is in dB, i.e., a logarithmic scale, as opposed to the linear scale in Figure 9.5, thus exaggerating the leakage effect

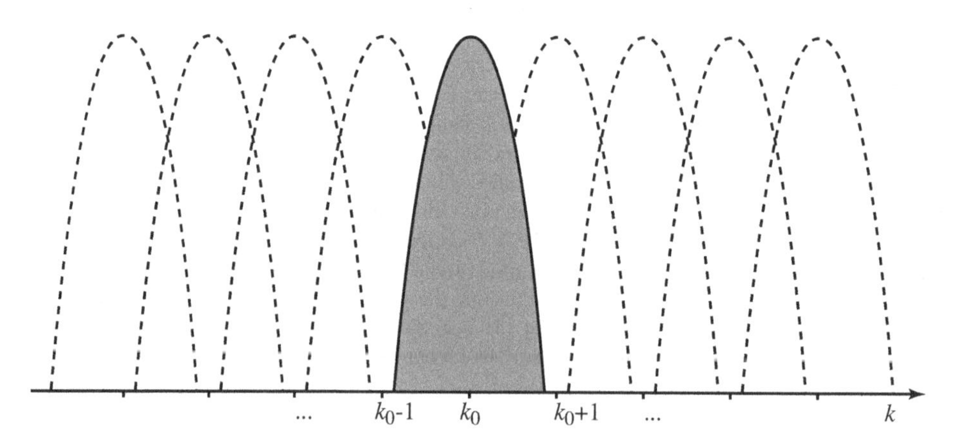


**Figure 9.8** The picket-fence effect. Each value in the discrete spectrum corresponds to the signal's RMS value after bandpass filtering. If we study a tone located between two frequencies it will be attenuated by the filter shape, and we will obtain a value which is too low

**Table 9.2** Some useful figures for the windows compared in Section 9.3.9. The windows are defined in the text. The amplitude correction factor is defined in Section 9.3.9 and the normalized equivalent noise bandwidth, NENBW, is defined in Section 9.3.9

| Window | First side lobe [dB] | Sidelobe falloff [dB/oct.] | Ampl corr. [−] | NENBW [bins] | Max. ampl. error [%] |
|---|---|---|---|---|---|
| Rectangular | −13.3 | −6 | (1) | 1 | −36 |
| Hanning | −31.5 | −18 | 2 | 1.5 | −15 |
| ISO flattop | −84 | – | 1 | 3.77 | −0.01 |
| Enh. flattop | −87.9 | – | 1 | 3.77 | −0.01 |

a number of parallel bandpass filters, where each filter is centered at the frequency $k$ (or $k \cdot \Delta f$ if we think in Hz). Each filter illustrated in Figure 9.8 is the main lobe of the window, which is of course a simplification; to be exact we really have to take into account the side lobes of the window as well.

This method of looking at the DFT is reminiscent of viewing the true spectrum through a picket fence and therefore it is sometimes called the *picket-fence effect*, or sometimes *scalloping* (which is an English word meaning something going up and down in a wavelike manner). Note that the picket-fence effect is also analog with the method of measuring a spectrum with octave band analysis that we discussed in Section 9.2. As was mentioned in Section 9.1.1 this principle is the *only* (nonparametric) way to measure or compute spectral content.

The picket-fence effect is a good illustration of what happens with the estimated amplitude of a sine which is located between two frequency lines; it will be attenuated by as much as the filter shape has decreased at the frequency of the sine. We can use this illustration for different windows to find the 'maximum amplitude error', which is tabulated for some common windows in Table 9.2.

## 9.3.9 *Time Windows for Periodic Signals*

As we saw in Section 9.3.7, we obtain leakage when calculating the DFT of a sinusoid with a non-integer number of periods in the observed time window. This error is caused by the fact that we truncate the true, continuous signal, which was also evident from the DFT overview in Section 9.3.2. By using a weighting function other than the rectangular window used in Section 9.3.7, we can reduce the leakage, and thus this amplitude error too. This process is called *time-windowing* and is illustrated in Figure 9.9 which illustrates the window effect on the same 51 Hz sine as in Figure 9.5.

The time window used in Figure 9.9 is called a *Hanning window* and is one of the most common windows used in FFT analysis. The effect of the window is that it reduces the transients in the periodic repetition of the time signal, although it is not very intuitive how this could improve the result. We will show, however, that we can estimate the amplitude much better than with the rectangular window.

The resulting spectrum after windowing in Figure 9.9 (d) is markedly sharper than that in Figure 9.5, but the peak does not have the correct value. We will soon present the Hanning window in some detail, but first we will see how to correct for this error.

### Amplitude correction of window effects

The result in Figure 9.9 (d) shows a peak that is considerably lower than the expected amplitude of the sine, which is 1. This is due to the fact that the Hanning window removes information in the signal, as is evident from the windowed function in Figure 9.9 (c). To see how to compensate for this window, we

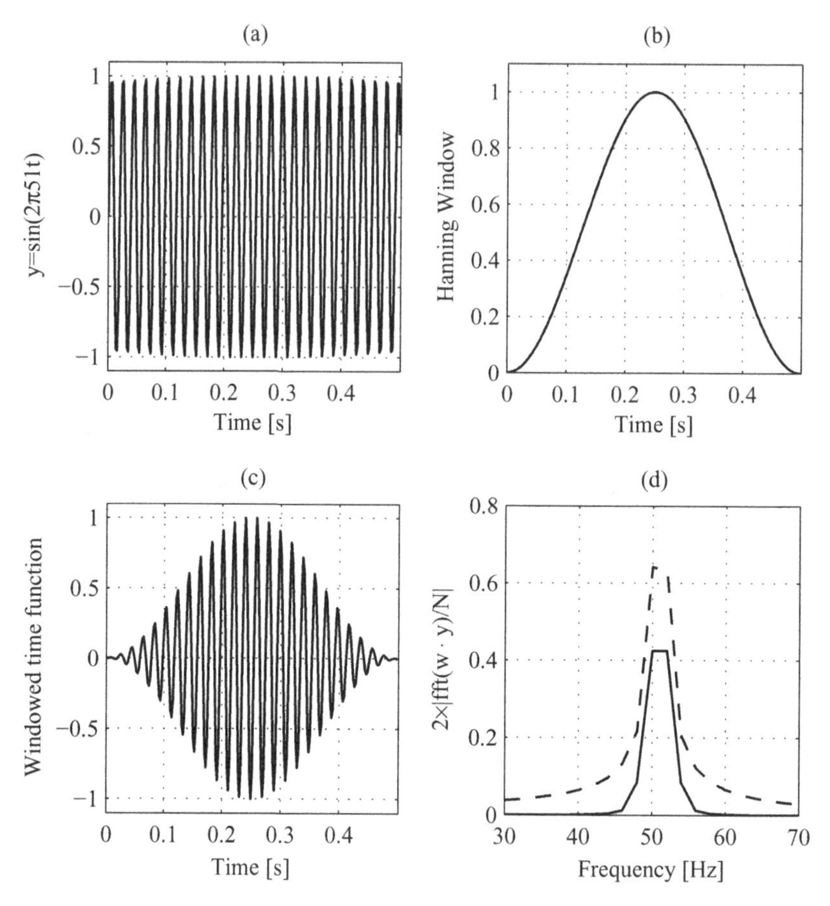


**Figure 9.9** Illustration of time-windowing with a Hanning window. The window reduces the jumps at the ends of the repeated signal. In (a) the signal is shown. In (b) is shown the Hanning window, in (c) the result of the multiplication of the two. In (d) the result of calculating the spectrum with the Hanning window (solid) and without (dashed) is shown. The spectrum now has much less leakage, but the amplitude is wrong, which will be addressed in Section 9.3.9

look at the effect of windowing a complex sine wave with a frequency which coincides with frequency line $k$, that is,

$$x_k(n) = Ae^{j2\pi f_k t} = Ae^{j2\pi kn/N} \tag{9.21}$$

when evaluated on the sample times, and for some arbitrary frequency $f_k = k\Delta f$. We now define a time window $w(n)$ and calculate the DFT for the same frequency line (all other DFT outputs will, of course, be zero). We get

$$X(k) = A\sum_{n=0}^{N-1} w(n)e^{j2\pi kn/N}e^{-j2\pi kn/N} = A\sum_{n=0}^{N-1} w(n) \tag{9.22}$$

which shows that the amplitude of the complex sine is scaled by a factor which is the sum of the window coefficients $w(n)$. Since we have already gotten accustomed to dividing the DFT by $N$, the *amplitude correction factor*, $A_w$, that we should use is the ratio of $N$ and the sum of the window coefficients, i.e.,

$$A_w = \frac{N}{\sum_{n=0}^{N-1} w(n)} \tag{9.23}$$

which, for the Hanning window is exactly 2, as we will see below. The way to apply this factor, is thus that we calculate a scaled spectrum $X(k)$ as

$$X_w(k) = \frac{A_w}{N} \sum_{n=0}^{N-1} x(n)w(n)e^{-j2\pi kn/N} \tag{9.24}$$

which will result in a double-sided spectrum with approximately half of the amplitude if $x(n)$ is a sine or cosine, at each of the positive and negative frequencies, respectively. We can then make it single-sided by the procedure discussed in Chapter 8.

### Power correction of window effects

In addition to the effect on the amplitude of periodic components, there is another effect from the window, which produces an incorrect *power* of the spectrum. This effect is due to the fact that the window bandwidth is not unity. Therefore, when applying Parseval's theorem to windowed spectra, the *equivalent noise bandwidth*, denoted $B_e$, has to be taken into account. We will use this factor in several occasions in Chapter 10 and will therefore derive it here.

We define the equivalent noise bandwidth as the width of a rectangular filter (in Hz) with the same height as the window spectrum squared at zero frequency, and which passes the same power as the time window in question. To derive the formulation for this bandwidth, we need a relation which we are going to develop in Section 13.3 for the output PSD of a filter. If we have a filter with frequency response $H$ and a double-sided input PSD, $S_{xx}$, then the output PSD, $S_{yy}$, is

$$S_{yy} = S_{xx} |H|^2 . \tag{9.25}$$

We now assume that we have a filter with a rectangular frequency response with height $h = W(0)$, the same height as the window spectrum, and width $B_e$. Furthermore, we assume that we have an input PSD limited within $\pm f_s/2$ which has a constant value of $1/f_s$ so that the total power is unity (PSD times frequency range). The output power of this filter will be

$$P_r = \frac{1}{f_s} B_e W^2(0) \tag{9.26}$$

where the square in $W^2(0)$ comes from the fact we are looking at power and not amplitude, so we have to take the square of the frequency response according to Equation (9.25).

We now look at the window as a filter with frequency response $W(f)$. The total output power of this filter, with the constant input PSD of $1/f_s$ will be

$$P_w = \frac{1}{f_s} \int_{-f_s/2}^{f_s/2} |W(f)|^2 \, df \tag{9.27}$$

which, using Parseval's theorem, can be calculated as

$$P_w = \frac{1}{f_s} \int_0^T w^2(t) dt \tag{9.28}$$

where we replace the infinity limits with 0 and $T$ because the window $w(t)$ is zero outside these limits. Observing this equation, we see that it is the Fourier transform of $w^2(t)$ evaluated at $f = 0$ where the exponential term is unity. Thus we can use Equation (9.18) to compute this equation directly from the samples of the window, as

$$P_w = \frac{\Delta t}{f_s} \sum_{n=0}^{N-1} w^2(n). \tag{9.29}$$

In the same manner we find the peak power gain of the window, $W^2(0)$, computed directly from the samples of the window, to be

$$W^2(0) = \left[ \Delta t \sum_{n=0}^{N-1} w(n) \right]^2. \tag{9.30}$$

Setting up the equality $P_r = P_w$ we now get

$$\frac{\Delta t}{f_s} \sum_{n=0}^{N-1} w^2(n) = \frac{1}{f_s} B_e \left[ \Delta t \sum_{n=0}^{N-1} w(n) \right]^2 \tag{9.31}$$

and by using the fact that $\Delta t = 1/(N\Delta f)$, this gives the *equivalent noise bandwidth*, ENBW, $B_e$, in [Hz], as

$$B_e = \frac{N\Delta f \sum_{n=0}^{N-1} w^2(n)}{\left[ \sum_{n=0}^{N-1} w(n) \right]^2}. \tag{9.32}$$

It is more useful in many cases to use the *normalized equivalent noise bandwidth*, NENBW, which we will denote $B_{en}$. The NENBW is simply the equivalent noise bandwidth divided by $\Delta f$, or

$$B_{en} = \frac{N \sum_{n=0}^{N-1} w^2(n)}{\left[ \sum_{n=0}^{N-1} w^2(n) \right]^2} \tag{9.33}$$

which for a Hanning window is exactly 1.5. The importance of this relation will be apparent in Chapter 10.

**Comparison of common windows**

There are many time windows which have been developed to optimize various properties. Many noise and vibration analysis software packages thus include a large variety of different windows from which to choose. Good overviews of different windows are found in Harris (1978) and Nuttall (1981), except for the later-developed flattop window which we will present below. For noise and vibration analysis, there is little use of this variety of windows, as the result for most windows, from a practical standpoint, is almost identical. We shall therefore examine only two windows, the Hanning and flattop windows, in more depth, as these two windows are sufficient for our purposes. We will also include the rectangular window for comparison purposes, which is the window that is effective if we do not multiply our data by any other window.

Leakage can be explained by either of two different principles. First, referring back to the convolution results in Figure 9.7, it is seen that leakage is produced in the convolution when the sine tone coincides with a side lobe. Thus reducing the side lobes will reduce leakage. There are two properties of the window that can be adjusted for this purpose: the first side lobe level, and the asymptotic falloff of higher side lobes. These properties can, however, only be reduced at the expense of increasing the width of the main lobe.

The second way to explain leakage, is by focusing on the periodic repetition of the windowed time sequence and its derivatives. It turns out that the higher derivatives of the window that are continuous in the periodic repetition, the higher the side lobe falloff will be. Of course, both these explanations are producing the same end result, but sometimes the one is more appropriate to explain a feature of the window than the other.

A second reason for wanting lower side lobes (including the first side lobe) is also worth mentioning. A real signal will, of course, in most cases be expected to have more than the single sine tone we used to illustrate the convolution process. When calculating the spectrum at one of the tones, say at $k = 0$, those other tones will be weighted by the window further away from $k = 0$ and summed into the spectrum value. For closely spaced tones, for the same reason, it is thus essential that already the first side lobe is as low as possible.

The Hanning window is probably the most common window used in FFT analysis. It is defined by half a period of a cosine, or alternatively one period of a squared sine,

$$w_H(n) = \sin^2\left(\frac{\pi n}{N}\right) = \frac{1}{2}\left(1 - \cos\left(\frac{2\pi n}{N}\right)\right) \tag{9.34}$$

for $n = 0, 1, 2, \ldots, N - 1$.

It should be particularly noted that the window defined in Equation (9.34) is periodic in the time window as described in Section 9.3.4. This means that the first value is zero, but the last value is identical to the second value, and not zero. Already Harris (1978) noted that this was often overlooked, and sadly, more than 30 years later this is still true. Thus, for example, the Hanning windows defined in MATLAB/Octave are, by default, not periodic in this way, although they can be made so by an option to the command **hann**. For large blocksizes the error is not very large, but for small blocksizes the error can be considerable, see Problem 9.1.

The Fourier transform of the Hanning window is really simple. It turns out that there are only three nonzero values in $W(f)$, namely $W(-1) = W(1) = -1/4$ and $W(0) = 1/2$. The minus sign is often neglected, which leads to erroneous phase, but does not affect the magnitude. Due to this simple spectrum, in early days of FFT/DFT, it was common to convolute by the Hanning window in the frequency domain instead of multiplying in the time domain. Due to the fast computers of today this has become less common.

The first side lobe of the Hanning window is approximately $-31.5$ dB below the main lobe maximum and the asymptotic falloff is $-18$ dB/octave. The Hanning window's Fourier transform has a main lobe that is wider than the rectangular window. This means that the picket-fence effect discussed in Section 9.3.8 results in a maximum error on the amplitude of a sine of approximately $-15\%$ (compared with $-36\%$ for the rectangular window). Note that the amplitude error is always negative or zero, since we scale the window so that a tone on a frequency line is correct, which means the main lobe maximum in the spectrum of the window is unity.

The amplitude error of 15% for the Hanning window is of course unacceptably large in many cases, for example when we want to measure the amplitude of a sinusoidal signal for calibration purposes. In that case, the *flattop window* may be utilized. The flattop window is not actually a uniquely defined window, but a name given to a group of windows with similar characteristics. The development of flattop windows was dominated by various companies manufacturing FFT analyzers in the 1980s, which has resulted in few publications specifying the coefficients of such windows. There is, however, an ISO standard, (ISO 18431-1: 2005) in which the coefficients of a relatively good flattop window are published. This window uses a formulation suggested by Nuttall (1981) where the window is built up by a sum of several sines. We formulate it in a slightly different way than the original, namely as

$$w(n) = \sum_{k=0}^{K} a_k(-1)^k cos(2\pi kn/N) \tag{9.35}$$

where the coefficients $a_k$ are real constants. For the ISO flattop window, as well as most other popular flattop windows, $K = 4$ is used. The ISO flattop window coefficients are $a_0 = 1, a_1 = 1.933$, $a_2 = 1.286$, $a_3 = 0.388$, and $a_4 = 0.0322$. This window has a first side lobe of approximately $-84$ dB, which essentially makes it unnecessary to worry about the falloff rate. The main benefit of the flattop window is its very flat main lobe characteristics, which give a maximum amplitude error of less than 0.1%, and makes the window well suited to estimate sine amplitudes.

Actually the ISO window is not very well optimized, and hence Tran *et al.* (2004) have published modified coefficients that reduce the first side lobe to $-88$ dB without essentially affecting the main lobe width. We will use this excellent flattop window in our comparison below. Other examples of flattop windows can be found in Reljin *et al.* (2007) which also provides some more insight into these windows. In Figure 9.10 the three windows, rectangular, Hanning, and flattop, with their Fourier transforms are shown for comparison. In Table 9.2 some useful figures for the three windows we have discussed are presented.

There is a price to pay for the decreased amplitude uncertainty when we use time windows. The price is in the form of increased frequency uncertainty, which occurs because the better the amplitude uncertainty, the wider the main lobe of the spectrum of the window. Therefore, if we measure a sinusoid with a frequency that matches one of our spectral lines, then the peak will become wider than if we had used the rectangular window. The flattop window, which has the best amplitude uncertainty, also has the widest main lobe, which, of course, is a direct result of the picket-fence effect. This trade-off is related to the bandwidth-time product which is explained more in relation to errors in PSD computation with windowing in Section 10.3.5.

Figure 9.11 (a) shows the result of the DFT of a sinusoid which exactly matches a spectral line after windowing with the Hanning window and with the flattop window, respectively. As shown in the figure, the Hanning window results in three nonzero spectral lines, while the flattop window gives 9 nonzero spectral lines. With this broadening of the peaks, there is a larger uncertainty of the exact frequency of the tone. Even with windows other than the rectangular, we get some leakage when the sinusoid's frequency does not coincide with a spectral line, as seen in Figure 9.11 (b), but at much lower levels than with the rectangular window. On the other hand we get a broadening of the peak as illustrated in Figure 9.11. The flattop window, because of its large main-lobe width, should only be used when it is known that the spectrum does not contain any closely spaced tones. The Hanning window should, therefore, be used as a standard window, since it provides a good compromise between amplitude accuracy and frequency resolution.

Finally in this section, we will present two more windows which are not recommended for spectral estimation, but which are nevertheless important to present, for different reasons. First, we present the Bartlett, or triangular window, which enters into our discussion on PSD estimation in Chapter 10, and second, the half sine window, which has recently been proposed for frequency response estimation (Antoni and Schoukens 2007) as an alternative to the Hanning window.

The Bartlett, or triangular window, is defined by a triangular shape,

$$w(n) = \begin{cases} 2n/N, & n = 0, 1, \ldots, N/2 \\ 2(N-n)/N, & n = N/2+1, N/2+2, \ldots, N-1 \end{cases} \tag{9.36}$$

which, like our other windows is periodic in its repetition. This window is actually the convolution of a rectangular window with itself, and is therefore of interest in spectral estimation, see Section 10.4.

The half sine window is defined as half a sine

$$w(n) = sin(\pi n/N) \tag{9.37}$$

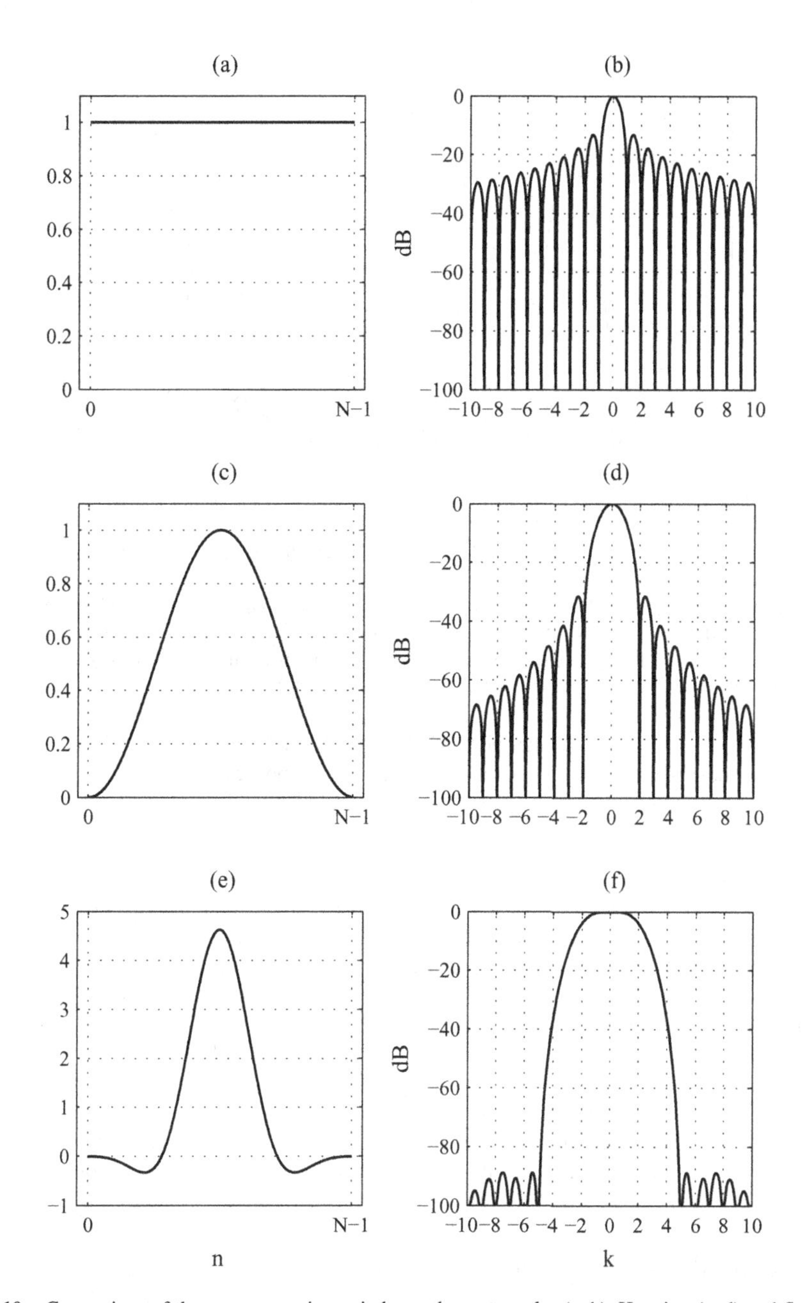


**Figure 9.10** Comparison of three common time windows; the rectangular (a, b), Hanning (c, d) and flattop (e, f) windows in time domain (left) and frequency domain (right). The Hanning window's first zero is situated at $k = 2$, which means that for a sinusoid situated between two frequency values in the DFT, the convolution with the window spectrum will make the value for $k = 0$ in Figure 9.7 be attenuated much less than for the rectangular window. For the flattop window almost no attenuation occurs. The uncertainty in amplitude therefore decreases when using windows with a periodic signal. On the other hand, with wider main lobes, the spectral peaks are broadened, which results in increased frequency uncertainty, see below

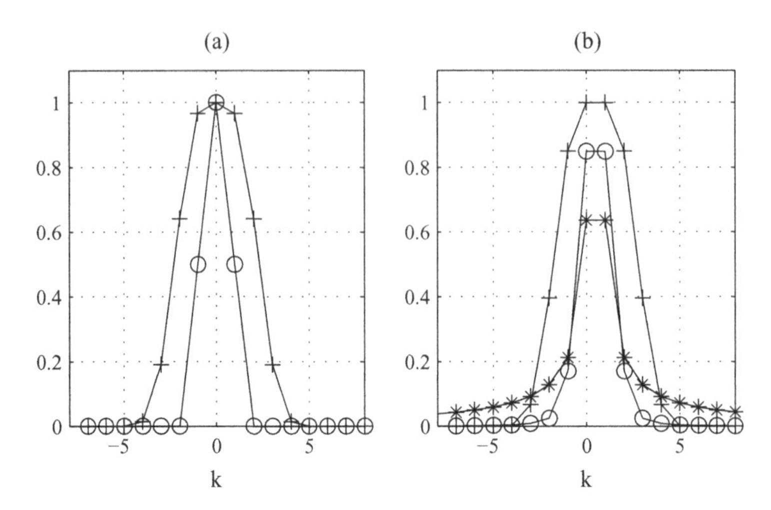


**Figure 9.11** The widening of the frequency peak is the price we pay to get a more accurate amplitude. In (a), the scaled DFT result is shown for a sine with amplitude 1 and frequency that matches the spectral line marked $k = 0$, for both Hanning (solid, circles) and flattop (solid, plus sign) windowing. With the flattop window the peak is much wider than with the Hanning window. The rectangular window is omitted as it would simply have one value of 1 at $k = 0$ in this case with no leakage. In (b) similar results for the case when the sine frequency is exactly midway between frequency lines $k = 0$ and $k = 1$ are shown, including the result for the rectangular window (solid, asterisks)

for $n = 0, 1, \ldots, N - 1$. This is a rather poor window for spectral analysis, but in Chapter 13 we will show that it has some advantages for estimation of frequency responses.

**Frequency resolution**

From the above discussion about widening of frequency peaks, it is clear that with a certain frequency increment, $\Delta f$, one may not, after the DFT computation, be able to distinguish two sinusoids, separated in frequency by only one, or a few, spectral lines. Two closely spaced sine tones can potentially result in one peak in the DFT result. For this reason we should differentiate between frequency increment and *frequency resolution*. Frequency resolution usually implies the smallest frequency difference that is possible to discern between two signals, while the frequency increment is the distance between two frequency values in the DFT computation, that is, $\Delta f$. Frequency resolution depends on the window, while the frequency increment depends only on the measurement time, $T$.

There is no exact frequency resolution for a particular window. How close two sinusoids can be in frequency, in order for the spectrum to still show two peaks, depends on the width of the window's main lobe, but also on where between the spectral lines the two sine waves are located, and on their respective amplitudes. In Problem 9.3, we will look into some aspects of frequency resolution.

## *9.3.10 Time Windows for Random Signals*

The window's influence on a random signal is quite different than the effect on periodic signals described in Section 9.3.9 due to the fact that random signals have continuous spectra, as opposed to periodic signals having discrete spectra. The result of the convolution between the continuous Fourier transform of the window and the random signal is therefore more complicated to visualize. If we recall that convolution implies that the qualities of both signals are 'mixed,' we can understand that the window will introduce a 'ripple' in the random signal's spectral density. At the same time, we get a smoothing of the spectral density, due to the influence of the main lobe. For narrow peaks in the spectrum, for example if we measure vibrations on resonant systems with low damping as discussed in Chapter 5, we get an undesired widening of the resonance peaks. More on these bias errors will be discussed in Section 10.3.4.

The qualities most important for the influence of the window when determining spectral densities are the width of the main lobe (which creates a broadening of sharp peaks) and the height of the side lobes (which create leakage). The lower the side lobes are, the less influence we get from nearby frequency content during convolution. The flattop window should never be used for random signals, because of its wide main lobe. The most common window is the Hanning window and there is little use of any other time window for PSD calculation of random signals on noise and vibration signals.

### 9.3.11 Oversampling in FFT Analysis

If we use a blocksize of $N$ time samples, the DFT results in half as many, that is, $N/2 + 1$ positive frequency (usable) spectral lines, as discussed in Section 9.3.5. Since the analog anti-aliasing filter is not ideal, but has some slope above the cutoff frequency, see Section 11.2.2, we cannot sample with a sampling frequency which is exactly $2 \cdot B_{max}$, the bandwidth of the signal. In data acquisition systems for noise and vibration analysis, a common oversampling factor is 2.56, although slightly lower oversampling factors (down to approximately 2.2) are sometimes used in the later generation systems, see Chapter 11. Thus, we can only use the discrete frequency values up to $k = N/2.56$ as the remaining frequency lines will be contaminated by aliasing.

Due to this fact, many data acquisition and analysis systems for noise and vibration analysis throw away the upper frequency values above $k = N/2.56$. Although this was reasonable many years ago when memory storage was expensive, today it is unfortunate because it means that some information is lost and that performing inverse DFT results in less accuracy than would be possible if the upper frequency values had been stored. In Table 9.3 some common number of spectral lines stored for different blocksizes are tabulated.

### 9.3.12 Circular Convolution and Aliasing

In some cases we want to compute the multiplication of two time functions or two spectra, for example to produce the filtering of a signal by multiplying the Fourier transform of the signal with a frequency response. When manipulating spectra like this it is essential to understand the *circular convolution* property of DFT, and how to avoid this, usually unwanted, property.

We have learned that for the continuous Fourier transform, a multiplication in one domain is equivalent to a convolution in the other domain. Note that this is the kind of convolution we discussed in

**Table 9.3** Typical blocksizes and corresponding numbers of stored spectral lines (unfortunately) used by many noise and vibration analysis systems. It should be emphasized that it is better to keep all frequency values up to $k = N/2$, the Nyquist frequency

| Block size | # of spectral lines |
|---|---|
| 256 | 101 |
| 512 | 201 |
| 1024 | 401 |
| 2048 | 801 |
| 4096 | 1601 |
| 8192 | 3201 |

Sections 9.3.7 and 9.3.9, even though in conjunction with the DFT, because we were dealing with continuous signals.

If, however, two DFT results, $X_1(k)$ and $X_2(k)$, are multiplied, due to the circular nature of the DFT, the equivalent relationship in the time domain is a circular convolution, denoted by $\circledast$, so that

$$y(n) = \text{IDFT}[X_1(k)X_2(k)] = x_1(n) \circledast x_2(n) \tag{9.38}$$

which is not equal to the ordinary convolution. We will not present the mathematics applying to this, but restrict the presentation with an example followed by a simple solution to avoid this phenomena, that will suffice for our purposes.

**Example 9.3.1** *Assume we have two signals (sequences) $x_1(n) = x_2(n) = 1$, for $n = 0, 1, 2, 3$. The ordinary convolution of these sequences (which can be obtained in MATLAB/Octave by the* **conv** *command) is the sequence*

$$x_1 * x_2 = 1, 2, 3, 4, 3, 2, 1 \tag{9.39}$$

*However, if we try to run the following MATLAB/Octave code*

```
x3 = ifft(fft(x1).*fft(x2));
```

*we obtain the sequence*

$$x_3(n) = x_1 \circledast x_2 = 4, 4, 4, 4 \tag{9.40}$$

*which is certainly not what we want. So how can we avoid this? The simple way is to add zeros to the two sequences $x_1(n)$ and $x_2(n)$, by redefining them as the sequences $1, 1, 1, 1, 0, 0, 0, 0$. The result of the IDFT of this sequence, which can easily be calculated with a small modification to the code above, namely*

```
x3 = iff(fft(x1,8).*fft(x2,8));
```

*i.e., by specifying a twice as large FFT blocksize N to the* ***fft*** *command. The result will now be*

$$x_3(n) = 1, 2, 3, 4, 3, 2, 1, 0 \tag{9.41}$$

*which solves the problem. This is one of the applications of* zero padding, *which we will look at a little closer in the next section.*

*End of example.*

The cyclic convolution can be seen as *aliasing* either in the time domain or the frequency domain, as it moves energy from one place to another in either domain (the opposite domain to where the multiplication took place).

### 9.3.13 Zero Padding

Zero padding is a technique where the DFT (or IDFT) is computed on a sequence that has been extended by zeros at the end. It is frequently used in various signal processing tasks. We saw an example in the previous section where it was used to produce an ordinary convolution instead of a circular convolution, when convolving two time domain signals by a frequency domain (DFT) multiplication. This is an important use of zero padding which we will use for estimating correlation functions in Section 10.4. It should be noted that zero padding does not add any information to the data, so the spectrum will not contain any information not already there.

Another use of zero padding is to compute spectra with a finer frequency resolution than the data sequence admits, particularly for transients, as we discussed in Section 9.3.6. In this case the zeros are added to the time sequence before the DFT. This procedure works correctly, if the time sequence is transient in the original time window with length $N$, i.e., it has already died out before we add extra zeros. The resulting spectrum will be a spectrum with finer resolution than if only the length of the sequence had been used. This technique was used to produce the spectra of the time windows in, e.g., Figure 9.10 and is correct because the time windows are assumed to be zero outside the measurement.

Zero padding is often also used in the frequency domain, before computing an IDFT. In this case the procedure produces interpolation in the time domain.

In early work on spectral estimation, several suggestions to use zero padding were published. Therefore it is not uncommon to still find suggestions to use zero padding in spectral estimation procedures. This is, however, due to a misconception as zero padding (in the time domain) only produces an interpolation in the spectrum between the spectrum values obtained without zero padding (Kay and Marple 1981). Higher-frequency resolution can only be accomplished by increasing the actual measurement time and zero padding in spectral estimation should be avoided.

## *9.3.14 Zoom FFT*

In many commercial systems for noise and vibration analysis, there is an option allowing the DFT (FFT) frequency bins to be concentrated to a limited frequency range $f_{\min} \le f \le f_{\max}$. This is often referred to as *zoom FFT analysis*, and is based on the Fourier transform property 8 in Table 2.2, which is repeated here for the reader's convenience. The transform pair states that if a time signal $x(t)$ has a Fourier transform $X(f)$, then the time signal $x(t)e^{j2\pi at}$ has the Fourier transform $X(f-a)$. Thus, by multiplying the measured time signal by the exponential term $e^{j2\pi at}$, which is easily done digitally, the spectrum of the signal is translated down to frequency $f-a$. We will show the procedure to follow to use this digital zoom capability by an example where we have a signal sampled with $f_s = 10$ kHz, which thus has a frequency range of $0 \le f \le 5$ kHz. We wish to zoom it in the frequency range $1900 \le f \le 2100$ Hz, and use a 1024 sample FFT.

1. Define the center frequency $f_c = 2000$ Hz, in the middle of the requested frequency range $1900 \le f \le 2100$ Hz.
2. Multiply the entire time signal by $e^{j2\pi f_c t}$. Note that this produces a complex signal. This will shift the frequencies so that the requested range is now $-100 \le f \le 100$ Hz.
3. Apply a lowpass filter of bandwidth $B = 100$ Hz, to each of the real and imaginary parts of the frequency shifted signal.
4. Extract every 25th sample ($5000/200 = 25$) from these lowpass filtered signals.
5. Put the real and imaginary part of the decimated signal back as a complex signal.
6. Apply the 1024 sample FFT to this complex signal.
7. Shift the negative frequencies to the lower half of the spectrum (e.g., by the MATLAB/Octave **fftshift** command).

It should be noted that zoom FFT processing does not violate the important relationship that the measurement time required for the time block is $T = 1/\Delta f$. This is a result of the fact that we decimate the signal in step 4, so we have to use 25 times longer data (in the original signal) than the 1024 samples we are using in the FFT.

An alternative way of obtaining exactly the same frequency resolution would be to use a blocksize of $N = 25 \cdot 1024$ samples. With the large possible FFT blocksizes possible in, for example, MATLAB or Octave, there is little use for zoom FFT for noise and vibration analysis (in the sense of post processing data). In some real-time applications, however, the zoom FFT still has its place.

## 9.4 Chapter Summary

We started this chapter with a discussion of nonparametric versus parametric methods for frequency (or spectral) analysis. The methods are called nonparametric if they do not require any *a priori* information about the model of the data (or form of the spectrum, if you like).

Nonparametric methods are the most common methods for spectrum analysis of noise and vibration signals, because of their reliability and generality. The most common type is FFT/DFT analysis, and octave and fractional octave analysis also belongs to the group of nonparametric frequency analysis methods.

The discrete Fourier transform, DFT, and the fast algorithm to compute it, the fast Fourier transform, FFT, were then presented and discussed in some detail. The DFT, $X(k)$ of a signal, $x(n)$, is defined by

$$X(k) = \sum_{n=0}^{N-1} x(n)e^{-j2\pi kn/N} \tag{9.42}$$

for $k = 0, 1, \ldots, N - 1$. The inverse DFT, IDFT, is similarly defined by

$$x(n) = \frac{1}{N} \sum_{k=0}^{N-1} X(k)e^{j2\pi nk/N} \tag{9.43}$$

for $n = 0, 1, \ldots, N - 1$.

The frequencies of the DFT, $k\Delta f$, are located at $f = k/T$, where $T = N\Delta t$ is the measurement time. Thus the frequency increment is $\Delta f = 1/T$.

We noted that the DFT is using the orthogonality properties of sines and cosines to extract any frequency content from $x(n)$ around each of the frequency lines, $k$. Each of these frequency lines come from a multiplication of $x(n)$ by a cosine (for the real part of $X(k)$), and a sine (for the imaginary part of $X(k)$) with $k$ periods within the measurement time $T = N\Delta t$.

We then noted that the DFT is periodic in both time and frequency, i.e., $X(k + N) = X(k)$ and the inverse DFT gives us a signal for which $x(n + N) = x(n)$. This can be interpreted such that the DFT gives us the spectrum of the periodic repetition of the measured signal $x(n)$.

As for the continuous Fourier transform, the symmetry properties of the DFT results in spectra of real signals $x(n)$ that have the properties,

$$\text{Re}\,\{X(-k)\} = \text{Re}\,\{X(k)\} \tag{9.44}$$

and

$$\text{Im}\,\{X(-k)\} = -\text{Im}\,\{X(k)\} \tag{9.45}$$

i.e., the real part of $X(k)$ is even, and the imaginary part is odd. This fact can be used to save some space by discarding the upper $N/2 - 1$ values of $X(k)$ because these values can then be recreated using the symmetry equations. It is important to note, however, that the value $X(N/2)$, (i.e., the 513th value for a blocksize of 1024, for example) is a real number which is not repeated in the symmetry.

Due to the truncation of continuous signals when computing the DFT, for such signals an error called leakage will affect the spectrum $X(k)$. Leakage in spectrum estimates can be reduced by using a time window, and we presented the Hanning window as the favored standard window, and the flattop window for the particular situation when we want to measure the amplitude or RMS value of periodic components accurately. The time windows typically trade frequency resolution for amplitude accuracy; the wider the main lobe of the window, the better amplitude accuracy we get for a periodic component, but at the expense of a broadening of the spectrum peak.

The leakage effects can be divided into three parts:

- a broadening of peaks due to the main lobe – an interpolation effect
- the additive effect of spectral content outside a frequency bin onto the DFT result at that frequency bin – true leakage
- a signal-to-noise deterioration caused (essentially) by the ratio of the measured (true) signal and the RMS of extraneous noise, both calculated inside the main lobe.

## 9.5 Problems

Many of the problems following are supported by the accompanying ABRAVIBE toolbox for MATLAB/Octave and further examples which can be downloaded with the toolbox. If you have not already done so, please read Section 1.6, and follow the instructions to download this toolbox together with example files.

**Problem 9.1** *Create a sampled time signal of the signal* $x(t) = 3\cos(2\pi 20t) + 5\cos(2\pi 40t)$ *using the following parameters:*

*Sampling frequency,* $f_s = 512$ *Hz,*

*Number of samples,* $N = 1024$.

*Calculate a spectrum using the DFT with Hanning window (use the ABRAVIBE toolbox* ***ahann*** *command), and scale it for correct amplitudes. Create a correct frequency axis and plot the spectrum, and make sure the cosines appear at the right frequencies.*

*Note: If you have done everything right, the signal should be periodic in the time window. You should have no leakage.*

*Rerun the example replacing the window with the default MATLAB/Octave* ***hann*** *window, which is not periodic, see Section 9.3.9. Observe the differences.*

**Problem 9.2** *Calculate the RMS value of* $x(t)$ *defined in Problem 9.1 analytically, using the orthogonality criteria of cosines. Then use the signal to calculate the RMS value of* $x(t)$ *in the time domain and verify that they are the same.*

**Problem 9.3** *Create a time signal of the signal* $x(t) = 3\cos(2\pi f_1 t) + 5\cos(2\pi f_2 t)$ *using the following parameters:*

*Sampling frequency,* $f_s = 512$ *Hz,*

*Number of samples,* $N = 1024$.

*Investigate the results of a scaled spectrum of* $x(n)$ *using Hanning window, with different frequencies* $f_1$ *and* $f_2$ *which are exactly on frequency lines, and midway between frequency lines. How close can they be and still be clearly separated?*

**Problem 9.4** *Repeat Problem 9.3 with flattop window instead (use the ABRAVIBE toolbox command* ***aflattop****, or make a MATLAB/Octave function using the window defined in Section 9.3.9).*

**Problem 9.5** *Use MATLAB/Octave to calculate the (regular) convolution between the following two sequences, with the command* ***conv*** *and by FFT, as explained in Section 9.3.12, and verify you get the same result.*

```
x1 (n) = 1 1 1 1 1 1 1 1
x2 (n) = 1 2 3 4 3 2 1 1
```

## References

Antoni J and Schoukens J 2007 A comprehensive study of the bias and variance of frequency-response-function measurements: Optimal window selection and overlapping strategies. *Automatica* **43**(10), 1723–1736.

Bendat J and Piersol AG 2010 *Random Data: Analysis and Measurement Procedures* 4th edn. Wiley Interscience.

Cooley J, Lewis P and Welch P 1967 Historical notes on the fast Fourier transform. *IEEE Trans. on Audio and Electroacoustics* **15**(2), 76–79.

Cooley JW and Tukey JW 1965 An algorithm for machine calculation of complex Fourier series. *Mathematics of Computation* **19**(90), 297–301.

Cooley JW and Tukey JW 1993 On the origin and publication of the FFT paper - a citation-classic commentary on an algorithm for the machine calculation of complex Fourier-series. *Current Contents/Engineering Technology & Applied Sciences* (51-52), 8–9.

Harris FJ 1978 Use of windows for harmonic-analysis with discrete Fourier-transform. *Proceedings of The IEEE* **66**(1), 51–83.

Heidemann MT, Johnson DH and Burrus CS 1985 Gauss and the history of the fast Fourier-transform. *Archive for History of Exact Sciences* **34**(3), 265–277.

ISO 18431-1: 2005 *Mechanical Vibration and Shock – Signal Processing – Part 1: General Introduction*. International Organization for Standardization, Geneva, Switzerland.

Kay SM and Marple SL 1981 Spectrum analysis - a modern perspective. *Proceedings of The IEEE* **69**(11), 1380–1419.

Newland DE 2005 *An Introduction to Random Vibrations, Spectral, and Wavelet Analysis* 3rd edn. Dover Publications Inc.

Nuttall AH 1981 Some windows with very good sidelobe behavior. *IEEE Transactions on Acoustics Speech and Signal Processing* **29**(1), 84–91.

Oppenheim AV and Schafer RW 1975 *Digital Signal Processing*. Prentice Hall.

Oppenheim AV, Schafer RW and Buck JR 1999 *Discrete-Time Signal Processing*. Pearson Education.

Proakis JG and Manolakis DG 2006 *Digital Signal Processing: Principles, Algorithms, and Applications* 4th edn. Prentice Hall.

Reljin IS, Reljin BD and Papic VD 2007 Extremely flat-top windows for harmonic analysis. *IEEE Transactions on Instrumentation And Measurement* **56**(3), 1025–1041.

Thrane N 1979 The discrete Fourier transform and FFT analyzers. Brüel & Kjær Technical Review 1.

Tran T, Claesson I and Dahl M 2004 Design and improvement of flattop windows with semi-infinite optimization *The 6th International Conference on Optimization : Techniques and Applications*, Ballarat, Australia.

# 10

# Spectrum and Correlation Estimates Using the DFT

In software for noise and vibration analysis, the FFT/DFT is typically used to estimate spectra as we mentioned in Chapter 9. Depending on the type of signal being analyzed, different types of spectra are recommended, as explained in Chapter 8. In this chapter we will discuss how spectra of these different types of signals should be scaled and interpreted.

Much of the available literature on spectral analysis, or spectrum estimation, is rather theoretical and it is often difficult to see, for example, how to obtain correctly scaled spectra. It is also difficult to find information on how to select a suitable spectrum estimator for a particular measured signal. In this chapter we will therefore present details about scaling spectra as well as giving detailed descriptions of how to select a proper spectrum type for different signals.

The discussion of Welch's estimator for auto and cross-spectral density in Section 10.3.2 is particularly thorough, as this method, despite the fact that it is virtually the only method available in commercial noise and vibration analysis software, is not described comprehensively in many other textbooks. In Section 10.6 at the end of the chapter, we also discuss some practical aspects of performing proper frequency analysis for some typical situations.

More general background theory on the topics of this chapter can be found in (Bendat and Piersol 2010; Newland 2005; Wirsching *et al.* 1995) and in the references at the end of the chapter.

## 10.1 Averaging

Many measurement signals contain random noise, either because the signal is random, or because it is periodic or transient, but contains random contaminating noise. In such cases several spectra are often averaged, frequency by frequency, to reduce the random error, $\varepsilon_r$, (see Section 4.2.2) of the spectrum estimate. This is often referred to as *ensemble averaging* or *frequency domain averaging*. Another form of averaging, *time domain* averaging, is sometimes used for certain types of signals. This type of averaging will be discussed in Section 11.3.4.

An illustration of the segment based processing used for frequency domain averaging in spectrum analysis software for noise and vibration analysis, is shown in Figure 10.1. The entire time signal is divided into $M$ segments which are independently used for a DFT calculation of each data segment. The squared magnitude value of each DFT result for each frequency value, $k$, of several subsequent spectra are averaged. In the case of deterministic (periodic or transient) signals, typically a few, say, 3–10 averages

*Noise and Vibration Analysis: Signal Analysis and Experimental Procedures,* First Edition. Anders Brandt.

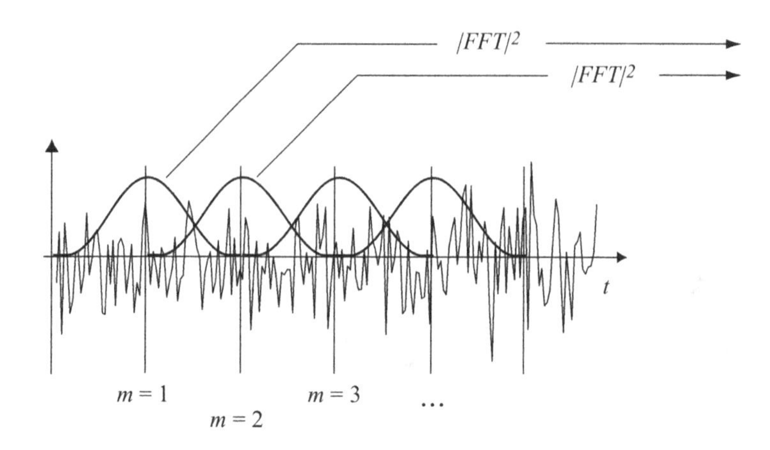


**Figure 10.1** Illustration of segment-based processing used for frequency domain averaging. The data are divided into a number of segments, possibly overlapping as in the figure, where 50% overlap is illustrated. Each segment of data is typically windowed, and then a DFT is performed on the windowed data segment. Finally, the magnitudes squared of the DFT results are averaged for each discrete frequency separately

are usually necessary, while several hundred averages may be necessary for signals of random nature, see Section 10.3.5. Perhaps it should be pointed out here, that although this is the typical implementation in current software for noise and vibration analysis, in Section 10.3.6 a different approach to spectrum estimation will be presented.

When averaging spectra, the squared magnitude values of the DFT have to be averaged, because we want the averaged result to give a correct RMS value (see Section 10.2.1). Spectra of transient signals can be averaged if several transients can be captured in a repeatable fashion, for example if the measurement is triggered by a level trigger.

*Overlap processing* is sometimes used in the averaging process, as indicated in Figure 10.1, particularly when computing spectral densities of random signals. Each time sample is then used more than once, so that the final average will contain more averaged DFT results from the same time data than if no overlap of the segments was used. The reason this gives a better result is essentially that the time window used prior to the DFT calculation removes some information in the data at the ends, where the window approaches zero. The amount of overlap that should be used depends on the time window. With the Hanning window, 50% overlap is usually seen as optimal, see Section 10.3.5.

## 10.2 Spectrum Estimators for Periodic Signals

For periodic signals, theoretically we use the Fourier series to describe the frequency contents. In practice, however, it is more common to use a scaling of spectra so that a peak in the spectrum is equivalent to the RMS value of a periodic component at that frequency. This is due to the fact that the amplitude of a signal is only defined for a single sine. As soon as we have more than one frequency component in a signal, it is more relevant to look at the RMS value. In most software for noise and vibration analysis, there is a choice for scaling spectra of periodic signals either to RMS or amplitude (the latter usually called 'peak' scaling) but it is strongly recommended to use RMS scaling. Due to the availability of different scaling, it is essential to document the scaling type used, as correct interpretation of the spectrum is otherwise impossible. This is usually done in the units of the plot, for example by typing 'Acceleration, [m/s$^2$ RMS]' or similar. In the following discussion we will assume RMS scaling consistently.

A periodic signal is deterministic, which means that, theoretically, a single DFT should give the correct spectrum. However, as measurement signals are almost always contaminated by extraneous noise from sensors and data acquisition equipment, or often from the vibration source itself being a combination of periodic and random contributions, it is often necessary to average a few spectra to get stable spectrum values. As mentioned in Section 10.1, the averaging is always done on squared magnitude values of the DFT, to keep RMS scaling consistent. In Section 10.6.4 we will discuss how to treat signals which contain contributions of both random and deterministic signals, for example where the main power in the signal is random, but where some periodic components are included.

## 10.2.1 The Autopower Spectrum

The *autopower spectrum* is a spectrum scaled to the square of the RMS value, the *mean power* or *mean square value*, of the signal at each frequency. This spectrum thus has square units, that is, if we measure voltage, the units of the autopower spectrum becomes [$V^2$]. We have already seen in Chapter 9 how to scale the DFT to yield correct amplitudes. Thus we have a suitable estimator for a single-sided autopower spectrum, $A_{xx}$, using $M$ averages, by

$$\hat{A}_{xx}(k) = \frac{S_A}{M} \cdot \sum_{m=1}^{M} \left|\hat{X}_{w,m}(k)\right|^2 \tag{10.1}$$

for $k = 0, 1, \ldots, N/2$, for some constant $S_A$ to provide correct scaling. Note that we use the 'hat' (^) to denote it is an estimate of the true spectrum as we discussed in Chapter 4. In Equation (10.1), $\hat{X}_{w,m}(k)$ is the windowed DFT of segment $m$, according to Equation (9.5), where the segmentation is done as was illustrated in Figure 10.1, either with or without overlap processing.

It follows from the discussion in Section 9.3, that the entire process to obtain a DFT scaled to RMS for one segment is

$$X_w = \frac{A_w}{N\sqrt{2}} \sum_{n=0}^{N-1} x(n)w(n)e^{-j2\pi kn/N} \tag{10.2}$$

where $A_w$ is the amplitude correction factor of the window, $w(n)$, from Equation (9.23), $N$ is the blocksize, and $\sqrt{2}$ scales to the RMS value. $X_w$ is a double-sided spectrum. We recall that this scaling results in double-sided peaks of half the RMS value each of a periodic component. When we square this spectrum, we will obtain a quarter of the RMS on each frequency bin which equals half the RMS value squared on each bin, instead of the RMS squared as we wish. We thus need to put an additional factor 2 in $S_A$ to get half the RMS squared on each positive and negative frequency. Finally, we add a factor 2 to make the spectrum single-sided, for all $k$ except the DC bin. Thus we find that the scaling constant $S_A$ for a correctly scaled autopower spectrum is

$$S_A = \begin{cases} \dfrac{2A_w^2}{N^2}, & k \neq 0 \\ \dfrac{A_w^2}{N^2}, & k = 0 \end{cases} \tag{10.3}$$

because we always have to treat the DC frequency line separately.

The autopower spectrum has some disadvantages; first of all, it is squared, which is not very intuitive for periodic components. We would rather like to have linearly scaled values of each periodic component for practical interpretation of periodic signals. Second, it is also very easy to confuse autopower spectrum with (auto) power spectral density for random signals. It is therefore not recommended to use this spectrum, but rather to take the square root of it, which results in the linear spectrum, see for example (ISO 18431-1: 2005).

### 10.2.2 Linear Spectrum

For reasons mentioned at the end of the last section, it is recommended to use the *linear spectrum* for periodic signals. With RMS scaling as described in Section 10.2.1, this spectrum is also referred to as *RMS spectrum*. It is more intuitive to interpret a periodic signal by directly presenting the RMS value of each of its periodic components, and not the square of those values. It should be especially noted that the averaging process when computing linear spectra, however, is still done on the autopower spectrum, as in Equation (10.1), and the square root is taken after the averaging is finished. We denote the linear spectrum by $X_L(k)$ to stress the fact that it is similar to the DFT, but must still be distinguished from it. The linear spectrum, or RMS spectrum, estimator is thus defined by

$$\hat{X}_L(k) = \sqrt{\hat{A}_{xx}(k)}. \tag{10.4}$$

The procedure for computing linear spectra of periodic signals is as follows.

1. Remove the mean of the time signal $x(n)$,
2. divide the total signal into $M$ segments, if averaging is wanted,
3. select a window and calculate the windowed FFT, $X_{w,m} = \text{DFT}[x(n)w(n)]$, of length $N$, of each segment,
4. calculate the magnitude squared of each $X_{w,m}$ from step 3,
5. for each frequency, make an average of all $X_{w,m}$, for $m = 1, 2, \ldots, M$,
6. calculate the scaling factor $S_A$ by Equation (10.3) and multiply the average from the previous step by this scaling, and
7. finally, compute the linear spectrum $\hat{X}_L(k)$ by taking the square root of the scaled autopower spectrum.

In addition to the steps in the list above, there are some practical aspects to estimate a relevant linear spectrum of a periodic signal. We will discuss this in Section 10.6.2.

### 10.2.3 Phase Spectrum

The linear spectrum, $X_L(k)$, in Equation (10.4) has no phase information since it is based on the absolute value (squared) of the DFT, and also because a single signal does not contain phase information since phase is a relative concept. Sometimes, e.g., for computing operating deflection shapes, ODS, as discussed in Section 16.6, however, we need the phase information relative to a reference channel. In such cases, a *phase spectrum* with phase relative to a reference channel (signal) can be used. The way to obtain an average phase relationship between two signals, is to calculate the cross-power spectrum of a signal $x(n)$ with respect to a reference signal $v(n)$, and take the phase from this averaged estimate. The cross-power spectrum for a periodic signal, $x(n)$, relative to a reference signal $v(n)$ is calculated by

$$\hat{A}_{xv}(k) = S_A \sum_{m=1}^{M} X_{w,m}(k) V^*_{w,m}(k) \tag{10.5}$$

where the scaling constant $S_A$ is the same as in Equation (10.3), and $V^*_{w,m}(k)$ is the complex conjugate of the DFT of the reference signal $v(n)$ of windowed segment $m$, to which the phase is related. The phase $\phi_{xv}$ of this cross-power spectrum is then added to the linear spectrum $X_L$, to form a *phase spectrum*

$$\hat{X}_P(k) = \hat{X}_L(k) e^{j\hat{\phi}_{xv}(k)} \tag{10.6}$$

where

$$\hat{\phi}_{xv}(k) = \angle \hat{A}_{xv}(k) \tag{10.7}$$

where $\angle$ denotes phase angle in [rad/s]. The absolute value of the phase spectrum is thus equal to the linear spectrum in Equation (10.4). You should note that the phase angle in Equation (10.7) comes from the mean value of the phase difference between both signals. This implies in principle that the signal $x$ is viewed as the output from a linear system, where $v$ is the input, see Chapter 13. The phase spectrum is the recommended spectrum estimator for ODS when the signals are periodic, see Section 16.6.

## 10.3 Estimators for PSD and CSD

We are now going to describe several estimators for autospectral and cross-spectral density for random signals. The history of estimators for PSD and CSD goes back to the days before the FFT algorithm. In those days the most suitable way to obtain a PSD estimate was to first compute a correlation function, and then use the Wiener–Khinchine relations discussed in Section 8.3.1, because correlation functions could be computed reasonably well, but spectra (DFT) were expensive due to the lack of the FFT algorithm. This method is called the Blackman–Tukey method, (Blackman and Tukey 1958a,b) and it will be discussed briefly in Section 10.3.6, although we will use a more concurrent method, namely the smoothed periodogram.

The development of the FFT algorithm soon made the Blackman–Tukey method obsolete, as (Welch 1967) published a method more suitable for use with low-memory computers. The Welch method is the method usually implemented in commercial noise and vibration analysis software because it is particularly well suited for computer processing. We will however later in this section discuss some advantages of reviving the Blackman–Tukey method (or more precisely, the similar smoothed periodogram method).

### *10.3.1 The Periodogram*

The basis for most FFT based PSD estimation is the *periodogram*, $\hat{P}_{xx}(k)$, which is simply the magnitude squared of a (long) DFT of the entire time signal, $x(n)$, scaled by the blocksize, i.e.,

$$\hat{P}_{xx}(k) = \frac{1}{L}\left|\sum_{n=0}^{L} x(n)e^{-j2\pi kn/N}\right|^2 \tag{10.8}$$

if we assume the length of the entire signal $x(n)$ is $L$ samples. The periodogram is an estimate of the PSD, but a very poor one. It can be shown (Bendat and Piersol 2010) that the periodogram is an *inconsistent* estimator of the PSD $S_{xx}$, i.e., it does not approach the true PSD even if we make $L$ very large. It is, however, the building block of most PSD estimators.

Similar to Equation (10.8) we can define a cross-periodogram as

$$\hat{P}_{yx}(k) = \frac{1}{L}\sum_{n=0}^{L} y(n)x^*(n)e^{-j2\pi kn/N}. \tag{10.9}$$

The periodogram can be shown (Bendat and Piersol 2010) to have a normalized random error of unity, i.e.,

$$\varepsilon_r[\hat{P}_{xx}] = \frac{\sigma[\hat{P}_{xx} - P_{xx}]}{P_{xx}} = 1 \tag{10.10}$$

which is independent of the length $L$, which is the problem with the periodogram. It is therefore an *inconsistent* estimator. Also note that the random error is independent of $k$, which means the random error is equally large for all frequency lines, which will be useful later.

It should be noted that the inappropriateness of the periodogram only applies to random signals $x(n)$. For periodic signals, the periodogram is a very good estimator (with suitable scaling) of autopower

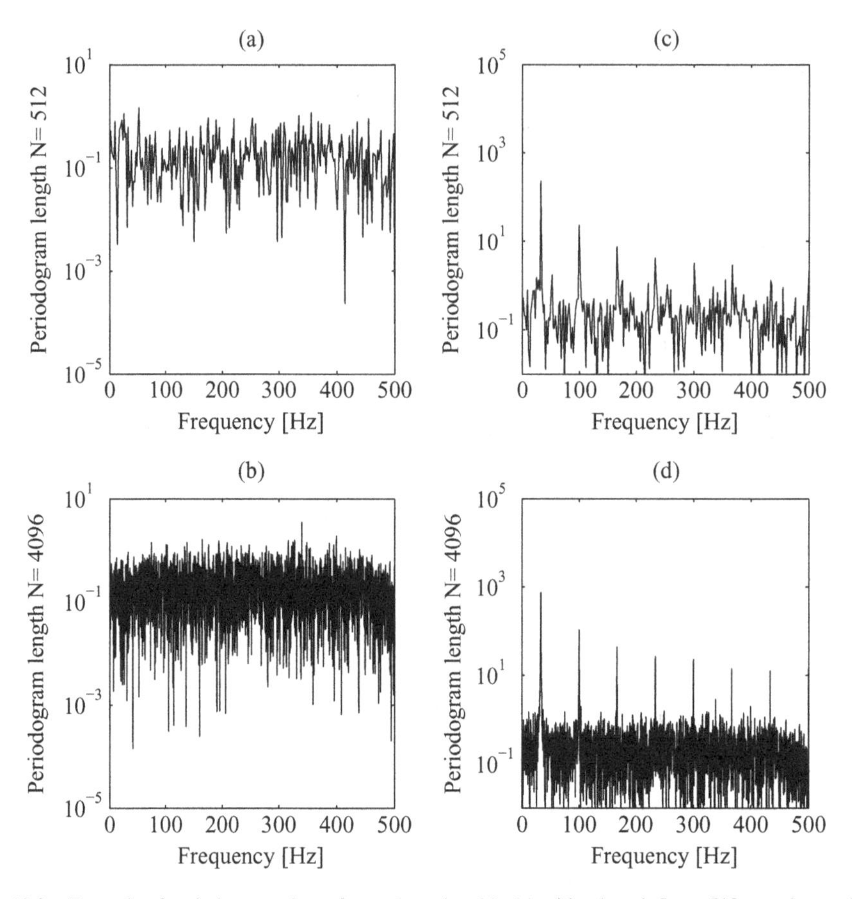


**Figure 10.2** Example of periodogram plots of a random signal in (a) with a length $L_1 = 512$ samples, and in (b) with $L_2 = 8192$ samples. As illustrated by the plots, the longer periodogram is not more stable than the short, which shows the periodogram is an inconsistent estimator. In (c) and (d) periodograms with the random signal plus a periodic signal (square wave) are shown, using the same lengths $L_1$ and $L_2$, respectively. As can be seen from these two plots, the longer periodogram makes the periodic signals stand out more, which illustrates that the periodogram (or long DFTs) can be good for finding periodic components hidden in noise

spectrum, as shown in Figure 10.2 where two periodograms of lengths $L_1 = 512$ and $L_2 = 8192$ samples are shown for a pure random signal, and for a random signal plus a sine signal. In the figure it is seen that the two periodograms in Figure 10.2(a) and Figure 10.2(b) of the random signal, with $N = 512$ and $N = 8192$ samples respectively, do not stabilize at all between the shorter and longer time signal. Actually, the longer the blocksize, the wilder the behavior of the periodogram. For the periodic signal in Figure 10.2(c) and Figure 10.2(d), however, which have the same data lengths, $N = 512$ and $N = 8192$, respectively, as the results in (a) and (b) in the same figure, we see that the periodic components stand out more from the background noise for the longer blocksize.

For periodic signals, we do not actually calculate the periodogram, of course. Rather, we use a long blocksize, $N$, to calculate the linear spectrum described in Section 10.2.2. If plotted in logarithmic $y$-scale, such a spectrum often reveals the resonances of a structure.

### 10.3.2 Welch's Method

The most common method for computing PSD and CSD is a method called Welch's method, (Welch 1967), which is based on an average of shorter, windowed periodograms. This method is virtually the only method available in commercial noise and vibration analysis systems. Welch's method is based on dividing the signal $x(n)$ into $M$ segments, each of length $N$, as was shown in Figure 10.1. Each segment is windowed before the DFT is calculated, and the modified periodogram (magnitude squared of the windowed DFT) is averaged for each frequency line. Usually the time blocks are overlapped, which decreases the random error of the PSD estimate, as we will show in Section 10.3.5.

Assuming we average $M$ windowed DFT spectra denoted $X_{w,m}$ as before, the power spectral density using Welch's method, $\hat{G}^W_{xx}(k)$, is computed by

$$\hat{G}^W_{xx}(k) = \frac{S_P}{M}\sum_{m=1}^{M} X^*_{w,m}X_{w,m} = \frac{S_P}{M}\sum_{m=1}^{M}\left|X_{w,m}(k)\right|^2, \quad k = 1, 2, \ldots, N/2 \tag{10.11}$$

where the scaling constant is to be determined. We choose the scaling factor $S_P$ so that the area under the function equals the square of the RMS value of the time function, which can be seen as an alternative definition of PSD instead of using the Wiener–Khinchine relation. $X_{w,m}(n)$ in Equation (10.11) is the DFT of the windowed segment for average number $m$. Thus, if $x_m(n)$ is segment number $m$ of the measured signal and $w(n)$ is the window used, usually the Hanning window, then

$$X_{w,m}(k) = \mathrm{DFT}\left[x_m(n)\cdot w(n)\right] = \sum_{n=0}^{N-1} x_m(n)w(n)e^{-j2\pi kn/N}. \tag{10.12}$$

For two different signals, an input signal $x(n)$, and an output signal $y(n)$, the discrete version of the cross-spectral density, CSD, that we described in Section 8.3.1, $\hat{G}^W_{yx}$, using Welch's method, is estimated in a similar fashion as

$$\hat{G}^W_{yx}(k) = \frac{S_P}{M}\sum_{m=1}^{M} Y_{w,m}X^*_{w,m}. \tag{10.13}$$

Note that the complex conjugate in Equation (10.13) will in effect mean that the phase angle of $X$ is subtracted from that of $Y$, which is natural if $x$ is thought of as the input signal to a linear system, and $y$ is the output.

### 10.3.3 Window Correction for Welch Estimates

We must now find the scaling factor $S_P$ in Equation (10.11) so that we can interpret the area under the PSD as the square of the RMS value of the signal $x(n)$. The first correction is to scale the function by dividing by the frequency increment, $\Delta f$, to make the PSD a *density function*. However, this will not be sufficient, as can be directly seen in Figure 9.6. This figure shows that, provided we have used the scaling factor $S_A$ for the autospectrum as in Equation (10.1), the peak value in the autopower spectrum already corresponds to the mean square value of the sinusoid (since we do not have leakage). Because of the window's main lobe width, we obtain a widening of the frequency peak, so that with the Hanning window we have two more nonzero spectral lines, besides the 'correct' value at the center. Thus, integrating (summing) the values in Figure 9.6 will naturally yield too large a result.

The factor we have to correct the spectrum estimate with, is the normalized equivalent noise bandwidth, $B_{en}$, as defined by Equation (9.33), see for example (Bendat and Piersol 2010; Harris 1978). This effectively means that we have to divide the scaling factor $S_A$ for the autopower spectrum in Equation (10.3) by the equivalent noise bandwidth, $B_e = B_{en}\Delta f$ to produce the PSD scaling $S_P$. We thus obtain the scale factor for Welch's PSD (or CSD, of course) estimate to be

$$S_P = \begin{cases} \dfrac{2A_w^2}{N^2 B_e}, & k \neq 0 \\ \dfrac{A_w^2}{N^2 B_e}, & k = 0 \end{cases} \tag{10.14}$$

where the factor $A_w$ is the amplitude correction factor from Equation (9.23).

To calculate the single-sided PSD based on $M$ averages, Equation (10.11) is therefore used with $S_P$ from Equation (10.14). Again note that for $k = 0$, $S_P$ is different (half the value). Also, note especially that $S_P$ in Equation (10.14) is the same as the scaling constant $S_A$ for the autopower spectrum of Equation (10.1), except for the division by the equivalent noise bandwidth, $B_e$. In many systems for noise and vibration analysis, consequently the conversion between different spectra is treated as a scaling chosen for the display, and not as different measurement functions, as the computational procedure for the autopower spectrum and PSD are the same.

The (single-sided) cross-spectral density, CSD, between two signals $x(n)$ and $y(n)$, $G_{yx}(k)$ where the signal $x$ is regarded as the reference (or input) signal, is computed by Equation (10.13) using the scaling constant $S_P$ from Equation (10.14).

Now that we have discussed the estimators for spectral density estimates, a few words need to be said about the signal processing involved in PSD/CSD calculations. In early work on spectral estimation, ideas about zero padding and detrending of signals where discussed. These ideas seem to be prevailing in some texts, and we therefore must investigate if it is a good idea to use these features. The idea of zero padding was discussed in Section 9.3.13 where it was concluded that it should not be used for spectral estimation, because it results in confusing results (i.e., lures you to believe there is a frequency resolution which there really is not).

Detrending means removing a possible trend in data (by linear regression analysis) and the idea of using this comes from knowledge about problems that can occur around DC (sensor drift, etc.). In effect, detrending is a highpass filtering operation. In some literature on spectrum estimation it is popular to recommend detrending each segment of data prior to windowing. Applying such detrending on each segment in Welch's method is, at best, dubious. If data contains an unwanted low-frequency drift, it is better to first apply a highpass filter to the entire signal, and thus remove the low-frequency content, by the methods described in Section 3.3.2 prior to calculating the spectral density. As mentioned before, it is a good idea to remove the mean of the signal before processing, but it is important to remove the mean of the entire signal, and not process the segmented data individually.

### *10.3.4 Bias Error in Welch Estimates*

When computing PSDs experimentally, we approximate a continuous function with discrete 'bars', as illustrated in Figure 10.3. This approximation leads to a bias error, which decreases with decreasing frequency increment, $\Delta f$.

A more rigorous understanding of the bias error can be found by noting that the multiplication of the measured signal by the time window (or, if no explicit time window is used, the rectangular window) corresponds to convolution in the frequency domain. A complication compared with the convolution discussed in Section 9.3.9 for the DFT, is that Equation (10.11) involves the square of the window spectrum, and not the window spectrum itself. It can be shown relatively easily, however, that the double-sided PSD estimate, $\hat{S}_{xx}(f)$ can be written as (Schmidt 1985a)

$$\hat{S}_{xx}(f) = \int_{-f_s/2}^{f_s/2} S_{xx}(u)\,|W(f-u)|^2\,\mathrm{d}u \tag{10.15}$$

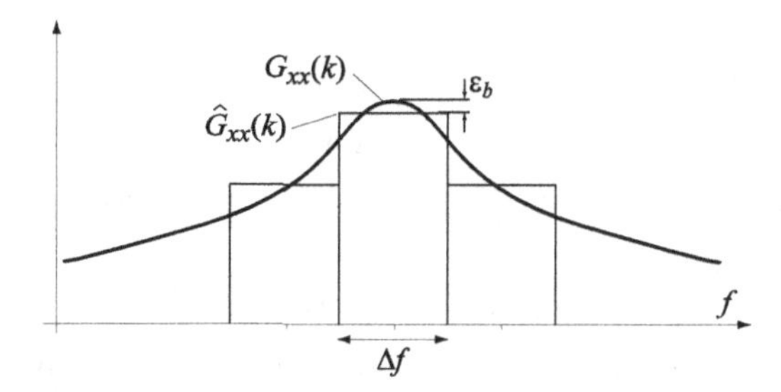


**Figure 10.3** Illustration of the bias error in estimating PSD by approximating the continuous function with bars of constant width, each bar (frequency line) having the same area (mean square of the signal) as the area under the continuous PSD within the same frequency interval

i.e., as the convolution of the true PSD and the magnitude squared of the window Fourier transform. The bias error resulting from the convolution in Equation (10.15) will, of course, give a broadening of a peak in the PSD corresponding to a resonance. In Figure 10.4 an example of such a PSD is shown, zoomed in around the resonance of an SDOF system. It can be seen that the bias error is negative at the resonance frequency, and positive on both sides of it. This distortion of the shape of the PSD can cause serious errors if we try to estimate the damping of the system by using the resonance bandwidth as defined by Equation (5.39).

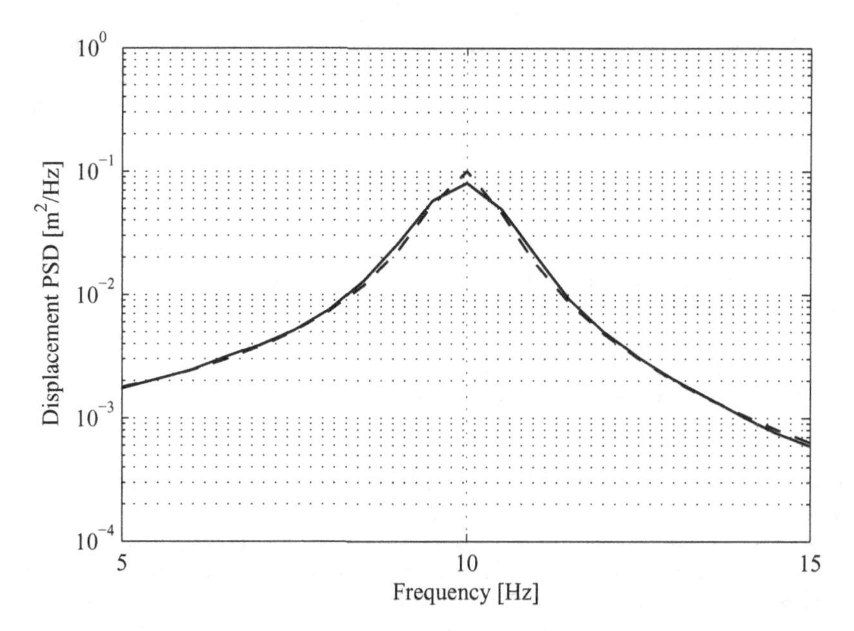


**Figure 10.4** Autospectral density of a simulated vibration signal from an SDOF system with undamped natural frequency $f_r = 10$ Hz, and relative damping $\zeta = 0.05$, excited by band-limited white noise. For the PSD calculation, Welch's method was used with $N = 256$ samples, $f_s = 128$ Hz, a Hanning window and 500 averages with 50% overlap (solid). Overlaid is the true PSD (dashed) evaluated at the same discrete frequencies as the experimental PSD. The bias is clearly seen as an underestimation at $f_r$, and an overestimation at frequencies a little further away from $f_r$

By noting that the autospectral density is the Fourier transform of the autocorrelation, an equation analogous to Equation (10.15) can also be formulated in the time domain for the autocorrelation, by

$$\hat{R}_{xx}(f) = \rho_{ww} R_{xx}(\tau) \tag{10.16}$$

where $\rho_{ww}$ is the 'autocorrelation' of the time window given by convolution of the window by itself time reversed,

$$\rho_{ww}(\tau) = w(\tau) * w(-\tau). \tag{10.17}$$

Equation (10.15) can be used to calculate the bias of the PSD if we know the true PSD and the spectrum of the time window. The normalized bias error of the PSD estimate, is then given by

$$e_b[\hat{S}_{xx}] = \frac{\mathrm{E}\left[\hat{S}_{xx}\right] - S_{xx}}{S_{xx}} \tag{10.18}$$

as defined in Section 4.2.2.

For noise and vibration analysis, we are particularly interested in this bias error in the vicinity of resonances as discussed in Chapters 5 and 6, as we can expect the convolution in Equation (10.15) to yield the worst error at frequencies where there is large change in $\hat{S}_{xx}$. The bias error will be equal for single-sided PSDs, so we will now state the errors for such estimates because we use single-sided estimates in practice.

To understand what affects the bias error, Equation (10.15) can be expanded by a Taylor series. A commonly cited bias error was developed, for a rectangular time window, by Bendat and Piersol (2010) in their early work (we reference the latest edition here), as

$$\varepsilon_b \approx \frac{B_e^2}{24} \frac{G''_{xx}}{G_{xx}} \tag{10.19}$$

where $G''_{xx}$ is the second derivative of $G_{xx}$ with respect to frequency, and $B_e = 1$ for the rectangular window. This approximation was later shown to be a rough approximation compared with a better approximation given by (Schmidt 1985a,b), (which is cited in later work (e.g., Bendat and Piersol 2010)), for the particular case of the Hanning window, given by

$$\varepsilon_b \approx \frac{(\Delta f)^2}{6} \frac{G''_{xx}}{G_{xx}} + \frac{(\Delta f)^4}{72} \frac{G^{(4)}_{xx}}{G_{xx}} \tag{10.20}$$

where $G^{(4)}_{xx}$ is the fourth derivative of $G_{xx}$ with respect to frequency. A comparison of the equations shows that the error in Equation (10.19) becomes the first term in the error of Equation (10.20) if we set $B_e = 2\Delta f$ in Equation (10.19).

For an output signal of a mechanical system with a resonance, we know from Equation (5.39), which we repeat here for convenience, that the resonance bandwidth, $B_r$ is related to the undamped natural frequency, $f_r$ and the relative damping of the system, $\zeta_r$, by

$$B_r = 2 f_r \zeta_r. \tag{10.21}$$

We can anticipate that the second derivative of the PSD will be dependent on the damping, and thus we can relate the frequency increment $\Delta f$ to the resonance bandwidth, $B_r$.

To investigate the bias error approximation in Equation (10.20), we can compute the 'true' normalized bias error using Equation (10.15) and compare it with the approximation, using a Hanning window and an example SDOF system. In Figure 10.5 the result of a simulation using the first and both terms of Equation (10.20) and the 'true' bias error are shown. It can be seen that the approximate bias error is good, provided the second term in Equation (10.20) is included, and relatively good for practical purposes, even using only the first term. It can also be seen that using only one term overestimates the error. It can be shown that the second term vanishes when the ratio $B_r/\Delta f$ becomes large, i.e., when the frequency increment is small.

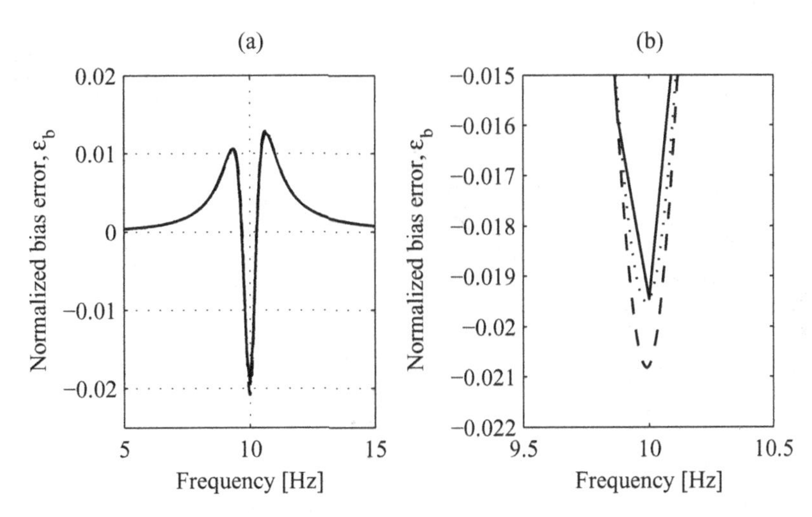


**Figure 10.5** Normalized bias error of a PSD calculated using a Hanning window on the output of an SDOF system with $f_r = 10$ Hz, and relative damping $\zeta_r = 0.05$ with a frequency increment of $\Delta f = 1/8$ Hz. The resonance bandwidth is $B_r = 1$ Hz, which means there are eight frequency lines within $B_r$. The bias error plotted with solid line is the result of using the theoretical formula for the convolution in Equation (10.15). The bias error calculated by Equation (10.20) with only the first term is plotted with dashed line. This is equal to the definition by Equation (10.19) using $B_e = 2\Delta f$. In dotted line the bias error using both terms in Equation (10.20) is shown. It can be seen in the figure that using only one term from Equation (10.20) overestimates the error slightly. This error is larger the smaller the ratio $B_r/\Delta f$ is. It can be concluded that Equation (10.20) with both terms gives an accurate estimate of the bias error

It is interesting to make a simulation on a known mechanical system using a PSD estimated by Welch's method. The result of such a simulation, using the method described in Section 6.5 to produce output noise of a system with known properties will now be presented. The normalized bias error around the resonance according to Equation (10.15) is plotted in Figure 10.6 for an example where the frequency increment $\Delta f = B_r/5$ for the same SDOF system used to produce Figure 10.5. To calculate the normalized bias error in the simulation, a fine resolution PSD was first computed with 100 000 averages to yield the 'true' PSD. Then a PSD with a frequency increment with the requested relation $\Delta f = B_r/5$ was computed with the same number of averages, to make the random error negligible (see Section 10.3.5). The error according to Equation (10.18) was then computed from the two estimated PSDs. From the figure it is clear that the bias approximation in Equation (10.20) is a good approximation. It should be stressed, however, that this error is only valid for the Hanning window, but as this window is the recommended window for PSD estimation this is not a major drawback.

From the previous discussion in this section, and the results in Figure 10.5, it is seen that the bias error is largest at the resonance frequency. We have also established that the bias error is related to the ratio $B_r/\Delta f$ which tells how many frequency lines reside inside the resonance bandwidth. In Figure 10.7 the maximum bias error (without sign, since it is always negative) as a function of the ratio $B_r/\Delta f$ is plotted. The most important result from this figure is that, we need very small $\Delta f$ to yield negligible bias errors. For 1% normalized bias error, as many as 12 frequency lines must be within the resonance bandwidth $B_r$.

**Example 10.3.1** *To illustrate the implications of the discussion of bias errors in this section, we will look at the frequency increment necessary to yield a bias error less than 1% on a system (e.g., a tall building) with an undamped resonance frequency of $f_r = 0.5$ Hz, and a relative damping of $\zeta_r = 1\%$.*

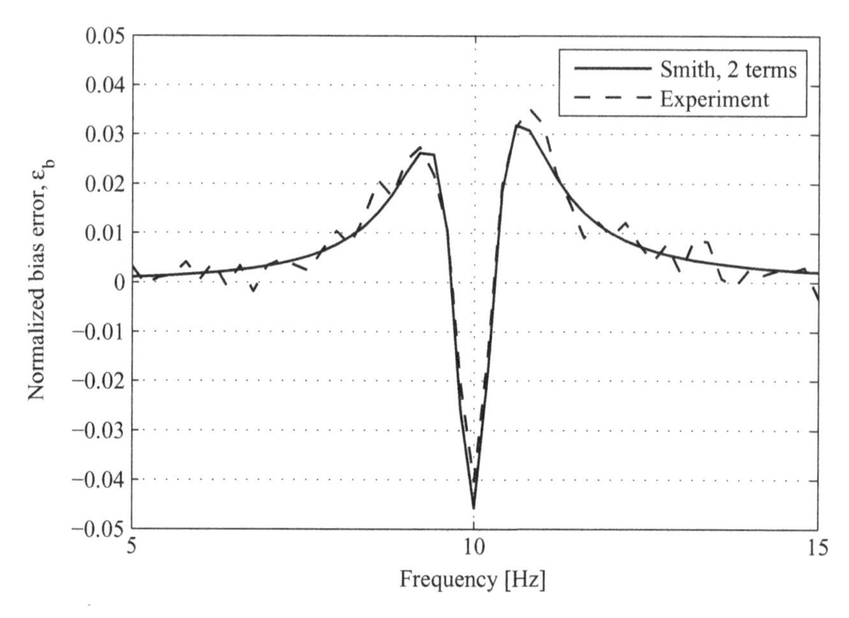


**Figure 10.6** Experimental bias error calculated on a PSD using the Hanning window and Welch's method, from simulated data of an SDOF system with an FFT frequency increment $\Delta f = B_r/5$. See text for details. In the figure it is shown that the experimentally obtained normalized bias error agrees very well with the theoretical expression of the bias error given by Equation (10.20). The variations in the experimentally obtained bias are due to the random error in the estimate of $G_{xx}$

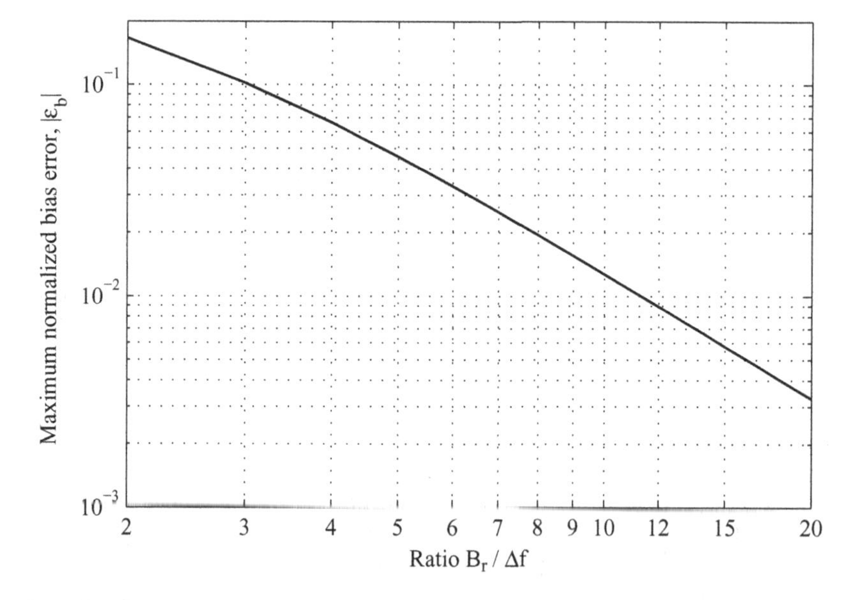


**Figure 10.7** Plot of the maximum normalized bias error for Welch's estimate of a PSD of the output of an SDOF system, as a function of the ratio $B_r/\Delta f$, the resonance bandwidth divided by the frequency increment. The error plotted is the magnitude of the maximum negative bias error at exactly $f = f_r$, the natural frequency of the SDOF system

*Equation (10.21) yields a resonance bandwidth $B_r$ of*

$$B_r = 2 \cdot 0.5 \cdot 0.01 = 0.01 \text{ Hz} \tag{10.22}$$

*which, combined with the results of Figure 10.7, which shows that $\Delta f < B_r/12$ for a maximum bias error of 1%, gives us*

$$\Delta f < \frac{B_r}{12} = \frac{0.01}{12} \approx 8.3 \cdot 10^{-4} \text{ Hz}. \tag{10.23}$$

*Since the block time of a DFT block is $T = 1/\Delta f$, this means that each block we acquire will take 1200 seconds, or 20 minutes. In the next section we will see that we also need many averages, which means that we may, in many cases, have to accept bias errors larger than 1% to get reasonable measurement times, at least when dealing with low frequency vibrations.*

*End of example.*

In practice we often do not know the resonance bandwidth of the resonances of the measured structure. An experimental procedure to investigate which frequency increment is required to achieve negligible bias error in such cases will be described in Section 10.6.3.

For cross-spectral density estimates using Welch's estimator in Equation (10.13), the bias error is caused by a convolution by the time window similar to Equation (10.15), i.e., the cross-spectral density estimate can be expressed as

$$\hat{S}_{yx}(f) = \int_{-f_s/2}^{f_s/2} S_{yx}(u)\,|W(f-u)|^2\,du. \tag{10.24}$$

This equation, however, only holds if there is no delay between the two signals $x$ and $y$. In cases where there is a known delay between the signals, the undelayed signal should be shifted so that both signals are synchronized. The bias errors for cross-spectral density estimates are then similar to the bias errors of autospectral estimates.

### *10.3.5 Random Error in Welch Estimates*

When calculating PSD, the random error is influenced by windowing and overlap processing (Nuttall and Carter 1982; Welch 1967). It is somewhat surprising that despite, Welch being very explicit about his method in his original paper, the random errors in PSDs estimated by Welch's method are commonly misunderstood, particularly when it comes to the random error when overlap processing is used. We will therefore treat this in some detail here. The presentation here will mainly follow the original work of (Welch 1967).

In order to understand what happens when computing a PSD using Welch's method, we first have to consider that we average, frequency by frequency, repeated windowed periodograms. Each periodogram consists of the magnitude squared of the DFT, see Equation (10.11). If we study one particular frequency line, $k$, from the consecutive periodograms, we can consider the periodogram values at that frequency line as a random variable. If we do not use any overlap processing, periodogram values will be (approximately, see Schoukens *et al.* (2006)) uncorrelated, and the random error relatively straightforward. When using overlap processing, however, the random variable will be correlated between averages that partly come from the same time data.

Let us first consider the *normalized variance*, the square of the normalized random error from Equation (4.7), for *one* periodogram in the averaging at *one* frequency line. This normalized variance will be unity from Equation (10.10). This result we will need later. Now consider an average based on $M$ dependent observations, $\phi_m$, of a variable $\phi$ which has a mean value of $\mu$, and a normalized correlation coefficient, $\rho(q)$

$$\rho(q) = \frac{\text{Cov}[\phi(k), \phi(k+q)]}{\text{Var}[\phi]} \tag{10.25}$$

where $q$ is the delay in samples. We now compute the variance of the average of $\phi_m$ and have that

$$\text{Var} = [\bar{\phi}] = E[(\bar{\phi} - \mu)^2] = E[\bar{\phi}^2] - \mu^2. \tag{10.26}$$

But

$$\begin{aligned} E\left[\bar{\phi}^2\right] &= E\left[\frac{1}{M}\sum_{p=0}^{M-1}\phi_{n+p}\frac{1}{M}\sum_{q=0}^{M-1}\phi_{n+q}\right] \\ &= \frac{1}{M^2}E\left[(\phi_n + \phi_{n+1} + \ldots)((\phi_n + \phi_{n+1} + \ldots)\right] \\ &= \frac{1}{M^2}\left[M\sigma_\phi^2 + 2(M-1)\sigma_\phi^2\rho(1) + 2(M-2)\sigma_\phi^2\rho(2) + \ldots\right]. \end{aligned} \tag{10.27}$$

If we now calculate the normalized variance of the variable $\phi$ we get

$$\varepsilon_r^2 = \frac{Var\left[\bar{\phi}\right]}{\sigma_\phi^2} = \frac{1}{M}\left(1 + 2\sum_{q=1}^{M-1}\frac{M-q}{M}\rho(q)\right). \tag{10.28}$$

If the variables $\phi_m$ are independent, i.e., $\rho(q) = 0$, then Equation (10.28) becomes $1/M$, which is what is expected for an average of $M$ independent variables. Because each periodogram has a unity normalized variance from above the result in Equation (10.28) is valid for a PSD computation, if we do not apply overlap processing.

It is now left to show what the correlation between the periodograms is, when using a particular degree of overlap with a particular time window. Welch showed that for Gaussian noise with constant spectral density within $\pm\Delta f/2$, this correlation, for a time window $w(n)$, with length $N$, and a delay of $D = \text{round}[N(1 - O/100)]$ samples if $O$ is overlap percentage, is

$$\rho(q) = \frac{\left[\sum_{n=0}^{N-1} w(n)w(n+qD)\right]^2}{\left[\sum_{n=0}^{N-1} w^2(n)\right]^2} \tag{10.29}$$

for $q = 1, 2, \ldots, M-1$. The result of this equation is put into Equation (10.28) to calculate the random error (or its square as in the equation). The numerator factor of the correlation coefficient comes from a multiplication of the window by a shifted version of itself (convolution, in effect). For an overlap percentage of up to 50%, $\rho(q)$ is thus only nonzero for $q = 1$, because the shifted window $w(n+2D)$, for $q = 2$ will be outside $w(n)$. For higher overlap percentages, $\rho(q)$ is nonzero for more values of $q$. For the Hanning window, even with 75% overlap, there are only three nonzero values of $\rho(q)$.

We will now summarize and illustrate the above mathematical results. When computing a PSD according to Equation (10.11), based on $M$ averages and *no overlap*, from Equation (10.10) we get a normalized random error of

$$\varepsilon_r\left[\hat{G}_{xx}(k)\right] = \frac{1}{\sqrt{M}}. \tag{10.30}$$

When using overlap processing, with the same number of averages (i.e., number of performed FFTs), $M$, we get a higher variance, because some of the values that have been used were correlated. But then we have not used all the samples of our previously used data; with 50% overlap we have only used (approximately) half as many samples, for example. With the same data we used with no overlap, we can increase $M$ when we use overlap. If we, for example, have a data set of 100 non-overlapping segments, then with 50% overlap and the same blocksize, we can perform $M = 199$ averages (if you think it would be 200, consider that the last block will be half a block to the right of the end of the data and thus cannot be used).

A useful term to introduce is the *equivalent number of averages*, $M_e$, i.e., the number of independent segments that would give the same variance. Then we have that with overlap

$$\varepsilon_r \left[\hat{G}_{xx}(k)\right] = \frac{1}{\sqrt{M_e}}. \tag{10.31}$$

Note that when we use overlap processing, then $M_e \leq M$, where $M$ is the number of overlapping segments used, i.e., the number of FFTs actually used in the averaging process. This is due to the fact that there is some correlation between the numbers we average.

From Equations (10.28) and (10.31) we can calculate the *relative increase* in the equivalent number of averages, $M_e$, compared with if we had not used overlap processing. Thus when we use a data set of $M_d$ (distinct, from Bendat and Piersol) non-overlapping segments and perform $M$ averages (FFTs), and thus $M > M_d$, then the relative increase is

$$\frac{M_e}{M_d} = \frac{M/M_d}{\left(1 + 2\sum_{q=1}^{M-1} \frac{M-p}{M}\rho(q)\right)} \tag{10.32}$$

where $\rho(q)$ is defined by Equation (10.29). This relative increase of the number of averages is independent of the blocksize, but is dependent on the time window and overlap factor used. In Figure 10.8 the relative increase is plotted for some common time windows, as a function of overlap percentage. Note especially that there is an increase in the equivalent number of averages even when using a uniform window!

**Example 10.3.2** *Let us illustrate the computation of normalized random error by an example. Assume that we have a data set with 102 400 samples. We select a blocksize of 1024 samples, which gives us 100 non-overlapped segments. We use a Hanning window and 50% overlap to compute a PSD. What will the random error be?*

*First, we calculate the correlation coefficient $\rho(q)$ for a Hanning window with 50% overlap according to Equation (10.29). We will have $M = 199$ overlapped segments, since the 200th segment will have half its block outside the data length. In our example then, $D = 512$, and the correlation coefficient for $q = 1$ becomes $\rho(1) \approx 0.02778$. Furthermore, $\rho(q) = 0$ for $q > 1$, since already a shift of $2D$ will have shifted the window outside itself.*

*We then use Equation (10.28) to sum up the normalized mean-square error and find that*

$$\varepsilon_r^2 = \frac{1}{199}(1 + 2 \cdot 0.02778) \approx \frac{1.05556}{199} \approx 0.00530 \tag{10.33}$$

*and the normalized random error is thus*

$$\varepsilon_r = \sqrt{0.00530} \approx 0.073. \tag{10.34}$$

*Alternatively, using the results of Figure 10.8 we could have observed that the relative increase in the number of averages is approximately 1.89. This means that the equivalent number of averages we should put into Equation (10.31) is 1.89 times 100, i.e., 189, which gives us a normalized random error of approximately $\varepsilon_r = 1/\sqrt{M_e} = 1/\sqrt{189} \approx 7.3\%$, the same result as above. Compare this with the*

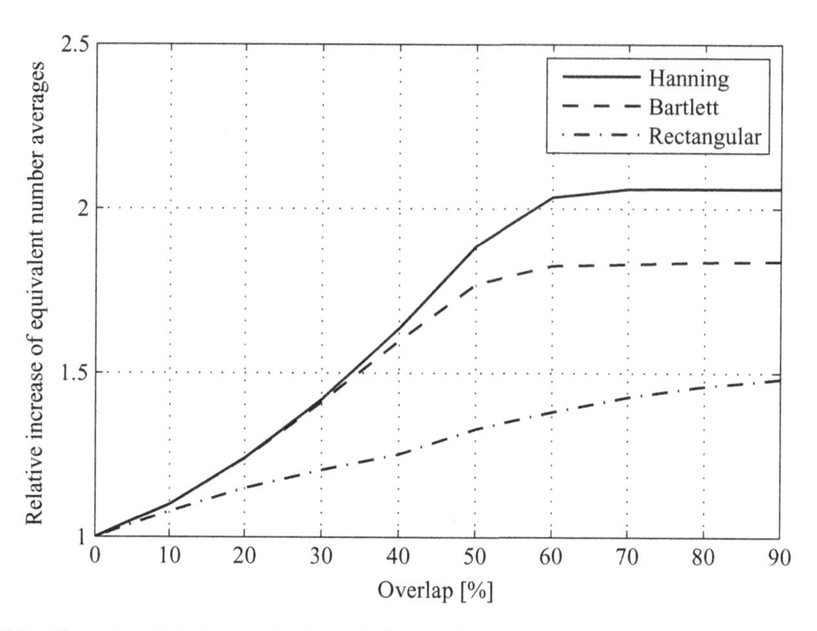


**Figure 10.8** Illustration of the increase in the equivalent number of averages as a function of overlap percentage when computing power spectral density based on a particular data set, using Welch's method. The plot assumes all data are used which means that the number of FFTs performed is increasing with increasing overlap. The equivalent number of averages is the number giving the random error as $\varepsilon_r = 1/\sqrt{M_e}$. For 50% overlap and Hanning window, $M_e = 1.89$, instead of 1.99 if all blocks had been statistically independent (because 100 independent segments leads to 199 overlapped segments with 50% overlap)

*normalized random error of 7.1% ($1/\sqrt{200}$) we would get if we used 200 averages and no overlap, which would require twice the amount of data! This shows that using 50% overlap with Hanning window produces almost entirely uncorrelated FFT blocks.*

*End of example.*

From Figure 10.8 and Equation (10.31) follows, that when we use overlap processing with increasing overlap percentage, at first we get a substantial reduction in the variance for increasing overlap, while above a particular overlap percentage, the additional gain is very small. For the Hanning window, the limit is 62.5% (Nuttall and Carter 1982), above which there will be no additional reduction in the variance. It is also clear from the figure that the increase from 50% to 62.5% overlap only gives approximately 8% less variance and thus approximately 4% less random error. The number of FFTs at the same time increases approximately from a factor of 2 to 2.7 times the number of independent segments. Therefore, 50% overlap percentage is usually considered optimal when using the Hanning window. The increase in the equivalent number of averages is approximately 1.89.

Thus, as the Hanning window and 50% overlap is what is recommended for PSD computations, it makes sense to plot the normalized random error for this combination, as a function of the number of averages (total number of FFTs, the number usually put into the FFT analysis software). This result is plotted in Figure 10.9 and provides a convenient way to find out the normalized random error when computing PSDs using Welch's method.

As we mentioned above, Welch assumed the noise to be Gaussian and flat between $\pm\Delta f/2$ in the deduction of the equation of the correlation between overlapped segments. This is a reasonable

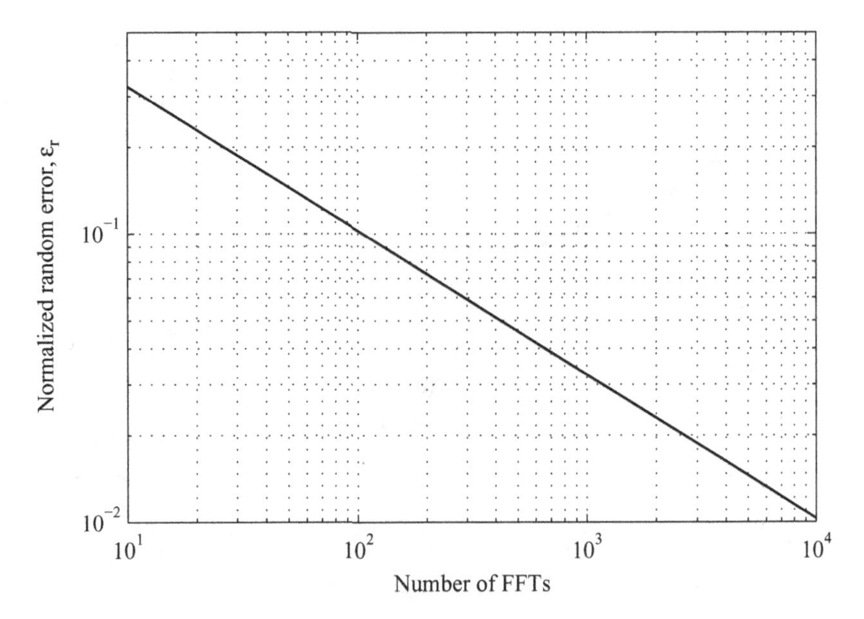


**Figure 10.9** The normalized random error for a PSD estimate using Welch's method, with 50% overlap and Hanning window, as a function of the number of (overlapping) FFT blocks in the averaging process. This number is what is normally entered into the FFT analysis software as 'number of averages'

assumption in all practical cases, regardless of the actual statistical properties of the noise, since the noise for narrow spectral bands will tend to be Gaussian. To prove that this assumption holds in real cases, in Figure 10.10 results from simulations of random errors using random noise from a simulated SDOF system for three different time windows (Hanning, Bartlett, and rectangular) are presented. As can be seen in the figure, simulation results agree very well with Welch's predictions. Simulations using uniformly distributed noise (which is far from Gaussian) give similar results. We can therefore use the random error as presented above with some confidence on real data.

For cross-spectral densities the random error is more complicated, because it depends not only on the number of averages and overlap percentage (and thus correlation between the averages) but also on the coherence, $\gamma_{yx}^2(f)$, (see Section 13.4) which is a measure of how much of the output $y$ that is linearly dependent on the input $x$. Thus, the random error of the magnitude of the cross-spectral density can be shown to be (Bendat and Piersol 2010)

$$\varepsilon_r[\hat{G}_{yx}] \approx \frac{\varepsilon_r[\hat{G}_{xx}]}{\sqrt{\gamma_{yx}^2}}. \tag{10.35}$$

This equation yields the result that when the two signals $x$ and $y$ are completely linearly dependent, and the coherence thus equals unity, the random error of the cross-spectral density equals that of the autospectral density.

### *10.3.6 The Smoothed Periodogram Estimator*

The 'original' method for computer estimation of power spectral densities was, as was mentioned in the introduction to this section, a method accredited to Blackman and Tukey (1958 a, b), which was

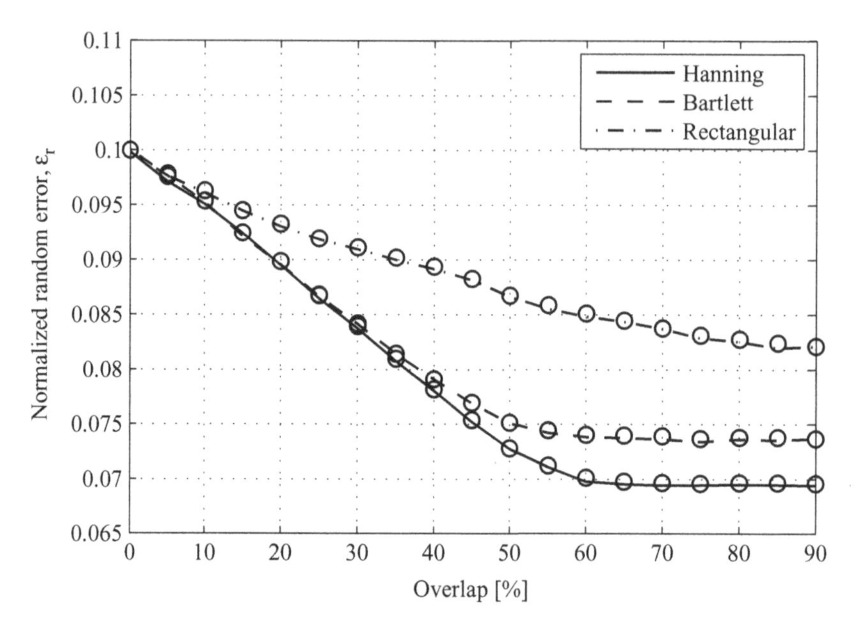


**Figure 10.10** Plot of experimentally obtained normalized random errors (lines) and theoretical values (rings) according to Welch's formula. Band-limited white Gaussian noise was passed through an SDOF system with relative damping of 1%. A PSD was computed using the output noise with different amounts of overlap, and for each case the normalized random error was computed. The figure shows a very good agreement between theory and experiment

superseded by Welch's method shortly after the publication of the FFT algorithm. The Blackman–Tukey method was originally formulated by taking the Fourier transform (DFT) of a windowed estimate of the correlation function (auto or cross). It had, however, been known already from Daniell (1946), that estimating PSDs can also be achieved by smoothing the periodogram in the frequency domain. With the increase in computer power and memory, the method of smoothing the periodogram has become more appealing, and it adds some advantages that we will present below, as an alternative to the popular Welch's method. The method is discussed in, for example, Bendat and Piersol (2010), Cooley *et al.* (1970) and Otnes and Enochson (1972).

The idea behind the smoothed periodogram is similar to the idea by Welch, i.e., to reduce the variance (random error) of the PSD estimate by averaging. However, in the smoothed periodogram method, the frequency lines around the frequency line $k$ to be calculated are used for the averaging. This essentially comes from the result of Equation (10.15) where we now calculate a periodogram using all data, which essentially produces an unbiased (low-bias) estimate of the PSD. The convolution in Equation (10.15) is then done on the periodogram, where we notice that convolving the periodogram with a (short with respect to the total data length $L$) smoothing window is the same, at a particular frequency line $k$, as a weighted average of the surrounding frequency lines.

It is practical to use a smoothing window $w_s$ with an odd length, so that it can be centered on the frequency line $k$ we wish to calculate. The smoothed periodogram estimator for an autospectral density with a rectangular smoothing window of length $L_s$ is thus

$$\hat{S}_{xx}^{SP}(k) = \frac{1}{L_s} \sum_{l=k-(L_s-1)/2}^{k+(L_s-1)/2} \hat{P}_{xx}(l) \tag{10.36}$$

where $\hat{P}_{xx}$ is defined by Equation (10.8) and $k$ (now corresponding to actual frequencies $kf_s/L$) are selected at suitable frequencies, see below. Similarly the cross-spectral density can be estimated by

$$\hat{S}_{yx}^{SP}(k) = \frac{1}{L_s} \sum_{l=-(L_s-1)/2}^{(L_s-1)/2} \hat{P}_{yx}(l) \tag{10.37}$$

where $\hat{P}_{yx}$ is defined by Equation (10.9).

We could, of course, include an arbitrary smoothing window in the summation in Equation (10.36) and Equation (10.37). Although this could reduce the bias in the estimator, it would also increase the random error. Some work has been done to try to find optimum smoothing operations (e.g., Hannig and Lee 2004; Stoica and Sundin 1999), but those methods normally lead to complicated algorithms. For noise and vibration analysis the rectangular smoothing window is recommended, and the experimental method described in Section 10.6.3 is recommended for minimizing the bias error of the estimate.

The smoothing window length can be considered almost equivalent to the number of averages in Welch's method, with the difference that the average in the smoothed periodogram (with our restriction) is a straight average, since we are using a rectangular smoothing window. An easy implementation of the smoothed periodogram, which becomes almost equivalent with Welch's estimator, is to choose a 'blocksize', $N$, being the number of frequency lines, equidistantly placed on the frequency axis. The smoothing window length then becomes the next lowest integer to $L_s = L/N$. To make the smoothing symmetric, it is also a good idea to make $L_s$ an odd number, centered on the frequency bin of the periodogram we want to calculate. The frequency lines $k$ of the estimated PSD or CSD can be chosen arbitrarily, for example at the frequencies $f_s/N$ where Welch's estimator would put them.

A further advantage with the smoothed periodogram method is that it is possible to choose the frequencies of the smoothed PSD with a logarithmic spacing, and simultaneously increase the smoothing window length exponentially as a function of frequency (so that the ratio of the smoothing window length and the frequency of the particular frequency line is constant). This is a relevant choice in many cases in vibration analysis, because we can expect the resonances to be broader (in Hz) at higher frequencies than at lower frequencies, according to Equation (10.21). Using a logarithmic frequency spacing allows for longer smoothing window length at high frequencies which gives a smaller random error where we do not need the high-frequency resolution. An example of this will be given in Section 10.6.3.

In principle Welch's method and the smoothed periodogram method of PSD/CSD estimation are versions of the same method. They asymptotically give the same result on stationary random signals. There is, however, a practical difference between the two methods which could be of great importance in the case of unwanted harmonics in random signals. To my best knowledge, this method has never been published, but the principle is very straightforward. As we showed in Figure 10.2, in the case of harmonics present in a random signal, the periodogram results in very sharp peaks for the harmonics. An efficient way to remove these harmonics, is thus to edit the periodogram by removing a few frequency lines around each harmonic, prior to applying the smoothing. The result is an accurate PSD with the harmonics removed. This method could be of special interest in operational modal analysis, where the presence of harmonics is often a problem, see e.g., Pintelon *et al.* (2008).

## *10.3.7 Bias Error in Smoothed Periodogram Estimates*

The bias error in estimates using the smoothed periodogram estimator is easy to find by Equation (10.15), using a boxcar weighting function $W(f)$ and convolving the true PSD with $|W(f)|^2$. As before, we are particularly interested in this bias error when estimating PSDs for vibration signals on resonant structures. The bias error then is similar to the bias error of Welch's estimator, and an example is shown in Figure 10.11.

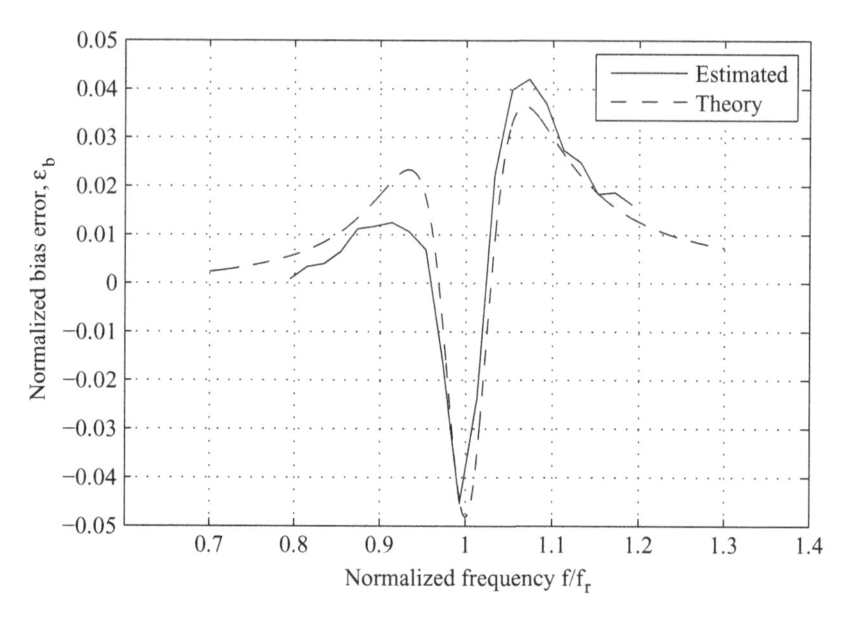


**Figure 10.11** Estimated normalized bias error of a smoothed periodogram estimate of PSD, around the natural frequency of an SDOF system. A rectangular smoothing filter was used for the PSD estimation. The ratio $B_r/\Delta f = 5$. The bias error at the resonance frequency is similar to the result for Welch's estimate in Figure 10.6 which is consistent with the fact that both methods lead to convolution in the frequency domain, although the convolution spectrum is (slightly) different

### *10.3.8 Random Error in Smoothed Periodogram Estimates*

Each spectral line in the periodogram is an approximately independent variable in a statistical sense, at least if the total number of samples in the DFT are much greater than the smoothing window length, $L \gg L_s$. Therefore the normalized random error can be calculated as

$$\varepsilon_r[\hat{G}_{xx}^{SP}] \approx \frac{B_{en}}{\sqrt{L_s}} \tag{10.38}$$

where $L_s$ is the smoothing window length and $B_{en}$ is the normalized equivalent noise bandwidth of the smoothing window from Equation (9.33), which is unity for a rectangular window as suggested to be used for the smoothed periodogram estimator in Section 10.3.6.

## 10.4 Estimator for Correlation Functions

When we want to estimate the autocorrelation of a signal, or cross-correlation between two signals, random or periodic, we want to have an unbiased and consistent estimator, i.e., an estimator which asymptotically approaches the true correlation function when enough data is used, and which has a variance that decreases as we increase the number of averages. Such an estimator will be presented here. As we discussed in Section 4.2.12, the autocorrelation function, $R_{xx}(\tau)$, is a special case of the cross-correlation function, $R_{yx}(\tau)$ when the two signals $x$ and $y$ are the same. We will therefore treat only the cross-correlation here.

Correlation functions can be estimated directly in the time domain by using the definition in Equation (4.32). This equation, however, is very computationally expensive, as we have to convolve the entire two sequences for each lag $\tau$. A much better estimator is obtained by a procedure which uses the Wiener–Khintchine relations in Equation (8.9), i.e., we first compute a PSD using FFT, and then take the inverse transform of this estimate.

The absolute value squared in the PSD estimators corresponds to a multiplication in the frequency domain with the spectrum and its complex conjugate. According to the discussion in Section 9.3.12, this corresponds to a circular convolution and not the ordinary convolution we wish to compute. Thus, as explained in Section 9.3.13, we need to use zero padding to avoid the circular convolution.

Furthermore, if we look at the formulation for the cross-correlation in Equation (4.32), it can be seen that an unbiased estimator based on $T$ seconds of data from $x(t)$ and $y(t)$, has to be formulated, in continuous time, as

$$\hat{R}_{yx}(\tau) = \frac{1}{T - |\tau|} \int_{-T/2}^{T/2} y(t)x(t-\tau)\mathrm{d}t \tag{10.39}$$

because at a particular lag $\tau$ we only have $T - |\tau|$ seconds of data of the two time signals which is overlapping.

To find a discrete estimator corresponding to Equation (10.39), we start by estimating a cross-spectral density using a rectangular window and Welch's method. It can be shown (Bendat and Piersol 2010) that this computation is equivalent to windowing the cross-correlation function by the 'autocorrelation' of the window from Equation (10.17). This 'autocorrelation' of a rectangular window is a Bartlett window, as described in Section 9.3.9. We thus have to compensate the estimator of cross-correlation by the reciprocal of the Bartlett window. The discrete cross-correlation estimator is thus defined by

$$\hat{R}_{yx}(m) = \begin{cases} \dfrac{N-r}{N\Delta t}\hat{R}_{yx}(r), & \text{for } r = 0, 1, \ldots, N-1 \\ \dfrac{r-N}{N\Delta t}\hat{R}_{yx}(r), & \text{for } r = N, N+1, \ldots, 2N-1 \end{cases} \tag{10.40}$$

where $\hat{R}_{yx}(r)$ is the IDFT of the special CSD estimate, $\hat{S}^{C}_{yx}(k)$, where the FFT is made with $2N$ blocksize, of which the last $N$ samples are zeros padded to the data, and a rectangular time window, i.e.,

$$\hat{S}^{C}_{yx}(r) = S_C \frac{1}{M} \sum_{m=0}^{M} Y_{mz} X^{*}_{mz} \tag{10.41}$$

where the spectra are zero-padded , and

$$X_{mz}(r) = \frac{1}{2N} \sum_{n=0}^{2N-1} x_m(n) \mathrm{e}^{-\mathrm{j}2\pi rn/N} \tag{10.42}$$

for $r = 0, 1, \ldots, 2N-1$, and where each segment $x_m(n)$ consists of $N$ samples, and $N$ zeros. This is very easily implemented in MATLAB/Octave as the **fft** command has an optional parameter with FFT size, so the actual zero-padding is done by the FFT command.

The scaling constant $S_C$ for the CSD estimator in Equation (10.41) is given by

$$S_C = \frac{2}{\Delta f} = \frac{2N}{f_s} \tag{10.43}$$

where the factor 2 comes from a compensation for the $N$ zeros. In addition to this, the obtained $\hat{R}_{yx}(r)$ should be FFT shifted, and the outer $N/2$ samples on each side removed to yield a final cross-correlation estimate of length $N$.

## 10.5 Estimators for Transient Signals

For transient signals, as mentioned in Section 8.4, the Energy Spectral Density, ESD, is the most common measurement function available in noise and vibration analysis systems. In general, no window should be used for transient signals, which means that there is no need for any correction factor; however, also see the comments in Section 10.5.1. Instead, the measurement time is adjusted, if possible, so that the transient signal both starts and ends at zero, to avoid leakage. As mentioned in Section 8.4, the ESD is essentially a PSD multiplied by the measurement time used (for one FFT block). Thus, the single-sided ESD is calculated as

$$\hat{G}_{xx}(k) = \frac{2T}{\Delta f} \cdot \left| \frac{DFT\{x(n)\}}{N} \right|^2 = 2(\Delta t)^2 \left|DFT\{x(n)\}\right|^2 \tag{10.44}$$

by exploiting the fact that $\Delta f = 1/T$ and that $T = N\Delta t$. The factor 2 should not be applied to the value for $k = 0$. In fact, Equation (10.44) is a discrete approximation of the magnitude squared of the continuous Fourier transform, where the time increment replaces the infinitesimal differential, and the ESD is converted to a single-sided spectrum. Performing such scaling, however, the DC value must be considered so that it is scaled properly.

As mentioned in Section 8.4, for transient signals one can also use a linear spectrum, called the transient spectrum, $T_x(k)$, which in the discrete form is estimated by

$$\hat{T}_x(k) = \Delta t \cdot |X(k)| \tag{10.45}$$

for $k = 0, 1, 2, \ldots, N/2 + 1$.

Equation (10.45) is a direct approximation of the continuous Fourier transform, which was the definition of the transient spectrum in Section 8.4. From a comparison between Equations (10.44) and (10.45) it follows also that the transient spectrum, $T_x$, besides scaling for single-sided spectra, is equal to the square root of the energy spectral density. Note that $T_x(k)$ is defined as double-sided, but usually displayed single-sided. This is particularly useful if the signal x(t) is a force, since then the DC value $T_x(0)$ is equal to the impulse of the force, which is seen from

$$\hat{T}_x(0) = \Delta t \cdot \sum_{n=0}^{N-1} x(n) \approx \int_0^T x(t)dt \tag{10.46}$$

where the second term follows directly from the definition of the DFT in Equation (9.5). The time increment times the sum in Equation (10.46) implies an estimate of the area under the curve $x(n)$ through a block summation (Riemann sum), which approximates the integral.

**Example 10.5.1** *As an example of a transient spectrum, we can study a half sine pulse, which is the form of, for example, a force pulse from an impulse hammer excitation, see Section 13.8. This type of pulse with length D and amplitude A is defined by*

$$x(t) = A \sin\left(\frac{\pi}{D}t\right) \tag{10.47}$$

*for $t \in [0, D]$, and zero outside this range. The impulse, that is, the area under this half sine is*

$$I_x = \int_0^D A \sin\left(\frac{\pi}{D}t\right) \mathrm{d}t = A\left[-\frac{D}{\pi}\cos\frac{\pi t}{D}\right]_0^D = A\frac{2D}{\pi} \tag{10.48}$$

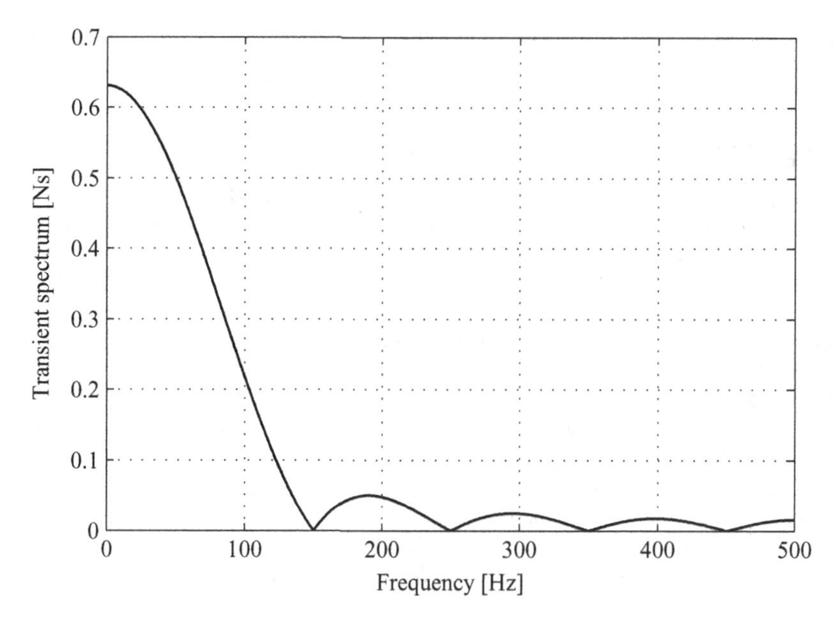


**Figure 10.12** Transient spectrum of a half sine pulse of amplitude 100 N and duration 10 ms. The impulse of the pulse is approx. 0.637 Ns which agrees with the transient spectrum value at frequency 0 Hz

*Figure 10.12 shows the transient spectrum of a half sine pulse with duration $D = 10$ ms and the maximum value $A = 100$ N. According to Equation (10.48) this pulse has an impulse of $2/\pi$, which is approximately 0.637 Ns. As seen in the figure, this value agrees with the transient spectrum value at 0 Hz.*

*End of example.*

### 10.5.1 Windows for Transient Signals

For transient signals, none of the windows discussed in Section 9.3.9 are normally used, since most transient signals are self-windowing, that is, they begin and end at zero. Actually, in most cases it would be devastating to use, for example, a Hanning window, since it would remove the most important part of the signal at the beginning (most vibration transients are exponentially decaying signals, as we saw in Chapter 5). Hence, no window is in general necessary, since the periodic repetition of the transient does not cause any transient at the time block beginning and end points.

For resonant structures with light damping, it can often happen that the signal does not die out before the measurement time is finished, resulting in spectrum leakage. This situation should be avoided by increasing the measurement time so that the transient dies out inside the time window. If this cannot be achieved an exponential window may be used but it is important to understand that the spectrum will be affected by this window. The exponential window has the form

$$w(n) = \mathrm{e}^{-an} \tag{10.49}$$

for $n = 0, 1, 2, \ldots, N - 1$. The constant $a$ is chosen so that the signal at the end of the measurement time is sufficiently attenuated so as not to give any significant leakage, i.e., so the windowed signal dies out before the measurement time is finished. If the measured signal is the impulse response of a mechanical system, the exponential window is equivalent to (has the same effect as) increasing the system damping.

We will discuss this more when we come to measuring frequency responses with impact excitation in Section 13.8.

## 10.6 Spectrum Estimation in Practice

We shall now present some guidelines for spectrum measurements with FFT analysis software in practice. We will use examples to explain the process. It might be good to carry out these examples if you have access to a system for FFT analysis or MATLAB/Octave, partly for practice and partly to gain increased insight into the method of operation of the analysis software.

In practice, as opposed to our theoretical treatment of the estimators in the previous sections of this chapter, time signals and spectra are usually plotted using physical units for time and frequency, respectively. Thus the $x$-axis of the following plots (which are all spectra) will be in the frequency variable $f = k\Delta f$.

Practical spectrum estimation issues that need to be discussed are most importantly the following.

1. Choice of the most suitable estimator,
2. choice of the most suitable time window,
3. choice of proper frequency increment, or blocksize, and
4. choice of the most suitable measurement length (or in many cases, to find the shortest acceptable time).

These issues will be discussed in the subsequent sections for various typical applications in noise and vibration analysis. There is also another issue, of course, which is to find the most suitable units of measurement. For vibrations, the question is often whether to look at acceleration, velocity, or displacement, as their respective frequency weighting in most cases will change the frequencies of 'highest vibration levels'. This is, however, a topic that depends on engineering judgments which are a little outside the scope of this book. To summarize briefly, the choice of units is dependent on the reason for the measurement. For example, noise emission from a structure is largely dependent on vibration velocity, so in cases where the reason for the vibration measurement is related to noise emission, it may be reasonable to look at vibration velocity.

It should be clear from the discussion so far in this chapter that the result of spectrum estimation depends on the frequency increment and time window, etc., chosen for the analysis. Once a particular frequency increment is chosen, for example, the trade-off between bias and random error is fixed. For this reason, the 'traditional' method of averaging spectra in real time is highly unsatisfactory. In many cases there is a need to be able to run frequency analysis with several different choices of, for example, frequency resolution. With the inexpensive storage capacity of modern computers, it is motivated to store the measurement signal so that it can be subsequently analyzed over and over again. I therefore strongly recommend that all signals are stored first on hard disk, which makes the frequency analysis much easier and safer (because you can run the analysis again, if needed).

An additional important point is the length of the recorded data. Many of the time domain processing procedures we discussed in Chapter 3 result in somewhat shorter data sequences after processing, for example resampling (where some thousand samples on both ends of data should be discarded after the processing), and integration and differentiation by filters, etc. It is therefore good practice to always measure more data than is anticipated to be needed. This increases the freedom of further processing prior to spectrum estimation without losing accuracy.

### *10.6.1 Linear Spectrum Versus PSD*

In order to motivate the use of the special spectra (autopower, or the recommended linear spectrum) for periodic signals we will first study what happens if we measure a sinusoid with linear spectrum

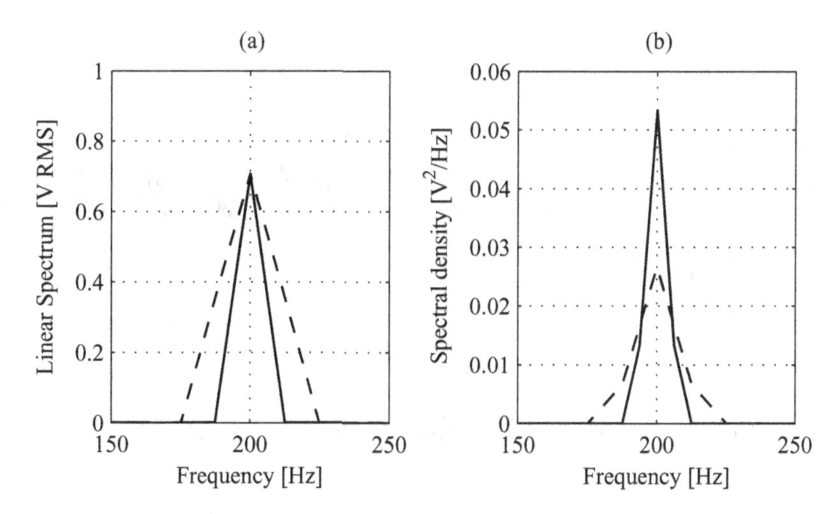


**Figure 10.13** Comparison between linear spectrum and PSD for sinusoid with amplitude 1 V and frequency 200 Hz. In (a) the linear spectrum of the sine is plotted with 6.25 Hz (solid) and with 12.5Hz frequency increment (dashed). In (b) PSDs with the same frequency increments as above are plotted. As seen in the figure, for the linear spectra the peak value is the same regardless of frequency increment, because the signal is periodic. See the text for an explanation. The frequency axes have been zoomed in to visualize the differences

versus power spectral density. In Figure 10.13 autopower and power spectral density with two different frequency resolutions (blocksizes) of a recorded sine signal are shown. The difference in level between the two PSDs in Figure 10.13 is explained by the fact that for a PSD it is not the level which is relevant to examine, but instead the area under the curve. Because the frequency peak with finer frequency increment is narrower, it must be higher so that the area under it is constant. Thus, it is not the choice of PSD that is 'incorrect' but the interpretation that is different than that of the linear spectrum. For periodic signals we usually want to be able to read out each periodic component's RMS value, and the linear spectrum is thus usually preferable.

In Figure 10.14, the result of a similar measurement on a random signal is presented in linear spectrum and PSD, respectively. Using reasoning analogous to the above, it is now the PSD that gives the easiest interpretation of the spectrum; for random signals it makes little sense to use linear spectrum.

### *10.6.2 Example of a Spectrum of a Periodic Signal*

As a first real measurement example, we shall study the spectrum from an accelerometer attached to a small fan, with constant rotation speed. The measured signal is shown in Figure 10.15. The acceleration signal in this case is periodic, since imbalance, etc., gives rise to one or more fluctuations per revolution. Since some contaminating noise can exist in the transducer signal, we may need to average the signal to obtain a correct spectrum. We can determine if this is necessary by examining the changes in a specific peak if we make successive measurements based on one time block. If the peak varies more than what we assume to correspond to the measurement uncertainty, then we must increase the number of averages until a sufficiently stable average is obtained.

Next we need to select a time window. For this type of signal a flattop window should be chosen if accurate RMS estimates are necessary. In many cases, however, the maximum inaccuracy of approximately 15% with the Hanning window is sufficiently accurate, and then the Hanning window offers the advantage of higher frequency resolution relative to the frequency increment, which means that the blocksize and thereby the measurement time of each time block can be kept shorter.

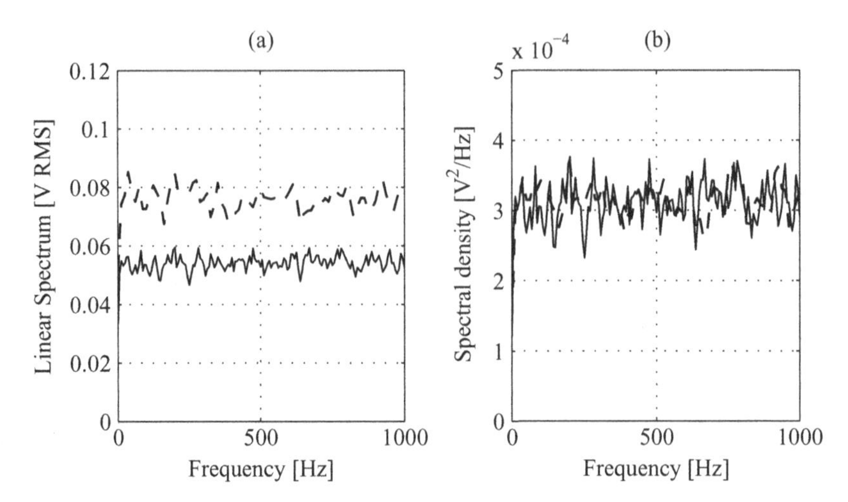


**Figure 10.14** Comparison between linear spectrum and PSD of a random signal. In (a) linear spectra with 6.25 Hz (solid) and 12.5 Hz frequency increment (dashed). In (b) PSDs with the same frequency increments are plotted. As seen in the figure, the value of the PSD is the same regardless of the frequency increment when the signal is random. See the text for an explanation

The data acquisition hardware may have to be set to a proper input range, and the frequency range selected. The frequency range should be set so that the desired number of harmonics are included. For 1800 RPM, which corresponds to 30 Hz, for this example we assume that 500 Hz, corresponding to 16 harmonics, is sufficient.

Once the frequency range is fixed, we adjust the blocksize to vary the frequency increment, $\Delta f$. The recommended procedure is to make a measurement with a coarse frequency increment, $\Delta f$, i.e., a slightly higher $\Delta f$ than we anticipate is sufficient. The resulting spectrum is then compared with spectra with

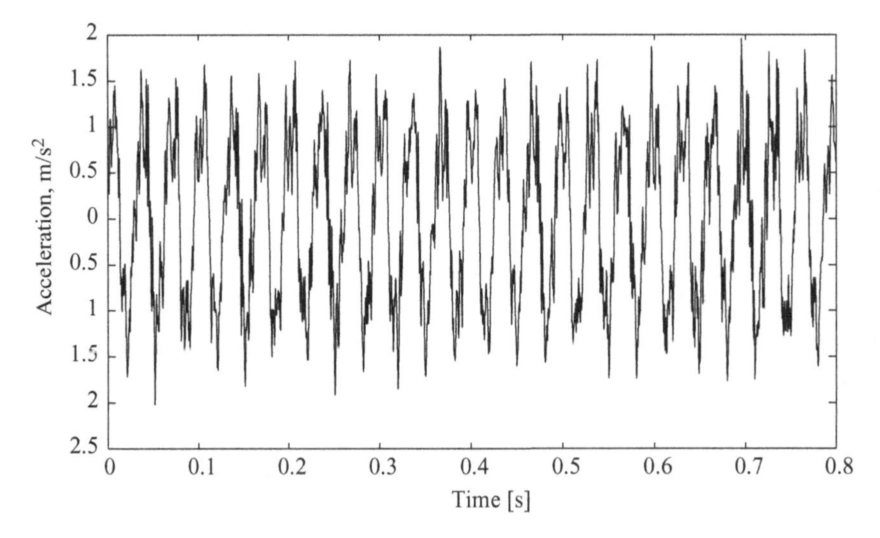


**Figure 10.15** Time signal from an accelerometer attached to a fan rotating with approximately 1800 RPM

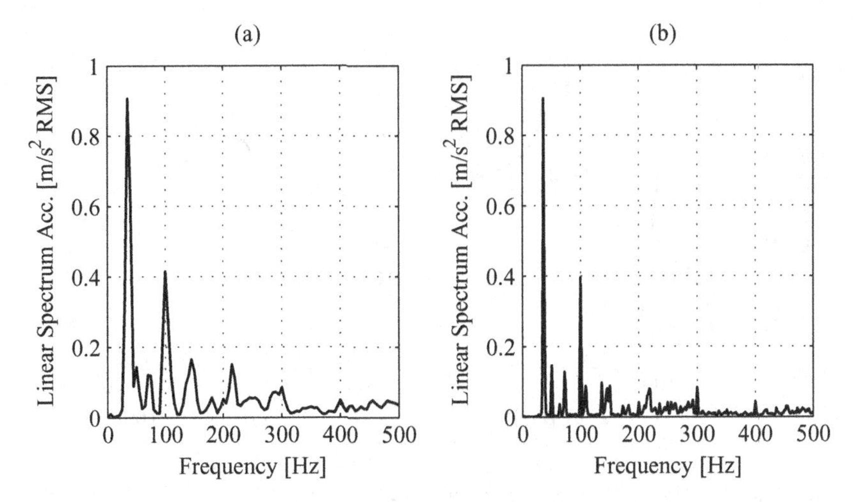


**Figure 10.16** Linear spectra of fan acceleration from Figure 10.15. Two different frequency increments: (a) 5 Hz and (b) 1.25 Hz were used in the frequency range 0–500 Hz. As seen in the figure, a frequency increment of 1.25 Hz is sufficient for the spectrum to decrease to near-zero between the periodic components (peaks), whereas the 5 Hz frequency increment is obviously too coarse

gradually smaller $\Delta f$ (larger blocksize) until the spectrum is clearly a line spectrum, i.e., the spectrum reaches near-zero values in between the peaks.

In Figure 10.16 two spectra from the fan measurement with frequency increment 5 Hz, and 1.25 Hz, respectively, are plotted. The result in Figure 10.16(a), with $\Delta f = 5$ Hz, shows that the spectrum does not reach near-zero values between the peaks, which is an indication that several spectral components are within the frequency resolution. With the finer frequency increment of $\Delta f = 1.25$ Hz in Figure 10.16(b), the details are enhanced and the spectrum looks like an anticipated spectrum of a periodic signal.

With repeated measurements without averaging (number of averages = 1) the spread in the highest periodic component peak, at approximately 60 Hz, was around 10%. This was judged as more than the measurement uncertainty, and therefore the number of averages was increased to 10. The result was a significantly more stable value, with a variation of approximately $\pm 1\%$.

Finally, an important point must be emphasized. When averaging spectra from rotating machines as in this case, it must be verified that the variation in rotation speed (in Hz) is smaller than the frequency increment. Otherwise, for several consecutive averages, the peak will not always match the same spectral line in the spectrum. If that is the case, naturally the average result will become too low, see also Section 12.8.

## *10.6.3 Practical PSD Estimation*

We shall also study a spectrum example for a random acceleration signal. For this purpose we use the simulation procedure that was described in Section 6.5 to generate data from a known mechanical system with three degrees of freedom, because this allows us to illustrate the resulting PSD and compare with the 'true' PSD. We let a bandlimited white random force excite the system and calculate the output acceleration.

As we described in Section 10.3.4, when measuring a PSD we must choose a sufficiently small frequency increment, so that the bias error is negligible. We therefore begin by setting the frequency range and frequency resolution, by guessing at first. The more knowledge we have before we begin the

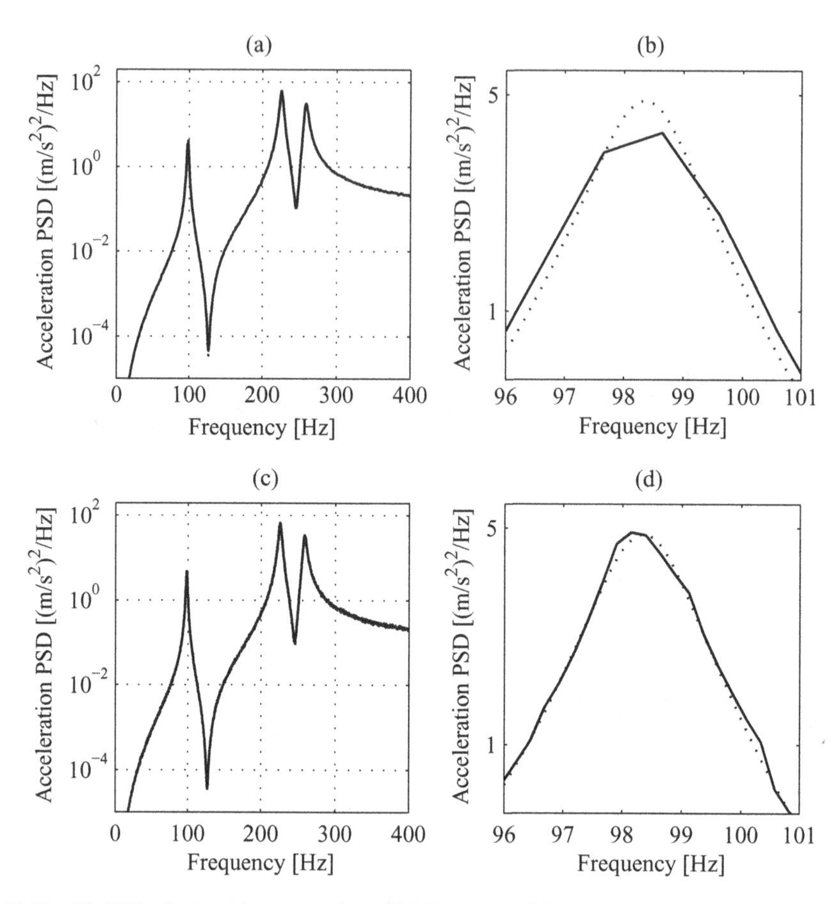


**Figure 10.17** The PSD of a signal from a simulated 3DOF system (solid) and the true PSD (dotted). The resonance bandwidth of the first mode is equal to $B_r = 2$ Hz. Two different frequency increments were used; in (a) 1 Hz and in (c) 0.25 Hz. In (b) and (d) the peak corresponding to the first mode in (a) and (c) respectively, has been zoomed in for closer inspection. As seen in the figure, with the higher frequency increment, there is a clear bias in the peak, whereas the bias is negligible with the finer frequency increment. The latter correspond to a ratio of $B_r/\Delta f \approx 8$ which, according to Figure 10.7 gives a normalized bias error of less than 2%. The PSDs were both calculated using the same data; for the PSD in (a) and (b) a total of 1000 FFTs with 50% overlap were used, and consequently for the PSD in (b) and (d) a total of 250 FFTs were used. The normalized random error in the latter PSD is therefore, according to Figure 10.9, approximately 6.5%

measurement the better off we are, naturally. In our case we begin with a 1000 Hz measurement range and 5 Hz frequency increment. We also choose 200 averages and 50% overlap to obtain a fairly small random error. The result of this measurement is shown in Figure 10.17(a) and (b), overlaid with the true PSD. The results of a measurement where the frequency increment has been decreased to 2.5 Hz (i.e., the blocksize was doubled) are shown in Figure 10.17(c) and (d), also overlaid with the true PSD. As seen in the figure, the frequency increment in the second estimate agrees well with the true PSD.

In a real measurement we do not have the luxury of being able to plot the true PSD overlaid with our estimate. So how can we assure we obtain a sufficient frequency resolution with negligible bias? The answer is, that we need to perform several PSD estimates with gradually decreasing frequency increment (increasing blocksize given that the sampling frequency is constant), until peaks due to resonances do not

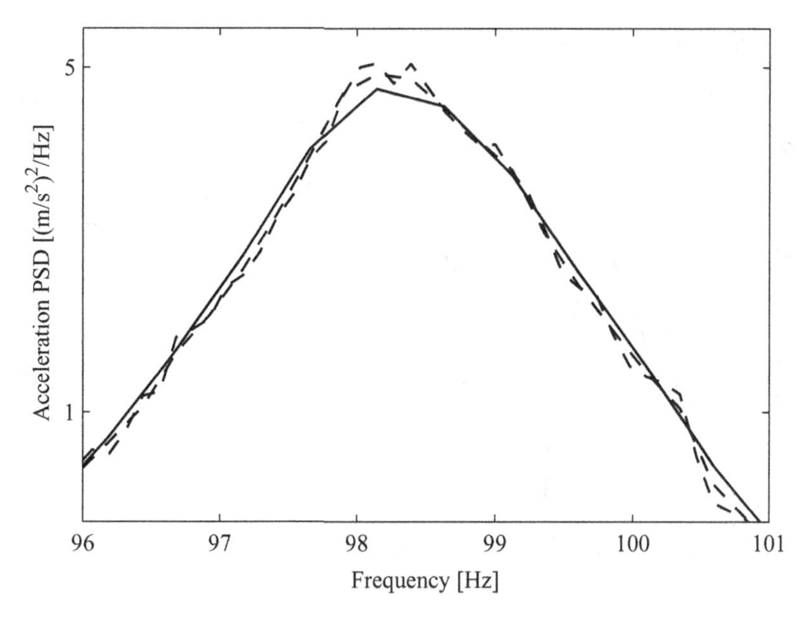


**Figure 10.18** PSD of the same signal as in Figure 10.17, with three different frequency increments; 0.5, 0.25, and 0.125 Hz. As evident from the figure, the two spectral densities with the finest frequency increment show negligible bias, and we can therefore conclude that the 0.25 Hz increment is sufficient. This also corresponds to the result of Figure 10.17 where we also had the correct PSD overlaid

increase. A result of three such estimates with 0.5, 0.25, and 0.125 Hz frequency increment are plotted and overlaid in Figure 10.18. To be able to do this type of analysis easily, it is convenient to have the time data stored, so that several analyses can be done on the same data.

The discussion about finding a frequency increment which produces a PSD with negligible bias error should perhaps be extended a little. It is not always necessary to obtain PSDs with this high precision. In many cases, the aim of estimating a PSD is not to make accurate analyses of the peaks in the PSD. In such cases a higher bias error can often be tolerated. Examples of such applications is for example PSD estimates to find a suitable test spectrum for vibration testing, where the PSD is usually used to build some average envelope from several PSDs in different measurement points. One should, however, always be sure to know when there is bias in the PSD, so the analysis procedure presented can still be used with success to understand how much bias error a particular measurement has.

As discussed in Section 10.3.6, an advantage in some cases with using the smoothed periodogram estimator, is that it can be implemented with a logarithmic frequency axis and smoothing window width, thus keeping a constant ratio of the smoothing window length and frequency. An example of this type of PSD estimate is shown in Figure 10.19 where it can be seen that the random error is smaller at higher frequencies than at lower frequencies.

### *10.6.4 Spectrum of Mixed Property Signal*

In many applications of noise and vibration measurements, the signal is not obviously either periodic or random. It is common to find vibrations with some periodic content and some random content. In such cases it is usually easiest to interpret estimated PSDs, particularly if the PSD plot is combined with a cumulative PSD plot, as discussed in Section 8.5. The RMS value of each periodic component is easy

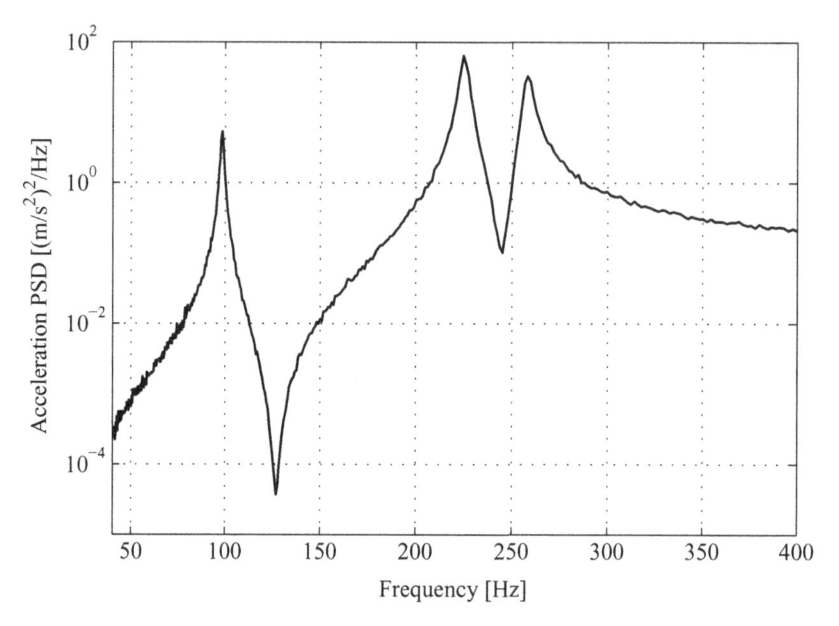


**Figure 10.19** PSD using the smoothed periodogram and a logarithmic frequency spacing and smoothing window length. The PSD is computed using 400 frequencies between 40 and 400 Hz, and an exponentially growing smoothing window length from 50 to 500 spectral lines. The random error can be seen as small fluctuations ('ripple'), particularly at low frequencies where it is highest

to read from the cumulated PSD, by taking the square root of the rapid increase at the frequency of the component. In Figure 10.20(a), a PSD of an example signal is shown. The signal is the same noise signal used for the PSD discussion in Section 10.6.3, with a sine with a frequency of 80 Hz added. The cumulated PSD in Figure 10.20(b) clearly shows the increase in cumulated mean square at 80 Hz, and it is easy to read out that the RMS of the sine is $\sqrt{4}$ m/s$^2$, i.e., 2 m/s$^2$ (using some cursor functionality, at least).

## *10.6.5 Calculating RMS Values in Practice*

In noise and vibration analysis it is very common to want to know the RMS value of a signal in a particular frequency range. For example, for a measured sound pressure, this RMS value in decibels relative to 20 $\mu$Pa, is the sound level in the particular frequency range. It is also common to apply some weighting characteristic to a spectrum and then calculate a weighted RMS value, for example acoustic A-weighting in the case of sound pressure. In this section we will see how to apply such weightings and how to calculate correct RMS values from different types of spectra.

## *10.6.6 RMS From Linear Spectrum of Periodic Signal*

Parseval's theorem (see Table 9.1) implies that a mean square summation in the time domain is equivalent to a magnitude square summation in the frequency domain. In order to compute the RMS value of a signal in a particular frequency range, it is thus suitable to perform the summation on an autopower

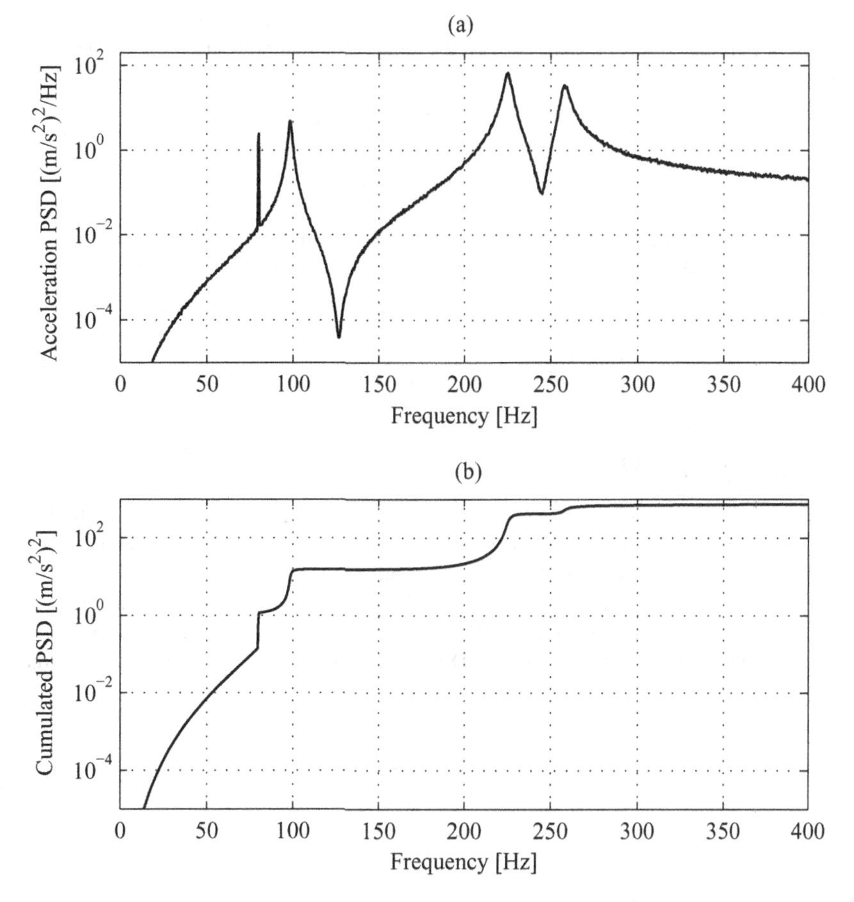


**Figure 10.20** PSD of a vibration signal consisting of random noise with a sine added at 80 Hz in (a). In (b) the cumulated PSD is plotted, which makes readings of RMS levels between any two frequencies easy, see Section 8.5

spectrum. If we use the preferred linear spectrum, we thus square each spectrum value to obtain the autopower spectrum prior to the summation.

From the discussion about broadening of spectrum peaks in Section 9.3.7, however, it should be clear that a straight summation of all the values in an autopower spectrum will yield a value which is too large, as can be seen in Figure 9.11. The peaks in an autopower spectrum are scaled so that the peak value, if there is no leakage, is exactly equal to the RMS value of the periodic component at a particular frequency. Since there is one additional value (in the case of a Hanning window and with no leakage) on each side of the peak, the sum of these three values will, of course, be larger than the true RMS. Indeed, for a Hanning window, the two side factors squared (because it is an autopower spectrum) are 1/4 of the center value, meaning that the squared sum will be a factor $1 + 2 \cdot 1/4 = 1.5$ too large. It turns out that this is no surprise, since the factor to use to compensate the overestimated value by, for an autopower spectrum is the normalized equivalent bandwidth, $B_{en}$, which, for a Hanning window, is $B_{en} = 1.5$. This is also obvious from the discussion on PSD window correction in Section 10.3.3, because the PSD is scaled so that a summation over frequency *is equal to* the mean square of the signal.

Using the normalized equivalent noise bandwidth, we can thus formulate an expression for calculating RMS values of periodic signals measured with the single-sided autopower or linear spectrum. For the autopower spectrum we have

$$x_{\text{RMS}}(k_1, k_2) = \sqrt{\frac{\sum_{k=k_1}^{k_2} \hat{A}_{xx}(k)}{B_{en}}} \tag{10.50}$$

where $k_1$ and $k_2$ are the spectral lines we wish to sum RMS values between. For a linear spectrum, we naturally replace $\hat{A}_{xx}(k)$ in Equation (10.50) with $\hat{X}_L^2(k)$ from Equation (10.4), i.e.,

$$x_{\text{RMS}}(k_1, k_2) = \sqrt{\frac{\sum_{k=k_1}^{k_2} \hat{X}_L^2(k)}{B_{en}}}. \tag{10.51}$$

### 10.6.7 RMS from PSD

For a PSD, the computation of the RMS value is even more straightforward. This spectrum is already scaled so that the area under the curve corresponds to the mean square of the signal, i.e., the square of the RMS value. Since the estimated PSD described above is usually measured with a relatively small $\Delta f$ to avoid bias error, it is hardly necessary to use a better method to calculate the area under the curve than simply summing the values of the PSD multiplied by the frequency increment. Therefore, a desired RMS value is determined for an estimated PSD in the frequency range between spectral lines $k_1$ and $k_2$ as

$$x_{\text{RMS}}(k_1, k_2) = \sqrt{\Delta f \cdot \sum_{k=k_1}^{k_2} \hat{G}_{xx}(k)}. \tag{10.52}$$

### 10.6.8 Weighted RMS Values

As mentioned in the introduction to this section, frequency weighting signals is very common in noise and vibration analysis. In acoustics, there are the common A-weighting and C-weighting (and the less common B- and D-weighting) characteristics that weigh sound pressure for better correlation with perceived sound levels. In analysis of the vibration effects on humans, there are a range of weighting characteristics specified by different standards, for example, (ISO 2631-1: 1997; ISO 8041: 2005). The idea of all such weighting characteristics is that there is some linear filter between the location of the measurement to the location of interest, for example, the acceleration level on the floor where a person is standing, to the vibration level of the abdomen. It should perhaps be mentioned that not all human vibration weightings are linear. In the case of nonlinear effects, however, the simple, linear weighting described here cannot be used, see for example (ISO 2631-5: 2004) for shocks applied to humans, where the particular interest is on the spinal loads – this leads to more complicated data processing.

In all cases of frequency weighting the principle is the same; there is a specified filter, for example the acoustic A- and C-weighting characteristics plotted in Figure 10.21. The measured signal is supposed to pass this filter, and the output unit (sound pressure in our example here) is to be computed – particularly the RMS value in a certain frequency range. The acoustic weighting curves are specified in the standard (IEC 61672-1 2005).

Frequency weightings are defined as filter characteristics, and in most cases they can be applied as time domain digital filters as described in Section 3.3.2. In many cases this is necessary because the

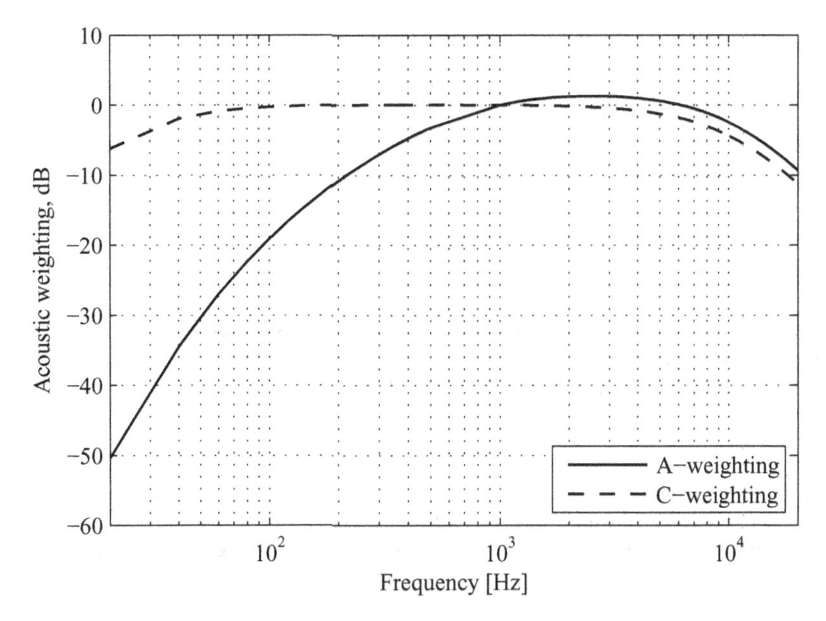


**Figure 10.21** Plot of the acoustic A- and C-weighting curves in dB. The curves are specified in the standard (IEC 61672-1 2005)

time signal should be analyzed with some particular time domain analysis. This is the case for acoustic analysis if particular sound levels with time constants are to be applied, see Section 3.3.5. Also many of the human vibration filters should be applied in the time domain, see for example (ISO 2631-1: 1997). When the signal is stationary and it is *only the total RMS level* which is to be calculated, however, the weighting can be applied in the frequency domain, and a much faster calculation can be done.

In order to apply a frequency weighting in the frequency domain, the principle is to compute the weighted spectrum, and then perform an RMS calculation as described in previous sections. This is an easy operation, although we have to consider if the scaling of the spectrum is linear (as for a linear spectrum), or squared, as in the case of autopower spectra or PSDs. In the case of a linear spectrum, if we assume we have a linear weighting function described by a frequency response $W_s(f)$, then the weighted linear spectrum $X_{Lw}(f)$ is

$$X_{Lw}(f) = \hat{X}_L(f)W_s(f) \tag{10.53}$$

and the easiest way to apply this weighting is, of course, to calculate the weighting frequency response $W_s$ on the same discrete frequencies, $f = k\Delta f$, where the estimated spectrum $\hat{X}_L$ is defined.

In the case of an autopower spectrum or a PSD, the frequency response $W_s$ has to be squared, so that the weighted PSD, for example, becomes

$$G_{xx,w}(f) = \hat{G}_{xx}(f)W_s^2(f) \tag{10.54}$$

where the weighting function, $W_s(f)$ again is calculated for the same discrete frequencies $k$ where $\hat{G}_{xx}$ is defined.

### 10.6.9 *Integration and Differentiation in the Frequency Domain*

In Section 5.3 we mentioned that differentiation in the time domain corresponds to multiplication of a frequency response by $j\omega$. Similarly, this can be applied directly on spectra. Thus, if we have measured a displacement signal, $u(t)$, and computed a linear spectrum $U_L(f)$, the spectrum of the velocity time signal, $v(t) = du/dt$ can be computed on the spectrum itself by

$$V(f) = \mathcal{F}[v(t)] = \omega U_L(f) \tag{10.55}$$

which follows straight from the properties of the Fourier transform in Table 2.2. The imaginary number j is omitted since $U_L$ is real.

If the signal $u(t)$ is a random signal and we have computed the spectral density $G_{uu}$ instead, however, since this spectrum is squared, we must take the (magnitude) square into consideration, and thus the spectral density of the corresponding velocity, $v(t)$ will be

$$G_{vv}(f) = \omega^2 G_{uu}(f) \tag{10.56}$$

where it should be particularly noted that we do not square the imaginary number j, of course, since a spectral density is based on the *magnitude* squared.

Similarly, integration in the frequency domain can be done by dividing by $j\omega$ in case of linear spectra, and by $\omega^2$ for squared spectra.

## 10.7 Multi-channel Spectral Analysis

In Chapters 14 and 15, we will use data for many channels sampled synchronously for estimation of multiple-input/multiple-output (MIMO) frequency response functions and principal components etc. It is therefore practical to introduce a matrix approach to spectra for general MIMO systems.

Let us assume that we measure a number of input signals, $x_1(t), x_2(t), \ldots, x_Q(t)$, and that we calculate the Fourier transform of each of these signals, which we denote $X_1(f), X_2(f), \ldots, X_Q(f)$. From consecutive blocks (records, frames) of spectra of each input signal, we average together single-sided autospectra, $G_{x_1,x_1}(f), G_{x_2,x_2}(f), \ldots$ and so on. We now simplify the notation by dropping the x's in the indexes and thus we define the auto (power) spectrum of $x_k(t)$ by

$$G_{kk}(f) = \mathrm{E}\left[X_k(f)X_k^*(f)\right] \tag{10.57}$$

where * denotes complex conjugate and E [] denotes expected value (in practice averaging). Autospectra are real and nonnegative functions. The simplified notation we will use will omit the variable $x$ in the index whenever the index stands for an input. We will also often drop the frequency variable $f$ for simplicity. Thus for example $G_{x_1,x_1}(f)$ will in most cases simply be denoted $G_{11}$.

We define the cross (power) spectrum of (reference) signal $x_q(t)$ with (response) signal $x_p(t)$, by

$$G_{pq}(f) = \mathrm{E}\left[X_p(f)X_q^*(f)\right] \tag{10.58}$$

where the 'reference' and 'response' should only be interpreted as from which signal to which signal we mean the cross-spectrum is defined; both signals here are, of course, input signals to the MIMO system. However, any two-channel function requires a reference and a response. Cross spectra are in general complex functions. Also note that the complex conjugation in Equation (10.58) corresponds to the phase of $X_q$ being subtracted from that of $X_p$, as would be expected if $x_p$ is the reference (input) signal, and $x_q$ is the response (output) signal.

Next, assume we also measure a number of output signals, $y_1(t), y_2(t), \ldots, y_p(t)$, simultaneously with the acquisition of the input signals, with corresponding fourier transforms $Y_1(f), Y_2(f), \ldots, Y_P(f)$.

Analogous with Equation (10.59), we then define the *input–output cross-spectra* of output signal $y_p$ with input signal $x_q$ by

$$G_{y_p,x_q}(f) = E\left[Y_p(f)X_q^*(f)\right]. \tag{10.59}$$

From Equation (8.13) we know that changing the order of the signals (indexes) corresponds to a complex conjugation of the cross spectrum, so that

$$G_{qp}(f) = G_{pq}^*(f) \tag{10.60}$$

which also follows straight from the definition in Equation (10.59).

Equation (10.60) is an important relationship in the discussion in Chapters 14 and 15. It should be mentioned here, to avoid confusion, that many textbooks define the order of the indexes in the reversed order compared with the one used here. However, we use the above notation in order to be consistent with standard matrix notation when we come to multiple inputs in later chapters.

### *10.7.1 Matrix Notation for MIMO Spectral Analysis*

When introducing multiple input and output signals in the later chapters, it will be useful to have a matrix notation for these signals. We therefore let the input spectrum vector $\{X(f)\}$ be a column vector of the instantaneous spectra (each FFT result in practice) of the input signals

$$\{X(f)\} = \begin{Bmatrix} X_1(f) \\ X_2(f) \\ \dots \\ X_Q(f) \end{Bmatrix}. \tag{10.61}$$

Similarly, the output spectrum vector $\{Y(f)\}$ is defined by

$$\{Y(f)\} = \begin{Bmatrix} Y_1(f) \\ Y_2(f) \\ \dots \\ Y_P(f) \end{Bmatrix}. \tag{10.62}$$

For the inputs, we define the input cross-spectrum matrix, $G_{xx}(f)$ as

$$[G_{xx}(f)] = E\left[\{X\}\{X\}^H\right] = \begin{bmatrix} G_{11} & G_{12} & \dots & G_{1Q} \\ G_{21} & G_{22} & \dots & G_{2Q} \\ \dots & \dots & & \\ G_{Q1} & G_{Q2} & \dots & G_{QQ} \end{bmatrix} \tag{10.63}$$

where $^H$ denotes *hermitian transpose*, i.e., a complex conjugation and transpose (see Appendix D). The expected value (averaging) operation is applied to each matrix element separately, and should be interpreted loosely: it needs to include the scaling factors we discussed for PSD and CSD estimators in Section 10.3. Note that the real autospectra $G_{11}, G_{22} \cdots$ etc. are found on the diagonal of $[G_{xx}]$, and that the off-diagonal elements of $[G_{xx}]$ contain complex cross-spectra between two input signals.

We further define the input-output cross spectrum matrix, $G_{yx}(f)$ similarly by

$$\left[G_{yx}(f)\right] = E\left[\{Y\}\{X\}^H\right] = \begin{bmatrix} G_{y1,x1} & G_{y1,x2} & \dots & G_{y1,xQ} \\ G_{y2,x1} & G_{y2,x2} & & G_{y2,xQ} \\ \dots & \dots & & \dots \\ G_{yP,x1} & G_{yP,x2} & \dots & G_{yP,xQ} \end{bmatrix}. \tag{10.64}$$

It should be noted, that the matrices in Equations (10.63) and (10.64) are dependent on frequency.

### 10.7.2 Arranging Spectral Matrices in MATLAB/Octave

In Chapters 13 and 14 we will see that the necessary spectral matrices for all analysis are $[G_{xx}]$, $[G_{yx}]$, and $[G_{yy}]$ defined in Section 10.7.1. However, there is an important limitation in that we will never have the need for any of the cross-spectra between any outputs, $y$. Thus we can limit $[G_{yy}]$ to contain only the autospectra between each signal $y_p$. We will now see how we can store such matrices conveniently in MATLAB/Octave.

As many commands operate on columns in MATLAB/Octave, for example **plot**, **fft**, and **std**, it is convenient to store spectral matrices with data in columns. This means, that the first index should be frequency. The second matrix index we will set to the output, or response, and the third index will be the input, or reference. With this convention the three matrices defined above will have the following size. We let $N_f$ here be the number of frequency lines (typically $N/2+1$ if $N$ is the blocksize), $D$ the number of responses, and $R$ the number of inputs (references).

- $[G_{xx}]$ will be a matrix with size $N_f \times R \times R$.
- $[G_{yx}]$ will be a matrix with size $N_f \times D \times R$.
- $[G_{yy}]$ will be a matrix with size $N_f \times D$ as we do not need to store any cross-spectra between any output signals.

**Example 10.7.1** *Write a MATLAB/Octave script which produces the 3D input cross-spectrum matrix in variable Gxx, if we have three input signals in three columns of a matrix in variable x in MATLAB/Octave. We assume we have a command* **acsd**, *producing a cross-spectral density between two vectors, and which will, of course, produce an autospectral density if the two vectors are the same. The following MATLAB/Octave lines then builds the 3D input cross-spectral matrix, $G_{xx}$, in variable (matrix) Gxx.*

```
[Nf,R]=size(x);
Gxx=zeros(Nf,R,R);   \% Allocate space
for r=1:R       \% x(:,r) treated as input
  for d=1:R     \% x(:,d) treated as output
    Gxx(:,d,r)=acsd(x(:,r),x(:,d),fs,N);
  end
end
```

*What if we want to plot this matrix? We will note that MATLAB/Octave cannot plot 3D matrices without some difficulty. So, some very useful commands are* **permute** *and* **squeeze**. *The former command rearranges a matrix in a different order. So, to produce a 2D matrix with only those spectra in the matrix Gxx which are related to the first reference i.e., Gxx(:,:,1), we can write*

```
A=permute(Gxx,[1 3 2]);
A=A(:,:,1);
semilogy(f,abs(A))
```

*which will produce the requested matrix. Since Gxx(:,:,1) is really a 2D matrix, we can also write*

```
A=squeeze(Gxx(:,:,1));
semilogy(f,abs(A))
```

*and whichever command is used is a matter of taste.*

*End of example.*

## 10.8 Chapter Summary

To summarize the contents of this chapter, we note some important points of (nonparametric) spectral estimation:

- all nonparametric spectral estimation methods in principle calculate RMS values of bandpass filters (filter bank theory),
- all FFT based spectrum estimators essentially use the magnitude squared of FFT/DFT results, or for cross-spectrum the product $(YX^*)$ between the DFT of each signal,
- averaging is done in the frequency domain, either as ensemble averages (Welch's method), or by averaging adjacent frequency lines (smoothed periodogram method),
- for easiest interpretation of spectra, it is important to choose an appropriate spectrum estimator (preferably linear spectrum, PSD, or transient spectrum).

For periodic signals it is most convenient to use the linear spectrum (also called RMS spectrum) defined in Section 10.2.2. For periodic signals the flattop window provides good amplitude accuracy, but requires larger blocksize (finer frequency increment) due to the wider main lobe of this window.

For random signals, we have discussed two estimators; the usual method implemented in noise and vibration analysis software is Welch's method, which is averaging windowed segments (blocks) of data. An alternative to Welch's method which can sometimes be attractive is the smoothed periodogram method, which instead uses one large FFT of the entire time signal, from which the magnitude squared (or product of $(YX^*)$) is calculated. This function, the periodogram, is then smoothed by using frequency lines around each frequency of interest to reduce the variance.

For mixed property signals we have discussed that the PSD should normally be used, and can be combined with a plot of the cumulated mean square value as a function of frequency, as in Figure 10.20. From the cumulated PSD plot it is easy to extract the RMS value of each periodic component.

For spectrum analysis of both periodic and random signals we have discussed that, to obtain correct spectra, a recommended procedure is to start with a coarse frequency increment, $\Delta f$, and gradually decrease $\Delta f$ (by increasing the blocksize) while observing the spectrum. For periodic signals, a line spectrum with distinct peaks with almost zero spectrum values between the peaks, is expected. Thus, a frequency increment is sought, for which each periodic component is clearly separated, as illustrated in Figure 10.16(b). For random signals, the aim of decreasing the frequency increment is instead to remove the bias of peaks due to resonances. Thus a $\Delta f$ is sought such that decreasing it further does not make the resonance peaks higher. This was illustrated in Figure 10.18.

For spectrum analysis of transient signals, the entire transient should be captured if possible. If the signal starts and ends with zero, it is 'self-windowing' and the transient spectrum should be calculated using the entire transient. For systems with low damping, sometimes an exponential window, defined in Equation (10.49), has to be applied to reduce the signal at the end to reduce leakage effects.

## 10.9 Problems

Many of the problems following are supported by the accompanying ABRAVIBE toolbox for MATLAB/Octave and further examples which can be downloaded with the toolbox. If you have not already done so, please read Section 1.6, and follow the instructions to download this toolbox together with example files.

**Problem 10.1** *Create a sine with an RMS value of 5 V and frequency* $f_0 = 20$ *Hz, with a sampling frequency of 1024 Hz, and a blocksize* $N = 1024$ *samples, in MATLAB/Octave.*

*(a) Perform each computation step described in Section 10.2.1 to obtain a single-sided autopower spectrum using a Hanning window. Make sure the level of the resulting autopower spectrum is correct.*

*(b) Calculate the linear (RMS scaled) spectrum from the result in (a).*

**Problem 10.2** *Repeat Problem 10.1 using a flattop window instead of the Hanning window.*

**Problem 10.3** *Use the entire linear spectrum of Problem 10.1 (b) to calculate the RMS value of the signal using the procedure in Equation (10.51). Compare the result numerically with the RMS value calculated from the time signal (for example using the MATLAB/Octave* **std** *command).*

**Problem 10.4** *Repeat Problem 10.3 on the result from Problem 10.2.*

**Problem 10.5** *Use the formulas in Section 10.3.2 to implement a MATLAB/Octave command to produce a PSD using 50% overlap and Hanning window. Create a Gaussian random signal with 102 400 samples using the MATLAB/Octave* **randn** *command and compute the PSD using your developed command with a blocksize of* $N = 1024$ *samples. Check that the RMS level calculated by Equation (10.52) equals the RMS calculated from the time signal.*

**Problem 10.6** *What is the normalized random error of the PSD estimated in Problem 10.5? Try to estimate this random error experimentally, by using the fact that the normalized random error is independent on frequency, i.e., use the standard deviation of the difference between the estimated and the true PSD as an estimate of the random error. How accurate is your estimated random error compared to the error given in Section 10.3.5?*

## References


Bendat J and Piersol AG 2010 *Random Data: Analysis and Measurement Procedures* 4th edn. Wiley Interscience.

Blackman RB and Tukey JW 1958a The measurement of power spectra from the point of view of communications engineering. 1. *Bell System Technical Journal* **37**(1), 185–282.

Blackman RB and Tukey JW 1958b The measurement of power spectra from the point of view of communications engineering. 2. *Bell System Technical Journal* **37**(2), 485–569.

Cooley JW, Lewis PAW and Welch PD 1970 The application of the fast fourier transform algorithm to the estimation of spectra and cross-spectra. *Journal of Sound and Vibration* **12**(3), 339–352.

Daniell PJ 1946 Discussion of 'on the theoretical specification and sampling properties of autocorrelated time-series'. *Journal of the Royal Statistical Society* **8** (suppl) (1), 88–90.

Hannig J and Lee TCM 2004 Kernel smoothing of periodograms under Kullback-Leibler discrepancy. *Signal Processing* **84**(7), 1255–1266.

Harris FJ 1978 Use of windows for harmonic-analysis with discrete fourier-transform. *Proceedings of the IEEE* **66**(1), 51–83.

IEC 61672-1 2005 *Electroacoustics - Sound level meters – Part 1: Specifications*. International Electrotechnical Commission.

ISO 18431-1: 2005 *Mechanical Vibration and Shock – Signal Processing – Part 1: General Introduction*. International Organization for Standardization, Geneva, Switzerland.

ISO 2631-1: 1997 *Mechanical vibration and shock – Evaluation of human exposure to whole-body vibration – Part 1: General requirements*. International Organization for Standardization, Geneva, Switzerland.

ISO 2631-5: 2004 *Mechanical vibration and shock – Evaluation of human exposure to whole-body vibration – Part 5: Method for evaluation of vibration containing multiple shocks*. International Organization for Standardization, Geneva, Switzerland.

ISO 8041: 2005 *Human response to vibration – Measuring instrumentation*. International Organization for Standardization, Geneva, Switzerland.

Newland DE 2005 *An Introduction to Random Vibrations, Spectral, and Wavelet Analysis* 3rd edn. Dover Publications Inc.

Nuttall A and Carter C 1982 Spectral estimation using combined time and lag weighting. *IEEE* **70**(9), 1115–1125.

Otnes RK and Enochson L 1972 *Digital Time Series Analysis*. Wiley Interscience.

Pintelon R, Peeters B and Guillaume P 2008 Continuous-time operational modal analysis in the presence of harmonic disturbances. *Mechanical Systems and Signal Processing* **22**(5), 1017–1035.

Schmidt H 1985a Resolution bias errors in spectral density, frequency response and coherence function measurements, i: General theory. *Journal of Sound and Vibration* **101**(3), 347–362.

Schmidt H 1985b Resolution bias errors in spectral density, frequency response and coherence function measurements, iii: Application to second-order systems (white noise excitation). *Journal of Sound and Vibration* **101**(3), 377–404.

Schoukens J, Rolain Y and Pintelon R 2006 Analysis of windowing/leakage effects in frequency response function measurements. *Automatica* **42**(1), 27–38.

Stoica P and Sundin T 1999 Optimally smoothed periodogram. *Signal Process.* **78**(3), 253–264.

Welch P 1967 The use of fast fourier transform for the estimation of power spectra: A method based on time averaging over short, modified periodograms. *IEEE Trans. On Audio and Electroacoustics* **AU-15**(2), 70–73.

Wirsching PH, Paez TL and Ortiz H 1995 *Random Vibrations: Theory and Practice*. Wiley Interscience.

# 11

# Measurement and Analysis Systems

There are many commercial systems for measurement and analysis of noise and vibration signals. The basic design of these systems has its roots in the early FFT analyzers developed in the late 1960s and early 1970s, shortly after the publication of the FFT algorithm. Although hardware has gone through a revolutionizing development since then, the basic software design has been kept relatively intact, although today, of course, most systems consist of relatively inexpensive hardware and sophisticated software.

Early analysis of noise and vibration signals involved analog tape recorders and expensive computers for converting the analog signals to digital signals, and for performing frequency analysis. A breakthrough was made in the early 1970s when the first *FFT analyzers* were brought on the market. These bulky machines could, thanks to the FFT algorithm, compute spectra in real-time. At that time, memory was expensive, so in most cases time data was thrown away after computing the FFT, and only the average spectrum end result was kept.

The basic design of modern FFT analysis systems was established already in these first analyzers. Today the tools for noise and vibration analysis usually consist of a relatively inexpensive hardware box in which the analog-to-digital conversion is done, and data is then transferred to a laptop computer for the remaining processing. Thus it is not appropriate to refer to current systems as *analyzers*. Rather, we will refer to modern analysis systems for noise and vibration analysis as *FFT analysis systems*, or *noise and vibration analysis systems*. The software architecture for noise and vibration analysis systems has been kept close to the original design in the sense that systems of today are usually still based on processing *blocks* of data, see Section 11.3. The main difference between a modern system and the original analyzers is that in modern systems, time signals can usually be stored onto a hard disk or other storage media for subsequent analysis. As we mentioned in Chapter 10, this is a recommended procedure, as frequency analysis sometimes has to be performed several times on the same time data to extract all information, and data quality analysis can only be performed on time data, see Section 4.4.

In the remainder of this chapter we will discuss some practical aspects of measuring signals for subsequent time domain or frequency domain analysis. We will discuss important hardware issues and how to interpret specifications of data acquisition hardware. We will also look at the basic design of systems available for noise and vibration analysis and discuss the demands a good noise and vibration measurement system has to meet. After a brief introductory overview of the design of general analysis systems, we will present hardware and software requirements and features, respectively.

*Noise and Vibration Analysis: Signal Analysis and Experimental Procedures*, First Edition. Anders Brandt.

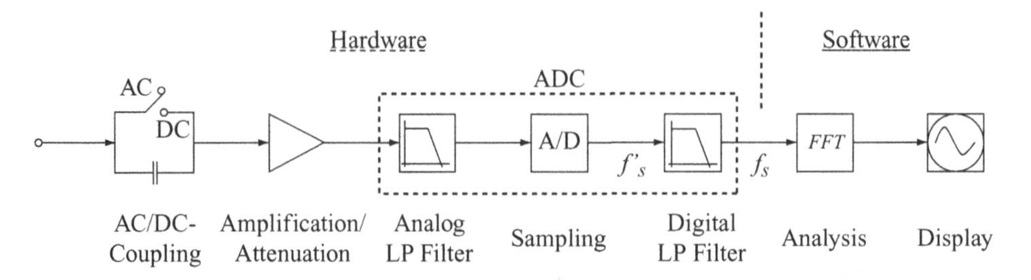


**Figure 11.1** Schematic diagram of an FFT analysis system. First comes a switch for AC/DC coupling, after which comes an amplifier/attenuator, which adjusts the signal to a reasonable level for A/D conversion. An analog filter, the anti-aliasing filter, follows, and then the A/D converter, and a digital lowpass filter. After this comes the processor in which the computations and analyses are carried out. Finally, the result can be displayed on a monitor. The dotted box surrounds components usually comprised in the sigma–delta A/D converter which is most common in modern measurement systems, see Section 11.2.2. The different steps are explained in the text below

## 11.1 Principal Design

A modern system for analysis of noise and vibration signals typically consists of an external hardware box in which the analog signals are conditioned and sampled digitally. The hardware box is usually controlled by software, usually implemented on a laptop computer, in which the signals are processed and stored. A schematic illustration of the main parts of a typical measurement system is shown in Figure 11.1.

The hardware usually consists of some signal conditioning electronics such as IEPE current feed for sensors with built-in impedance conversion as discussed in Chapter 7, or microphone power supply, etc. The signal conditioning is not illustrated in the schematic diagram in Figure 11.1, but will be discussed briefly in Section 11.2. Then, as illustrated at the very left of Figure 11.1, there is an AC/DC switch and an amplifier, or attenuator. Following this, the signal passes the analog-to-digital converter (ADC), indicated inside the dotted box in the figure, which in most cases today consists of an analog anti-aliasing filter, the actual A/D conversion, and a subsequent digital lowpass filter. The digital signal after the digital lowpass filter is transferred to the PC where it is processed and displayed, as illustrated on the right-hand side of Figure 11.1.

There is an indicated boundary (vertical dotted line) between hardware and software in Figure 11.1. This is to indicate those parts of the data acquisition and processing that are typically implemented in the (external) hardware box, and PC software, respectively. This does, of course, not mean that the parts implemented in hardware are not seen in the software. The hardware is *controlled* by the data acquisition and analysis software, so you will find menu choices in the software which control the different parts of the hardware illustrated in Figure 11.1.

Details of each part of the hardware and software will be discussed in Sections 11.2 and 11.3, respectively.

## 11.2 Hardware for Noise and Vibration Analysis

There are many hardware products on the market today which are dedicated to noise and vibration measurement. The main reason for this is, that there are some special demands on, particularly, dynamic range, cross-channel match, and signal conditioning (current supply) for IEPE sensors which need to be fulfilled for a high-quality noise and vibration measurement system. It is, of course, in principle possible to build a system based on general ADC components, but once all necessary parts are put together, the price will likely be higher than using dedicated hardware. The special demands on a good system for noise and vibration signals, require highly accurate antialiasing filters, matched between the channels, and electronics with relatively low noise floor, and those components are very expensive. The introduction of *sigma–delta* ADC technology in the 1990s (see Section 11.2.2), which is used in almost all dedicated hardware for noise and vibration signals today, has essentially led to very high performance at a relatively low cost per channel. This means that it is very difficult to match the price/performance of these dedicated systems by building a system based on more general ADC components.

## 11.2.1 Signal Conditioning

Many systems for noise and vibration analysis have built-in signal conditioning for typical transducers in vibration analysis and acoustics. Inputs with current supply for piezoelectric transducers of IEPE type, as discussed in Section 7.3, are now available in almost all noise and vibration systems. In addition, there are systems with direct inputs for microphones, charge amplifiers for piezoelectric transducers, etc. Since the signal conditioning is dependent on which transducer type is used, it has been omitted from the general schematic illustration in Figure 11.1, but is located before the input on the left-hand side of the figure.

Immediately after the signal conditioning, or sometimes inside the signal conditioning stage, there is a circuit for AC or DC coupling. Choosing AC coupling will force the DC component to be removed from the input signal, which is desired when the AC component we want to analyze is relatively small in comparison with the superimposed DC component. If we try to A/D convert the entire signal, we will obtain worse dynamic range, see Section 11.2.2, in the measurement than necessary, since the A/D converter's measurement range must be based on the maximum value in the signal, that is, the AC plus the DC component. Most noise and vibration signals should preferably be acquired with AC coupling, since we are in most cases only interested in the dynamic part of the signal. There are, however, some cases where the vibration signal has a natural nonzero mean, or where very low frequencies are of interest, which can justify acquiring the signal with DC coupling.

The capacitor in the AC coupling circuit, together with the input resistance of the next electronics stage, will form a high pass (HP) filter. Typically this is specified as a cutoff frequency in Hz by the manufacturer. For low-frequency measurements this can be problematic and care should be taken when measuring frequencies below, say, 20 Hz or so. Also see Section 7.3.1 for a discussion on low frequency characteristics of transducers.

After the AC/DC circuit in Figure 11.1 there is an amplification/attenuation stage, with the purpose of scaling the input so that it has a suitable voltage range for the lowpass filter and A/D converter. This stage will be touched upon in the section on A/D conversion and dynamics.

## 11.2.2 Analog-to-digital Conversion, ADC

A/D conversion (analog to digital conversion) is a collective term for conversion from a continuous (analog) signal to discrete (digital) samples. A/D conversion can be split into two parts: (i) discretization of amplitude, so-called *quantization*, and (ii) sampling.

### Quantization and dynamic range

By quantization we mean the effect that each continuous amplitude (voltage) value is converted to a (binary) number with a fixed number of bits (one bit is one binary digit). After this conversion process the amplitude resolution is naturally fixed, which results in a limited *dynamic range*, see Figure 11.2.

Dynamic range is defined as the ratio between the largest and smallest number that can be represented simultaneously after the A/D conversion. Since every extra bit in a binary number gives twice as many

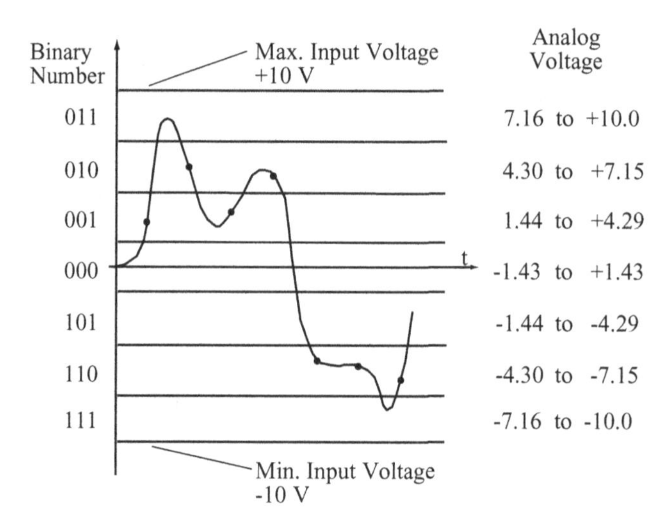


**Figure 11.2** Schematic illustration describing the principle of quantization during A/D conversion. In the illustration, three bits are used in the conversion, which is clearly too few for most applications, but is used here for the sake of simplicity. A continuous (analog) value at the input is converted to a digital sample, indicated by a filled ring, with a certain number of binary digits. The marked samples in the figure would give the binary sequence (001, 010, 001, 010, 110, 110 110). This limitation in resolution gives rise to limited dynamic range, see text

possible values (intervals), as a first approximation we can calculate the dynamic range, $D$, expressed in dB, after A/D conversion using $p$ bits, as

$$D \approx 6 \cdot p \text{ [dB]} \tag{11.1}$$

because a factor of 2 is equivalent to 6 dB ($20 \cdot \log_{10}(2) \approx 6$).

Besides the dynamic limitation due to quantization, the total dynamic range of the analysis system will usually also decrease due to the signal-to-noise ratio in the input amplifier if the noise floor of the amplifier is higher than the dynamic range due to quantization. The total dynamic range is typically around 100–110 dB for modern 24-bit measurement systems. A more rigorous treatment of dynamic range limitations due to quantization (e.g., Oppenheim and Schafer 1975) leads to a dynamic range due to quantization which is approximately 1.5 dB less than that given by Equation (11.1), but this correction is normally negligible compared with the limitation due to the amplifier's signal-to-noise ratio. Therefore, we refrain from this treatment here.

As the alert reader you are, you may have noticed that a possible digital value is missing in Figure 11.2. This is because we have used a representation where the leftmost bit stands for the sign (+ or −), the so-called *two's-complement representation*, the most common representation used in measurement equipment (ADCs and signal processors). As the aim of this chapter is not to describe how to design low-level software, we need not worry about the binary representation.

### Setting the measurement range

In order to obtain maximum dynamic range in a measurement, it is essential to adjust the amplifier/attenuator in the analyzer so that the measured signal exploits as much as possible of the total input range of the A/D converter. This is done by adjusting the input amplifier or attenuator, as the reference for the A/D converter is normally set to a fixed reference voltage for reasons of long-term stability. (The ADC reference voltage needs to be insensitive to aging of electronic components, temperature changes, etc.) If the input signal contains a DC component, it is recommended to use AC coupling of the input signal so that the DC component disappears, as mentioned above. We will now illustrate the effect of changing full scale range with an example.

**Example 11.2.1** *Use MATLAB/Octave to simulate a measurement of a sine signal with 1 V RMS level with two different full scale voltages, 2 and 20 V, respectively, using an ADC with 16-bit resolution. Calculate linear spectra of the two signals, and plot and compare the results.*

*MATLAB/Octave includes functionality to use integer arithmetics. This can be used to 'simulate' a real-case measurement situation the following way. To compute the linear spectra we use the* ***alinspec*** *command from the accompanying toolbox.*

```
fs=1000;                  % Sampling frequency
N=2048;                   % FFT block size
fsine=67;
t=(0:1/fs:(N-1)/fs)';       % Time axis
y=sqrt(2)*cos(2*pi*fsine*t);         % Create the sine
Scale=double(intmax('int16')/2);    % This makes 2V = intmax
ys1=double(int16(Scale*y));  % Truncate product to 16 bits
[Y1,f]=alinspec(ys1/Scale,fs,ahann(N),1,0); % Linear spec.
Scale=double(intmax('int16')/20);   % This makes 20V = intmax
ys2=double(int16(Scale*y));
[Y2,f]=alinspec(ys2/Scale,fs,ahann(N),1,0); % Linear spec.
```

*In Figure 11.3, the results of the code above are plotted for comparison. As can be seen, the effect of using an inappropriate full scale range is that the noise floor of the measurement is greatly increased (by a factor of ten just as the full scale range was increased by a factor of ten). This shows the importance of optimizing the input range.*

*End of example.*

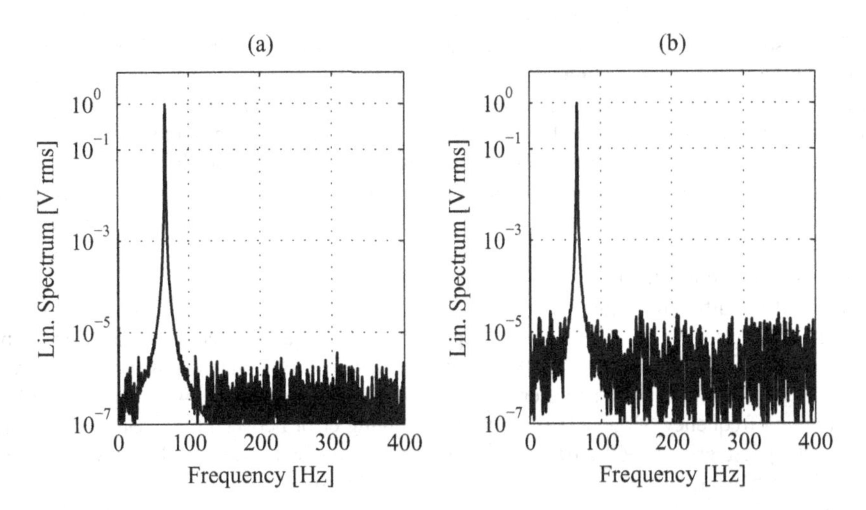


**Figure 11.3** Plot for Example 11.2.1. The plots illustrate the importance of selecting an appropriate full-scale range for dynamic measurements. In the figure, two spectra of a sine signal of 1 V RMS voltage (1.4 V amplitude), discretized using, in (a) a full-scale range of 2 V, and in (b) a full-scale range of 20 V. The ADC resolution was 16 bits. As can be seen in the figure, the background (discretization) noise is increasing with approximately 20 dB (10 times) with the higher full-scale range

It is also appropriate to warn against *overloading* the input electronics, which occurs if the measurement signal exceeds the full-scale range of the instrument. Overload is completely devastating to frequency analysis (and in most cases time domain analysis as well), as the spectrum of an overloaded signal differs from that of the original signal. Therefore, it is vital for all measurements that we can detect when overload occurs. This must be performed both in the input amplifier and in the A/D conversion itself. Because we have a lowpass filter before the A/D conversion, it is not possible to avoid the fact that overload can occur in the analog part, but it will be impossible to discover in or after the A/D conversion. Hardware should therefore be designed such that overload is indicated if it occurs before the ADC as well as if it occurs in the A/D conversion process.

In recent years, 24-bit A/D converters have become predominant in FFT analysis systems. The dynamic range of the ADC is then approximately 144 dB, which is much greater than the total dynamic range, which is rarely larger than approximately 100 dB, due to cross-channel talk and the noise floor of the input amplifiers etc. The dynamic range of most transducers for vibration measurements is approximately 90–100 dB. Thanks to the larger dynamic range of the ADC, we can avoid overloading the analysis system by setting the measurement range a few times above the measured signal's voltage without the total dynamic range of the measurement being compromised.

### Sampling accuracy

The second aspect of A/D conversion is the sampling, as we can of course only carry out the above quantization at specific points in time. It is important that the samples for all channels in a multi-channel measurement system are sampled simultaneously and that the time interval between samples, $\Delta t$, is accurate. To fulfill these requirements, the A/D converter should contain so-called sample-and-hold circuits, which are circuits that temporarily 'holds' the analog voltage at very precise time instants, so that the ADC can convert this constant value. Sample-and-hold circuits can usually be added to commercial ADC boards, and a system for noise and vibration must contain such circuits. Commercial FFT analysis systems always contain these circuits whereas many inexpensive ADC boards do not, but can usually be equipped with optional sample-and-hold circuits. The price for the sample-and-hold circuits can exceed the price of the ADC board.

The DFT is sensitive to the fact that the time increment, $\Delta t$, between all samples is identical. Figure 11.4 illustrates what happens to the dynamic range of a spectrum of a sinusoid when sampled with a normally-distributed random error in the sampling instances, with a standard deviation of $10^{-4} \cdot \Delta t$. The signal's spectrum was subsequently calculated, which should be a peak at the sinusoid's frequency, with low values, under the dynamic range, for other frequencies. Because of the error in time intervals between samples, the dynamic range is considerably limited. In many cases of vibration analysis, high dynamic range is necessary, for example to measure frequency response. Especially note that in our example, if the sampling frequency was, say, 1 kHz, then the standard deviation of the time error in our example would correspond to only 100 ns.

The result if the different channels are sampled at different instances in time, is a phase difference between the channels. This error is frequency dependent and can be analyzed by considering the period of a sine at a particular frequency as 360°. If the error between two channels is $T_\epsilon$ seconds, and the frequency is $f_0 = 1/T_0$, then the phase error is $360 \cdot T_\epsilon / T_0$. If, for example, we sample a 100 Hz sine with a sampling frequency of $f_s = 1kHz$, and the sampling error between two channels is 0.1%, i.e., 1 $\mu$s, then the phase error is $360 * 10^{-6}/10^{-3} = 0.36°$. Phase match between channels will be further discussed in Section 11.2.4.

### Anti-alias filters

In order to ensure that the spectrum measured by an FFT analysis system does not contain aliasing errors, the signal must, in most cases, be lowpass filtered to remove all frequencies above the Nyquist frequency (half the sampling frequency), see Section 3.2. This filter must clearly be an analog filter, as

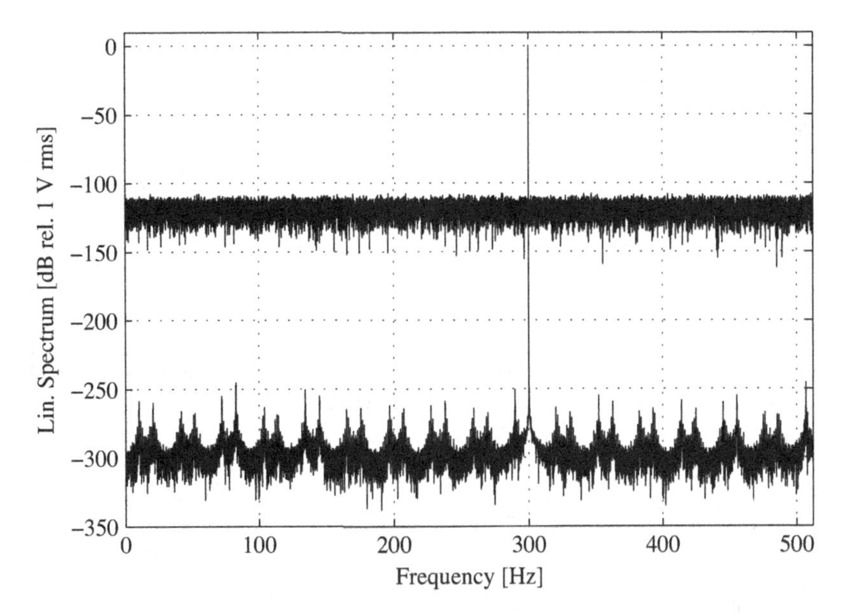


**Figure 11.4** An irregular sampling frequency (spreading of time delay $\Delta t$) gives rise to limited dynamic range. The lower curve shows the numerical dynamic range and the upper curve shows what happens if we introduce a random error in the sampling times $\Delta t$ with a standard deviation of $10^{-4} \cdot \Delta t$

it appears before the A/D converter. Because anti-aliasing filters, for physical reasons, cannot cut off all frequencies immediately above the cutoff frequency, but rather have a slope above the cutoff frequency, we must also have a 'safety margin' between half the sampling frequency and the cutoff frequency. In early equipment for noise and vibration analysis, a standard oversampling factor of 2.56 was established, as illustrated in Figure 11.6. In more recent systems utilizing sigma–delta ADCs, however, this factor is sometimes somewhat smaller, say down to approximately 2.2.

Following the A/D converter in Figure 11.1 there is a digital lowpass filter. This filter applies a lowpass filtering and subsequent decimation of the higher sampling frequency ($f_s'$ as illustrated in Figure 11.1) used by the ADC to the lower sampling frequency, $f_s$, requested by the user. Its function is to reduce some of the drawbacks with the analog anti-aliasing filter, especially its phase characteristics, as discussed in Section 3.3.2. An illustration of the characteristics of a typical anti-aliasing filter in an older analysis system without sigma-delta ADC is illustrated in Figure 11.5. This type of design with a digital decimation filter following the ADC has been predominant in systems for noise and vibration analysis since the mid 1980. If the analysis system is designed this way, the A/D converter's sampling frequency is fixed, usually at the system's highest frequency range. When we choose a lower measurement range, the digital filter is used to lowpass filter the measured signal, and at the same time it removes some samples, that is, decimates the data, so that the data after the digital filter has a sampling frequency of (typically) 2.56 times the highest frequency of interest.

An advantage of this process is that the digital filter can be designed with better characteristics, such as ripple in the pass-band, stop-band attenuation, and not least linear phase, compared to analog filters. Besides these advantages, the digital filter will have the same characteristics for all channels, giving good cross-channel match, which is important for cross-channel analysis, for example when estimating frequency response. Furthermore, the lowpass filtering process enhances the effective dynamic range, since each output sample of the filter is essentially an average of several input samples, which reduces the variance (noise level). There is, however, also a potentially severe cause of problems with this type of ADC design, if the measured signal contains frequencies above the Nyquist frequency, which will be discussed in Section 11.2.3.

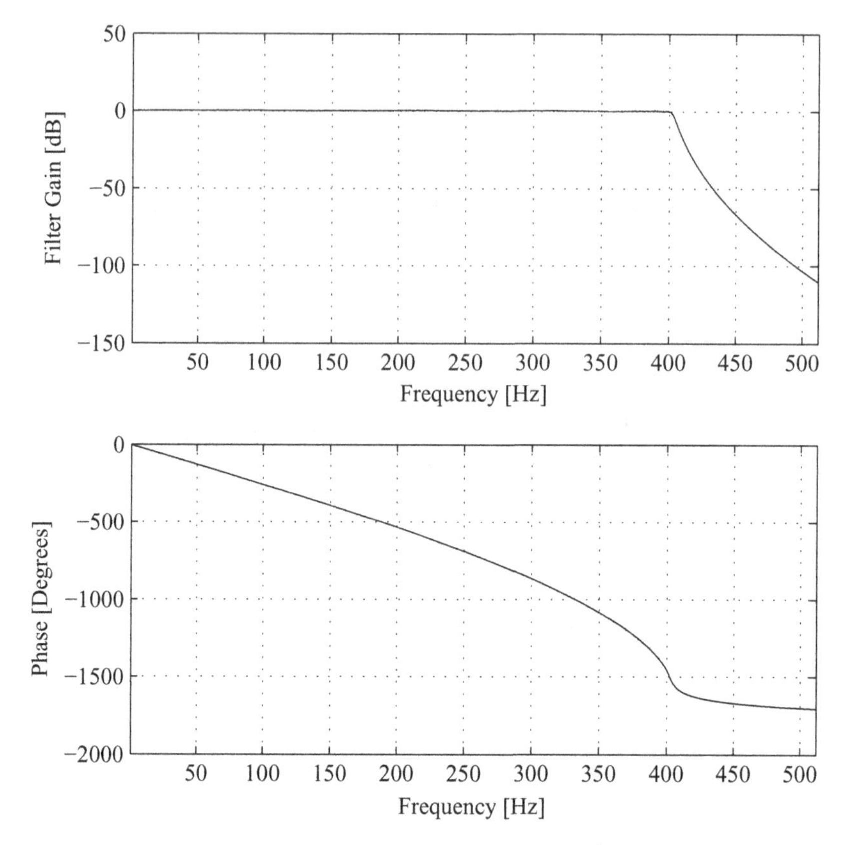


**Figure 11.5** Typical antialiasing filter. Because of the filter's non-ideal characteristics, the cutoff frequency, $f_c$, needs to be set lower than half the sampling frequency. It is typically set to $f_s/2.56$ in FFT analysis systems, for historical reasons, which approximately corresponds to $0.8f_s/2$. In the figure, the cutoff frequency is 800 Hz, which gives a sampling frequency of $2.56 \cdot 800 = 2048$ Hz. Note the non-linear phase characteristics, which is discussed in Sections 3.3.2 and 11.2.2

### Sigma–delta ADCs

In recent years, an A/D converter type called the *sigma–delta converter* (sometimes, more appropriately, called delta–sigma converter) has become popular in FFT analysis systems. It works on the same principle as the process just mentioned, but has a considerably higher sampling frequency than the original analysis systems using the principle described in Section 11.2.2, and consequently the digital lowpass filter decimates the data even more, which allows for even higher-quality data. A further advantage of the high sampling frequency is that a lower-order filter can be used for the analog anti-aliasing protection which introduces less problems with the nonlinear phase characteristics.

The sigma–delta ADC builds on the so-called one-bit converter (Higgins 1990; Proakis and Manolakis 2006) which in turn is based on delta modulation. This technology, developed for use in audio equipment, results in ADCs with very good performance at a very low cost. Typical sampling frequencies used in sigma–delta ADCs are in excess of 6 MHz, although at this frequency rate the signal has only a single bit resolution. Using the decimating digital filter following the ADC, however, the effective resolution is increased to 24 bits or more. The only restriction with this type of ADC is that, since the technology is developed for audio applications, the possible frequency range is limited; nominally to the upper audible

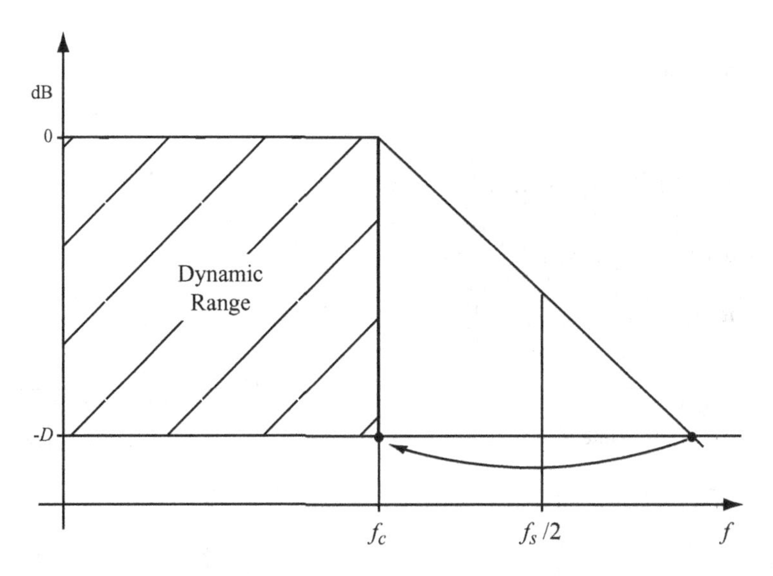


**Figure 11.6** Schematic illustration of the sampling frequency in relation to the cutoff frequency of the anti-aliasing filter. How large the ratio must be depends on the slope of the filter above the cutoff frequency, and the wanted dynamic range, $D$. The 'standard' in FFT analysis systems has become an oversampling factor of 2.56, that is, $f_s = 2.56 \cdot f_c$. This factor is a result of the fact that typical anti-aliasing filters had a slope of approximately 120 dB/octave. In some more recent systems utilizing sigma–delta ADCs, this factor is slightly reduced, see text for details

frequency of 20 kHz. Overclocking sigma–delta ADCs, some manufacturers are currently marketing sigma–delta ADCs with upper frequency range of over 200 kHz (usually with reduced bit resolution).

The digital filter following the ADC in sigma–delta converters should be a linear-phase filter, see Section 3.3.2, if time domain analysis such as transient analysis, for example, is the aim of the measurement. For price/performance reasons, however, not all manufacturers build this type of filters into measurement systems for noise and vibration analysis. You can therefore not be sure that systems with sigma–delta ADCs are suited for transient analysis without checking data sheets.

### *11.2.3 Practical Issues*

Measurement systems designed with digital filters after the ADC as illustrated in Figure 11.1 have a potential cause of problems which is illustrated in Figure 11.7. If the measured signal has substantial frequency content outside the measurement range, but below the analog antialiasing filter's cutoff frequency, and this 'outside signal' is larger than the signal of interest, the A/D converter can be overloaded, even though the measured signal looks fine on the screen. The reason is that what is observed on the screen is the output of the digital filter, which in the case under discussion will remove the high-frequency signal causing the overload. This situation can be perplexing if one is not aware that it may occur. Thus, if seemingly inexplicable overload indications occur, the measurement frequency range should be set to its highest value and the signal from the A/D converter should then be investigated. If the signal with the higher frequency is very large in relation to the signal of interest, and the dynamic range too low to allow the entire signal to be acquired and subsequently downsampled to the frequency range of interest, the only solution may be to use an external analog lowpass filter between the transducer and the measurement system to remove the higher frequency content. It should be said, however, that with 24-bit ADCs, due to their extremely large dynamic range, in many cases the situation can be solved by recording the signal with a high frequency range, and subsequently downsample to the requested frequency range.

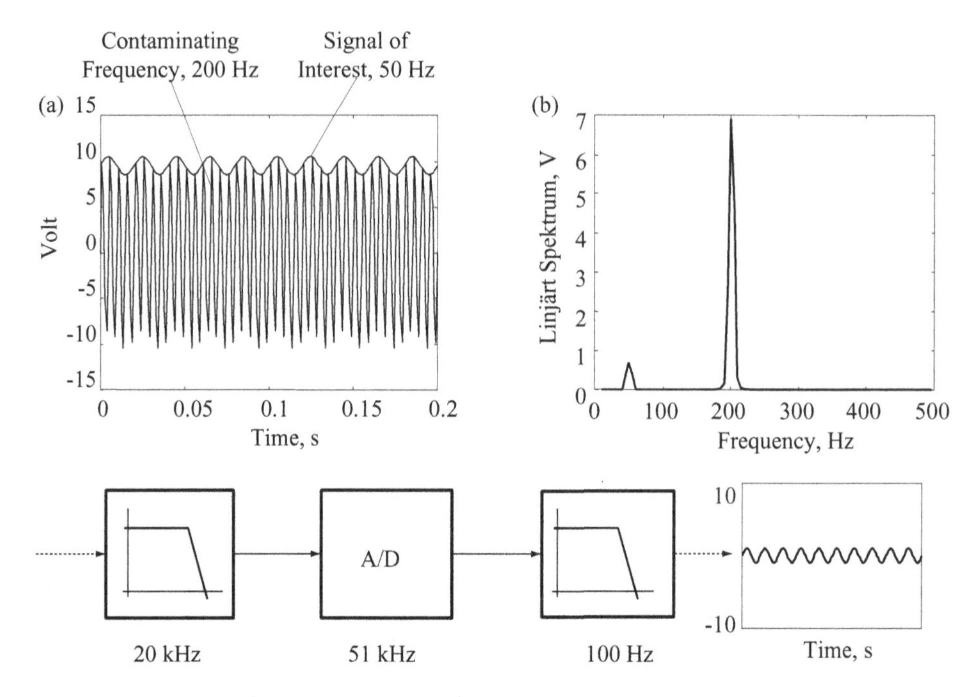


**Figure 11.7** Description of the case where overload can occur, even though the signal on the screen seems to be smaller than the A/D converter's input range. The overload occurs because the A/D converter sees the signal at 200 Hz, while we (in this example) on the screen only see up to 100 Hz. The A/D converter's amplitude range is assumed to be $\pm 5$ V

In summary, we shall list a few points about sampling and A/D conversion, which should be observed when using an FFT analysis system.

- Set the measurement range so that it approximately matches the measured signal's level. When using modern analysis systems with an A/D converter with 24 bits or more, a larger safety margin can be used, as few transducers have as great dynamic range as the A/D conversion. As IEPE sensors (see Section 7.3) are the most common sensors, it is worth pointing out here, that these signals have 5 V full range.
- Potential overloading of the input is fatal for spectral analysis. Therefore, watch out for overload indications and do not forget to check the entire measurement chain, including transducers, signal conditioning, and the ADC.
- An FFT analysis system is suitable to use for analyzing the spectrum of a signal. The time data seen in an FFT analysis system can normally only be used as an indicator that there is a signal, but not to see what the signal looks like, since the oversampling is insufficient. (However, some analysis systems specially designed for time domain analysis have the ability to use higher oversampling ratio so that time analysis can be performed.) Alternatively, the signal can be recorded and subsequently lowpass filtered.
- If you have an indication of overload, even though the signal looks smaller than the full scale, increase the frequency range and check if you might have a signal at a higher frequency which is overloading the A/D converter. If the dynamic range is insufficient after the A/D converter is adjusted to the higher-signal, then an analog lowpass filter is necessary between the sensors and the FFT analysis system to remove the higher frequency signal.

### 11.2.4 Hardware Specifications

When evaluating hardware for data acquisition systems, there are some very important specifications that need to be understood. In this section we will discuss some of the most important specifications, and also discuss how they can be easily measured. The main reason for this discussion is that some of the specifications often given for ADCs are theoretical values. It is important to understand the total performance of a system, particularly if one wants to build a system from individual components.

When testing specifications, it is important to correctly treat the channels which are *not* tested. These channels need to be shortened, usually best done by connecting 50 Ω BNC terminating plugs to the input BNC connectors. The reason for this is that open channels can easily pick up 50/60 Hz line power components which can ruin the measurements.

At this time it can be appropriate to discuss the importance of *calibrating* your measurement system. The importance of calibration to make traceable measurement cannot be over emphasized. A measurement system should be calibrated regularly, and the recommended time between calibrations is usually specified by the manufacturer of the system.

In the following subsections, the most important specifications of dynamic measurement systems will be discussed. Some relatively easy checks that can be done in the lab will be described, as it is good practice to, once in a while, check your measurement system to ensure that nothing has changed since the last calibration.

**Absolute amplitude accuracy**

The absolute accuracy of measurement systems for noise and vibration signals is usually good to within 1% or better. This can seem rough compared with, for example, a digital voltmeter, but we have to remember that the absolute accuracy in a dynamic measurement system is frequency dependent, and the worst case is specified. As sensors for noise and vibration analysis are usually specified within ±5%, one percent absolute accuracy of the instrument is rather acceptable. To measure the absolute accuracy, a high-quality sine generator must be used, which is usually only possible in calibration labs.

**Anti-alias protection**

As mentioned in Sections 3.2.1 and 11.2.2, it is essential that a system for noise and vibration analysis is equipped with analog anti-alias filters which ensure that a frequency component showing up in a spectrum is definitely due to frequency content at that frequency, and not due to a higher frequency, aliasing as a lower-frequency component. This means that the anti-alias filters must attenuate all frequencies above the Nyquist frequency by more than the dynamic range (as discussed in Section 11.2.2).

The anti-alias protection can be investigated by attaching a sine generator with an amplitude slightly less than the voltage range (full-scale voltage) of the channel to be tested. Then, instantaneous spectra are studied while increasing the frequency of the sine tone up above the Nyquist frequency. Once the frequency of the sine tone comes above the measurement frequency range, there should be no tone sticking up through the noise floor of the spectrum. If there is aliasing, the aliasing tone will sweep down in the spectrum, as the actual frequency is swept up between $f_s/2$ and $f_s$, where it (if still visible) will turn and start sweeping up.

With some low-cost anti-aliasing filters there are cases where some frequencies can come across with virtually no attenuation at all. With sigma–delta ADCs, it can be difficult to anticipate which these frequencies are. A large frequency range above the Nyquist frequency should therefore be tested.

**Simultaneous sampling**

As mentioned in Section 11.2.2, it is important that systems for noise and vibration analysis include sample-and-hold circuits for simultaneous sampling. This should be checked in the data sheet, as it is difficult to measure experimentally. This is particularly important if you build your own system, as dedicated systems in this field normally include such circuits.

### Cross-channel match

Many applications in noise and vibration analysis involve estimation of two-channel functions such as frequency response, coherence, and cross-correlation functions. In estimates involving two channels, any difference in amplification or phase characteristics between the two channels will be added to the cross-channel estimate. The *cross-channel match* is therefore an important specification of a measurement system. For multi-channel systems, it should be specified as the worst case, between the two channels in the system with the largest mismatch.

The cross-channel match is measured simply as the frequency response between any two channels. To investigate it, two channels are therefore connected to the same source by using a (usually random noise) signal from a signal generator to both inputs. The frequency response between the two investigated channels is then estimated using the procedures we will discuss in Chapter 13.

The cross-channel match is typically far superior on systems using sigma–delta ADCs compared with any other design. Modern measurement systems are often within $\pm 0.1\%$ in amplification (the magnitude of the estimated FRF), and less than $0.5^\circ$ in phase. The phase deviation usually becomes gradually worse closer to the upper frequency limit (cutoff frequency of the digital lowpass filters).

### Dynamic range

The *dynamic range* of measurement systems is a very important measure. Unfortunately, this term is sometimes slightly abused, so it is important to understand what is meant by the manufacturer when interpreting specifications. Dynamic range is a measure of the ratio of the largest and the smallest signal that can be resolved in a particular measurement, usually given in dB. High dynamic range is essential for frequency analysis, especially for frequency response measurements as these functions often contain very large dynamic range.

Thus, with this definition, it is a measure of the noise floor of the instrument, relative to the full scale range. As there are some other parameters, particularly cross-channel talk (see below) which can reduce the *total* dynamic range in a measurement, it is very important to know how the dynamic range is defined. Unfortunately, many manufacturers are not very clear about this.

It is quite common, particularly for ADC boards, that the quantization noise is specified as dynamic range. This can, however, easily be seen, as it is typically specified as 6 dB per bit ADC resolution. This, as obvious from Section 11.2.2, should not be confused with the dynamic range as we have defined it.

In most cases in modern data acquisition hardware, the dynamic range is limited by spurious narrow band signals (tones) which originate from clock frequencies, etc. in the hardware above the random noise floor of the instrument. In such a case, the dynamic range can be measured rather easily by shortening the input, for example by connecting a 50 Ω BNC terminating plug to the input connector. The spectrum will now contain only the spurious 'noise', since the input is shortened, and the ratio of the highest peak in the spectrum relative to the full scale range, is the dynamic range. After setting the input voltage range to, say, $A$ [V], a linear spectrum is measured with a few averages to obtain a stable spectrum of the background 'noise' (which in this case is not really noise as we assume it to be periodic). The dynamic range, $D$, in dB, is then computed by reading the root mean square (RMS) value of the highest peak in the spectrum, $X_{\max}$, and using the equation

$$D = 20 \log_{10} \left[ \frac{A/\sqrt{2}}{X_{\max}} \right] \tag{11.2}$$

where $A/\sqrt{(2)}$ is the RMS of the full-scale voltage.

If the dynamic range is instead limited by random noise, i.e., no tones are appearing above the noise floor, the dynamic range can alternatively be measured by calculating the RMS value of the signal of the shortened channel (i.e., the noise floor), $V_{RMS}$, and then calculating the dB ratio by

$$D = 20 \log_{10} \left[ \frac{A/\sqrt{2}}{V_{RMS}} \right]. \tag{11.3}$$

**Cross-channel talk**

Another very important property of measurement systems is the *cross-channel talk*. This measure, usually specified in dB, is measuring how much a sine tone connected to any channel is coming across on another channel. This factor is actually often limiting the effective dynamic range of many measurement systems since, if the cross-talk attenuation is less than the dynamic range (noise floor) of the instrument, it is impossible to tell if a low-level spectral component appearing on a particular channel comes from the signal at that channel, or if it is cross-talk from another channel, polluting the channel in question.

The cross-channel talk is measured by connecting a sine, with an amplitude close to the full-scale range, to one of the channels, and measuring a linear spectrum on all other channels, usually set to the same full-scale range. In the spectrum of the adjacent channel, at the frequency of the sine, the sine is usually detected, and the cross-channel talk is defined as the ratio of the measured amplitude and the true amplitude of the sine, on the tested channel and on the channel where the sine tone is connected, reported in dB. It is important to terminate all channels but the one where the sine is connected, when performing this measurement.

### *11.2.5 Transient (Shock) Recording*

In Chapter 3 we discussed that time domain analysis often requires that the signal is recorded with linear-phase filters. Many measurement systems for noise and vibration analysis are not designed with this in mind, although some systems are. It should be particularly noted that not all systems with sigma–delta ADCs are designed with linear-phase digital filters, due to price/performance issues. Before using a noise and vibration analysis system for recording transients for time domain processing, the system specification should therefore be carefully checked.

A solution if the system is not designed with linear-phase digital filters following the ADC, a solution could be to record the transients at the maximum sampling speed. Linear-phase digital filters can then be applied to the signals in post-processing, either in the software provided with the system, if available, or by exporting data, for example to MATLAB/Octave. This procedure usually works if the bandwidth of the transient is much lower than the bandwidth of the measurement system.

## 11.3 FFT Analysis Software

In this section we will discuss how modern FFT analysis software is designed, and typical applications for which such software is designed. A modern FFT analysis system is usually based on a laptop computer with software in which all analysis is done. There are many turnkey solutions on the market which provide all the functionality needed for typical applications in noise and vibration analysis. Most of these systems use a design established in the 1970s with the first FFT analyzers. The main thing characterizing these systems is that they are designed to process data block wise, as will be described in Section 11.3.1. Before proceeding with this, we should mention a few words about an alternative way of working.

As mentioned several times earlier in this book, there are some disadvantages with online processing of noise and vibration signals, where time data are discarded once fed to the FFT processor. The main disadvantage is that, as we know from Chapter 10, spectrum analysis is always a compromise between frequency resolution and variance or amplitude accuracy (depending on whether the signal is random or deterministic). Therefore, it is not uncommon to want to analyze the same data several times, for example with different frequency increment, or with different time windows. For this reason, it is good practice to record time signals for subsequent processing with different measurement settings.

Another important issue is that of *quality assurance* of the measured data, which is particularly important after field measurements. As we discussed in Section 4.4, such analysis can only be done in the time domain, and this is a strong reason to record time signals.

A third point is that there are some analysis procedures, for example estimating spectral densities by the smoothed periodogram method, which are not at all block based, but rely on performing one FFT on the entire data, and then smooth this FFT result in the frequency domain. This estimator is usually not found in FFT analysis software because it does not fit the block processing architecture, however the estimator is sometimes to be preferred over the more common Welch estimator.

These points make it attractive to record time data in most cases. This is also indeed possible with most analysis systems available on the market, which include the possibility to export data to, for example, MATLAB/Octave, to perform non-block-based processing, if the analysis software itself does not support it.

In the remaining sections of this chapter we will discuss the typical architecture of most commercial systems for noise and vibration analysis, as this is inevitably what you are most likely using.

### *11.3.1 Block Processing*

Before we continue with details about analysis hardware and software, we should take some time to discuss the principal operation of FFT analysis software. The FFT analysis system measures time data block wise, meaning that a certain number of samples, a block (or frame) of, for example, 1024 samples, is acquired. As soon as this acquisition is made, the block with data is transferred, normally to RAM memory, where the computations take place. The FFT algorithm starts the computation, while samples continue to stream into the now empty buffer. When the FFT computation is finished, the result, the *instantaneous spectrum*, is moved to a new memory location where it is accumulated in average operations and possibly being displayed, see Figure 11.8.

It may happen that the FFT computation takes more time than it takes to acquire the next block of time data. In that case, data acquisition halts until the FFT computation is complete and the instantaneous spectrum is moved on to the averaging buffer. That is to say, the samples that are A/D converted in the meantime are lost. This usually causes no problems if the measured signal is stationary. In certain special applications it can, however, be important that time data are not lost. Therefore, many manufacturers specify what is called the *real-time bandwidth*, which is usually given as the highest bandwidth (analysis range) which can be analyzed, with a given blocksize, without any data loss. Usually, such maximal performance is obtained without the accumulated spectrum being displayed, as displaying always costs some performance. It should be added that between the A/D converter and the FFT processor, there is usually a certain buffer to facilitate the communication necessary when the data are transferred internally.

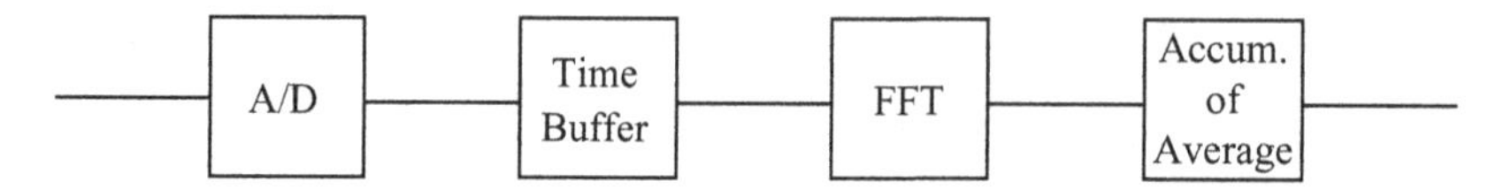


**Figure 11.8** Illustration of the principles of the memory buffers of an FFT analysis system. Time data are buffered in memory right after A/D conversion, which is used for trigger functions, etc. The FFT process uses a certain memory location, and stores the result in the final average accumulation

Therefore, one can usually carry out a number of averages without the limitation above coming into play, as long as data can fill this buffer. It is only when many averages are made, in some cases several hundred, that this limitation is noticed.

### 11.3.2 Data Scaling

Most FFT analysis software includes the possibility of using the sensitivity of the transducers used, so that the data are presented directly in the unit measured. If the software is well designed, there is also a field for setting an amplification factor for an external amplifier between the transducer and the analysis system. In this way, one can avoid loading the brain with what the net result is of a combination of transducer sensitivity and amplification; something experience tells us is more difficult than one may think.

As we discussed in Section 7.3.3, most IEPE sensors can be ordered today with a built-in memory circuit in which all necessary information about the sensor is stored. If the measurement systems supports reading this information, it is a good step forward to safe and reliable measurements, as entering wrong scale factors for sensors is a common cause of error in vibration measurements.

### 11.3.3 Triggering

All FFT analysis software packages have a trigger function, that is, a start function, which can be used in at least three ways. Although the terminology differs between manufacturers, the following should be helpful to understand the trigger function in most analysis systems. Most software packages have only a common level trigger. It may be possible to also set a certain hysteretic effect, that is, when the trigger level is reached, a certain level above and/or below this level must be reached before a new trigger can be activated. In general, one may set the trigger level and pulse edge (positive or negative).

In addition to the trigger level, the block processing can usually be controlled by setting a trigger mode condition according to the following. This setting may have different names, but is almost always available in noise and vibration analysis systems.

- *Free run* means that the trigger function is not activated. The analysis system then treats the incoming data as they arrive from the A/D converter.
- *First frame* or *continuous trigger* means that the trigger function is only used to begin the measurement, but afterwards the data are treated as with free run. This function is used if one wants the measurement to begin when the input signal exceeds a certain level.
- *Every frame* or *transient mode* means that the trigger level must be fulfilled for each time block. This function is used to synchronize data in some way, for example when analyzing transients, where a number of transients are to be acquired and a spectrum calculated for each transient. An additional feature is often found with this type of trigger, which gives the ability to check the data for each block, before they are added to the average. This is useful for impact excitation, which is described in Section 13.8.

In some applications, a built-in signal generator in the FFT analysis system is used to control a shaker or loudspeaker. In the case that the excitation signal is transient, a special trigger function called *source triggering* or similar, is sometimes used. This implies that every time block for the input signals is synchronized with the transient signal sent out from the signal generator. See Sections 13.9 and 14.4 for details about excitation signals.

When the input signal is triggered at a certain level, it is also essential to be able to set a pre-trigger, so that no part of the signal is cut off. If we imagine a transient as in Figure 11.9, which we trigger at

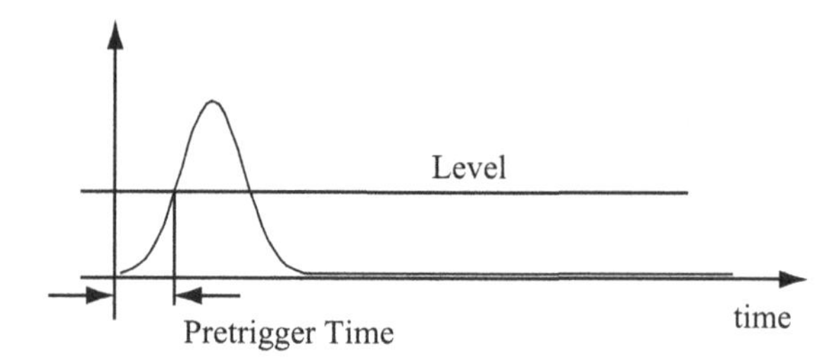


**Figure 11.9** Illustration of pre-trigger. A part of the signal before the triggering instant is acquired, so that the entire pulse can be analyzed

some level along the positive pulse edge, without pre-triggering we would only obtain the part of the transient that came after the triggering point. To avoid cutting the signal like this, with the pre-trigger, we set a time interval before the trigger that is also included in the acquired time block. This presumes, of course, that the time data are successively stored in memory in a so-called FIFO buffer (first in first out). Similarly, in some cases one may like to have a delay after the triggering point, before data are acquired in order to, for example, compensate for a time delay in the physical system being measured.

### *11.3.4 Averaging*

Most of the time, spectrum analysis involves an averaging process in the frequency domain, known as *frequency (domain) averaging*, as explained in Chapter 10. FFT analysis software thus allows this form of averaging to be selected. Usually there is also a choice of *time domain* averaging. These two forms of averaging are seemingly similar, but have quite different implications.

Frequency averaging decreases the variance in each spectral line if there is random noise between the estimated instantaneous spectra (each scaled and squared FFT result). Time domain averaging, on the other hand, is a completely different process, which can be used to improve the *signal-to-noise ratio* of deterministic signals. This is, however, only possible in two different cases,

1. for repeated transients with noise added on each transient, using a well-defined triggering condition, a time average could remove the random noise and result in a 'clean' average where the transient appears without the noise, or
2. for signals which are periodic exactly in the time window used for averaging, after some (many) averages only harmonic components which are periodic within the measurement block (window) remain in the averaged signal.

The second averaging case can be used either in order tracking applications with synchronous sampling, see Chapter 12, or to improve estimates of frequency response functions when using periodic excitation signals, see Section 13.12.

In most analysis software there is also a choice of averaging type which can usually be either 'linear' (sometimes called 'stable'), 'exponential', 'peak hold', or an averaging type called 'interrupted' or similar. The first choice produces a linear average where each value is equally weighted in the averaging process; this is the standard form of averaging. Exponential averaging can be used for spectral averages to produce a result similar to old analog analyzers. The exponential average is an average where the result is formed by weighting the most recent spectrum by $1/2$ and older spectra with a series $1/4$, $1/8$, .... This produces an average similar to the result of an analog analyzer. It is mostly used for monitoring purposes where spectra are monitored continuously. The interrupted average is used in conjunction with impact testing which will be described in Section 13.8. The averaging type means that the data acquisition is stopped after acquisition of the time block, and usually the user can manually select to use the current time block in the averaging process, or reject it. It can only be used if each data block is triggered. Peak hold, finally, is an averaging type which, for frequency domain averaging, keeps the maximum value of all averaged spectra, at each frequency. It can, for example, be used for conservative measurements of periodic signals with some contaminating noise, or to track (slowly) sweeping sines. You should look in the documentation of your analysis software for more details on how the different averaging types are used in your particular software.

### *11.3.5 FFT Setup Parameters*

The parameters needed for spectrum or frequency response estimation are blocksize, overlap factor (usually in percent of the blocksize), frequency range (or sampling frequency), and number of averages to perform. The number of averages specified is always the actual number of FFTs that will be performed. In order to calculate the random error for a spectral density estimate, with particular overlap percentage and number of averages, the formulas from Section 10.3.5 can be used. The only method available for spectral density estimates in commercial software packages for noise and vibration analysis is usually Welch's method (see Section 10.3.2).

As we know from Chapter 9, the relations between sampling frequency and blocksize on the one hand, and the frequency increment on the other hand, is that

$$\Delta f = \frac{f_s}{N} \tag{11.4}$$

where $\Delta f$ is the frequency increment of the DFT, $f_s$ is the sampling frequency, and $N$ is the blocksize.

## 11.4 Chapter Summary

This chapter has included an overview of some of the most important concepts relating to hardware and software for measurement and analysis of noise and vibration analysis. We have presented many factors which explain why, in most cases, dedicated hardware designed for noise and vibration measurements has better price/performance than systems built by standard components such as A/D boards, etc. The main reason for this is that a good noise and vibration measurement system puts high demands on the precision in A/D conversion, input dynamic range, etc. This makes modern measurement hardware based on the sigma–delta ADC superior.

Some important specifications for a good measurement system for noise and vibration analysis are:

- anti-aliasing filters with high cross-channel match,
- sample-and-hold circuits for simultaneous sampling and accurate cross-channel phase match,
- low noise floor for high dynamic range necessary for spectra and, particularly, frequency response measurements,
- high number of bits resolution, typically 24 bits is standard today.

## 11.5 Problems

Many of the problems following are supported by the accompanying ABRAVIBE toolbox for MATLAB/ Octave and further examples which can be downloaded with the toolbox. If you have not already done so, please read Section 1.6, and follow the instructions to download this toolbox together with example files.

**Problem 11.1** *If you have a commercial measurement system, go through your system documentation and find the specifications presented in this chapter so you know the limitations of your system.*

**Problem 11.2** *Assume you are going to measure an acceleration with an IEPE accelerometer. We assume the sensor is well chosen so that it gives close to maximum output voltage (full range). The accelerometer has a scaling constant of 100.4 mV/g. Which settings should you set for:*

1. *Input voltage range?*
2. *Sensitivity in mV/EU if your measured unit is* $m/s^2$*. (EU is commonly used for Engineering Unit)?*

*If the acceleration you measure is 34* $m/s^2$*, what will the voltage from the accelerometer be?*

**Problem 11.3** *Assume you have a microphone connected to a microphone power supply, which is then connected to a channel on your measurement system. The microphone has a sensitivity of 38 mV/Pa, and the power supply is set to a gain of 60 dB. What is the maximum sound pressure level in dB SPL that you can measure, if the maximum voltage from the power supply and in your measurement system is 10 V? (Note dB SPL means dB relative to 20* $\mu$*Pa.)*

## References

Higgins RJ 1990 *Digital Signal Processing in VLSI*. Prentice Hall.
Oppenheim AV and Schafer RW 1975 *Digital Signal Processing*. Prentice Hall.
Proakis JG and Manolakis DG 2006 *Digital Signal Processing: Principles, Algorithms, and Applications* 4th edn. Prentice Hall.

# 12

# Rotating Machinery Analysis

When analyzing vibrations and sound (noise) from rotating machines, a special type of frequency analysis, usually called *order tracking* is often used. This analysis is based on tracking RMS levels of time-varying sine tones resulting from the periodic forces acting on the machine. In this chapter we shall study how such analysis is carried out, while we will not study the different mechanisms behind vibration problems in rotating machines more than briefly. The analysis methods we will focus on in this chapter are predominantly used in the automotive and aerospace industries, and on power plant generators, etc. Other areas where rotating machinery analysis is commonly used are, for example, in vibration monitoring applications in process industry, and for balancing turbine engines and many other machines. The subject is vast and can fill several books, see for example Wowk (1991).

The procedures for order tracking that we describe in this chapter are not commonly found in textbooks. A good source for a more comprehensive discussion is the dissertation thesis by Blough (1998).

## 12.1 Vibrations in Rotating Machines

Two properties are of particular interest in the analysis of rotating machinery. As always in vibration analysis, structural resonances (modes) are of interest because they amplify vibrations. Secondly, however, on rotating machines, vibrations directly or indirectly caused by the rotation itself, are also of interest because they can become large without any resonance amplification. These latter vibrations are caused by, for example, imbalances, axle deformation or misalignment, defects in bearing races, defects in teeth on gears, etc. Each of these sources of vibration produce vibration at a particular factor times the rotational speed of the machine. Rotational speed dependent vibrations in rotating machines can, of course, occur at a frequency where the structure has a resonance, which can often cause very high vibration levels and sometimes disaster.

A factor times the rotational speed is called an *order*, where the rotation speed is referred to as *order 1*, two times the rotation speed is *order 2*, etc. Orders do not need to be integer numbers; we can have order 2.5 or 3.938 etc. The methods for analyzing vibrations in rotating machines which we are focusing on here, are mainly based on measuring the amount of noise or vibration due to either an order or a resonance frequency. The order number of the dominant vibration can often be used to deduce from where the vibration originates. If, for example, we have a gearbox with gear ratio 1:2.3 and we have high vibration levels at order 4.6, the problem is related to something that happens twice per rotation of the output shaft.

*Noise and Vibration Analysis: Signal Analysis and Experimental Procedures*, First Edition. Anders Brandt.

## 12.2 Understanding Time–Frequency Analysis

Most analysis of rotating machines is based on investigating the vibrations during a *speed sweep*, where the machine is either run up from a low to a high RPM (revolutions per minute), or run down from a high to a low RPM. The time data measured during the speed sweep are divided into smaller segments, each of which is processed by FFT to find the spectrum. It is evident that we are here talking about a special case of nonstationary signals, as the frequency content of the signal is changing. Therefore, it is important to understand some fundamental properties of this type of signals.

First, we need to have a model of the type of signals encountered. The assumption made in analysis of rotating machines is usually that the signal comprises a number of sine tones with some *instantaneous frequency* which changes with time, with each sine tone having a time-varying amplitude (and thus RMS level). We can thus formulate the signal as

$$x(t) = \sum_{k=1}^{N_O} A_n(t)\cos(\omega_n(t)t + \phi_n) \tag{12.1}$$

where $A_n(t)$ is a time-varying amplitude of the $n$-th order component, $\omega_n(t)$ is the angular frequency of the same component, $\phi_n$ is the phase angle offset of the tone, and $N_o$ is the number of order-related components in the signal.

The concept of instantaneous frequency can be understood by observing that for a stationary sine tone, with constant amplitude and frequency, we have

$$x(t) = A\cos(\omega t) = A\cos(\Phi(t)) \tag{12.2}$$

from which it can be seen that the (constant) angular frequency is

$$\omega = \frac{\mathrm{d}}{\mathrm{d}t}(\omega t) = \frac{\mathrm{d}}{\mathrm{d}t}(\Phi(t)). \tag{12.3}$$

From this it seems reasonable to define the instantaneous (angular) frequency of any signal

$$x(t) = A(t)\cos(\Phi(t)) \tag{12.4}$$

by

$$\omega_i(t) = \frac{\mathrm{d}}{\mathrm{d}t}(\Phi(t)) \tag{12.5}$$

where we have used the index $i$ on $\omega_i$ to indicate 'instantaneous'.

When we are analyzing a time-varying signal, there is an intrinsic problem related to the *bandwidth–time product*, BT product, discussed in Section 9.1.1. The instantaneous frequency is not possible to estimate instantaneously, due to the fact that we need to measure the signal during some time to be able to compute a spectrum. Furthermore, the frequency resolution we obtain using a particular measurement time, is inversely related to the measurement time. So, the finer frequency resolution we choose, the poorer time resolution we obtain, and vice versa. This results in an ambiguity when we are analyzing speed sweep signals and it is very important to understand this limitation.

To illustrate the BT product limitation, in Figure 12.1 two so-called *spectrograms* of a recorded microphone signal during a sweep from approximately 1000 to 6700 RPM is shown with two different FFT blocksize (the data will be described in Section 12.9). A spectrogram is a plot of a number of spectra with a certain blocksize versus time. On the left-hand side, in Figure 12.1(a), the frequency resolution is coarse and the time resolution thus fine, whereas in the right-hand spectrogram in Figure 12.1(b), the opposite situation is shown. Comparing the two spectrograms it is obvious how the change in time–frequency resolution affects the result.

As a consequence of the time–frequency limitations it should be realized that from any time–frequency analysis, there is not one, true, answer to the question of what the spectrum content of a time-varying

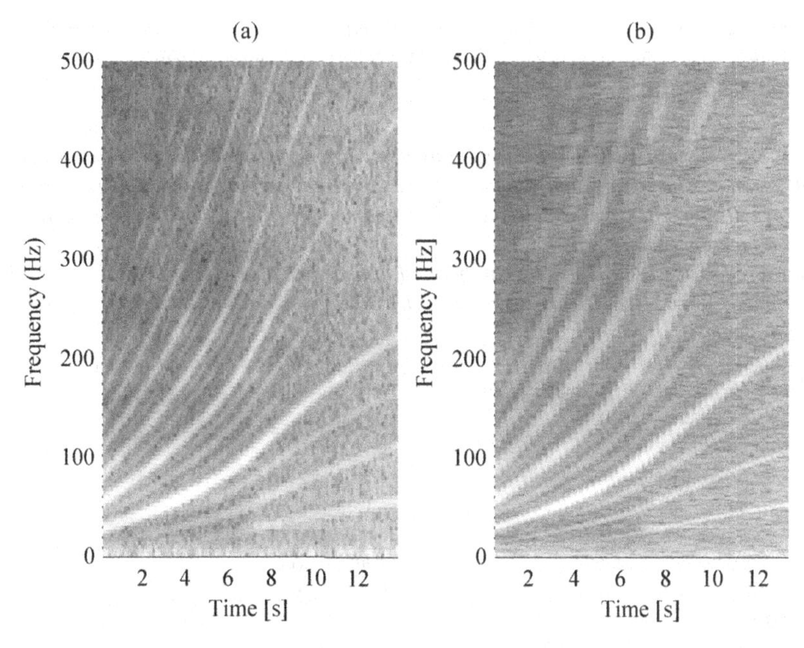


**Figure 12.1** Two spectrograms of a recorded microphone signal during a sweep from approximately 1000 to 6700 RPM (see Section 12.9 for details about the signal); in (a) the frequency increment was approximately 4 Hz, and in (b) approximately 1 Hz. It can be seen in (a) that the large frequency increment yields a relatively fine time resolution, whereas in (b) the properties are reversed [Data courtesy of Prof. Jiri Tuma]

signal is – it depends on how it is observed. In most cases of speed sweep analysis, the speed of the sweep therefore has to be kept fairly slow, so that changes in amplitude of order components do not occur too fast. In many cases it is a good idea to standardize some experiment and analysis parameters, such as frequency resolution and speed sweep rate, to facilitate comparisons between different measurements.

## 12.3 Rotational Speed Signals (Tachometer Signals)

In most cases of rotating machinery analysis, a rotation speed transducer, a *tachometer*, is connected to the rotating machine to measure the RPM. Such a transducer is usually either optical or inductive. In either case it produces some form of pulse signal where the time between the pulses is related to the rotation speed, $v_r(t)$, expressed in rotations per minute (RPM) as

$$v_r(t) = \frac{60}{N_p(t_2 - t_1)} \tag{12.6}$$

where $N_p$ is the number of pulses per revolution, and $t_1$ and $t_2$ are the time instances of two pulses. The estimated RPM readings are very important for the analysis of rotating machinery, as we will see later in this chapter. Many FFT analysis systems designed for rotating machinery analysis therefore have functionality to ensure that the rotation speed measurement is accurate. For example, lowpass filtering is often included to remove high-frequency disturbances, averaging to decrease random error, and limiting the change in the estimate of $v_r(n)$ between nearby $n$ (slew rate limitation).

The RPM as a function of time can be rather easily computed from a measurement of the tacho signal. Usually, the tacho signal is recorded on a measurement channel simultaneously with the accelerometer and/or microphone signals, so that the tacho signal is sampled synchronously with the signals to be analyzed. To illustrate how the RPM as a function of time can be computed, we will start by an example where a sine sweep is generated in MATLAB/Octave, which is then processed to obtain the RPM–time profile.

**Example 12.3.1** *Generate a simulated run-up signal of an engine increasing in RPM from 600 to 6000 RPM in 30 seconds. The signal should contain the fundamental and first two harmonics, with constant RMS levels of 1, 0.5, and 0.25, respectively. Use the fundamental frequency of the same signal to compute the RPM as a function of time.*

*The following MATLAB/Octave code can be used to generate a sweeping sine. Note that 600–6000 RPM corresponds to 10–100 Hz. The code generates a tacho signal in variable* ***tacho*** *and the signal with orders 1, 2, and 3, in variable* ***x***.

```
fs=2000; % Sampling frequency
T=30; % Time duration
f0=10; % Start freq.
f1=100; % End freq.
t=(0:1/fs:T)'; % Time axis
Sq2=sqrt(2); % Fundamental amplitude
x=Sq2*chirp(t,f0,t(end),f1);    % Fundamental
tacho=x;                        % Tacho signal
x=x+0.5*Sq2*chirp(t,2*f0,t(end),2*f1);  % Add order 2
x=x+0.25*Sq2*chirp(t,3*f0,t(end),3*f1); % Add order 3
```

*The first second of the generated signal is plotted in Figure 12.2.*

*We now use the tacho signal to extract the instantaneous RPM at each positive zero crossing, since our tacho signal will be a pure sine wave. In analysis systems it is common to be able to set both the*

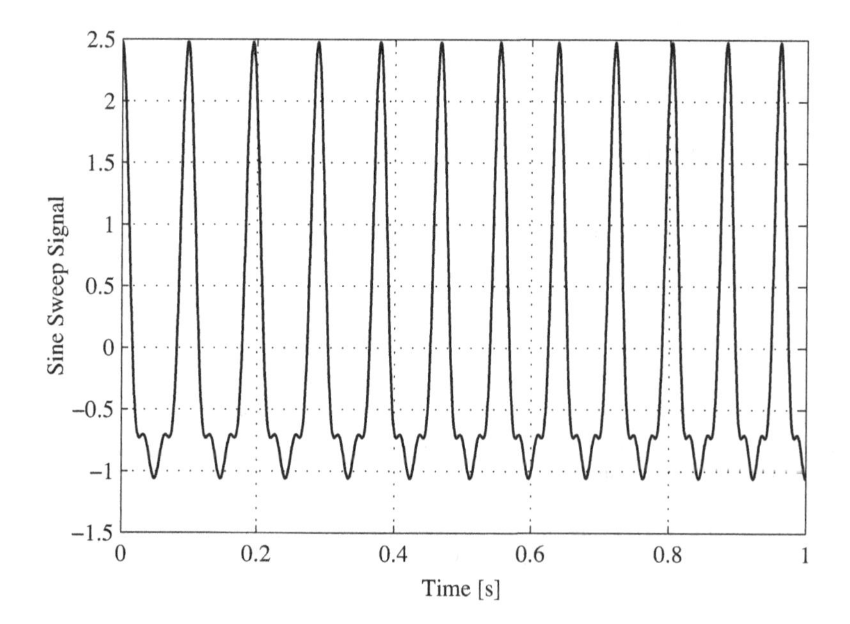


**Figure 12.2** First second of sine sweep with orders 1, 2, and 3 from Example 12.3.1

*triggering level and the slope, where the tacho pulses are counted. An example of MATLAB/Octave code to generate the RPM–time profile is as follows.*

```
% Produce +1 where signal is above trigger level
% and -1 where signal is below trigger level
TLevel=0;
xs=sign(tacho-TLevel);
% Differentiate this to find where xs changes
% between -1 and +1 and vice versa
xDiff=diff(xs);
% We need to synchronize xDiff with variable t from the
% code above, since DIFF shifts one step
tDiff=t(2:end);
% Now find the time instances of positive slope positions
% (-2 if negative slope is used)
tTacho=tDiff(find(xDiff == 2));
% Count the time between the tacho signals and compute
% the RPM at these instances
rpmt=60/PPR./diff(tTacho); % Temporary rpm values
% Use three tacho pulses at the time and assign mean
% value to the center tacho pulse
rpmt=0.5*(rpmt(1:end-1)+rpmt(2:end));
tTacho=tTacho(2:end-1); % diff again shifts one sample
```

*The code above produced the variables **rpmt** (temporary RPM, see below), with the instantaneous RPM at time instances in variable **tTacho**, using the time instances where the tacho signal crossed the trigger level with positive slope. In most applications, it is preferable to have an estimate of the instantaneous RPM corresponding to each sample in the sampled signals. To get this we can interpolate the estimated RPM values onto the time axis of the original, sampled signals, with the following line of code.*

```
rpm=interp1(tTacho,rpmt,t,'linear','extrap');
```

*The RPM as a function of time generated by the above procedure is plotted in Figure 12.3 (a). It is apparent that the estimates have some 'noise'. This comes from the uncertainty of $\pm\Delta t/2$ in the time instance of each trigger position which produces 'quantization noise'. In order to reduce this error, we can add a smoothing filter (see Section 3.3.3) before the last interpolation. The result of using 10 trigger instances to average over, is plotted in Figure 12.3 (b). As can be seen in the figure, this produces a more stable and reliable RPM–time profile.*

*End of example.*

The instantaneous RPM estimate has drawn some interest in the literature (Blough 1998; Fyfe and Munck 1997; Saavedra and Rodriguez 2006; Vold and Leuridan 1993). The most reliable way to improve the accuracy of the RPM–time profile, however, is to sample the tacho signal with a higher sampling frequency, thus reducing the uncertainty in the trigger occasions. Some commercial systems for noise and vibration signals therefore include special tacho input channels with this feature.

## 12.4 RPM Maps

The typical analysis procedure in rotating machinery analysis, is to make a run-up or a coast-down of the machine or engine, during which the time signals are recorded, or analyzed in real time. A run-up

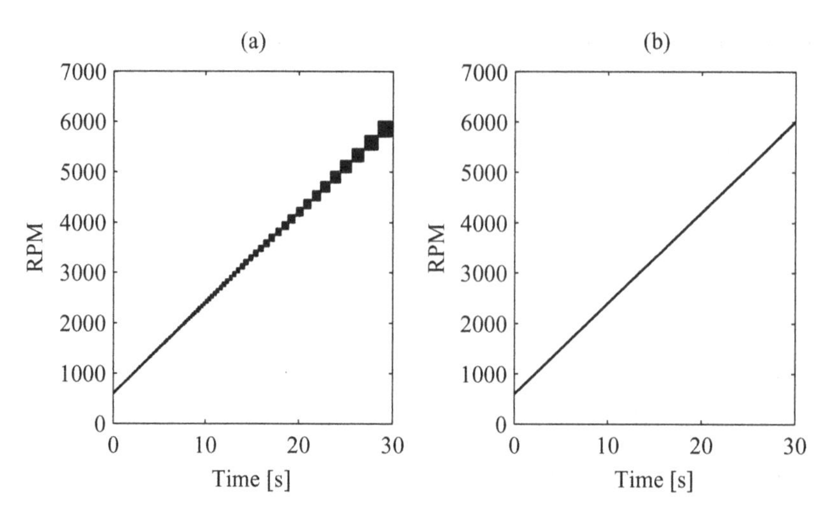


**Figure 12.3** RPM–time profile from Example 1, in (a) without smoothing prior to the interpolation, and in (b) with smoothing prior to the interpolation

is when the rotational speed of the machine or engine is increased from a low RPM to a high RPM, for example from 800 to 5500 RPM of an automobile engine. On some machines, for example electrical generators, the RPM cannot be smoothly swept up, for example because some machines are made to operate at a constant RPM (related to the power line frequency). In such cases, it is common to shut the drive of the machine off, and measure the vibrations during the coast-down of the machine. Another example where coast-down is often done, is on gearbox analysis, where a run-up and a coast-down will find problems on different sides of the gear teeth.

In the analysis of the time signal measured from the run-up or coast-down, the time signal is typically divided into short segments, for each of which an instantaneous spectrum is calculated, sometimes called a short-time Fourier transform, or STFT. Each of these spectra is 'stamped' by the RPM during the corresponding time segment either by taking the mean RPM from the RPM–time profile, or in some cases by taking the RPM at the end of the time block. The set of spectra thus produced is often referred to as an 'RPM map' which can be plotted in several formats.

### *12.4.1 The Waterfall Plot*

A common way to plot RPM maps is the *waterfall plot*. This is a three-dimensional diagram with frequency on the $x$-axis, amplitude on the $y$-axis and rotation speed on the $z$-axis. In this type of plot, order-related spectrum components, which occur at locations proportional to the rotation speed, will be visible as peaks on a straight line, and structural resonances will often be visible as peaks at fixed frequencies. In the diagram it can thus be seen which peaks are highest, at which speed the maximum occurs, and if they are caused by resonances or rotation-speed-dependent phenomena.

Figure 12.4 shows a waterfall diagram of the sine sweep from Example 12.3.1. The highest peak in this vibration signal is following order one, and orders two and three are also clearly seen.

### *12.4.2 The Color Map Plot*

A disadvantage with the waterfall plot is that sometimes smaller peaks can be difficult to distinguish. An alternative to the waterfall diagram which utilizes modern computer graphics is the so-called color map

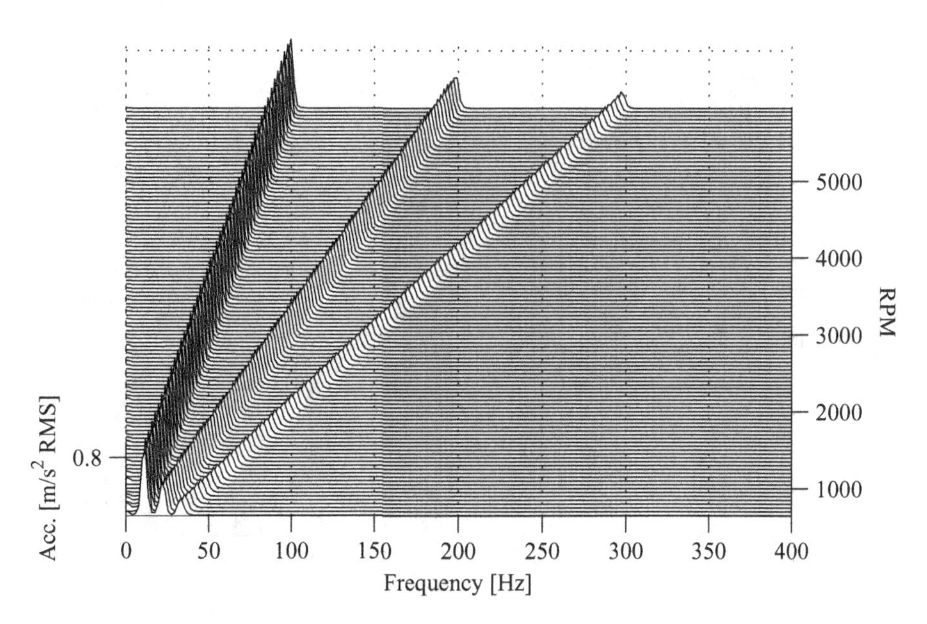


**Figure 12.4** Waterfall diagram of the sine sweep from Example 12.3.1. Orders one, two, and three are clearly seen

plot. An example of this type of plot is shown in Figure 12.5, although the print of this book only allows gray scale to be printed. In the color map plot, it can often be easier to differentiate the peaks, especially when an order meets a resonance frequency. A common way to obtain more details in the color map is to plot the logarithm of the RPM map, which enhances lower peaks.

## 12.5 Smearing

From the RPM map discussed in Section 12.4, so-called *order tracks* are often produced. An order track is an extracted RMS level of a particular order, versus RPM. To see how we can extract this information accurately , we need to discuss the concept of *smearing*. As discussed in Section 12.2 the signals encountered in order tracking analysis are nonstationary. The spectrum of the signal is thus not constant during the measured time block, which violates the DFT assumption. An assumption made in order tracking analysis is that the change in frequency for a tone we are studying is small during the acquisition of the short time data block used for each FFT. Smearing is an effect of the DFT which occurs if the frequency of the signal is changing during the duration of the time block.

Analysis of the smearing effect shows that it is dependent on the time window used and the amount of frequency change during the time block, relative to the frequency increment, $\Delta f$. To illustrate the smearing effect, in Figure 12.6(a), the result of a spectrum computation using a Hanning window, on a sinusoid swept linearly upwards in frequency, is shown. Three spectra are shown in the figure, for no frequency change, and a frequency change of $2\Delta f$ and $4\Delta f$, respectively. As can be seen, the smearing effect causes the peak to decrease, whereas frequency bins on both sides of the peak are increased.

A better window to use for this type of analysis could be the flattop window, which is often recommended for periodic signals. In Figure 12.6 (b) the spectrum of the same signal as in Figure 12.6, but with a flattop window instead of the Hanning window, is plotted. It can be seen that the flattop window has the peculiar effect that the smearing causes a peak which is higher than the true value. As evident

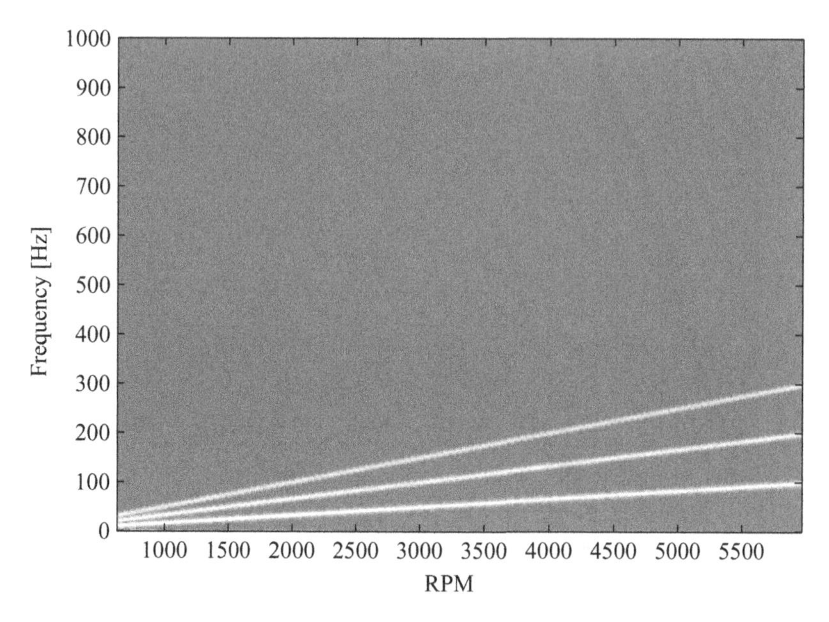


**Figure 12.5** Example of color map plot, converted to gray scale. The color map is a view of the waterfall diagram from above. On the $y$-axis is frequency, on the $x$-axis is rotation speed. The spectral peaks are coded in different colors, the higher the vibration amplitude, the larger the symbol. In this type of diagram, it is often easier to distinguish different orders and resonance frequencies, some of which may be hidden in the waterfall diagram. The gray scale print here does not give the figure full credit, see the problem section for examples of producing this type of plot in MATLAB/Octave

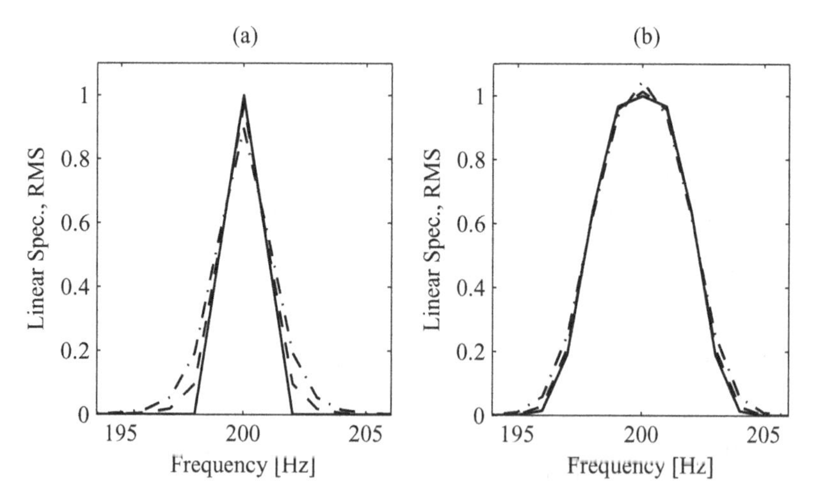


**Figure 12.6** Linear spectrum of a sinusoid with RMS value of 1 and a frequency which sweeps from $f_0 - k\Delta f/2$ to $f_0 + k\Delta f/2$ for $k = 0, 2$, and 4. In (a) a Hanning window was used before the spectrum computation, and in (b) a flattop window. It can be seen that the the peak decreases and the spectrum on both sides of the peak increases, an effect called smearing

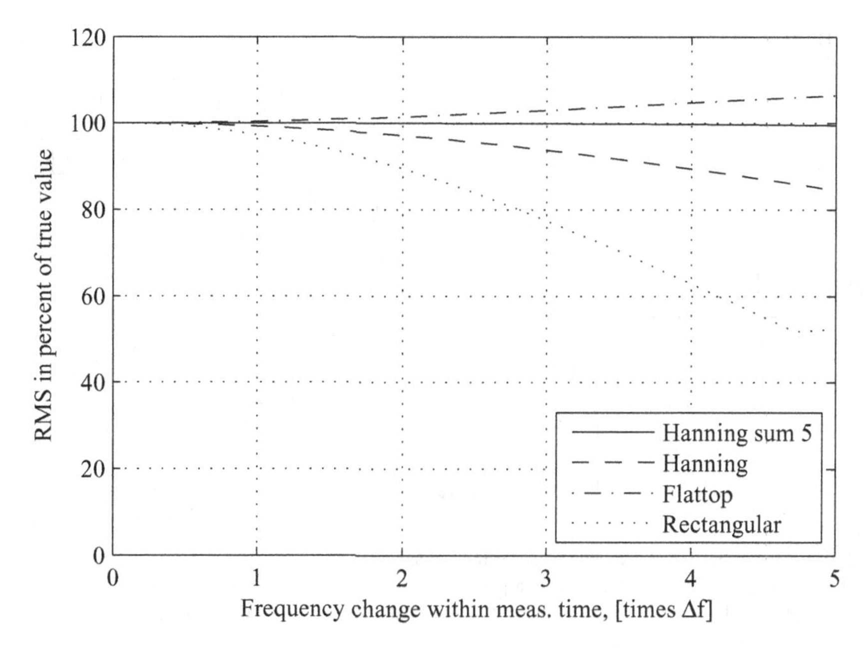


**Figure 12.7** Smearing effect for different windows. The graph shows the error in the RMS value in %, when measuring the linear spectrum of a sinusoid with linearly increasing frequency within the time block. The $x$-axis shows how many DFT spectral lines the signal's frequency changes by during the time of the FFT block. It is evident from the figure that the rectangular window is very sensitive to frequency variation in the measured signal, while the flattop window (ISO window, see Section 9.3.9) is less sensitive. The flattop window has the quality that the smearing error is positive up to a relatively large frequency alteration in the measured signal. The best method is, however, to use a Hanning window, and sum five frequency bins centered on the peak, as shown in solid. For comparison the result of using a rectangular window is also shown

from the plot, the error is less than for the Hanning window due to the greater width of the main lobe of the flattop window. However, the overestimation of the RMS level is, in most cases, an unwanted effect. As will be shown next, we can also considerably reduce the error in the order track by utilizing a different estimation approach.

There is a third alternative, which is to use a Hanning window and to sum the RMS value using several DFT bins around the peak frequency as described by Equation (10.51). It turns out that this alternative, using five frequency values, centered around the peak, is a much better alternative, which is due to the fact that Parseval's theorem is valid also for sweeping sines. This means that the RMS level calculated over the entire frequency range will always be identical to the 'true' RMS level calculated in the time domain. Also in a narrower frequency range around a single spectral component, the summed RMS level is very close to the true RMS value. In Figure 12.7 the results of using this method with a summation over 5 frequency bins is plotted versus the total frequency change of the sine in number of frequency increments, $\Delta f$. In the same figure the results of taking the peak of the spectrum with Hanning, flattop and rectangular windows are also plotted. As can be seen in the figure, the error in RMS level is negligible when using the frequency summation with Hanning window, compared with using the peak for any of the windows. Using a Hanning window and summing five frequency lines around a peak to obtain the RMS level is thus the preferred method for order tracking with fixed sampling frequency.

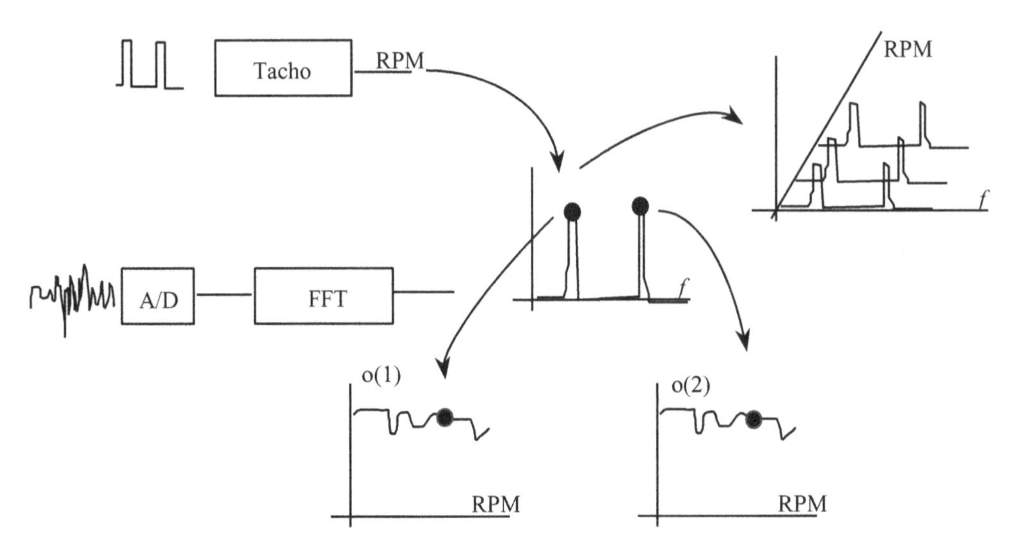


**Figure 12.8** Illustration of the principle of order track diagrams. While the time data are collected, the rotation speed is measured. Spectra are computed with a particular RPM increment and stored in an RPM map. From this RPM map, the RMS values for the orders of interest are then extracted and plotted in order track plots

## 12.6 Order Tracks

When interpreting how different rotational-speed-dependent components contribute to vibration or sound levels, two-dimensional diagrams called order tracks are often used. These diagrams are calculated from the RPM map by extracting information about the RMS value of an order component versus the RPM. A principal illustration of the process of generating order tracks as well as RPM maps is shown in Figure 12.8.

Using the process described in Section 12.5 of summing five frequency lines after applying a Hanning window in the spectrum computation, for the three first orders of the sine sweep in Example 12.3.1, results in the order tracks plotted in Figure 12.9. Since the RMS levels of all tones are constant during the sweep, the result is a check that the order tracking procedure works.

When computing the RMS level of an order by summing five frequency lines as recommended in Section 12.5, the 'filter' will have constant frequency bandwidth, i.e., a constant bandwidth in Hz. Sometimes it can be better to use a constant order bandwidth for the summation which can be easily accomplished by increasing the number of frequency lines the RMS summation is made over, with increasing RPM. The main reason for using constant order bandwidth is that the smearing effect gets worse for higher orders since the change in frequency increases with the order. If order one changes, for example, by one frequency increment, $\Delta f$, inside the FFT block, then order $o_n$ will change by $o_n \Delta f$. When using constant order bandwidth, with the recommended procedure of using a Hanning window, the order bandwidth must be set so that it corresponds to at least five frequency lines at the lowest RPM.

## 12.7 Synchronous Sampling

In order to avoid the smearing effect discussed above, so-called synchronous sampling can be used instead of sampling the data with fixed sampling frequency as has been assumed so far. In the past this was done by intricate analog electronics devices called phase-locked loop circuits. With the fast processing power currently available, today it is always accomplished using a resampling technique

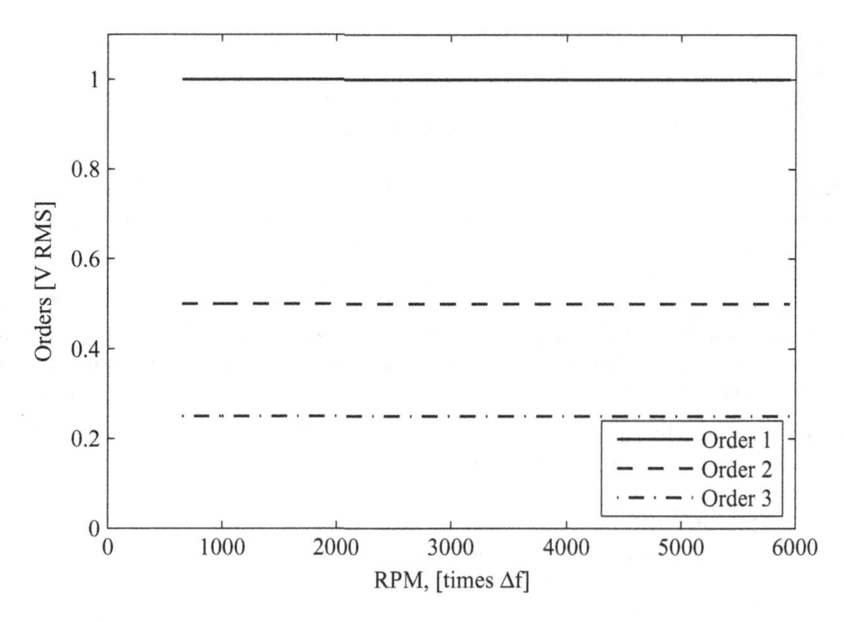


**Figure 12.9** Order tracks. The first three orders of the sine sweep used in Example 12.3.1 are plotted versus RPM. The RMS level of each order was computed by summing five frequency bins in the spectrum after applying a Hanning window. The levels of 1, 0.5 and 0.25 corresponds to the values used in the example

which will be presented in this section. This technique was developed and patented by Hewlett-Packard, (Potter 1990a,b) in the 1980s and has become the industry standard for synchronous sampling of rotating machinery signals.

The aim of synchronous sampling is to sample the vibration or noise signal at equal *angles* along each cycle (one revolution being one cycle), instead of equidistantly in time. The samples after synchronous sampling are usually said to be in the *order domain*, or *angle domain*. With for example 16 samples per revolution of the engine (or shaft), a signal will be obtained which is periodic with constant period length, if the $x$-axis is scaled in cycles. The DFT of this signal, will in turn have peaks at locations corresponding to the harmonics of this basic cycle (one rotation of the engine), which are exactly the orders, or fractions of them, see Section 12.7.1.

Before going into the DFT process of synchronously sampled data, we should look at how to obtain samples synchronous with the angle of the engine. We will describe two approaches here (i) Using one tacho pulse per revolution as synchronization pulses, and (ii) using the RPM–time profile we already computed in Section 12.3. We assume that we have the tacho signal recorded together with the vibration signal we wish to resample, in two separate records.

If we have a tacho signal with one or more pulses per revolution, we can easily obtain a trigger once per revolution by simply dropping all but one pulse per revolution. We can also assume that the RPM does not change substantially during one cycle, so we approximate the speed as constant during each cycle (this can, of course, be refined, but the first approximation used here usually works well). This is a reasonable assumption since there is always inertia in the machine preventing rapid changes of the rotational speed. We thus place the samples at equidistant times between the two tacho pulses. We now do this for each cycle (tacho pulse pair) during the run-up, and obtain time instances where we should have sampled the signal, had we not used fixed sampling frequency during the recording. If the number of samples per revolution is an integer factor times the number of tacho pulses per revolution, it is, of course, not necessary to remove the tacho pulses to have one pulse per revolution. The resampling can then be done with higher accuracy using all tacho pulses.

The next step is now to resample the vibration signal onto this new time axis. It turns out that this can be done arbitrarily accurately by simply using an upsampled vibration signal, and then using a linear interpolation algorithm. In most cases, 10 or 20 times oversampling gives enough accuracy. The procedure will be illustrated next, by a MATLAB/Octave example.

**Example 12.7.1** *Assume we have measured a vibration signal and a tacho signal with four pulses per revolution with a regular data acquisition system for noise and vibration analysis. Resample the signal with 16 samples per cycle.*

*We showed how to obtain the time instances of each tacho pulse in Example 12.3.1, in the variable* ***tTacho*** *on page 267 (before the last interpolation onto the final time axis!). We now start with this signal and add the following MATLAB/Octave code. Note that we could use all tacho pulses here, since we have four tacho pulses per revolution and 16 samples per cycle, which means we want exactly 4 samples per tacho pulse. To make the example more general, however, we will use only one pulse per revolution.*

```
tTacho=tTacho(1:PPR:end); % Pick out every 4th pulse
ts=[]; % Synchronous time instances
SampPerRev=16;
for n = 1:length(tTacho)-1
    tt=linspace(tTacho(n),tTacho(n+1),SampPerRev+1);
    ts=[ts tt(1:end-1)];
end
% Now upsample the original signal 10 times (to a total
% of approx 25 times oversampling).
x=resample(x,10,1);
fs=10*fs;
create a time axis for this upsampled signal
tx=(0:1/fs:(length(x)-1)/fs);
% Interpolate x onto the x-axis in ts instead of tx
xs=interp1(tr,x,ts,'linear','extrap');
```

*The reason for the code inside the* ***for*** *loop is that the new sampling instances should be at* ***SampPerRev*** *evenly spaced points between the two tacho pulses. The last sample, however, should be 'one sample before' it reaches the next tacho pulse, to obtain a continuous signal.*

*If we want a new time axis for this new variable,* ***xs****, we obtain it by the extra line*

```
tc=(0:1/SampPerRev:(length(xs)-1)/SampPerRev;
```

*which will produce an x-axis in cycles. The result of resampling the sine sweep from Example 12.3.1 using the code above is shown in Figure 12.10 for some cycles.*

*End of example.*

An alternative to the above method, is to use the RPM–time profile already established in Example 12.3.1. If we divide the instantaneous RPM by 60, we obtain the instantaneous frequency, $f_i$ in Hz. The integral of this frequency with respect to time, is immediately the 'angle' in parts per revolution (if we multiplied by $2\pi$ we would get it in radians, but for our purpose it is the straight derivative we want) We thus have that the 'instantaneous angle', $A_i$ is

$$A_i = \int_0^t f_i(t)\mathrm{d}t. \tag{12.7}$$

If we now want, for example, 16 pulses per period, we simply interpolate the values in $A_i$ onto a new $x$-axis being the fractions 0, 1/16, 2/16.... This procedure will now be illustrated by a MATLAB/Octave example.

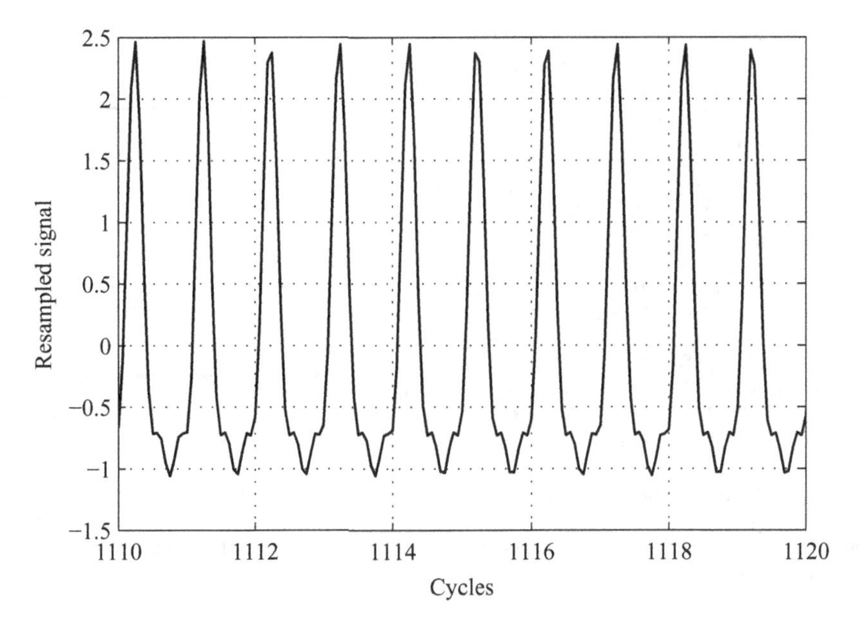


**Figure 12.10** Sine sweep signal from Example 12.3.1 resampled using the tacho signal as discussed in Example 12.7.1 for cycles 1110 to 1120. The signal appears versus the $x$-axis of 'cycles' as a stationary sine

**Example 12.7.2** *Use the computed RPM–time signal in variable* **rpm** *from Example 12.3.1 to resample a signal using 16 pulses per revolution.*

*We obtain the following MATLAB/Octave code.*

```
% Calculate the inst. angle as function of time
% (in part of revolutions, not radians!)
Ainst=dt*cumsum(rpm/60);
% Find every 1/SampPerRev of a cycle in Ainst
minA=min(Ainst);
maxA=max(Ainst);
Fractions=ceil(minA*SampPerRev)/SampPerRev:1/SampPerRev:maxA;
% New sampling times
tt=interp1(Ainst,t,Fractions,'linear','extrap');
```

*The synchronous sampling is then done as in Example 12.7.1, by upsampling the signal and interpolating it onto time instances in variable* **tt**.

*End of example.*

Of the two methods described here to synchronously resample a signal, the first method, using the tacho pulses, is usually a little more accurate because the tacho pulses are well defined in time, whereas the RPM–time profile usually needs some smoothing to obtain stable RPM values and therefore introduces some uncertainty in the instantaneous RPM.

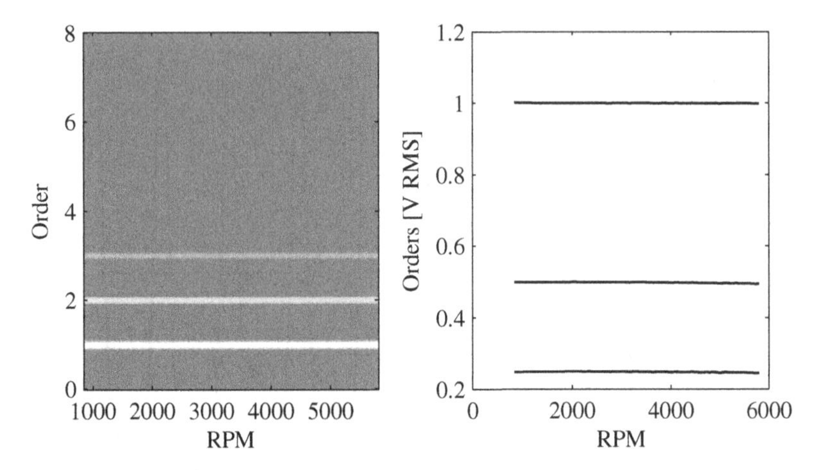


**Figure 12.11** RPM map, in (a), and the first three orders of the sine sweep from Example 12.7.1, in (b), after resampling the signal using 16 samples per revolution. The RPM map now has order on the $y$-axis and the order components appear along straight horizontal lines. The order tracks in (b) are usually of higher accuracy than those obtained by fixed sampling frequency as described in Section 12.6

### *12.7.1 DFT Parameters after Resampling*

We will now look at the DFT parameters and their interpretation on synchronously sampled signals. We recall from Chapter 9 that each frequency line, $k$, in the DFT, corresponds to a sine with $k$ periods in the time window. This means that if we want an order resolution, $n_o$, of for example 1/8th order, we need to sample eight periods of the synchronously resampled signal (so that the first order is on spectral line eight). Furthermore, the highest order we want to calculate, has to be below frequency line $N/2$, if $N$ is the blocksize. If we denote this maximum order by $O_{\max}$, then we can calculate the necessary blocksize for the DFT by

$$N = 2O_{\max}n_o. \tag{12.8}$$

For resampled signals, if the resampling was perfect, no time window should be necessary prior to computing the DFT. In practice, however, it is recommended to use a flattop window to get the best possible RMS level accuracy, even if the resampling process has failed slightly at some instance. An RPM map and corresponding order tracks of the first three orders of the sine sweep from Example 12.7.1 are shown in Figure 12.11.

## 12.8 Averaging Rotation-speed-dependent Signals

Sometimes the instantaneous values extracted in the procedures previously mentioned in this chapter do not produce stable values. This can be caused by noise in the transducer, or, e.g., by irregularities in the combustions of a combustion engine, etc. To obtain more stable values, averaging can be applied several ways.

In general, averaging is not necessary to produce the RPM map, as the accuracy in RMS levels is not critical to the purpose of the RPM map plot. The best approach of averaging to produce less variance in order tracks is to average several adjacent RMS estimates in the order track. This is equivalent to applying a smoothing filter (see Section 3.3.3) to the order track calculated as described in Section 12.6. This is the recommended procedure to obtain more stable order tracks.

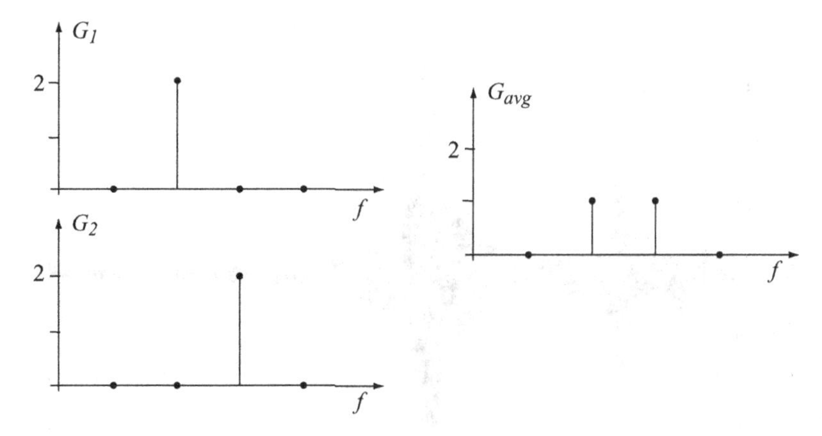


**Figure 12.12** Illustration of the error that occurs if different instantaneous spectra in the averaging process contain a spectral peak at different frequency bins. As illustrated, this produces an erroneous average spectrum. Averaging on signals from rotating machinery can usually only be done in the order domain, after synchronous resampling of the signals

Another approach is to average several consecutive spectra to produce each spectrum in the RPM map. This requires that the run-up or coast-down is very slow so that the spectrum peaks match the same spectral line in each spectrum included in the average, see also Section 10.1. Otherwise an error will arise as illustrated in Figure 12.12. In practice this type of averaging can only be successfully applied to synchronously resampled data and sometimes to constant RPM measurements, although in the latter case great caution has to be used to ensure the RPM is constant enough to produce spectrum peaks on the same spectral line in each instantaneous spectrum.

## 12.9 Adding Change in RMS with Time

So far in this chapter we have demonstrated the order tracking techniques with a sine sweep with constant RMS level across the run-up. This means that we have not taken the BT product into full consideration, as there was no change in RMS level with time. We will therefore now add some change to the RMS level of each order, to see how that affects the accuracy of the order tracking. To do this, we use the forced response simulation method described in Section 6.5, and let our sine sweep be the force input to an SDOF system with a natural frequency of 50 Hz (corresponding to 3000 RPM), and 2% damping. In Figure 12.13 the time signal of the vibration signal from the simulation is shown.

The resulting order tracks of the first order for fixed sampling frequency and for synchronous resampling using the tacho pulses, respectively, are shown in Figure 12.14. In Figure 12.14(b), where the plot is zoomed in around the RPM range where the first order passes the natural frequency of the SDOF system, it can be seen that with fixed sampling frequency and FFT blocksize of 1024 and 2048 samples, there is an error in the estimated order versus RPM. With synchronous sampling, however, the order is tracked with correct RMS level at the peak. It should be noted that the fixed sampling frequency could perhaps have been used with a shorter blocksize, giving a better accuracy, but using synchronous sampling is 'safer'.

It has sometimes been argued that fast run-ups can not be correctly tracked, even using synchronous sampling. This is not entirely correct, however. Using commercial systems where the synchronous sampling is done in real time, due to the reduced accuracy of the resampling process necessary to fulfill the real-time requirements, the resampling can sometimes fail. In post-processing, however, the accuracy

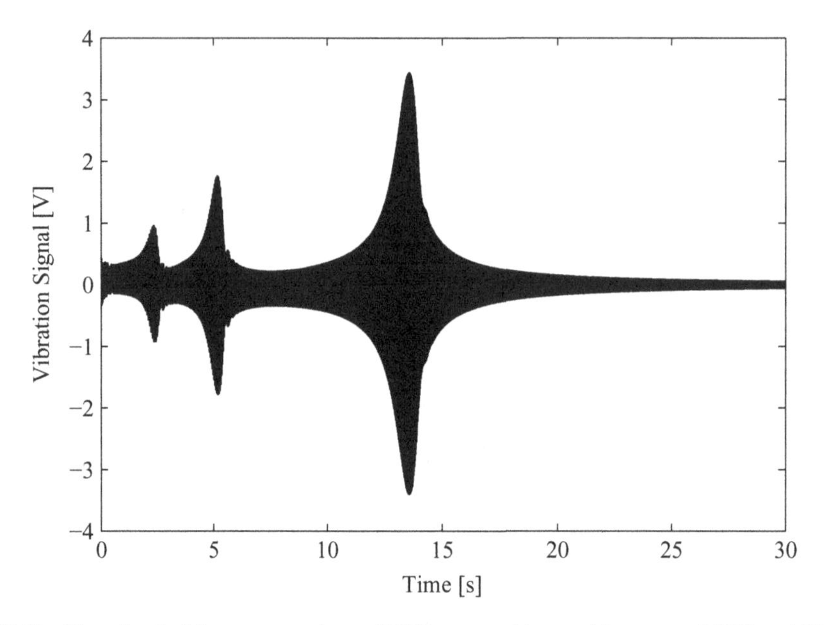


**Figure 12.13** Time signal of sine sweep passing an SDOF system with natural frequency of 50 Hz and 2% relative damping. When the first, second, and third order passes the natural frequency there is an amplification of the amplitude. This happens in the reverse order, of course, so that the third order passes the natural frequency first

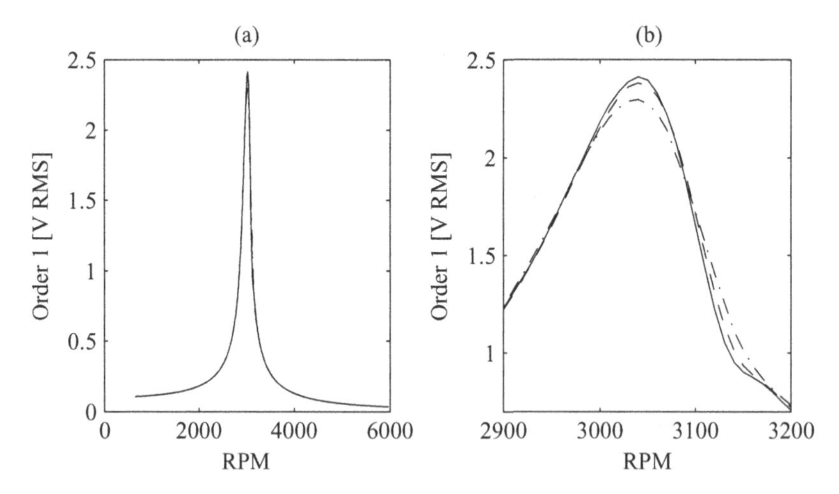


**Figure 12.14** Order track for first order of sine sweep which has passed an SDOF system, using the simulation method described in Section 6.5. Solid line: first order using synchronous sampling; dashed line: fixed sampling frequency with blocksize $N_1 = 1024$ samples; and dash-dotted line: fixed sampling frequency with blocksize $N_2 = 2048$ samples. In (a) the order track over the entire RPM range is shown, and in (b) the RPM range around where the first order passes the natural frequency of the SDOF system is zoomed in. It can be seen in the zoomed plot that the order tracks using fixed sampling frequency do not give correct estimates of the RMS level when the change rate is large. With synchronized sampling, however, the estimated RMS level reaches the true value of approximately 2.4 V, which corresponds to the maximum amplitude of the time signal in Figure 12.13

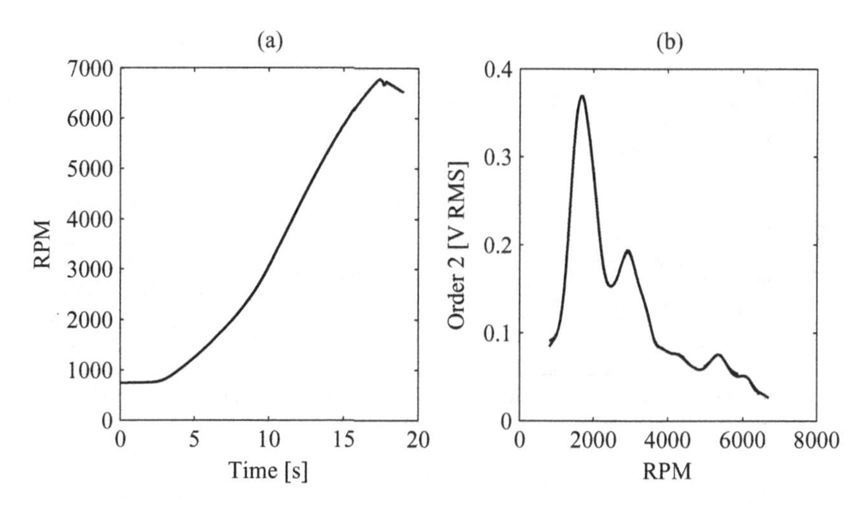


**Figure 12.15** Results of a fast run-up of a car during approximately 14 s. In (a) the RPM–time profile is shown, and in (b) order 2 is tracked by fixed sampling frequency (solid line) and by synchronous sampling (dashed line). As can be seen, in this case both fixed sampling frequency and synchronous sampling gives similar results. Some smoothing was applied to get this result [Data courtesy of Prof. Jiri Tuma]

can be increased by using higher oversampling, as is seen in Figure 12.15, where a very fast run-up is shown. The signal in this case was a microphone positioned near the exhaust of a car during a full-throttle acceleration in third gear, taking approximately 15 seconds.

The BT-product limitation is essentially the only limitation to tracking fast run-ups, regardless of the method used. This means that the accuracy of tracking a rapidly increasing RMS level versus RPM as in Figure 12.15, depends on the bandwidth of the tracking filter. If this filter bandwidth can be set high enough, then the slope of the rapid increase can be accurately tracked. If, however, there is a close order, for example, which will be included in the bandwidth of the tracking filter, then the problem is unsolvable using an FFT-based approach, and the only way to obtain accurate order tracks is to make a slower run-up, or to use a parametric method such as Prony's method described in Section 12.10.

Using very fast run-ups can pose another problem, which needs to be discussed. Assuming there are resonances which causes amplification of the order of interest, during a very fast run-up the frequencies can sweep through the resonance so fast that the mode is not fully excited, i.e., the amplitudes do not increase as much as they would if the run-up speed had been slower. This has to do with the transient behavior of the mechanical system, as discussed in Chapter 5. Therefore, in most cases it is better to use slow sweep rates, as was recommended earlier in this chapter.

**Example 12.9.1** *In this example we are going to look at some results of a run-up analysis on a diesel car. The acceleration on the driver's seat foundation, in a car equipped with a four cylinder diesel engine, was measured during a run-up from approximately 1400 to 3000 RPM. In Figure 12.16, the results are shown as an RPM map from an analysis using fixed sampling frequency, an RPM map after resampling the acceleration signal synchronously with the rotation, an RPM–time profile, and the orders of interest for this example. In the last plot, the overall level computed from the fixed sampling frequency analysis is plotted together with orders 2 and 25 of both analyses (fixed and synchronous sampling). The results for order 2 are almost identical and no differences can be seen. For order 25, however, it is evident that the synchronous sampling generates more accurate results.*

*For a four-cylinder four-stroke engine, order 2 will be dominant because there are two combustions per revolution of the engine. This is clearly seen in the RPM maps in Figure 12.16 (a) and (b). Order 2 is*

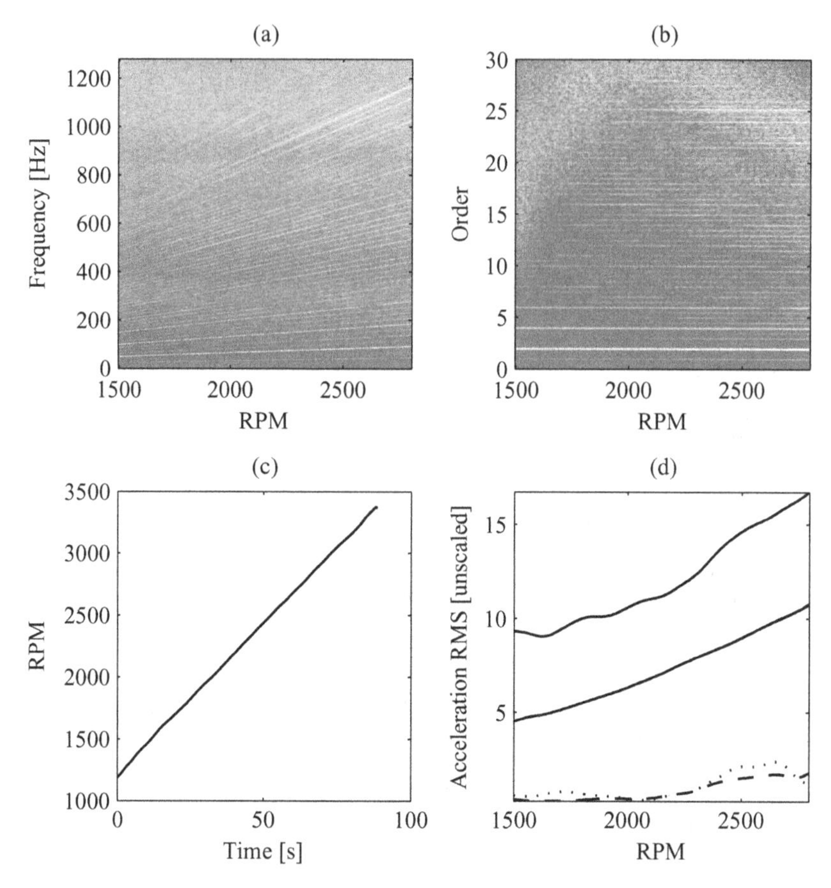


**Figure 12.16** Results of a run-up analysis of a diesel engine used for Example 12.9.1. In (a) an RPM map using 4096 samples blocksize is shown; in (b) an RPM map of synchronously resampled data using 64 samples per revolution and a maximum order of 32; in (c) the RPM–time profile used for the RPM map in (a) as well as for resampling the data; and in (d) the overall level based on the fixed sampling frequency data, order 2 from fixed frequency (solid) and resampled data analysis (dash-dotted), and order 25 from fixed frequency (dashed) and resampled data (dotted). Order 2 from both analyses are indistinguishable. See Example 12.9.1 for details. [Data courtesy of SAAB Automobile AB]

*dominated by shaft unbalances at low RPMs and of the mass inertia of the piston and combustion forces at higher RPMs, so it should increase approximately linearly with RPM. As is seen in Figure 12.16 (d), order 2 is increasing approximately linearly with RPM, which indicates there are no resonances in the structure in the frequency range of order 2, i.e., approximately 23–55 Hz.*

*In addition to order 2, a large number of higher orders are visible. Order 25 seems to be higher than most orders in the RPM map in Figure 12.16 (d). This is a driveline frequency, due to some gearbox ratio (since 25 is not a multiple of 2) and we select it for processing simply to illustrate the difficulty of tracking higher orders using fixed frequency sampling (at least with fixed frequency bandwidth). Thus order 25 is plotted in Figure 12.16 (d), near the bottom of the plot. The differences between the order from fixed sampling and synchronous sampling are clearly seen, and are due to the inaccuracy of the tracking using fixed sampling frequency.*

*To compute the results in Figure 12.16, the RPM–time profile in Figure 12.16 (c) was first computed by the procedure described in Section 12.3. To compute the RPM map in Figure 12.16 (a), the RPM–time profile was used to find the occasions for every 10th RPM, and a linear spectrum was computed using 4096 samples blocksize and a Hanning window. The orders from this RPM map were then computed by summation over 5 frequency lines as described in Section 12.6. The overall level, i.e., the total RMS level at each RPM, in Figure 12.16 (d) was computed by summing the RMS level of each spectrum, as described by Equation (10.51).*

*For the synchronous sampling, the original acceleration signal was resampled synchronously using the RPM–time profile, as described in Example 12.7.2, using a maximum order of 32 (i.e., 64 samples per revolution). Then an RPM map of the resampled signal was computed, as in Figure 12.16 (b), again using every 10 RPM, and an order resolution of 1/32-th of an order, i.e., the blocksize, according to Equation (12.8) was 2048 samples. A flattop window was used for the synchronous resampling RPM map, and thus the orders from this RPM map were obtained by tracking the peak of every order.*

*End of example.*

## 12.10 Parametric Methods

All the methods so far described in this chapter have been non-parametric methods, i.e., methods not using any *a priori* information about the signals. As we have discussed previously in this book, the advantage with nonparametric methods is that they give reliable results without the need for any assumptions about the data. The orders tracked by the methods described so far in this chapter can be slightly wrong, due to bandwidth–time restrictions, or because the synchronous resampling fails due to, for example, tachometer errors. But RMS levels which are completely absent from the data cannot accidently appear, which is the case with some parametric methods.

Some parametric methods, for example Prony's method (see e.g., Proakis and Manolakis 2006; Vold *et al.* 1988) have been suggested for order tracking. However, there has been little success using such methods, which is probably due to the fact that the models underlying vibration signals are not easily defined. There are almost always vibration artifacts due to the fact that any machine is more complicated than any reasonable model can include. Therefore, this type of parametric methods usually fails on noise and vibration data.

More recently, however, a parametric method which is less prone to errors from the model assumption, usually referred to as the *Vold–Kalman filter method*, has been developed, (Pan and Wu 2007; Pelant *et al.* 2004; Tuma 2004, 2005; Vold and Leuridan 1993; Vold *et al.* 1997). This method uses adaptive bandpass filters whose center frequencies are controlled by the instantaneous RPM–time estimate described in Section 12.3. Although originally suggested for fast run-ups, with the results obtained in Section 12.9 it is clear that fast run-ups can be tracked using post processing synchronous sampling. The Vold–Kalman technique is still limited by the BT–product – the narrower the bandpass filters, the longer time constants the filters have, and thus the slower they can react to rapid changes in RMS level in an order, see for example Brandt *et al.* (2005).

The Vold–Kalman technique offers two other advantages, however, not available with non-parametric methods; first, it can be used to filter out time signals corresponding to order-related components, and second, it can be used for multiple RPM signals with crossing orders. The first of these points is very attractive in sound quality applications, where the Vold–Kalman method offers the possibility to listen to a summation of the order-related components of, for example, the noise from an engine. The second point is very important in many applications on for example automatic gearboxes and on turbines, where typically several independent rotational speeds are causing orders, related to the different rotating shafts, to cross. With Vold–Kalman filters these crossing orders can be split into independent orders.

Another technique which has recently been proposed is a method based on the Gabor transform, (Pan *et al.* 2007; Qian 2003; Shao *et al.* 2003). This method is computationally attractive although there is some disadvantage with virtual (nonexistent) components which can cause difficulties in the interpretation of results.

Any parametric method for order tracking requires more experience from the user to obtain reliable results, compared with the non-parametric, FFT–based methods. Thus, in practice it is often useful to first calculate order tracks using either fixed sampling frequency or synchronous sampling, and based on those results, for example Vold–Kalman filters can be defined and used for higher accuracy.

## 12.11 Chapter Summary

In this chapter we have presented a method commonly used for analysis of rotating machines, *order tracking*. This technique assumes that the noise or vibration signal from a rotating machine consists of a number of RPM-dependent sine waves with variable amplitude and phase, and with a frequency which is a constant factor (the *order number*) times the RPM of the engine. Thus, order one corresponds to the rotational frequency of the machine, order two is the first harmonic (twice the frequency of order one), etc.

Analysis of rotating machines by order tracking are usually made either during a run-up of the machine, where the RPM is increased from a low to a high RPM, or during a coast-down, where the machine is slowing down from a high RPM to a low RPM. In any case, the analysis is done the same way by dividing the signal into short time segments, and by computing an FFT of each segment. The map of spectra versus RPM can be plotted in a 3D waterfall diagram or a color intensity map, in which the orders and resonance frequencies of the machine can be identified.

Order tracking is a technique where nonstationary signals are analyzed. The concept of the bandwidth–time product limitation for frequency analysis is therefore important to understand. Using a particular bandwidth of analysis, the time resolution is limited in such a way that changes in for example RMS level versus RPM, which are of interest in order tracking, can only be correctly estimated if they are slow enough. This means that it is best to use relatively slow run-up speeds in order to get reliable estimates of order tracks.

From the RPM map, order tracks can be extracted, and the preferred method is to use a Hanning window in the FFT computations, and then sum the RMS level around a particular order of interest by using at least five frequency lines. This reduces the effect of smearing, as discussed in Section 12.5.

To further reduce the effect of smearing, the signal originally sampled with fixed sampling frequency can be resampled into the angle domain, where the signal is sampled at a fixed number of samples per revolution of the engine. This produces a signal with seemingly constant 'frequency', with an $x$-axis of cycles. The DFT of this signal has peaks at fractions of the orders, as was discussed in Section 12.7.1.

Parametric techniques, such as the now popular Vold–Kalman filter method, can offer advantages when orders are crossing in machines with two or more independently rotating parts, or when time signals of each order are of interest in sound quality applications. Other parametric techniques, such as Prony's method, have not proven to be reliable for most rotating machinery applications.

## 12.12 Problems

Many of the problems following are supported by the accompanying ABRAVIBE toolbox for MATLAB/Octave and further examples which can be downloaded with the toolbox. If you have not already done so, please read Section 1.6, and follow the instructions to download this toolbox together with example files.

**Problem 12.1** *Create a simulated run-up signal using MATLAB/Octave with the following properties: Start RPM: 800 RPM Stop RPM: 5800 RPM Sweep time: 60 seconds Order one RMS level: varying* *sinusoidally between 1 and 2 V, and with one period during the speed sweep. Order two RMS level: varying sinusoidally between 0.5 and 8 V, with two periods during the speed sweep. Use a sampling frequency of four times the highest frequency in the signal.*

*Also create a tacho signal being a sine with constant amplitude during the speed sweep.*

*Extract the RPM–time profile with a time resolution corresponding to the sampling frequency of the signal.*

**Problem 12.2** *Create an RPM map of the signal in Problem 12.1 and use it to extract the first and second orders. Compare the order tracks with the 'true' values and experiment with the blocksize of the FFT. (Use the preferred method of using a Hanning window and five frequency lines to sum the RMS level presented in Section 12.6). Determine a blocksize that gives good results.*

**Problem 12.3** *Resample the signal in Problem 12.1 using 8 samples per revolution, and calculate order tracks for the first and second orders, using an FFT resolution of 1/16th order. Compare the tracked orders with the results of Problem 12.2.*

## References


Blough J 1998 *Improving the Analysis of Operating Data on Rotating Automotive Components* PhD thesis University of Cincinnati, College of Engineering.

Brandt A, Lagö T, Ahlin K and Tuma J 2005 Main principles and limitations of current order tracking methods *Sound and Vibration* **39**(3), 19–22.

Fyfe KR and Munck EDS 1997 Analysis of computed order tracking. *Mechanical Systems and Signal Processing* **11**(2), 187–205.

Pan MC and Wu CX 2007 Adaptive vold-kalman filtering order tracking. *Mechanical Systems and Signal Processing* **21**(8), 2957–2969.

Pan MC, Liao SW and Chiu CC 2007 Improvement on Gabor order tracking and objective comparison with Vold–Kalman filtering order tracking. *Mechanical Systems and Signal Processing* **21**(2), 653–667.

Pelant P, Tuma J and Benes T 2004 Vold-kalman order tracking filtration in car noise and vibration measurements *Proc. Proc. 33rd International Congress and Exposition on Noise Control Engineering, INTER-NOISE, Prague, Czech Republic.*

Potter R 1990a A new order tracking method for rotating machinery. *Sound And Vibration* **24**(9), 30–34.

Potter R 1990b Tracking and resampling method and apparatus for monitoring the performance of rotating machines.

Proakis JG and Manolakis DG 2006 *Digital Signal Processing: Principles, Algorithms, and Applications* 4th edn. Prentice Hall.

Qian S 2003 Gabor expansion for order tracking. *Sound and Vibration* **37**(6), 18–22.

Saavedra PN and Rodriguez CG 2006 Accurate assessment of computed order tracking. *Shock and Vibration* **13**(1), 13–32.

Shao H, Jin W and Qian S 2003 Order tracking by discrete Gabor expansion. *IEEE Transactions on Instrumentation and Measurement* **52**(3), 754–761.

Tuma J 2004 Sound quality assessment using Vold–Kalman tracking filtering *Seminar, Instruments and Control, Ostrava, Czech Republic.*

Tuma J 2005 Setting the passband width in the Vold–Kalman order tracking filter *Proc. 12th ICSV, Lisbon, Portugal.*

Vold H and Leuridan J 1993 High resolution order tracking at extreme slew rates *Proc. SAE Noise and Vibration Conference*, Traverse City, MI Society of Automotive Engineers.

Vold H, Crowley J and Nessler J 1988 Tracking sine waves in systems with high slew rates *Proc. 6th International Modal Analysis Conference*, Kissimmee, FL, pp. 189–193.

Vold H, Mains M and Blough J 1997 Theoretical foundation for high performance order tracking with the Vold–Kalman tracking filter *Proc. 1997 Noise and Vibration Conference, SAE*, vol. 3, pp. 1083–1088.

Wowk V 1991 *Machinery Vibration: Measurement and Analysis*. McGraw-Hill.

# 13

# Single-input Frequency Response Measurements

It is common in many noise and vibration applications to compute frequency response functions from measurements. In most cases, the structure is excited by known (measured) forces applied either by an impulse hammer or by a shaker. In some cases, frequency response functions are measured between response signals which are due to natural excitation by for example wind or traffic loads on buildings or bridges. The latter is common, e.g., when measuring transmissibilities for operating deflection shape measurements, see Section 16.6.

Common reasons that one may wish to measure frequency response functions between a force and a response signal are, for example, to determine

- natural frequencies and relative damping (e.g., in experimental modal analysis, see Section 16.7)
- point stiffness to be used in analytical models
- noise transfer paths (NTF) for *noise path analysis* (NPA), which is a method of studying sound paths from, for example, the engine mounts to the driver's ear in a vehicle, see Chapter 15. NPA has so far mainly been used within the automotive industry, but is increasingly being applied for other products.

In this chapter, we shall present two different ways of measuring frequency response, namely through excitation with an impulse hammer and with a shaker, respectively. We limit the discussion in this chapter to systems with one input and one output, so-called single-input/single-output (SISO) systems. In Chapter 14 we will extend the concept to general systems with many inputs and many outputs. Much of the theory in this chapter was developed in the early days of noise and vibration analysis. A good source for a more detailed discussion is Bendat and Piersol (2010).

Estimation of frequency response turns out to be rather complicated in terms of all the errors involved, because, as we will see, there are errors due to the spectral estimation as well as errors due to unwanted noise in the measured force and response signals. We will therefore make extensive use of simulations in this chapter, rather than real measurements. This allows us to focus on one of the sources of error at the time, whereas in real measurements all errors occur at once, and are unknown. In order to make good measurements of FRFs, it is essential to understand all the errors involved, and to be able to determine when or if they are occurring in a particular measurement.

*Noise and Vibration Analysis: Signal Analysis and Experimental Procedures,* First Edition. Anders Brandt.

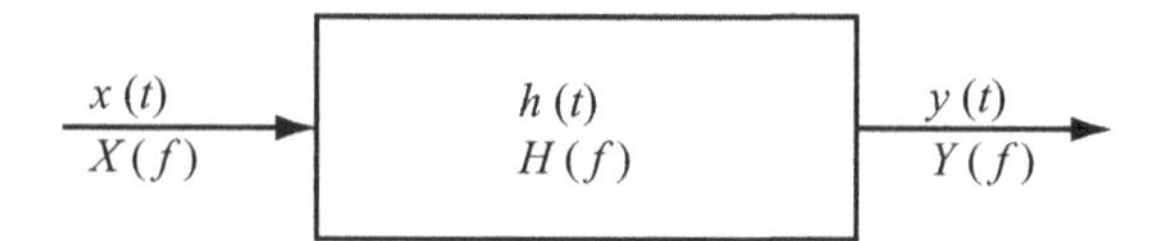


**Figure 13.1** Linear time-invariant system with input $x(t)$, output $y(t)$, impulse response $h(t)$ and frequency response $H(f)$

## 13.1 Linear Systems

We recall from Section 2.6, that a system with an input signal $x(t)$ and an output $y(t)$ is considered to be *time invariant* if the coefficients of the system of differential equations which describe the system do not change with time. In practice, this means that the system parameters, for example mass, damping, and stiffness, do not change during the time we study (measure) the system. Further, a system is *linear* if it fulfills two criteria: (i) the system is *additive*; and (ii) the system is *homogeneous*.

Real-life systems are rarely completely linear. During all measurements of frequency response we therefore need to investigate whether or not the conditions of linearity are fulfilled. If they are not, then many of our common assumptions fail, and the methods presented in this chapter will not be appropriate. How to check the linearity of mechanical systems will be discussed in Section 16.7.2.

## 13.2 Determining Frequency Response Experimentally

We shall now see how the frequency response can be estimated experimentally, under the assumption that the system, $H(f)$, is linear and time-invariant. We thus assume that we have a system as in Figure 13.1, with input $x(t)$ and output $y(t)$. Recall from Section 2.6 that we may write the output as the convolution of the input with the system's impulse response, $h(t)$, i.e.,

$$y(t) = x(t) * h(t) = \int_{-\infty}^{\infty} x(u)h(t-u)\mathrm{d}u. \tag{13.1}$$

Recall also that we can carry out the computation of Equation (13.1) in the frequency domain, which is undeniably more attractive, since in that case we obtain the equation for the spectrum $Y(f)$ as

$$Y(f) = X(f) \cdot H(f). \tag{13.2}$$

The input is thus amplified and phase-shifted at each frequency independently of other frequencies.

In real-life measurements, particularly in the field of noise and vibrations, we often have no hope of being able to measure $x(t)$ and $y(t)$ without at least one of these signals being contaminated by extraneous noise from the sensor and input electronics in the measurement system. However, we can assume that the contaminating noise is uncorrelated with the input and output signals. A general model for an actual system where extraneous noise is added to the signals we measure is illustrated in Figure 13.2. A discussion about the contaminating noise on the input and output signals is found in Section 13.7.

### *13.2.1 Method 1 – the $H_1$ Estimator*

In order to find a method which estimates the frequency response of the linear system, we shall first make a simplification of the system in Figure 13.2, which consists of setting the noise at the input to $m(t) = 0$. We thus assume that we can measure the input $x(t)$ without any extraneous noise, but that the output contains noise. Furthermore, we assume for the moment that the input signal, $x(t)$ is random noise. In Section 13.7.1 we will extend the discussion to input signals with other properties.

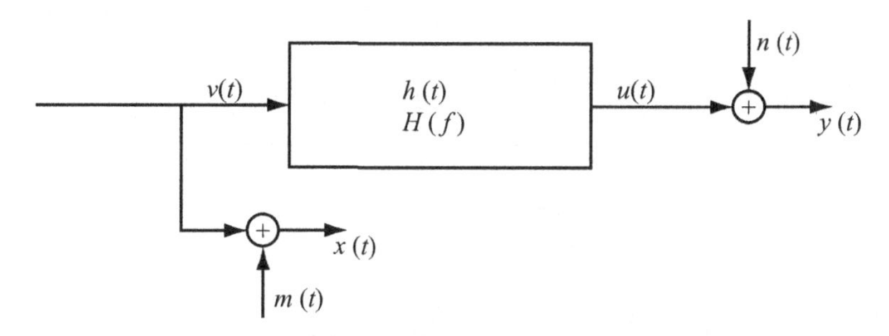


**Figure 13.2** Linear system with extraneous noise contaminating both the input and output signals

For the system in Figure 13.3 we can formulate the spectrum of the output as

$$Y(f) = X(f)H(f) + N(f). \tag{13.3}$$

Next, we multiply both the left-hand and right-hand sides in Equation (13.3) by $X^*(f)$, i.e., the complex conjugate of $X(f)$, and obtain

$$X^*(f)Y(f) = X^*(f)X(f)H(f) + X^*(f)N(f). \tag{13.4}$$

If we take the expected values (i.e., experimentally we make an average) of the left-hand side and each term on the right-hand side separately, and scale the results in a suitable way to obtain PSDs as descried in Section 10.3, we obtain, from Equations (10.11) and (10.13), the cross- and auto-spectral densities

$$G_{yx}(f) = G_{xx}(f)H(f) + G_{nx}(f). \tag{13.5}$$

The last term in Equation (13.5), the cross-spectral density between the input and the noise, approaches zero when we average, since these signals are uncorrelated. We thus obtain the so-called $H_1$ *estimator* of $H(f)$, as

$$\hat{H}_1(f) = \frac{\hat{G}_{yx}(f)}{\hat{G}_{xx}(f)} \tag{13.6}$$

which is the equation used in FFT analysis systems for estimating $H(f)$ from measurements of $x(t)$ and $y(t)$ if it is assumed that the contaminating noise in the input signal is negligible. As before, we use the symbol ˆ (hat) to denote that we are dealing with estimated functions. Note specially in Equation (13.6), that, because $G_{xx}(f)$ is real, the phase angle in $H_1(f)$ comes exclusively from the phase angle of the cross spectrum, $G_{yx}(f)$.

The $H_1$ estimator in Equation (13.6) is a *least squares* estimate of the system $H(f)$, see for example Bendat and Piersol (2010). If there is noise added to the input, and thus the model we use for the $H_1$ estimator is wrong, then the estimated $H_1$ will be *biased*. This will be discussed in Section 13.4. Furthermore, as we will see in Section 13.5.1 the $H_1$ estimator is always biased due to the limited frequency resolution, although this error can be made arbitrarily small by increasing the blocksize.

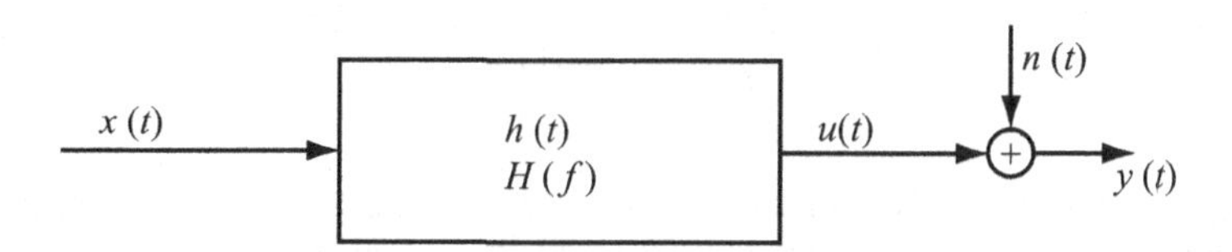


**Figure 13.3** Linear system with noise only at the output

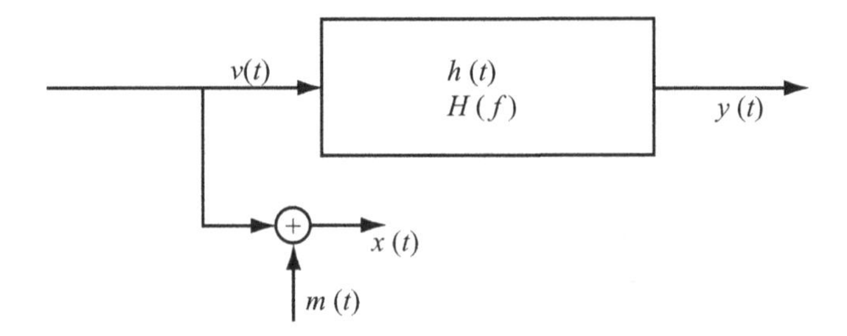


**Figure 13.4** Linear system with noise only at the input

**Example 13.2.1** *Write a MATLAB/Octave script which calculates the $H_1$ estimator of a single-input/single-output system, using spectral densities in variables* `Gxx` *and* `Gyx`.

*We assume the variables are column vectors as described in Section 10.7.1. The $H_1$ estimator is then simply the element-wise division of the cross-spectral density and the autospectral density of $x$.*

```
H1=Gyx./Gxx;
```

*End of example.*

## *13.2.2 Method 2 – the $H_2$ Estimator*

In some cases it may be the case that the preceding assumption, that the noise at the input is negligible, is not reasonable. For example, when we measure the accelerance of mechanical systems with shaker excitation, then near the natural frequencies of the structure, the acceleration signal (output) is large, but the force signal (input) is often small because the structure is weak, see also Section 13.9. In that case it is more reasonable to assume that the dominating extraneous measurement noise is present in the input signal, $x(t)$, as illustrated in Figure 13.4.

To find an estimator for this case, from Figure 13.4 we obtain the following relationship

$$Y(f) = [X(f) - M(f)]\, H(f). \tag{13.7}$$

The trick this time is to multiply Equation (13.7) by the complex conjugate of the output spectrum, $Y^*(f)$, which gives

$$Y^*(f)Y(f) = H(f)\left[G_{xy}(f) - G_{my}(f)\right]. \tag{13.8}$$

We now take the expected value (average) like we did for the $H_1$ estimator above, and scale properly, and obtain

$$G_{yy}(f) = H(f)\left[G_{xy}(f) - G_{my}(f)\right]. \tag{13.9}$$

Similar to the $H_1$ case above, the cross-spectrum, $G_{my}(f)$ will approach zero when we average, if we assume that the noise, $m(t)$, is uncorrelated with $v(t)$ and therefore with $y(t)$. We thus obtain the so-called *$H_2$ estimator* as

$$\hat{H}_2(f) = \frac{\hat{G}_{yy}(f)}{\hat{G}_{xy}(f)} \tag{13.10}$$

Note in Equation (13.10) that $G_{xy}(f) = G^*_{yx}(f)$ (see e.g., Equation 8.13), that is, the phase of $G_{xy}(f)$ is equal to the phase of $G_{yx}(f)$ with opposite sign. Since $G_{xy}(f)$ in Equation (13.10) is in the denominator while $G_{yx}(f)$ is in the numerator in Equation (13.6), we see that the phase of $H_1$ is equal to the phase of

$H_2$. It should also be noted that if the $H_2$ estimator is missing in the FFT analysis software, it can easily be computed by switching place between $x$ and $y$ and then inverting the resulting $H_1$ estimate, i.e.,

$$^2\hat{H}_{yx} = \frac{1}{^1\hat{H}_{xy}}. \tag{13.11}$$

where we introduce the left superscript '1' in $^1\hat{H}_{yx}$, etc., to indicate that we are using specifically the $H_1$ estimator. This nomenclature will be used whenever it is not obvious from the context which particular estimator we are discussing.

Like the $H_1$ estimator in the case of noise on the output, the $H_2$ estimator is also a least squares estimate of $H(f)$. In case the model assumption is wrong, the estimate $\hat{H}_2$ is biased, which will be discussed in Section 13.4. It is also, like the $H_1$ estimator, always biased due to the limited frequency resolution, although this error can be made arbitrarily small by increasing the blocksize, see Section 13.5.1.

### 13.2.3 Method 3 – the $H_c$ Estimator

Now, what if we are taking the noise on both input and output into consideration? There has been several attempts at solving this system, including the $H_v$ (Rocklin *et al.* 1985), and $H_s$ (Wicks and Vold 1986) estimators. In order to get an unbiased estimate in the case of noise on both input and output, it turns out that this is only possible with some *a priori* information about the input and output noise (or their ratio), or, by knowing one signal without any contaminating noise at all. In Section 14.1.6 we will look at the common $H_v$ estimator which is an estimator trying to address this problem. It should also be emphasized that the discussion here applies to random input signals; in Section 13.12 we will show that we *can* remove the bias due to both input and output noise if we use a periodic excitation signal (or at least make it arbitrarily small).

Here, we will look at another, rather ingenious, estimator, namely the $H_c$ estimator, which assumes that there is a signal which can be measured without contaminating noise. In the case of shaker excitation of structures, the electrical output signal of the signal generator is typically chosen. This estimator was originally proposed by Goyder (1984), and further developed by Mitchell and Cobb (1987).

In Figure 13.5, the concept of the $H_c$ estimator setup is illustrated. We have a signal $v(t)$ which we assume we can measure without any contaminating noise. In addition we have the force signal, $x(t)$, and a response signal, $y(t)$, both of which are measured with contaminating noise, $m(t)$, and $n(t)$, respectively.

The $H_c$ estimator is now based on the fact that, using the $H_1$ estimator, we know from Section 13.2.1 that we can estimate the two systems, $^1\hat{H}_{yv}$ and $^1\hat{H}_{xv}$, indicated in Figure 13.5 without bias due to the

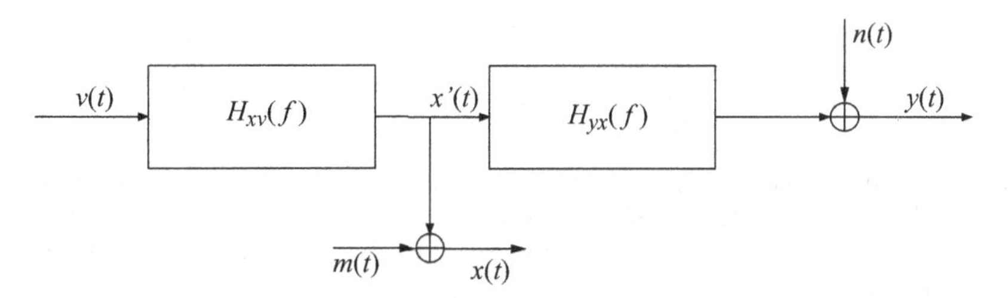


**Figure 13.5** Illustration of the model assumption for the $H_c$ estimator. The estimator is based on the assumption that there is an input signal, $v(t)$, which can be measured without any contaminating noise, and which is located before the measured input to the system. In mechanical applications, this is typically the voltage output of the signal generator

extraneous noise, if we just make appropriately many averages. But we now have the simple relation that the system we are seeking, $H_{yx}$ is given by

$$H_{yx} = \frac{Y(f)}{X(f)} = \frac{Y(f)/V(f)}{X(f)/V(f)} = \frac{H_{yv}}{H_{xv}} \tag{13.12}$$

which means that we can define the $H_c$ estimator by

$${}^{c}\hat{H}_{yx}(f) = \frac{{}^{1}\hat{H}_{yv}}{{}^{1}\hat{H}_{xv}}. \tag{13.13}$$

In case we make enough averages so that the $H_1$ estimators in Equation (13.13) are unbiased, then the $H_c$ estimator is also unbiased. This is at the expense of an extra input channel, however, which is probably a reason why this estimator is not very common in commercial software. A further disadvantage with this estimator is that it relies on the frequency responses involved being linear, whereas for shakers used for structural excitation, we know they are not linear, which can cause uncertainty in the $H_c$ estimator.

## 13.3 Important Relationships for Linear Systems

We shall now study some important relationships which apply to linear systems. If we measure the input $x(t)$ to a linear system with frequency response $H(f)$, the complex conjugate of Equation (13.2) is

$$Y^*(f) = X^*(f)H^*(f). \tag{13.14}$$

Multiplying Equation (13.2) by Equation (13.14) term by term, taking the expected value of each term, and scaling properly, we obtain the important relation

$$G_{yy}(f) = |H(f)|^2\, G_{xx}(f) \tag{13.15}$$

This equation is a useful relation to express the PSD of the output of a linear system. Note that Equation (13.15) is valid for ideal linear systems without contaminating noise on the measured signals, as in Figure 13.1.

We now move to the system with noise only at the output, in Figure 13.3. If we start with the signals summed at the output we have that

$$Y(f) = U(f) + N(f). \tag{13.16}$$

We form the complex conjugate of this equation and multiply it by itself, obtaining

$$Y^*Y = U^*U + N^*N + U^*N + UN^* \tag{13.17}$$

where we have left out the frequency variable for the sake of simplicity. Taking the expected value of each term, and scaling properly, we obtain

$$G_{yy} = G_{uu} + G_{nn} + G_{nu} + G_{un}. \tag{13.18}$$

Equation (13.18) expresses a very important relationship, namely that *if we sum several signals, the power spectral density of the summed signal is in general not equal to the sum of each signal's PSD.* Instead, we must take into account all of the cross spectral densities between the signals included in the summation.

If the signals $u$ and $n$ are *uncorrelated* (independent), however, which they are in our case for the system with noise at the output, we obtain the special case

$$G_{yy} = G_{uu} + G_{nn}. \tag{13.19}$$

## 13.4 The Coherence Function

In the case of existing input noise, $m(t)$, not included in the assumption for the $H_1$ estimator, the magnitude of the estimate $\hat{H}_1$ will be less than or equal to the true value of $H(f)$. This can easily be observed by the fact that when $m(t)$ is not zero, then $\hat{G}_{xx}$ will be measured with this noise and thus

$$E\left[\hat{G}_{xx}\right] = G_{vv} + G_{mm} \tag{13.20}$$

according to Equation (13.19). Because the $\hat{G}_{xx}$ term is found in the denominator of the $H_1$ estimator in Equation (13.6), $|\hat{H}_1|$ will become *smaller than or equal to* the true $|H(f)|$ due to this bias error. In a similar fashion, because the $H_2$ estimator has a factor $G_{yy}$ in the numerator which will be affected similarly with $G_{nn}$, $|\hat{H}_2|$ will always be *greater than or equal* to the true value $H(f)$, and equal only when $n = 0$. In both cases it is also necessary that we have made many averages in Equation (13.6) and (13.10), respectively, so that the cross-spectrum terms, including the extraneous noise in Equations (13.5) and (13.9), respectively, become zero. Thus, we may conclude that the true value of $|H(f)|$ fulfills

$$\left|\hat{H}_1(f)\right| \leq |H(f)| \leq \left|\hat{H}_2(f)\right|. \tag{13.21}$$

When we estimate the frequency response with the $H_1$ or $H_2$ estimators as above, we can simultaneously compute the *coherence function*, $\gamma^2(f)$ which is defined as the ratio between the $\hat{H}_1$ estimate and the $\hat{H}_2$ estimate, i.e.,

$$\hat{\gamma}^2_{yx}(f) = \frac{\hat{H}_1(f)}{\hat{H}_2(f)} = \frac{\left|\hat{G}_{yx}(f)\right|^2}{\hat{G}_{xx}(f)\hat{G}_{yy}(f)}. \tag{13.22}$$

Note that the coherence function in Equation (13.22) is defined as the *squared* function, $\gamma^2_{yx}$. From Equation (13.21) and the definition of the coherence function, it follows directly that

$$0 \leq \gamma^2_{yx}(f) \leq 1. \tag{13.23}$$

If $\gamma^2_{yx}(f) = 1$ then $\hat{H}_1 = \hat{H}_2$ which implies that we have no extraneous noise, and moreover that the measured output, $y(t)$, derives solely from the measured input, $x(t)$.

The coherence function is a quality measure of our estimated frequency response, regardless of which estimator we use. The coherence function drops below unity if there is contaminating noise on either the measured input signal, $x(t)$, or in the measured output signal, $y(t)$, or in both signals. In all three cases, there is a bias error in the determination of the frequency response in at least one of the estimators, $\hat{H}_1$, or $\hat{H}_2$, which follows from the fact that $\hat{\gamma}^2_{yx} = \hat{H}_1/\hat{H}_2$ and the properties in Equation (13.21). It the coherence drops because of noise $m(t)$ or $n(t)$ (or both, of course) is impossible to tell solely from observing $x(t)$ and $y(t)$. We will return to a more thorough discussion on the coherence function in Section 13.7, but first we need to establish some more relations.

## 13.5 Errors in Determining the Frequency Response

We will now look at some aspects of the errors involved in estimating frequency response functions using the $H_1$ and $H_2$ estimators for a single-input/single-output system. This is considerably more complicated than the discussion on spectrum estimation in Chapter 9, because the correlation between the input and output signals is coming into the error relations. Also, the estimators we use to estimate the auto- and cross-spectral densities change the relations for the bias and random errors involved.

It is important to understand that there are two completely different errors involved in the FRF estimates: (i) spectral analysis errors, and (ii) model errors. We have already discussed the model errors, which arise if, for example, we use the $H_1$ estimator, but there is input noise, $m(t)$. The errors we will look at in this section assume that the model is correct.

The spectral errors in the FRF estimates are further divided into two parts; the errors caused by the estimator itself, without any extraneous noise, $n(t)$ or $m(t)$ as illustrated in Figure 13.2, and then the errors caused by the extraneous noise, i.e., $n(t)$ if we look at the $H_1$ estimator, and $m(t)$ if we look at the $H_2$ estimator.

The most common estimator for the auto- and cross-spectral densities, used in all commercial systems to date, is Welch's method, as described in Section 10.3.2. A comprehensive analysis of the errors involved when using this method when the input signal is random noise, is found in Antoni and Schoukens (2007, 2009) and Schoukens *et al.* (2006). A thorough discussion on errors if the smoothed periodogram estimator is used is found in Bendat and Piersol (2010).

## 13.5.1 Bias Error in FRF Estimates

Similarly to what we discussed for spectral densities in, e.g., Section 10.3.4, the frequency response estimated with the $H_1$ estimator, even without the presence of contaminating noise, is biased due to the limited frequency resolution. In practical frequency response measurements, this bias error should be minimized by selecting a blocksize for the FFT by using the same procedure as presented in Section 10.6.3, i.e., by gradually increasing the blocksize until peaks do not increase in height when the blocksize is increased further.

The spectral analysis bias error has two parts; one caused by the fact that the measured output block is not actually caused exactly by the measured input block, and one caused by leakage. The first of these errors is illustrated in Figure 13.6 and is caused by the fact that when we truncate the signals, we measure

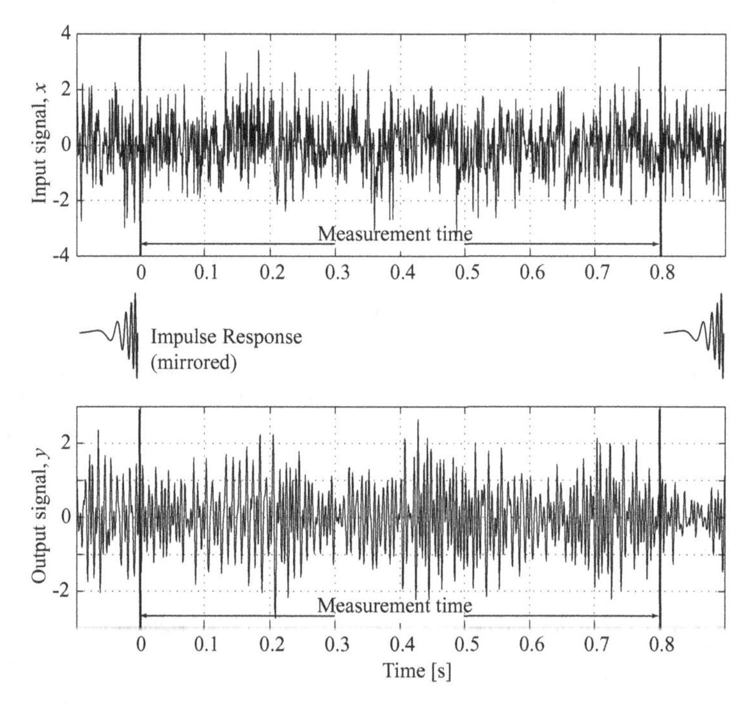


**Figure 13.6** Illustration of an arbitrary time block of the input and output signals in the averaging process to compute a frequency response function. A 'time mirrored' impulse response is added to the picture to illustrate the convolution of the input signal with the impulse response, which involves mirroring the impulse response before multiplying the input signal and summing, see Chapter 2. The measured output block is therefore not caused by (exactly) the measured input block, which causes an error. This error is minimized by increasing the measurement time so that it is much larger than the length of the impulse response

an output signal of the linear system, which is not totally caused by the input signal. To understand this effect, we must consider that the output signal is the input signal convolved with the input signal, which from Section 2.6.4 we know means that the impulse response is 'mirrored' (time reversed), then multiplied by the input signal, and the product is summed up. At the beginning of the output signal in a particular time block, there is, therefore, a region which actually depends on the input signal *prior* to the time block. At the end of the time block, there is correspondingly a part of the input signal that causes the output signal outside (after) the time block. This error is often erroneously called leakage, although it is not really the same as the previously mentioned effect called leakage in Chapter 9 which was due to the end effects of the blocks in the DFT process.

Once the blocksize is made much larger than the impulse response length, the remaining bias error is caused by leakage. An important result in Antoni and Schoukens (2007) is that the normalized bias error, $\varepsilon_b\left[\hat{H}\right]$, of a frequency response estimate is proportional to

$$\varepsilon_b\left[\hat{H}\right] \propto \frac{1}{N^2}\frac{H''(f)}{H(f)} \tag{13.24}$$

where $H''$ is the second derivative of the FRF with respect to frequency. First of all, we can conclude that the bias error is inversely proportional to the blocksize squared, which means it vanishes quickly as we make the blocksize large. In Section 13.10 we will illustrate how this bias error can be made negligible in measurements of a typical lightly damped system.

We recognize the ratio of the second derivative to the value of the FRF from the bias error expression for spectral densities in Equation (10.18). Since we usually measure FRFs of lightly damped systems in noise and vibration analysis applications, it is of particular interest to know the nature of the bias error on such systems. It turns out that the bias error, similarly to the spectral density bias error, essentially depends on the ratio of the resonance bandwidth and the frequency increment, i.e., the ratio $B_r/\Delta f$. A plot of a simulation result of the maximum bias error (located at the resonance frequency) of a frequency response of an SDOF system, is presented in Figure 13.7.

**Example 13.5.1** *Assume we have a truck frame with a first natural frequency of 1 Hz, and relative damping of 1%. Determine the necessary frequency increment so that the normalized bias error of an FRF, measured using random excitation, is below 1%. If you need to make 50 independent averages (without overlapping), how long will the measurement take?*

*The resonance bandwidth according to Equation (5.39) is* $B_r = 2\zeta_r f_r = 0.02$. *In order to have a bias error less than 1%, we need to have a frequency increment of maximum* $\Delta f_{max} = 0.1 B_r$ *according to the plot in Figure 13.7. Thus, we need to have a frequency increment of less than 0.002 Hz. Since the time of each FFT time block is* $T = 1/\Delta f$ *this means the measurement will take*

$$T_m = \frac{50}{0.002} = 25000s \approx 7\,hours. \tag{13.25}$$

*This example shows what a high price low-bias measurements of FRFs can come at!*

*End of example.*

## *13.5.2 Random Error in FRF Estimates*

As shown by Antoni and Schoukens (2007) and (Schoukens *et al.* (2006), the random error in estimates of frequency responses using Welch's method, has two parts; one caused by the extraneous noise, $n(t)$ in the $H_1$ case, and one caused by leakage, of which the latter has previously been unaccounted for. The total random error therefore depends on the time window as well as the number of averages and the overlap percentage, similarly (but not equivalently) with the random error of PSD estimates. In Antoni and Schoukens (2007) it was shown that the smallest random error possible, given a certain amount of data, is obtained by using a *half sine window* instead of the more common Hanning window, and with an

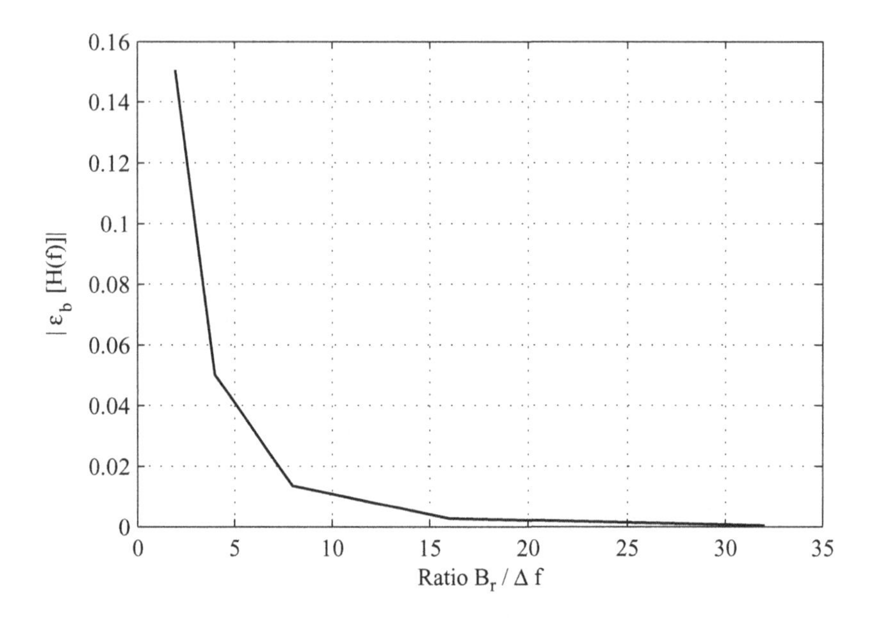


**Figure 13.7** Maximum normalized bias error of FRF estimate on an SDOF system, versus the ratio of the resonance bandwidth and the frequency increment, $B_r/\Delta f$. The plot is a result of a simulation of the maximum bias error, which occurs at the resonance frequency, using various blocksizes. Note that the plot shows the magnitude of the bias error; the bias error is actually negative at the resonance frequency. Furthermore, the plot was obtained by estimating the FRF using a half sine window and 67% overlap, see Section 13.5.2. The plot shows that in order to have a normalized bias error of less than 1%, we have to have at least ten frequency lines within the resonance bandwidth. It also shows, however as does Equation (13.24), that the bias error can be made arbitrarily small by increasing the blocksize

overlap percentage of 67% instead of the more common overlap of 50% used for spectrum analysis. The differences in results using a Hanning and a half sine window are not very large, but it is always good practice to use the best possible method. This will be further discussed and illustrated in Section 13.9. It should be noted that the half sine window is unsuitable for spectrum estimation, so if both spectra and frequency responses are to be calculated and used, it may be better to use the common Hanning window.

The random errors as defined by Antoni and Schoukens (2007) are rather complicated to interpret. For our purpose it is sufficient to provide an approximate equation for the random error in the magnitude of frequency response functions, given by Bendat and Piersol (2010). They give the random error of the frequency response between input $x(t)$ and output $y(t)$ for the $H_1$ estimator, in the absence of extraneous noise on the input, as

$$\varepsilon_r\left[\left|\hat{H}(f)\right|\right] \approx \sqrt{\frac{1-\gamma_{xy}^2(f)}{2n_d\gamma_{xy}^2(f)}} \tag{13.26}$$

where $n_d$ is the number of distinct averages, i.e., the number of averages without overlap processing. When using overlap processing, $n_d$ should be replaced by $M_e$, the equivalent number of averages, as given by Figure 10.8. This equation, although neglecting the random error due to leakage, is attractive in its simplicity, as it reveals that the best way to achieve a small random error is to make sure the coherence is near unity. If the coherence is not near unity, due to excessive output noise, the random error in the estimate can still be reduced by a (large) number of averages. A near unity coherence function should always be attempted to achieve when measuring frequency responses, as will be demonstrated in Section 13.9. In addition, it should be mentioned that in practice we usually also have other errors, such as extraneous noise on the input signal, which makes the precision of this equation good enough for practical use.

### *13.5.3 Bias and Random Error Trade-offs*

From Sections 13.5.1 and 13.5.2 it can be concluded that the bias error and random error are contradictory, as they were shown to be for spectral density estimates in Chapter 10. To reduce the bias error we need to use large blocksize, which (if we have a fixed amount of recorded data) reduces the possible number of averages. To reduce the random error we need to take many averages, which (under the premise of a fixed amount of data) makes the available blocksize less. We thus have to make a compromise between bias and random errors, if we have a certain amount of data. In practical FRF measurements, however, it is good practice to experimentally verify a sufficiently small bias error, prior to recording the data, and then to verify how many averages are needed to produce a sufficiently small random error. In many cases it is necessary to analyze the necessary measurement time prior to the measurement, so that enough averages can be made to get a reasonably low random error and bias error.

Instead of using Welch's method to compute the auto- and cross-spectral densities needed for the FRF estimator, it is, of course, possible to use the smoothed periodogram method described in Section 10.3.6. Actually, this estimator suffers a little less from the end block leakage than Welch's method, particularly for small blocksizes. However, the difference is so small that from a practical point of view it is equivalent to the Welch estimator. It can also be shown that the two estimators are asymptotically equivalent when the blocksize increases, (Antoni and Schoukens 2007).

## 13.6 Coherent Output Power

Assuming we have a system with noise on the output only, as in Figure 13.3, we will now look at some properties which allow us to compute the spectral densities of the output from the linear system, $G_{uu}$, and of the extraneous noise, $G_{nn}$. From Equation (13.15) and the expression for $H(f)$ with the $H_1$ estimator from Equation (13.6) we first have

$$\hat{G}_{uu} = \left|{}^1\hat{H}\right|^2 \hat{G}_{xx} = \frac{\left|\hat{G}_{yx}\right|^2}{\hat{G}_{xx}}. \tag{13.27}$$

Using the first and second term in this equation we can thus estimate the spectral density of $u(t)$, i.e., the part of $y(t)$ which comes from $x(t)$ through the linear system, even though we cannot measure $u(t)$.

By noting the similarity between Equation (13.27) and the definition of the coherence function in Equation (13.22), we also find that Equation (13.27) can alternatively be calculated as

$$\hat{G}_{uu}(f) = \hat{\gamma}^2(f) \cdot \hat{G}_{yy}(f) \tag{13.28}$$

which is the usual formula used to compute $\hat{G}_{uu}$.

$\hat{G}_{uu}(f)$ is called the *coherent output power (spectrum)*, since it stands for the part of $G_{yy}(f)$ which linearly derives from the input, $x(t)$, i.e., which is *coherent with* $x(t)$. This function can be used to identify noise sources (Bendat and Piersol 1993). However, it is crucial to remember that this formula is based on the system without noise at the input. In order to identify noise sources using Equation (13.28) it is thus important to find a 'clean' input, which can be measured without extraneous noise, see Section 15.3.

Using Equation (13.28) we can also express Equation (13.19) as

$$\hat{G}_{nn}(f) = \left(1 - \hat{\gamma}_{yx}^2(f)\right) \hat{G}_{yy}(f) \tag{13.29}$$

because $x$ (and therefore $u$) and $n$ are uncorrelated.

We can therefore also estimate the PSD of the contaminating noise, even though we cannot measure this signal. Again, this formula is valid under the assumption that no noise exists at the input of the system.

## 13.7 The Coherence Function in Practice

We will now find a few more relations which help in interpreting the coherence function. First of all we should stress that the coherence function is a result of the fact that there is a difference between the $H_1$ and the $H_2$ estimator results, i.e.,

$$\mathrm{E}\left[\frac{\hat{G}_{yx}}{\hat{G}_{xx}}\right] \neq \mathrm{E}\left[\frac{\hat{G}_{yy}}{\hat{G}_{xy}}\right]. \tag{13.30}$$

Also, it is apparent that the coherence function equals exactly 1 if we only make one average, since then

$$\hat{\gamma}_{yx}^2 = \frac{\left|\hat{G}_{yx}\right|^2}{\hat{G}_{xx}\hat{G}_{yy}} = \frac{(YX^*)(Y^*X)}{XX^*YY^*} = 1. \tag{13.31}$$

This should be interpreted such that the coherence is *undefined* when only one average is taken.

In practice, the coherence function requires more averages to obtain a small random error, than does the frequency response. There is no exact solution of the random error in coherence estimates, but Bendat and Piersol (2000) give an approximate equation where the normalized random error is given by

$$\varepsilon\left[\hat{\gamma}_{yx}^2\right] \approx \frac{\sqrt{2}(1-\gamma_{yx}^2)}{\left|\gamma_{yx}\right|\sqrt{n_d}} \tag{13.32}$$

where $n_d$ is the number of distinct averages, without overlap processing. It is worth pointing out that this error formula means that a very large number of averages is needed to keep the random error low when the coherence is low. For example, a random error of $\varepsilon \leq 0.02$, when the coherence is $\gamma_y^2x = 0.5$, requires 5000 distinct averages, whereas when the coherence is $\gamma_{yx}^2 = 0.8$, only 500 averages is needed to produce the same random error. With overlap processing, $n_d$ in Equation (13.32) should be replaced by $M_e$, the apparent number of averages from Figure 10.8. It is apparent that in general the error given by Equation (13.32) is larger than the random error of the magnitude in the FRF, given by Equation (13.26).

We will now look at the quantitative value of the coherence function, in the cases of noise on the output and input, respectively. If we assume a system with extraneous noise on the output only, combining Equation (13.19) and Equation (13.28) leads to the relation

$$\gamma_{yx}^2 = \frac{G_{uu}}{G_{yy}} = \frac{G_{uu}}{G_{uu}+G_{nn}} = \frac{1}{1+\frac{G_{nn}}{G_{uu}}}. \tag{13.33}$$

From this equation it follows that the coherence function deviates from unity when the extraneous output noise is not zero.

If, instead, we assume a system with noise on the input only, then according to Figure 13.4 we have that

$$G_{yx} = G_{yv} = {}^2HG_{vv} \tag{13.34}$$

which follows if we formulate the equation for $Y$ as a function of $V$, multiply by $X^*$ and then average over many blocks. But we also have, from Equation (13.27),

$$G_{yy} = \left|{}^2H\right|^2 G_{vv}. \tag{13.35}$$

If we combine the two last equations with Equation (13.22) we arrive at the relation

$$\gamma_{yx}^2 = \frac{\left|G_{yx}\right|^2}{G_{xx}G_{yy}} = \frac{\left|G_{yv}\right|^2}{\left(G_{vv} + G_{mm}\right)G_{yy}} = \frac{\left|{}^2H\right|^2 G_{vv}^2}{\left(G_{vv} + G_{mm}\right)\left|{}^2H\right|^2 G_{vv}} = \frac{1}{1 + \frac{G_{mm}}{G_{vv}}}. \tag{13.36}$$

Equation (13.36) shows how the coherence deviates from unity when there is noise (only) on the input. It should be noted that the coherence function is thus, by combining Equations (13.33) and (13.36), a result of the amount of total extraneous input and output noise in the measurement in the general case of noise on both input and output signals.

We can now summarize some possible reasons for the coherence function to deviate from unity:

- Either the noise $m(t)$ or $n(t)$, or both, are not negligible compared with the measured signals $x(t)$ and $y(t)$.
- Bias errors due to insufficient frequency resolution, as described in Section 13.5.1.
- The system $H(f)$ is (strongly) nonlinear or not time invariant.
- Bias error exist due to a time delay between the signals $x(t)$ and $y(t)$. This is explained in detail in Bendat and Piersol (2010).
- The model is incorrect, for example, $v(t)$ is not the only (correlated) input which contributes to $y(t)$. This is explained in detail in Bendat and Piersol (2010).

### *13.7.1 Non-random Excitation*

If the input signal $x(t)$ to a linear system is not random, which is often the case, as will be evident in the following sections, the above discussion is not entirely applicable. However, the coherence function can still be used as an indicator of the signal-to-noise ratio at each frequency bin. If the coherence function equals unity, it is always an indication that the signal-to-noise ratio is sufficient. It should be mentioned, however, that even if the coherence is unity, the estimated frequency response can be erroneous. This is the case, for example, if the *model is not true*, for instance if there is some other signal, correlated with the measured input signal, which adds to the measured output signal. An example of this case is found in Section 13.12.3.

## 13.8 Impact Excitation

We shall now study how the above methods are used to measure frequency responses on mechanical systems. We will start by the method of impact excitation (sometimes called impulse excitation), where an impulse hammer described in Section 7.7 is used to excite the structure under test. This technique was first developed in the 1970s, (Halvorsen and Brown 1977), and received some renewed interest in the work of Fladung (1994), who, by increasing the number of references has shown that impact testing in many cases can give high-quality results for modal analysis.

Impact excitation is easier to apply than shaker excitation, since the structure does not need to be loaded by a force transducer and mounted to a shaker. The method also has a few other advantages compared with shaker excitation, for example:

- the force is well defined in the striking direction without transverse forces (see the section on shaker excitation below),

- it is easier to excite higher frequencies (which frequencies are considered 'higher' naturally depends upon the measurement object).

The method, however, also has some disadvantages compared with shaker excitation, most importantly:

- the signal-to-noise ratio (SNR) is low compared with methods using continuous excitation signals, particularly because the method measures the free decay of the structure, which means the contaminating noise leads to smaller SNR the longer the time record is extended,
- the risk of exciting, and difficulties caused by, nonlinearities is higher, and
- the total energy in the excitation cannot be divided into multiple-input force signals, as it can with shaker excitation (see Chapter 14).

Impact excitation may be preferred for the above reasons when one or more of the following points weigh to its advantage.

1. It is desired to measure quickly and the requirements for precision are less important.
2. The structure is lightly damped and difficult to excite by the shaker at its natural frequencies.
3. It is desired to investigate suitable excitation locations for shaker excitation.
4. Relatively high frequencies are being studied.
5. The structure is heavy which requires large and expensive shakers (for example when studying bridges and other building structures).

Measuring frequency response using impact excitation can be carried out with relatively simple equipment. As a minimum, an FFT analysis system with two channels, an impulse hammer and an accelerometer, are required, as illustrated in Figure 13.8. With common modern data acquisition systems with multiple (more than two) channels, in general it is recommended to use several reference accelerometers to acquire more redundant data.

### *13.8.1 The Force Signal*

The basis of impact excitation is that a short force pulse has a broad spectrum – the shorter the pulse, the higher the frequencies in its spectrum, see Figure 13.9. The force spectrum also depends on the structure's hardness and its dynamic stiffness at the excitation point. The strength of the impact also plays a role (a harder impact in general produces higher frequencies). Thus, it is important to choose the right hammer and the right tip to obtain a 'good' force spectrum, i.e., a spectrum that does not drop too much within the desired frequency range. A total drop of less than 20 dB is usually considered good, but with the increased dynamic range in modern instruments, this can be stretched. As we will show later, the recommended procedure is to test the available dynamic range, so no assumptions need to be made with regards to the correct tip to be used.

In Figure 13.9 (a) it can be seen that particularly the shorter impact displays some ringing before and after the main impact signal. This is seen very often during impact testing and is an effect of the fact that the actual bandwidth of the force signal is higher than the bandwidth used for data acquisition. The bandwidth limitation does not negatively affect the measurement, as each frequency is independent of all other frequencies, see Chapter 2.

During excitation, the force from the hammer impact should normally be as small as possible, as nonlinearities are otherwise more likely to be excited. In the worst case the structure can be deformed at the impact point, resulting in nonlinearities which completely destroy the results. How much 'as small as possible' is, is naturally difficult to say in general. When exciting a bridge with a sledgehammer it could

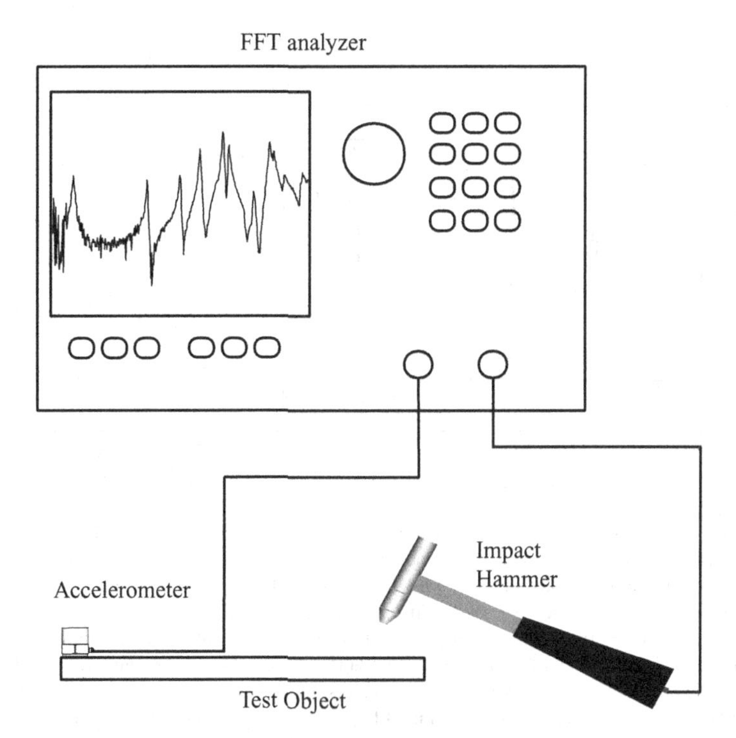


**Figure 13.8** Typical setup for measuring frequency response using impact excitation. An impulse force is applied to the test object with the hammer, causing a response measured by the accelerometer. The analysis system, here illustrated by an 'old-fashioned' FFT analyzer to appeal to all nostalgists, measures the signals and calculates the frequency response between force and response

be quite a lot, but as a rule of thumb, the impact should be carried out so that it feels soft but distinct. If the signal-to-noise ratio is not sufficient, then one may have to increase the force, see below. A better alternative, however, is often to select a more sensitive accelerometer.

The force signal always contains some noise from the transducer and (to much lesser degree in modern measurement hardware with very high dynamic range) electronics of the input channel. This noise is deteriorating the quality of the force spectrum, particularly after the impact has occurred, where the true force signal is zero. The majority of the noise can therefore be removed by applying a force window to the force signal, as illustrated in Figure 13.10. The force window is unity from the first sample in the time block, and a while after the impact has gone to zero, and then smoothly approaches zero, where it stays for the remainder of the block. Its function is to multiply the noise at the end of the force signal by zero to eliminate that noise. The DFT is sensitive to sharp edges, so the force window must be defined smoothly going from 1 to 0, so that the force spectrum is not getting distorted, for example by adding the upper part of a half sine with a duration of the same length as the part which is unity.

In most cases a careful examination of the measured force signal will show that there is a slight offset voltage in the signal, resulting from the input electronics. It is important to remove this offset prior to computing the spectrum of the force signal. Otherwise the offset will result in a distorted spectrum at low frequencies. The offset can be removed by using, for example, the first half of the samples in the pretrigger part of the force signal to calculate a mean value, which is then subtracted from the force signal. Alternatively, the first sample in the force signal can be subtracted from the force signal.

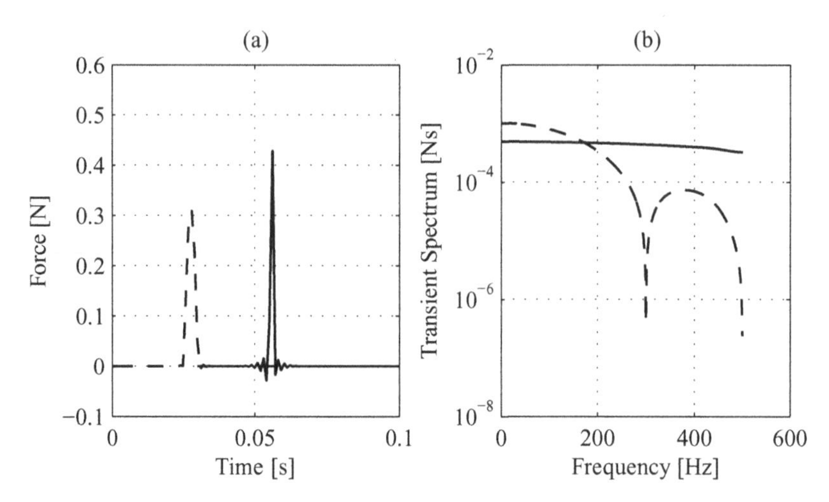


**Figure 13.9** (a) Shape of the force pulse of two different hammer tips; a hard (solid) and a softer tip (dashed). (b) Corresponding spectra (transient spectrum, see Section 10.5). The time scale for the impacts in (a) has been zoomed in, and the impacts have been separated in time for better viewing. A hard tip (solid line) gives a short impulse, which has a spectrum with high frequencies. The dashed line in (b) shows that the spectrum of the soft impact has zeros at certain frequencies, that is, there are frequencies with little energy. Normally, only frequencies in the main lobe are used for determining the transfer function. In (a) the shorter impact also shows some 'ringing' before and after the main pulse. This is an effect of lowpass filtering which is often seen during impact testing. The bandwidth of the pulse is in this case larger than the bandwidth used for data acquisition. This does not affect the quality of the measurement

## *13.8.2 The Response Signal and Exponential Window*

The response signal from an impact test is a free decay which approaches zero. The further away from the impact we get, the closer to the noise floor the response signal gets. This means that if the transducer noise is not negligible, there will be a deterioration in the signal-to-noise ratio the longer we make the measurement time. Multiplying the response signal with an exponential window defined by

$$w_e(t) = e^{-at} \tag{13.37}$$

with a suitably adjusted exponential constant, $a$, will suppress the upper part of the response signal and therefore improve the signal-to-noise ratio. The exponential window is acting as artificial damping on the structure, and will thus distort the frequency response. If modal parameter estimation is done on the measured FRF, the added artificial damping can be calculated and removed from the modal damping, as described in Section 13.8.4, provided the exponential window has been applied to both the force and the response signal. An example of a response signal, before and after application of an exponential window, is shown in Figure 13.11.

## *13.8.3 Impact Testing Software*

As for spectrum analysis which we discussed in Chapter 9, there has become a 'standard' for how impact testing is performed in commercial software for noise and vibration analysis. In this section we will start by explaining this standard technique, and after that, suggest some ways to improve it, using the power of modern computers. Unfortunately, the standard method was implemented in the days of

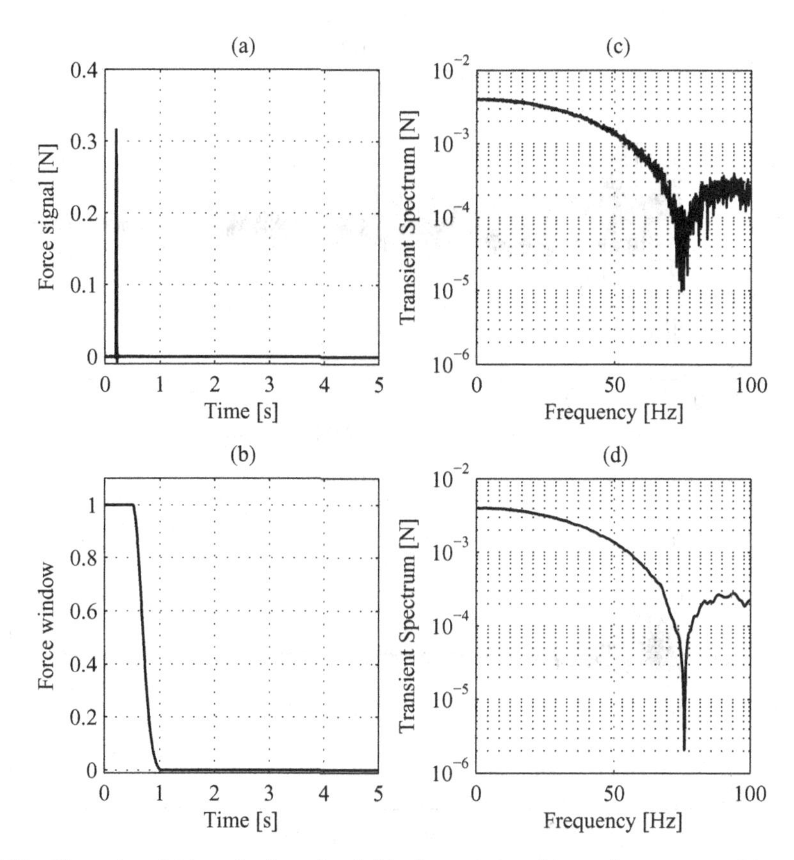


**Figure 13.10** Illustration of: (a) a noisy force signal; (b) a force window; (c), transient spectrum of the unwindowed force signal; and (d) transient spectrum of the windowed force signal. It is seen that the force window removes most of the noise in the spectrum

expensive storage space and limited computing capacity. Nevertheless, it is important to understand how commercial systems implement impact testing, in order to correctly use such systems.

The first setting is the trigger which is set with a trigger level and slope, and naturally should be set to trigger on the force signal. The data acquisition is triggered by each impact, after which $N$ samples are acquired. The software then computes the auto- and cross-spectra necessary for the FRF estimator, and adds them to a cumulated average. Many FFT analysis systems include the possibility of stopping after the time block has been collected, but before it is included in the average (called 'interrupted averaging' by some manufacturers), for a manual or automatic selection process where the user can select whether to include or reject the current impact in the averaging. Also, some manufacturers have overload and double-impact (see below) detection, which automatically prevents faulty hammer impacts from being included in the average. The typical procedure is to average in the frequency domain, but we will discuss the possibility of using time domain averaging in Section 13.8.6.

Once the trigger level is working, the pre-trigger has to be set. Usually the force should start a few hundred samples into the time block, or 5–10% of the blocksize. This ensures that the force time signal is not truncated at the beginning.

Once the trigger is set up and tested (which can sometimes be cumbersome because you need to know the full-scale range and approximately what the force level is to set the trigger correctly in percent of the

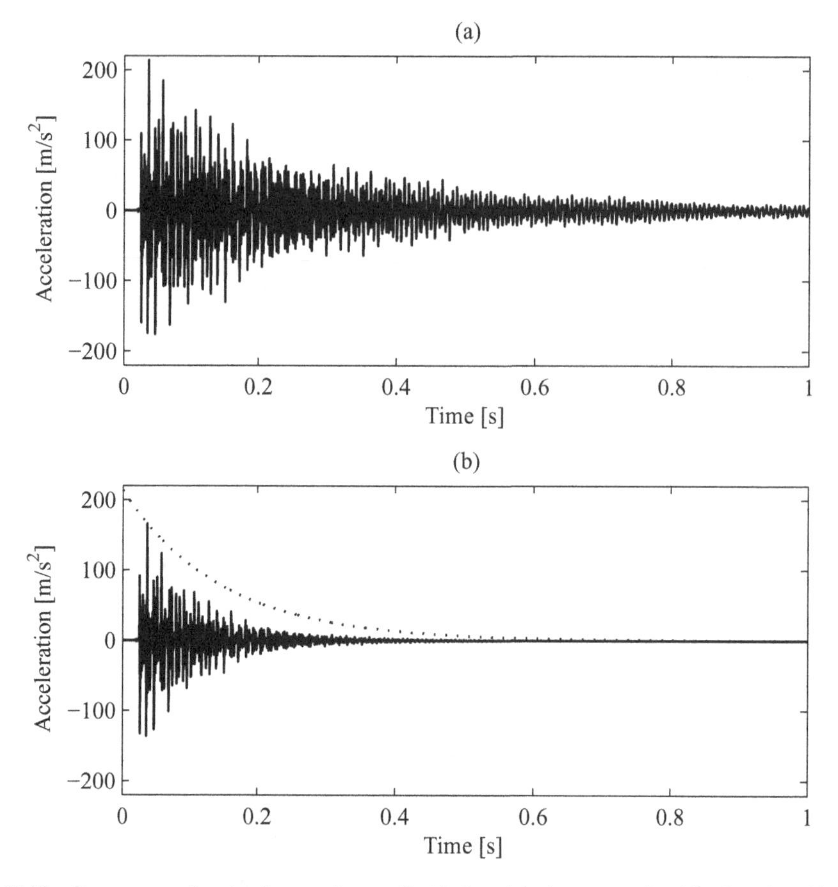


**Figure 13.11** Response acceleration from an impact. In (a) the original response signal is plotted, and in (b) the response after applying an exponential window with 0.001 as final value is plotted (solid) together with the exponential window (dotted). Note that the exponential window has been scaled to be visible in the figure, it always starts at 1

full scale range, which is most common), it is time to optimize the FFT parameters. We will here outline a procedure that simplifies this, and ends with optimal settings. We assume that we know the frequency range, which is usually given by the number of modes we want to study. Then we start with some default settings presented in Table 13.1.

**Table 13.1** Default settings for impact testing

| Frequency range | As high as the highest resonance (of interest) plus some margin |
|---|---|
| Window | None, uniform |
| Block size | 1024 or 2048 |
| Averaging | Stable, 3–5 measurements |
| FRF estimator | $H_1$ |

The next step is to set full-scale range for the A/D converters as described in Chapter 11. Make some impacts and see that you do not get any overloads. If there is a possibility to automatically reject overloaded blocks this can be recommended.

When the analyzer triggers as expected, the frequency spectrum of the force signal should be examined, and a proper tip should be selected for the impulse hammer. With an appropriate tip the force spectrum should drop at the end of the frequency range, but not so much that the signal-to-noise ratio is too bad (we will check this later, so you may need to go back to this point). The force spectrum should drop, because otherwise it includes higher frequencies, which causes the response signals to be higher than necessary, and increases the risk of exciting nonlinearities.

When the spectral content is acceptable it is time to set the optimal blocksize. The proper blocksize is given by the system properties as we discussed for spectrum analysis in Chapter 9; the frequency resolution should be set so that the bias error at each resonance of interest is minimal. It is investigated as mentioned in Section 13.5.1, by gradually increasing the blocksize and making new measurements, and comparing the FRFs. When the bias error is eliminated (i.e., small enough by engineering judgement), the peaks at the resonances do not increase with increasing blocksize (or decreasing $\Delta f$, really).

Once these fundamental settings have been established, what remains is to check if the signal-to-noise ratio (SNR) is sufficient, by looking at the coherence function. If there are dips in the coherence function, three actions can be taken; (i) a force window can be applied as explained in Section 13.8.1; (ii) an exponential window can be applied, as described in Section 13.8.2; and (iii) a harder tip can be selected, if the reason for the low SNR is that the force spectrum drops too much (this is seen by the fact that the coherence gets noisy at higher frequencies). The second point is illustrated in Figure 13.12. In this figure, the effect of adding an exponential window is illustrated. It is seen that the dip in the coherence, which coincides with an anti-resonance in the FRF, disappears when the exponential window is added. This is an indication that the dip in the coherence was a result of low signal-to-noise ratio in the accelerometer signal.

### *13.8.4 Compensating for the Influence of the Exponential Window*

The effect of the exponential window, that the 'apparent' damping of the structure increases, can be computed, assuming that the window is used on both the force and response signals. Thus, if the damping is later estimated using the FRF estimated including an exponential window, a correction factor can be applied to the estimated damping factor to obtain the true damping of the structure. This can easily be shown through the following equations.

The Laplace transform pair for multiplication by an exponential function is

$$e^{-at}y(t) \Leftrightarrow Y(s+a) \tag{13.38}$$

that is, multiplying by an exponential window implies a substitution of $s$ by $(s + a)$. Furthermore, we know from Chapter 5 that the transfer function of a simple SDOF system is

$$\frac{X(s)}{F(s)} = \frac{1/m}{s^2 + s2\zeta\omega_n + \omega_n^2} \tag{13.39}$$

If we multiply both force and response by an exponential window, this implies substituting all instances of $s$ by $(s + a)$, and we obtain

$$\frac{X(s)}{F(s)}\Big|_{s=s+a} = \frac{1/m}{s^2 + s2\zeta\omega_n + \omega_n^2}\Big|_{s=s+a} \tag{13.40}$$

In order to investigate how the exponential window influences damping, we shall now study the poles of Equations (13.39) and (13.40), since the Laplace variable $s$ clearly only exists in the denominator.

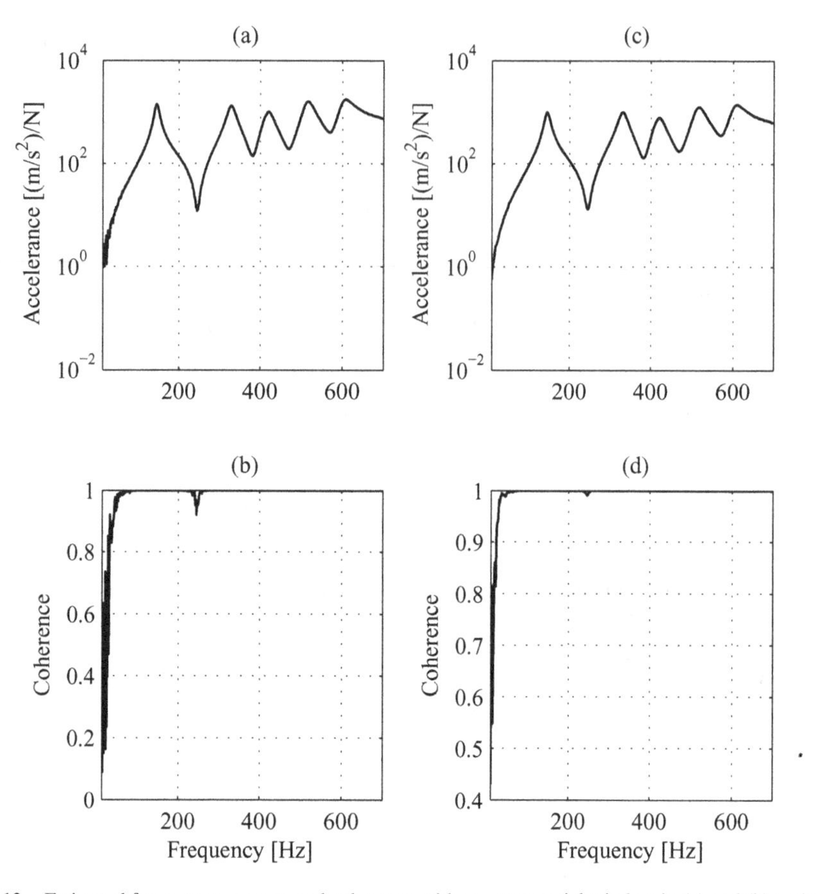


**Figure 13.12** Estimated frequency response and coherence with no exponential window in (a) and (b) and with an exponential window in (c) and (d). Five averages was used and the blocksize was $N = 2048$ samples. A force window was applied in both cases. In (b) it is seen that not using an exponential window results in a coherence function with a dip at approximately 250 Hz. The application of the exponential window removes this dip almost entirely, as shown in (d). This is an indication that the dip (which is coinciding with the antiresonance in the frequency response) was caused by noise in the accelerometer signal

The poles of Equation (13.39) are, as we know,

$$s_{1,2} = -\zeta\omega_n \pm j\omega_n\sqrt{1-\zeta^2} \text{ for } \zeta \leq 1. \tag{13.41}$$

Equation (13.40) requires a few steps of calculation, but finally the poles are

$$s_{1,2} = -(\zeta\omega_n + a) \pm j\omega_n\sqrt{1-\zeta^2} \text{ for } \zeta \leq 1. \tag{13.42}$$

Consequently, the exponential window only influences damping, that is, the real term in Equation (13.42). The increased damping can thus easily be compensated for if the variable $a$ used in the exponential window is known. If we denote the measured relative damping $\zeta_m$, and the undamped resonance frequency (in Hz) $f_r$, the corrected ('true') damping, $\zeta_c$, is

$$\zeta_c = \zeta_m - \frac{a}{2\pi f_r} \tag{13.43}$$

It should be noted that Equation (13.43) is only valid if the exponential window is applied to the force time signal as well as to the response signal. This is not very intuitive, and thus it is important to check that the software used correctly multiplies the force signal with an exponential window.

### *13.8.5 Sources of Error*

Theoretically, the impacts can vary without any effect on the estimate of the frequency response function. However, in reality, when the structure is not entirely linear, each impact must be equal. This criterion is not always easily fulfilled, as will be shown in Section 13.8.6.

If a double-impact occurs (one impact containing two or more hits of the hammer tip), which is difficult to avoid on some flexible objects that bounce back at the hammer, the phenomenon shown in Figure 13.13 occurs. The Fourier transform of the force is no longer nice and smooth, and this results in numerical errors in the computation of the frequency response. Examining the time signal and spectrum of the force, it is easy to discover double-impacts. Even better is when the analysis system has an automatic 'double-impact detection' functionality, which is an algorithm that automatically detects double-impacts and warns the user, or automatically ignores that impact in the averaging process.

Another error which can easily destroy an impact test occurs if one or more of the hammer impacts hit different locations on the structure. It is, of course, important that each impact hit the same point, as is easily understood from the relation between mode shapes and FRF appearance which we discussed in Section 6.4. How accurate one must be thus depends on the spatial wavelength of the mode shapes. If different points on the structure are excited with the different impacts during a measurement, the error that results is a bias error, which can be detected by the coherence function dropping below unity. Finally,

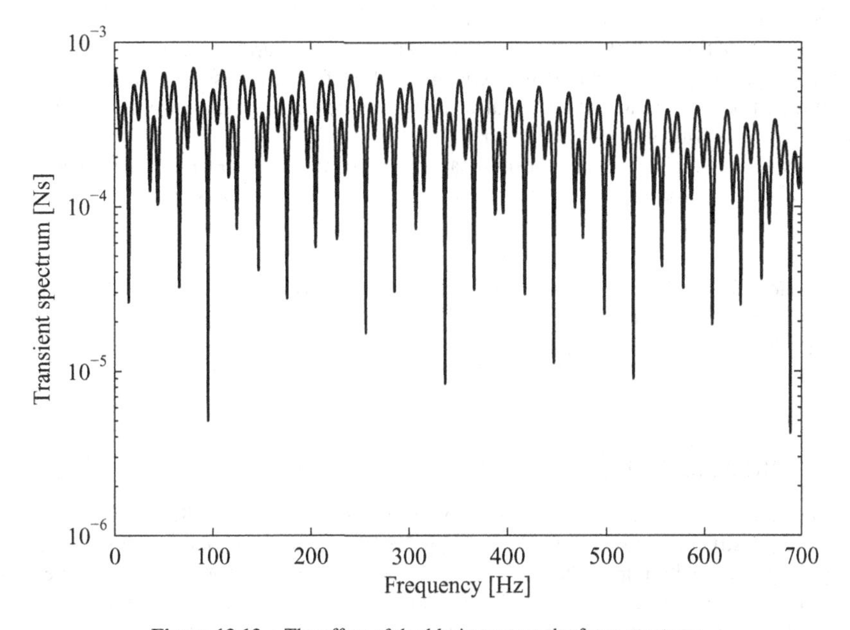


**Figure 13.13** The effect of double-impact on the force spectrum

as in all measurements, it is important to avoid overloading the sensors and inputs electronics, which destroys the results.

### *13.8.6 Improving Impact Testing by Alternative Processing*

So far we have used averaging in the frequency domain which is the most common way to perform impact testing. If each impact is nearly equivalent, time domain averaging can be used instead. When using manual impact excitation, in general the difference between individual impacts is often too large for time domain averaging to be valid, unless the operator of the impulse hammer is able to create very repeatable impacts. In practice, time domain averaging requires automatic impactors, which are available from some manufacturers. The advantage with using time domain averaging, is that it reduces the extraneous noise, and thus improves the signal-to-noise ratio, whereas frequency domain averaging only reduces the variance of the estimate. Time domain averaging in impact testing is discussed in for example Fladung *et al.* (1999).

In practical impact testing, with an imperfect operator of the impulse hammer, typically the individual impacts during the measurement of a particular point will differ by perhaps as much as $\pm 50-100\%$. This can significantly reduce the quality of the estimated FRF if the structure is slightly nonlinear (as many structures are). In Brandt and Brincker (2010) the advantages of recording the time signal during impact testing, and then post-processing the time signal, as opposed to the online processing discussed in Section 13.8.3, were discussed. The idea with this method is to set the measurement system up to record the signals from the impulse hammer and accelerometer during a time long enough to make five to ten impacts, with long enough time in between each impact to allow the maximum anticipated blocksize to be used in the later processing. After the measurement, software customized for post-processing the signal can then be applied to obtain the best possible FRF out of the available data, possibly discarding some of the impacts. Among the most important advantages with this technique are that

- the data acquisition becomes easier because no interaction is needed during the measurement (such as accept/reject of each impact, which is commonly required),
- the check for optimal blocksize can be made without remeasuring, since the data are available (assuming the impacts were spaced apart enough to allow for the necessary blocksize),
- the optimization of force and exponential windows can be made easier, since the data are available for repeated tests without remeasuring, and
- in the post-processing phase, only those impacts that produce a good FRF estimate need to be taken into the averaging process; for instance, the user can select only impacts which have approximately the same force level.

Unfortunately, few designers of commercial software packages have so far implemented impact testing software using this feature. It is easy, however, to implement the above mentioned processing in, for example, MATLAB/Octave. A typical example of the possible improvement measurement is shown in Figure 13.14.

## 13.9 Shaker Excitation

In many measurement applications it is preferable to mount a shaker (see Section 7.11) with a force transducer to the structure, instead of using an impulse hammer for excitation. The shaker has several advantages compared with impact excitation, most importantly that it, in general, gives a better signal-to-noise ratio since the excitation signal is active during a larger part of the measurement time. For a discussion about attaching the shaker and force transducer to the structure, see Section 7.5.

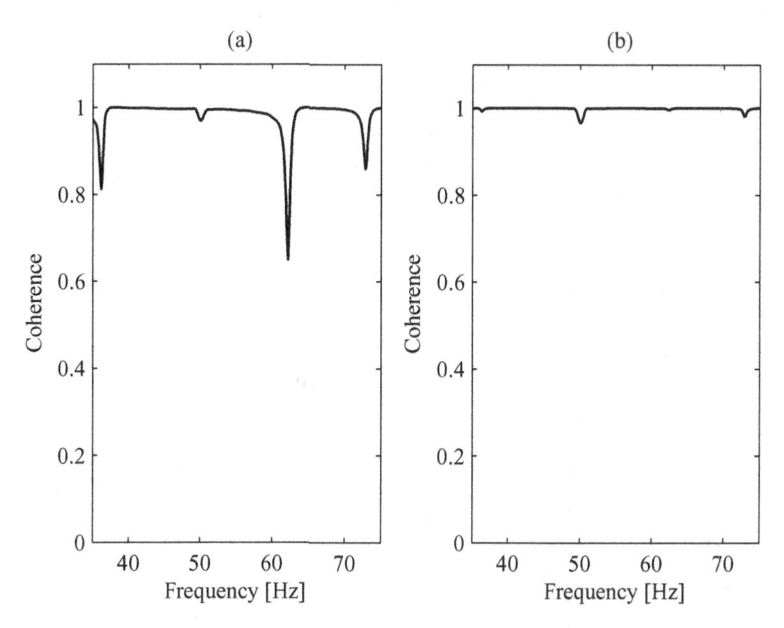


**Figure 13.14** Illustration of typical improvement of coherence on slightly nonlinear structure by selecting only force impacts with approximately equal level. The figure shows the coherence using five averages in (a), compared with using only the two impacts giving best coherence, in (b). The plots are from a measurement on a slalom ski

Different signals may be used to drive the shaker and can be divided into signals with continuous spectra and signals with discrete spectra. The former are either random signals or transient signals, or a combination of the two. Signals with discrete spectra are of course periodic in the measurement window (see Section 9.3.4) and have the advantage of concentrating the energy in the signal on those frequencies which we estimate with the DFT.

## *13.9.1 Signal-to-noise Ratio Comparison*

To consider the signal-to-noise ratio of different excitation signals, we will use the parallel filter approach discussed in Section 9.1.1 and consider a single spectral line, $k$. The (approximate, since we only consider the main lobe of the spectrum of the time window) result of a single DFT at a frequency line is illustrated in Figure 13.15, where the extraneous noise, $n(t)$, is also depicted. Two cases are of interest in this situation; either the excitation signal has a continuous spectrum, as is the case in impact testing and for pure and burst random excitation, or the excitation signal has a discrete spectrum, which is the case for the periodic excitation signals treated in Section 13.9.4 and 13.9.5.

In the first case, of an excitation signal with continuous spectrum, the value obtained at the spectral line of interest is a result of the true spectrum of the signal weighted by the spectrum shape of the time window (which in this case should be a rectangular window for impact testing or burst random, but a half sine window or Hanning window for pure random). The extraneous noise, if present, is also weighted by the time window spectrum shape, and added to the spectral line. The signal-to-noise ratio, is the ratio of the mean-square value of the weighted signal spectrum and the mean-square value of the weighted background noise spectrum, as depicted in Figure 13.15 (a) and (b). Furthermore, due to the weighting by the spectrum shape of the time window, with excitation signals with continuous spectra, there will always be some bias, although with long blocksizes, the bias can be made small.

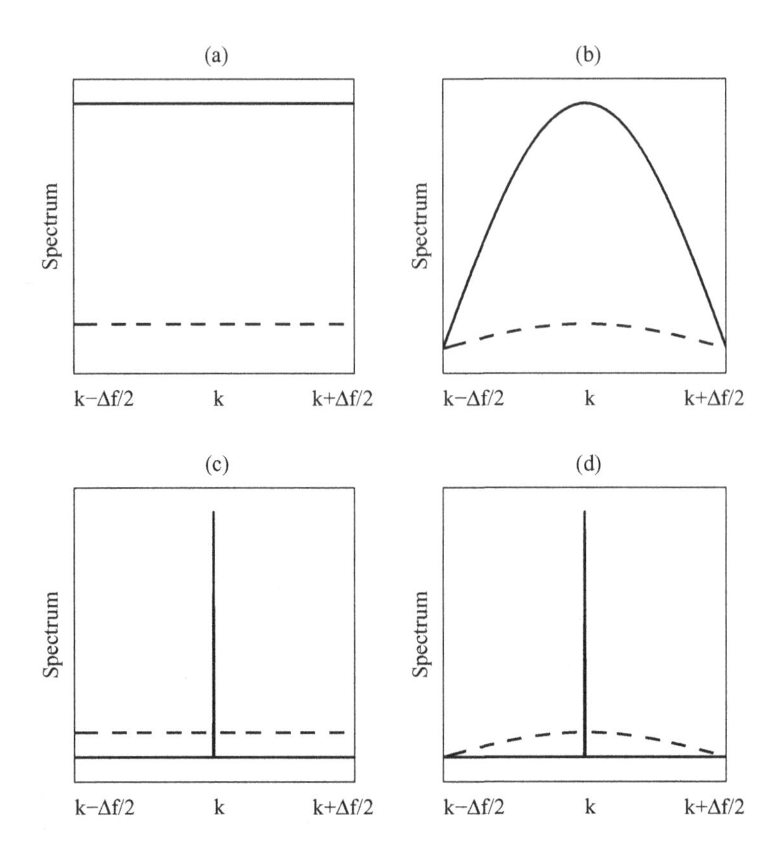


**Figure 13.15** Illustration of the signal-to-noise ratio at a spectral line for excitation signal with continuous spectrum (a and b) and with discrete spectrum (c and d). In (a) the true spectra of the excitation signal (solid) and the extraneous noise (dashed) are plotted; in (b) the same spectra are plotted after the spectrum shaping due to the effect of the time window are shown. In (c) and (d), the corresponding spectra are shown for an excitation signal with discrete spectrum. The signal-to-noise ratio is the mean square sum of the excitation signal divided by the mean square sum of the extraneous noise. As can be seen in the figure, by comparing (b) and (d), the signal-to-noise ratio will thus be higher for an excitation signal with discrete spectrum, since in this case the spectrum of the excitation signal is not affected by the window shaping

In the second case, we have a periodic excitation signal, with a period coinciding with the measurement time. In this case the frequency of the deterministic signal coincides with the spectral line, $k$, and thus, without the extraneous noise, we get a correct value of the spectrum at the spectral line, with no bias at all. If there is some background noise, this noise will be weighted by the time window spectral shape before being added to the spectral line. The signal-to-noise-ratio is thus larger in this case (provided the mean-square levels of the transient and periodic signals are the same within $\pm\Delta f/2$), as there is no weighting of the signal in the case of a periodic signal. This fact makes the signal-to-noise ratio superior when using periodic excitation signals compared to using non-periodic excitation signals.

### 13.9.2 *Pure Random Noise*

Pure random noise (or 'true random'), see Figure 13.16 (a), is a continuous, normally distributed noise signal. This signal has a continuous spectrum and since the signal is continuous in the time domain, it must be windowed by, for example, a half sine window when we carry out the frequency analysis. As we discussed in Section 10.3.3, any window but the rectangular gives rise to a widening of spectral peaks. Therefore, when measuring frequency response with low damping, which we normally have in structural dynamics, continuous noise gives poor resolution of the resonances. Because of this, pure random is often erroneously said to be completely inappropriate for exciting structures with low damping. Using pure random noise as excitation requires much larger blocksize for a particular bias error

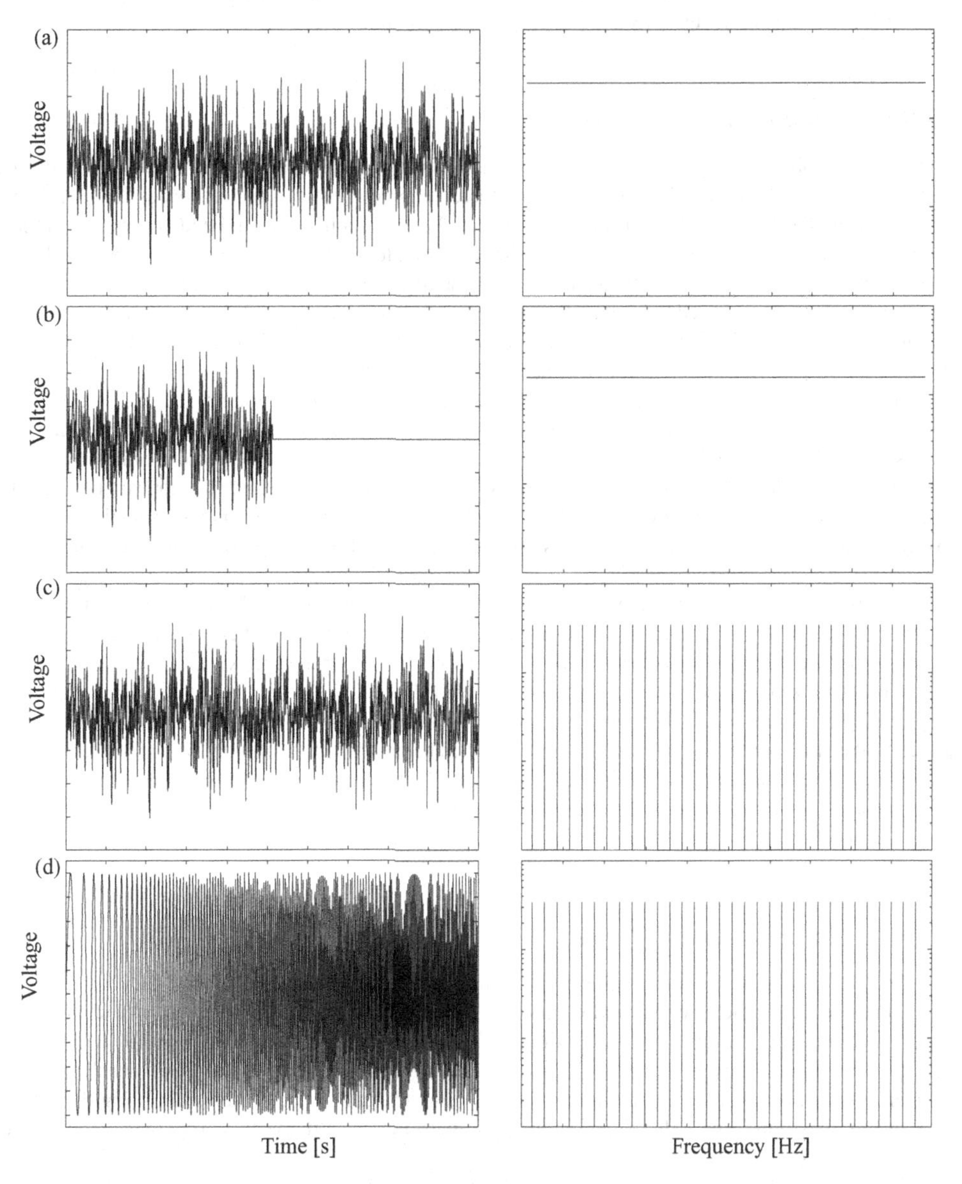


**Figure 13.16** Common excitation signals (left) for vibration excitation when measuring frequency response, and their respective (theoretical) spectra (right). (a) pure random noise, (b) periodic noise, (c) burst random noise, and (d) periodic chirp. See Sections 13.9.2–13.9.5 for descriptions of the different excitation signals

than other excitation signals, as we will show in the comparison between different excitation signals in Section 13.10. Therefore, it may be considered as an inappropriate excitation signal for shaker excitation, as the measurement will be possible to be done much more quickly, with equal bias, with one of the excitation signals presented in the next sections. However, it is important to understand that pure random can be used with (from a practical standpoint) the same bias, if only more data is recorded. In many cases with operating measurements, for example, the natural signals are random, and then good FRFs can very well be measured, but at the price of rather long measurement times, compared with what we need with other excitation signals when exciting a structure in the lab.

### *13.9.3 Burst Random Noise*

Burst random noise, see Figure 13.16 (b) is continuous noise which is turned off at a certain time during the measurement of each block, after which the force becomes zero and the responses (accelerations) die out during the remaining part of the block. Since all signals both begin and end at zero, no window should be used for this signal. It is relatively simple to create this type of signal within the hardware. The only requirement is that the data acquisition can be triggered when each burst block is sent out; the samples, however, do not need to be exactly synchronized. This is therefore the most common excitation signal for modal analysis available in commercial measurement systems. Yet another advantage is that it can easily be used for multiple-input estimation, see Chapter 14.

As mentioned above, burst random noise requires all signals to die out before the end of the block. Although this may sound easy, sometimes it is not so easy to accomplish, particularly if the rigid body modes of a freely supported structure have low damping. Sometimes, an improvement can then be made by adding some time between the bursts which is not measured. The leakage that appears because the response signals have not died out entirely at the end of the block is then mostly affecting the very low frequencies where the rigid body modes are located, and do not severely affect the frequency range of interest.

An often misunderstood fact is the effect of the interaction between the shaker and the structure during the end of the block, when the source is turned off. Depending on the amplifier type used to control the shaker, either the force is kept at zero (in the case of a current-controlled shaker), or the velocity of the shaker is kept at zero (in the case of a voltage-controlled shaker). In the first case, there is no interaction between the structure and shaker when the excitation signal is turned off, and thus the shaker is not affecting the structure. In the latter case, however, the shaker will add damping to the structure, which will come to rest more quickly than in the former case. This does not, however, influence the estimated FRF, because the force during the interaction is measured, and thus the linear relationship between the force and acceleration is correctly estimated.

### *13.9.4 Pseudo-random Noise*

Another way of achieving an excitation signal which does not require a time window is to make the excitation signal periodic. Pseudo-random noise, is, despite its name, a completely deterministic signal, which is furthermore periodic within the time window of the FFT. It can be created rather simply by setting the amplitude of each frequency line in a spectrum at a desired level (usually the same level for all frequency lines), and then adding a random phase to each frequency line. When calculating the inverse FFT of this spectrum, a periodic signal is obtained, with the desired spectral properties, as plotted in Figure 13.16 (c).

When using a periodic excitation signal, it is important to take into account that the structure has to achieve its steady-state condition before the data is acquired. The first few blocks after turning the shaker on will cause the structure to respond with a transient behavior, and this has to be 'waited out'. Usually, five to ten periods is enough to achieve a sufficient steady-state condition. On structures with very low damping, however, many more periods may have to be waited out before steady-state conditions apply.

Pseudorandom noise has a good signal-to-noise ratio, since all the signal energy coincides with the spectral lines of the DFT. For slightly nonlinear structures, however, the periodicity can be a problem, as these structures respond with nonlinear harmonics of the periodic frequencies.

A very important point with periodic excitation signals, which has very rarely been considered, is that the ideal averaging domain of these signals is *time domain* averaging, and not the 'normal' frequency domain averaging which we use for random signals. This was discussed in Phillips and Allemang (2003). Since the periodic excitation signal is entirely periodic in the time window, the extraneous noise can be *removed* by time domain averaging, thus achieving a bias-free estimate of the FRF, even in the case of noise on both the input and output signals. This does not seem to have been acknowledged in literature, but will be demonstrated in Section 13.12.

Sometimes, particularly on light structures, it is difficult to excite the structure near its natural frequencies where the force spectrum then shows dips. In such cases, pseudo-random allows shaping of the spectrum by setting the spectral lines to different values, sometimes called 'coloring' of the spectrum. This is not as easily done with other excitation signals, although it is possible with most signals.

The main drawback with pseudo-random noise is that it requires more sophisticated hardware, because it requires synchronization between the signal generator and the data acquisition input channels to assure the periodicity in the time window. Furthermore, the signal generator has to include a digital-to-analog converter, DAC, which is more expensive than a random generator. With the rapid cost reduction in modern hardware, DACs are becoming increasingly more available, however.

### *13.9.5 Periodic Chirp*

Another common periodic excitation signal is the periodic chirp signal, which gets its name from the sound it produces, provided its frequency range is appropriate. This signal consists of a sinusoid, which is continuously swept through the frequency frequency range of interest during each block, see Figure 13.16 (d). The advantage of using this type of signal compared with the random signals above is that the signal-to-noise ratio is the best, since the crest factor is lower than for pseudo-random. If the disturbance noise is negligible, then it suffices with one average, although as with the noise signals above, a few averages should in general be used to obtain a good result. The fact that the chirp signal consists of a sinusoid can also be advantageous if one knows that the structure is nonlinear and it is desired to have an excitation signal that always has the same amplitude. The noise signals, however, are often better when one has a nonlinear structure and wants to measure a linear approximation of the system, because the noise signals have amplitude distribution such that all amplitudes are randomly mixed.

Since the periodic chirp signal is a periodic signal within the time window, time domain averaging should be applied, see the discussion in Section 13.9.4. The drawback with pseudo-random, that it requires more sophisticated hardware than pure random and burst random, applies also to periodic chirp excitation.

### *13.9.6 Stepped-sine Excitation*

Stepped-sine is a completely different measurement method from those described above. All of the above methods, from impact excitation to shaker excitation with the excitation signals mentioned in Sections 13.9.2–13.9.5, are examples of *broadband excitation*, i.e., where the excitation signal contains frequencies within a wide frequency band. Stepped-sine excitation implies that we instead allow a sinusoidal excitation signal to step up in frequency. At each frequency, steady-state conditions are reached, after which the input and output signals are measured and the amplitude and phase relationships are determined between the output and the input force, before stepping the frequency to the next test frequency.

Stepped-sine is often done using the FFT and by setting the sampling frequency so that exactly an integer number of periods are measured and leakage is avoided. Stepped-sine is thus a slow method, but it has the advantage of being able to cope with very low signal-to-noise ratios. The signal-to-noise ratio is the highest possible, which makes it possible to use relatively low levels (since all the signal power is concentrated at one frequency, while the extraneous noise is spread over all frequencies).

In some systems supporting stepped-sine excitation it is possible to control either the excitation force or the response signal using feedback so that either the force or response signal is held constant. This is particularly useful when studying nonlinear structures, but can often be vital also on measurements on lightly damped structures, to avoid excessive vibration levels (by controlling the response, in this case).

## 13.10 Examples of FRF Estimation – No Extraneous Noise

In this section we will show some examples of estimation of frequency response using some different excitation signals which will illuminate the many practical aspects described theoretically in the preceding sections. We will first use an SDOF system simulated by the method described in Section 6.5, and with no extraneous noise. This is essential to understand the inherent errors in the estimates of FRFs using the various excitation signals. For the simulations, we set the natural frequency of the SDOF system to 100 Hz and the relative damping to 1%. The frequency range is selected to 1024 Hz (sampling frequency 2048 Hz), with additional ten times oversampling in the simulation, to reduce the error due to the simulation to a negligible level. Furthermore, the 3 dB bandwidth of the resonance for the SDOF system is 2 Hz (see Equation 5.39).

### *13.10.1 Pure Random Excitation*

We start by exciting the SDOF system with pure random noise. We make 149 averages (50 independent blocks), with a half sine window, and 67% overlap, which gives optimum bias and random error, as discussed in Section 13.5. The resulting FRFs and coherence functions for three different blocksizes are shown in Figure 13.17. In the zoomed-in plots on the right-hand side it is seen that the largest blocksize gives a very small bias error. The blocksizes used for the results shown in Figure 13.17 were 2048, 4096, and 16384 samples. The ratio of the 3 dB bandwidth of the SDOF system to the frequency increment is thus 2, 4, and 16, respectively.

As mentioned before, it is often said that pure random cannot be used to estimate FRFs for lightly damped systems due to the large bias caused by the time window. As the results in Figure 13.17 shows, this is not true. However, the price for a low bias error is a large amount of data; in this case we have used $50 \cdot 16384 = 819200$ samples for the largest blocksize. As the next sections will show, there are more economical methods of estimating bias-free FRFs, if we select a more appropriate excitation signal. Still, this example shows that for natural data of random nature, it is quite possible to estimate good FRFs.

Another interesting result in Figure 13.17 is that there is a random error in the estimate. At some spectral lines, this error makes the estimate larger than the true value. It should be noted that there is no contradiction between this and the remark in Section 13.4, which stated that the $H_1$ estimate is always less than the true value, as this remark referred to the bias error due to *input noise*, in the case of the $H_1$ estimator.

### *13.10.2 Burst Random Excitation*

As mentioned in Section 13.9.3, burst random noise is the most commonly used excitation signal for mobility and accelerance measurements on mechanical structures. With this excitation signal, the burst length is adjusted so that the response signal completely dies out before the end of the time block, and no

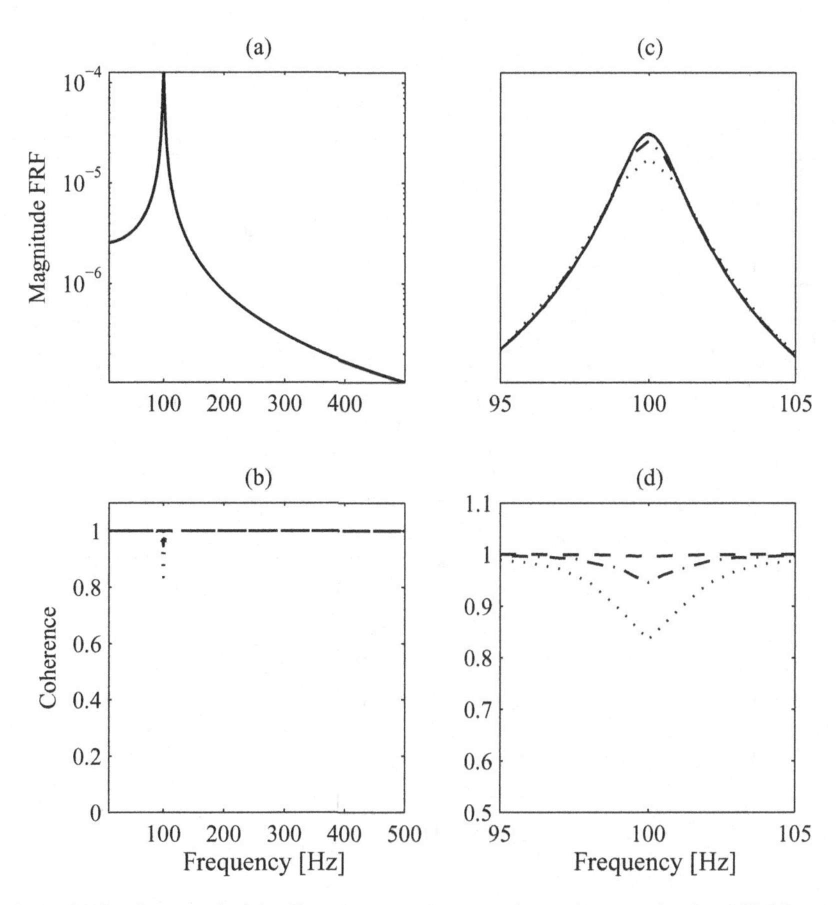


**Figure 13.17** FRF estimates using the $H_1$ estimator and pure random noise on a simulated SDOF system with $f_n = 100$ Hz, and $\zeta = 0.01$. In (a) and (b) the magnitude of the FRF and coherence, respectively, estimated using 2048 samples blocksize (dotted), 4096 samples (dash-dotted), and 16384 samples (dashed) are plotted. In (c) and (d) the same plots are zoomed in for a detailed look around the natural frequency. The true FRF is plotted in solid in (a) and (c). As can be seen, with the largest blocksize the bias error is negligible, and the coherence is very near unity ($\min[\gamma^2(f)] = 0.996$ at 100 Hz for the blocksize of 16384 samples)

window is then used in the FFT processing. In Figure 13.18 (a) and (b) the resulting FRF and coherence for two different burst lengths, 75% (too long) and 25% (sufficiently short) are plotted, using 2048 samples blocksize. As can be seen in Figure 13.18 (b), the dip in the coherence at the natural frequency vanishes when the leakage disappears.

The number of averages for burst random have to be large enough to produce a small random error. At least 50–100 averages should be used, and the number has to be higher the more extraneous noise is present. In the example we have used 50 averages, comparable to the case for pure random in the previous section. In addition, as we mentioned before, it may be necessary in practice to add some idle time between the bursts, in order for rigid body motion to entirely die out. Note also, that since the spectrum of the burst random signal is continuous, there will be a bias error if the blocksize is not much larger than the length of the impulse response.

As shown in Figure 13.18 (c), the bias error is almost zero even with a relatively low blocksize compared with the blocksize necessary for pure random excitation. If the FRF is going to be used

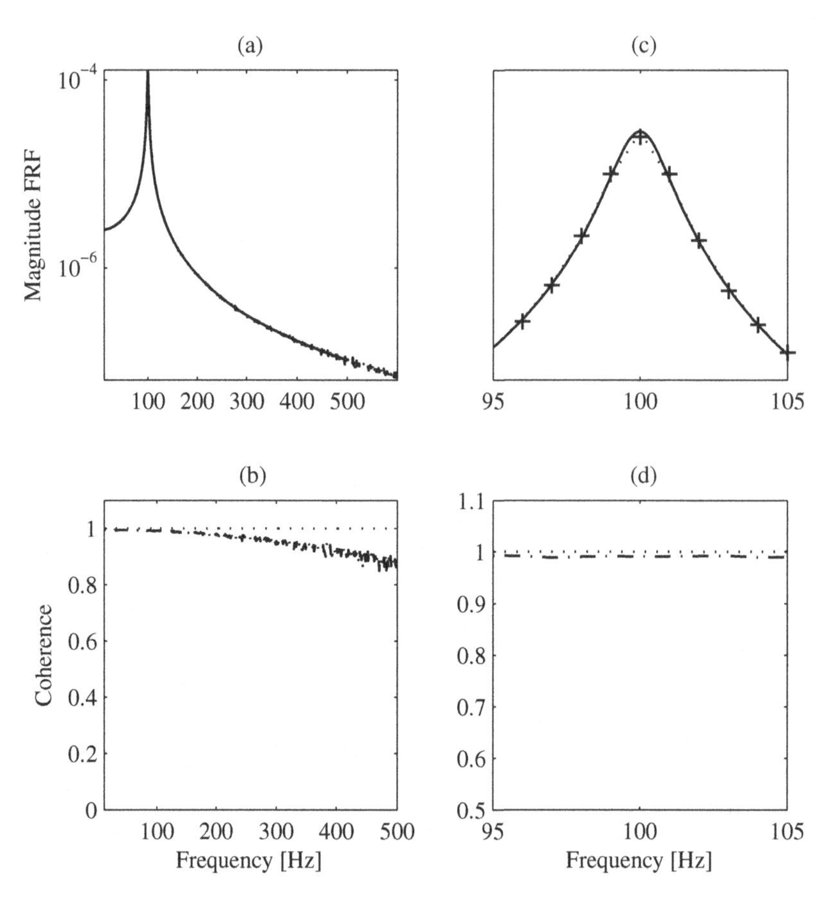


**Figure 13.18** Plots of magnitude of FRF and corresponding coherence for two cases of burst random; 75% burst length (dotted), and 25% burst length (dash-dotted, or '+' sign), with a blocksize of 2048 samples. With the leakage produced by the larger burst length, in (c) it can be seen an effect where the coherence drops at higher frequencies, and in (b) there is a clear bias at the natural frequency. With the low burst length this leakage error is minimal and the frequency lines are very close to the true FRF (solid). However, it should be remembered that with any excitation signal with continuous spectrum there is always some bias

for experimental modal analysis curve fitting, many algorithms can use spatial information, as well as frequency information, to achieve high accuracy in modal parameter estimates, so this low frequency resolution is sufficient. In other cases, however, where the FRF is to be used for other purposes, it may still be necessary to have a high resolution of the FRF, which may necessitate higher blocksize. It should be pointed out that the amount of data we have used for the burst random case is $50 \cdot 2048 = 102400$ samples, which is an eighth of what we used for the pure random excitation.

## *13.10.3 Periodic Excitation*

Without any extraneous noise, the result of using periodic random or periodic chirp excitation would be similar to using burst random, although with periodic excitation the bias error is exactly zero when no extraneous noise is present. Another difference is, of course, that we could use time averaging with periodic excitation signals and that we must allow the structure to enter a steady-state response before starting to acquire data. In Section 13.12 we will show some results of periodic excitation in the presence of extraneous noise on both input and output.

## 13.11 Example of FRF Estimation – with Output Noise

We will now add some noise to the output signal to illustrate the effect on the FRF estimate and on the coherence. To illustrate this point, we add a degree-of-freedom to our mechanical system and simulate a two-degree-of-freedom system with a second natural frequency of 200 Hz and damping of 0.01. This produces an antiresonance at approximately 158 Hz, as illustrated in Figure 13.19 (a). We add a constant

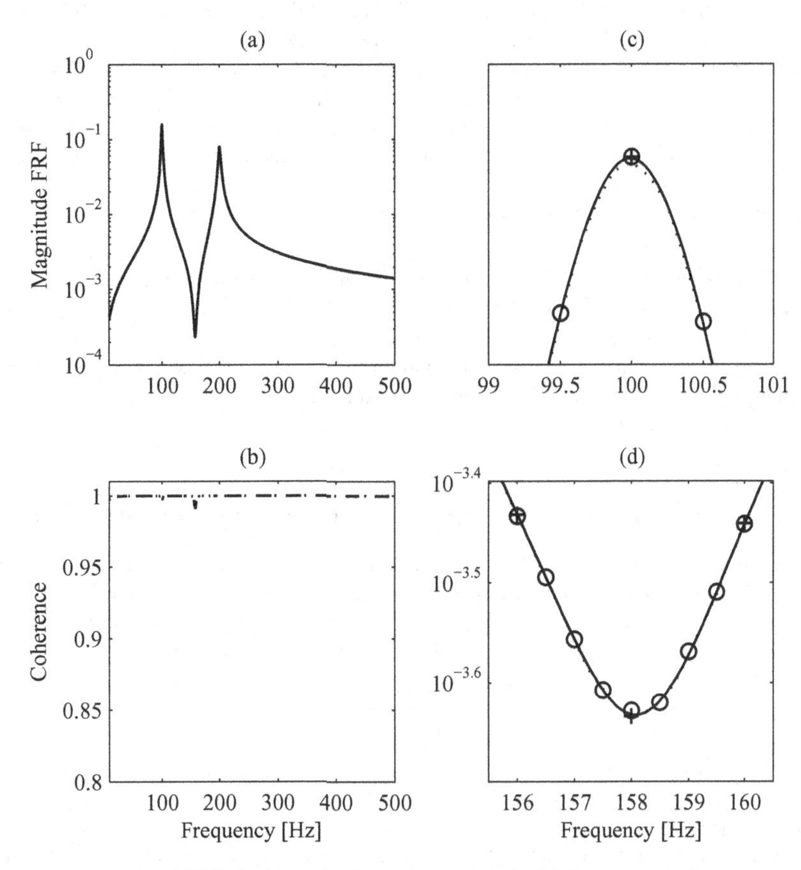


**Figure 13.19** Magnitude of FRFs in (a), and coherence functions in (b) for two excitation cases; pure random excitation with blocksize of 16384 samples (dotted), and burst random with blocksize of 4096 samples blocksize and 25% burst length (rings or dash-dotted). The output signal was contaminated by uncorrelated noise with a signal-to-noise ratio of 100 times (40 dB). The true FRF is plotted in solid. In all cases, 50 blocks of time data were used. In (c) and (d), the FRF estimated by a pseudo-random excitation signal, with 1024 samples (+ sign or dashed) is shown together with the other estimates, in a frequency region around the first natural frequency (in c), and around the antiresonance (in d). Although all three excitation signals have a small bias error, the plots illustrate the improvement in bias error achieved by a periodic excitation signal, in this case pseudorandom noise, although a periodic chirp would yield similar results

noise signal with an RMS level of $10^{-2}$ times the RMS level of the output signal of the system, and with constant PSD over the frequency range. This is a rather large extraneous noise level which we use to illustrate clearly the result of extraneous noise. The signal-to-noise ratio is, of course, worst where the response signal is small, as the extraneous noise has a continuous spectral density, and the most difficult part of the FRF to estimate, is therefore around the antiresonance. In Figure 13.19 the FRF (in a) and coherence (in b) results are shown for pure random, burst random, and pseudorandom excitation. On the right-hand side, a small frequency region around the natural frequency (in c) and the antiresonance (in d), are shown, to more clearly see the bias error in this region. As evident from Figure 13.19 (b), the noise at the output causes a dip in the coherence at the antiresonance.

As seen in Figure 13.19 (a) and (b), the coherence is still good at the natural frequencies of the 2DOF system. At the antiresonance, however, there is some random error, with pure random and burst random, but not for pseudorandom, but no evident bias error. The bias error is averaged away, since we are using the $H_1$ estimator, and the noise is in the output signal, which is the assumption for the $H_1$ estimator. The random error will diminish as we increase the number of averages.

The result from the pseudo-random excitation signal in Figure 13.19 is more remarkable and requires some comments. To produce the results in this example, we have used time domain averaging, which, as mentioned in Section 13.9.4 reduces the noise in both signals, $x$ and $y$. Using time domain averaging on all blocks of data, yields the coherence estimate invalid, as it would be based on only one average. It is therefore not plotted in Figure 13.19 b).

Two main conclusions can be drawn from this example:

- Any excitation signal can be used to estimate the frequency response with a small bias error, using the $H_1$ estimator in the case of extraneous noise in the output signal. (And the same is valid for the $H_2$ estimator in the case of noise in the input signal.)
- A periodic excitation signal such as pseudo-random (or periodic chirp, which is equivalent in terms of all relevant properties), is superior to any other excitation signal, due to its lack of bias error and better signal-to-noise ratio.

## 13.12 Examples of FRF Estimation – with Input and Output Noise

As a final example we will add extraneous noise also to the input signal. To make this example more realistic we will also shape the input force spectrum to illustrate a phenomenon common when exciting light structures. In such cases, often the force spectrum will have dips around the resonances of the structure, which is caused by the structure moving away from the shaker at the resonances. With this coloring of the force spectrum, typically the noise in the force transducer comes closer to the 'true' force signal at these dips in the force spectrum, causing the $H_1$ assumption to fail.

The results of a simulation similar to the simulation described in Section 13.11, but with colored force spectrum and with addition of noise also on the input signal, $x$, are plotted in Figure 13.20. In (a) and (b), the PSDs of the force and response signals are plotted (solid) together with the extraneous noise $m(t)$ and $n(t)$, respectively. Of course, in a practical situation we do not have access to the spectra of the extraneous noise, but it is useful to picture these noise sources as signals with relatively constant PSD as in the figure. When the force PSD is high near the antiresonance of the system, the signal-to-noise ratio is relatively high, and the assumption for the $H_1$ estimator is valid. Around the natural frequencies of the system, however, since the PSD of the force dips, the $H_1$ assumption is invalid. In Figure 13.20 (d) this can be seen to cause some (small in this case) bias in the estimates using pure random (dotted) and burst random (rings).

As mentioned in Section 13.2.3, there is no general estimator which can minimize the error in the frequency response in the case of both input and output noise, without *a priori* knowledge about the noise properties. This is valid for random noise input signals. However, as Figure 13.20 (d) shows, with

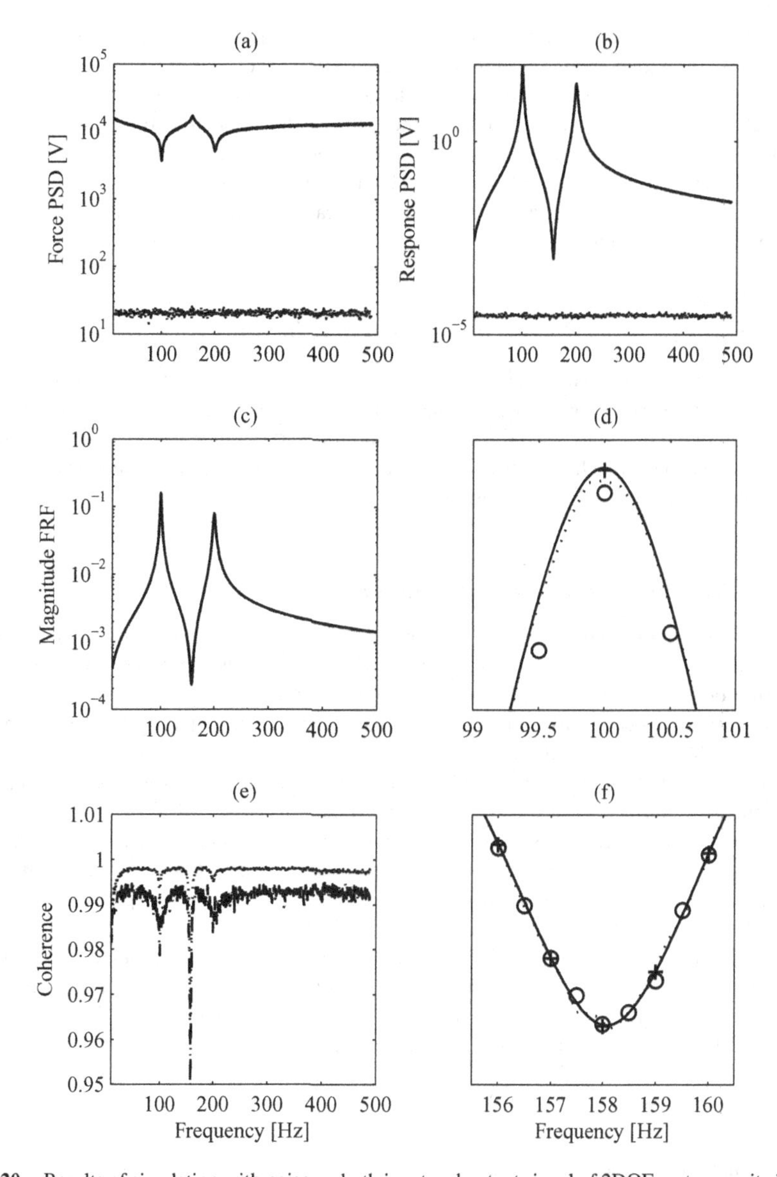


**Figure 13.20** Results of simulation with noise on both input and output signal of 2DOF system excited by colored force spectrum. (a) Force PSD (solid) and extraneous input noise (dotted); (b) response PSD (solid) and extraneous output noise (dotted); (c) Magnitude of FRFs of true system (solid), pure random excitation (dotted), burst random (dash-dotted), and pseudo-random with time domain averaging (dashed); (e) coherence functions of pure random (dotted), and burst random (dash-dotted); For the pure random excitation a blocksize of 16384 samples was used, for burst random 2048 samples and 25% burst length. In (c) and (d), the FRF estimated by the pseudorandom excitation signal, with 1024 samples (+ sign) is shown together with the other estimates, in a frequency region around the first natural frequency (in d), and around the antiresonance (in f). Although all three excitation signals have a small bias error, the plots illustrate the improvement in bias error achieved by a periodic excitation signal and time domain averaging, in this case pseudo-random noise, although a periodic chirp would yield similar results. Although not easily seen, the coherence for burst random in (e) is significantly worse than for pure random, due to the reduced signal-to-noise ratio with burst random signals

pseudorandom (plus sign), and with time domain averaging as we have used here, the estimated FRF is (almost) bias-free, provided enough averages have been made to 'clean up' the two input signals. This has, to my best knowledge, not been reported in literature, but is an important advantage with periodic excitation signals. It should also be noted that if the ordinary coherence function is wanted, it could be computed if a hybrid method is used. In such cases time domain averaging can be made on portions of data, after which each result of these portions is added together using frequency domain averaging. If 5–10 frequency averages are made, a coherence function can be estimated which can be used to assess the quality of the measurement.

### 13.12.1 Sources of Error during Shaker Excitation

A few practical aspects of shaker excitation need to be discussed. Attaching a shaker and making a correct measurement requires a great deal of caution and skills. It is very easy to introduce, for example, transverse forces into the stinger between the shaker and force transducer, which causes the force sensor to produce an erroneous signal, as discussed in Section 7.5. In the following sections some common checks and causes of errors will thus be discussed.

### 13.12.2 Checking the Shaker Attachment

The first thing which must be done, after attaching the shaker to the structure and choosing some initial settings for analysis, is to check that the stinger is not too flexible. This can be done by checking that the force spectrum does not show too high dynamic range (variation between its minima and maxima) in the frequency range of interest. How much the force spectrum may acceptably drop depends both on the transducer used and on the dynamic range of the analysis system. When the force spectrum drops too much it will be detected through a loss of coherence, and this should be considered particularly serious if the $H_1$ estimator is going to be used, as it will be biased in this case.

When the optimal measurement settings have been set, as discussed earlier in this chapter, a check of the shaker attachment should be carried out. As discussed in Section 6.4.6, a driving point frequency response between force and acceleration (velocity, displacement) always displays an anti-resonance between each resonance. This is equivalent to the fact that the imaginary part of the driving point FRF (real part in the case of mobility) will display peaks as either positive or negative, but never both. If the driving point FRF does not have this property, it is an indication that something is wrong with the coupling between the force sensor and the structure, or that the accelerometer is located too far away from the force sensor to provide 'one point', in cases where the accelerometer has to be mounted next to the force transducer. One should in such cases consider using an impedance head, see Section 7.6.

The causes of these deviations can include

- the force transducer is incorrectly attached (too soft),
- the accelerometer is incorrectly attached, or it sits too far from the force transducer to be seen as sitting at the 'same' position,
- the stingers are connected with transverse forces, or are unsuitably chosen, making the force transducer measure more than just the force in the desired direction.

Another check which should preferably be made is a check of *reciprocity*, as discussed in Section 6.4.2. Thus, if the shaker is intended to excite the structure at a certain point, a frequency response between the force in a *different* point, to an accelerometer in the future shaker location should first be measured. This requires either that the shaker is first mounted in this other point, or alternatively, if this is considered to be too much work, a separate measurement with an impulse hammer exciting the other point should be made. The FRF between this alternative force point and the accelerometer in the future force location

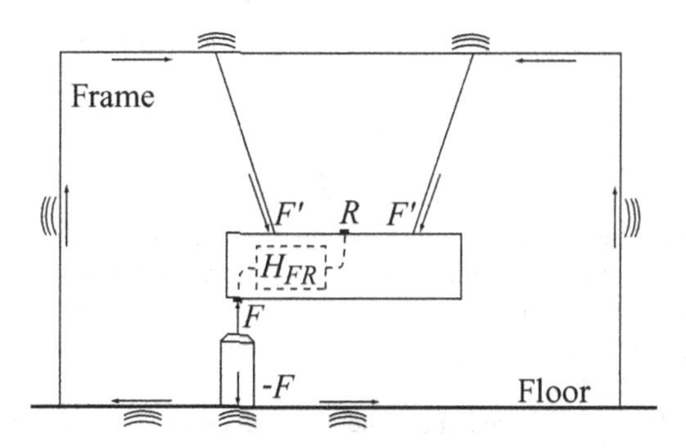


**Figure 13.21** Illustration of how forces correlated with the measured excitation signal can be transferred to the structure through the suspension and floor. The shaker's reaction force produces floor vibrations . These vibrations propagate through the frame suspending the structure, through the suspension springs. The structure is effected by forces $F'$ from the springs, which are highly correlated with the force $F$. This leads to bias error in the frequency response estimate between force $F$ and response $R$, which does not affect the coherence function making it more elusive. One should therefore always check that such forces do not exist by, for example, measuring the response signals with the shaker detached

is then measured and stored. After the shaker is then installed in its final location, an accelerometer is mounted in the point just excited by the force, and the FRF between these two points measured and stored. The two FRFs should be equal. The difficulty of attaching a shaker correctly is so large that this procedure should always be followed to ensure accurate measurements.

### *13.12.3 Other Sources of Error*

During shaker excitation, there is a potential error source which is very important to avoid, and which is illustrated in Figure 13.21. This error arises if the shaker causes vibrations which propagate through the structure through a different path than through the stinger and force transducer. These vibrations make the linear system, measured between the input force and the response, incorrect according to the model we assume. The input forces to the structure via the wall, floor, frame, and suspension as illustrated in Figure 13.21 are highly correlated with the force propagating through the stinger, which we measure with the force transducer. Therefore, this type of error is not discovered by the coherence function, since the measured response signal is correlated with the measured input. Sometimes, this type of error can be detected through measuring the response signals partly while the shaker is turned off and partly after it is turned on, but before the shaker is attached to the structure. After that, a comparison between the measured response signals is made. If there is no difference (easiest to see in the spectrum) then this is a good indication that no vibrations propagate through the structure via other paths than the desired one. See for example Ewins (2000) for more examples of situations like this.

## 13.13 Chapter Summary

This chapter has presented the fundamental theory and application of experimental frequency response estimation for single-input/single-output, linear systems. The two most common estimators for these systems are the $H_1$ and $H_2$ estimators. The $H_1$ estimator defined by

$$\hat{H}_1(f) = \frac{\hat{G}_{yx}(f)}{\hat{G}_{xx}(f)} \tag{13.44}$$

minimizes the bias due to extraneous noise present in the measured output signal. The $H_2$ estimator, defined by

$$\hat{H}_2(f) = \frac{\hat{G}_{yy}(f)}{\hat{G}_{xy}(f)} \tag{13.45}$$

instead minimizes extraneous noise present in the input signal. In case the input signal to the linear system is random noise, both these estimators are always biased, although the bias error decreases with increased blocksize. Thus with large enough blocksize, the bias error can be made negligible. Also, a relatively large number of averages usually have to be used, in order to get the bias error due to the extraneous noise to be small.

In addition to the FRF estimators, the coherence function is defined as the ratio of $\hat{H}_1$ to $\hat{H}_2$, or

$$\hat{\gamma}^2_{yx}(f) = \frac{\hat{H}_1(f)}{\hat{H}_2(f)} = \frac{\left|\hat{G}_{yx}(f)\right|^2}{\hat{G}_{xx}(f)\hat{G}_{yy}(f)} \tag{13.46}$$

where $0 \leq \gamma^2_{yx} \leq 1$. The coherence function is interpreted as the power of the output signal, $y(t)$ which can be explained by a linear relationship with the input signal $x(t)$. The main reason for a coherence deviating from unity, is that there is contaminating noise, either on the input, the output, or on both signals. There are, however, also a number of other situations where the coherence can differ from unity, which were listed in Section 13.7.

We also discussed that the optimal settings for the FFT processing for estimates of frequency response functions when the signals are pure random noise, is to use a half sine window and 67% overlap, if Welch's method for spectral density estimation is used. The FRF estimates will always be biased, but the bias error can be made negligible by increasing the blocksize until the peaks at the natural frequencies do not increase with higher blocksize. The random error in frequency response estimates is always nonzero, but usually small when there is limited amount of extraneous noise.

Two main techniques for measuring frequency response on mechanical systems were presented; (i) impact testing, where an impulse hammer is exciting the test structure, and (ii) shaker excitation, where a force sensor is attached, through a stinger, to a shaker. For impact testing, we showed that the noise in the input signal (force signal) can be almost eliminated using a force window. The $H_1$ estimator is therefore usually the best choice for impact testing. Furthermore, if an exponential window is used to improve the signal-to-noise ratio, it is increasing the apparent damping in the measured FRF. The amount of added damping can, however, be calculated from knowledge of the exponential factor of the exponential window, and the natural frequency of the system, see Equation (13.43).

For shaker excitation we have a range of excitation signals available. It was shown that the best excitation signals are the periodic signals, i.e., pseudorandom and periodic chirp. If these are not available in the measurement system, burst random can be used as an only slightly less efficient excitation signal in most cases (provided the noise in the sensors is reasonably low and provided the burst length is adjusted appropriately). All these excitation signals are self-windowing, i.e., they should be used without any time window in the FFT processing. We also discussed the possibility of averaging periodic excitation signals in the time domain instead of the more commonly used frequency domain averaging. This can potentially lead to removal of both input and output noise, although more averages than for frequency averaging should then be used.

The main advantage with the self-windowing signals, is that the bias error is small for smaller blocksizes than those which have to be used with pure random noise. This means that the measurement can be done much quicker with the same quality (bias error). Modern modal analysis algorithms allow relatively coarse frequency resolution, as they trade frequency resolution for spatial resolution.

## 13.14 Problems

Many of the problems following are supported by the accompanying ABRAVIBE toolbox for MATLAB/Octave and further examples which can be downloaded with the toolbox. If you have not already done so, please read Section 1.6, and follow the instructions to download this toolbox together with example files.

**Problem 13.1** *Which estimator ($H_1$ or $H_2$) should you use in an impact test? Explain why it is better than the alternative!*

**Problem 13.2** *Write a MATLAB/Octave script which creates a mechanical SDOF system with a natural frequency of 12.5 Hz and 2% damping, using appropriate commands from the accompanying toolbox. Set the sampling frequency to 200 Hz and use 100 (frequency domain) averages. Excite the system with random noise and determine which blocksize you should use as a minimum, to eliminate the bias error by running the script with larger and larger blocksize until the peaks do not become higher with a larger blocksize. Plot the magnitudes of the FRFs (with logarithmic y-axis) and coherence.*

**Problem 13.3** *Make a new script using part of the script from Problem 13.2 and replace the excitation signal by burst random noise. First adjust the burst length using a blocksize of 1024 samples until the measurement quality is good. Then check the maximum burst lengths (in % of the blocksize) you can have with blocksizes of 2048 and 4096, with good coherence.*

**Problem 13.4** *Define a 2DOF system with natural frequencies of 12.5 Hz and 25 Hz, and 2% damping. Use parts of the script in Problem 13.2 and add output noise. Rerun the script with different amounts of noise level. Check that you understand what happens at the natural frequencies as well as at the antiresonance.*

**Problem 13.5** *Use the script from Problem 13.4 and replace the excitation signal with burst random, and the same burst length you found out in the previous problem. Add the same amount of noise as in Problem 13.4. Run the script with different blocksizes and note what happens with the random error. Explain why!*

**Problem 13.6** *Using parts of the script from Problem 13.2 and the minimum blocksize obtained in that problem, compare the estimated FRF using half sine window with 67% overlap with an FRF estimated using Hanning window and 50% overlap. How much larger is the bias error with Hanning window?*

## References

Antoni J and Schoukens J 2007 A comprehensive study of the bias and variance of frequency-response-function measurements: Optimal window selection and overlapping strategies. *Automatica* **43**(10), 1723–1736.

Antoni J and Schoukens J 2009 Optimal settings for measuring frequency response functions with weighted overlapped segment averaging. *IEEE Transactions on Instrumentation and Measurement* **58**(9), 3276–3287.

Bendat J and Piersol A 1993 *Engineering Applications of Correlation and Spectral Analysis* 2nd edn. Wiley Interscience.

Bendat J and Piersol AG 2010 *Random Data: Analysis and Measurement Procedures* 4th edn. Wiley Interscience.

Brandt A and Brincker R 2010 Impact excitation processing for improved frequency response quality *Proc. 28th International Modal Analysis Conference*, Jacksonville, FL Society for Experimental Mechanics.

Ewins DJ 2000 *Modal Testing: Theory, Practice and Application* 2nd edn. Research Studies Press, Baldock, Hertfordshire, England.

Fladung W 1994 *The development and implementation of multiple reference impact testing* Master's Thesis University of Cincinnati.

Fladung W, Zucker A, Phillips A and Allemang R 1999 Using cyclic averaging with impact testing *Proc. 17th International Modal Analysis Conference*, Kissimmee, FL Society for Experimental Mechanics.

Goyder H 1984 Foolproof methods for frequency response measurements *Proc. 2nd International Conference on Recent Advances in Structural Dynamics*, Southampton.

Halvorsen WG and Brown DL 1977 Impulse technique for structural frequency-response testing. *Sound and Vibration* **11**(11), 8–21.

Mitchell L and Cobb R 1987 An unbiased frequency response function estimator *Proc. 5th International Modal Analysis Conference*, pp. 364–373, London, U.K.

Phillips AW and Allemang RJ 2003 An overview of MIM-FRF excitation/averaging/processing techniques. *Journal of Sound and Vibration* **262**(3), 651–675.

Rocklin GT, Crowley J and Vold H 1985 A comparison of H1, H2, and Hv frequency response functions *Proc. 3rd International Modal Analysis Conference*, Orlando, FL.

Schoukens J, Rolain Y and Pintelon R 2006 Analysis of windowing/leakage effects in frequency response function measurements. *Automatica* **42**(1), 27–38.

Wicks A and Vold H 1986 The Hs frequency response estimator *Proc. 4th International Modal Analysis Conference*, Los Angeles, CA.

# 14

# Multiple-input Frequency Response Measurement

In this chapter we will introduce identification of linear systems with multiple-input signals and multiple-output signals. Such systems are found in many applications in the field of structural dynamics and acoustics. For example, the noise in different parts of the interior of a car (outputs) are made up of sound from the engine, the road, wind-induced sound, etc. (inputs). In experimental modal analysis, it is common to use several shakers to excite the test structure, and simultaneously to measure a large amount of response signals. In these cases, it is inappropriate to use the SISO estimation methods we discussed in Chapter 13, but instead frequency response functions must be estimated using a multiple-input/multiple-output (MIMO) model.

As we will show in this chapter, the only results necessary for calculating all input/output relations for linear systems, are the autospectra of all signals (input and output), and cross-spectra between each input and all other signals (including other input signals). We will use the multi-channel spectral analysis concepts introduced in Section 10.7 for the discussion of estimators for MIMO FRF models.

The success of understanding MIMO FRF estimation relies on correctly understanding the different effects of excitation signals, signal-to-noise ratios, and the extraneous noise on input, or output signals, or both these signals. We will therefore make frequent use of simulations in this chapter, to illustrate each effect without the other effects present.

## 14.1 Multiple-input Systems

MIMO systems are defined such that each output is caused by a linear combination of all the inputs, and there are no causal relations between any of the outputs, see Figure 14.2 where a general MIMO system is depicted. If there are causal relations between any of the outputs then one or more output signals need to be redefined as an input signal. Thus there is a linear system between each input and each output, and there can, of course, be some uncorrelated, extraneous noise added to either the inputs or outputs, or both, as for the SISO system. The MIMO system can therefore be seen as a number of parallel multiple-input/single-output systems. The fact that we have several outputs is not an important issue, from a conceptual standpoint, since each output is independent of all the other outputs. The problem, instead, lies in the decoupling of the individual contributions in a particular output, to each of the inputs. We can thus concentrate on looking at the MISO system, at least from a conceptual point of view. It may

*Noise and Vibration Analysis: Signal Analysis and Experimental Procedures,* First Edition. Anders Brandt.

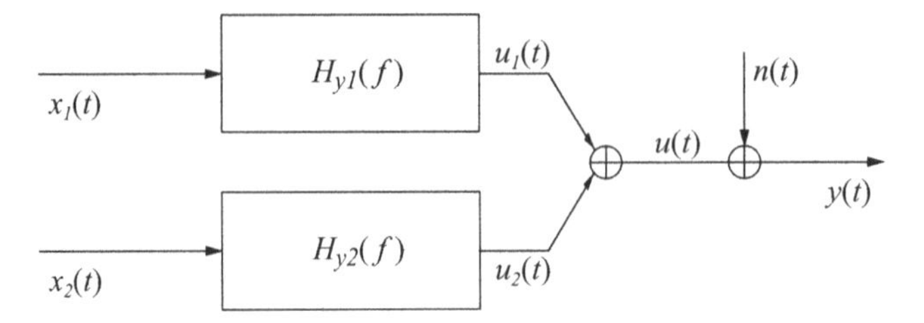


**Figure 14.1** 2-input/1-output system with noise on the output

be more computationally efficient to compute the entire MIMO system at once, but the understanding of the problem can be visualized by a MISO system.

### *14.1.1 The 2-input/1-output System*

Before we introduce the general multiple-input-multiple-output (MIMO) system, we will first consider a 2-input/1-output system, as depicted in Figure 14.1 in order to visualize the concept, because the matrix algebra we use in the general case becomes rather abstract.

In Figure 14.1, the output is caused by a linear combination of the two inputs, $x_1(t)$ and $x_2(t)$ which can have some mutual correlation, but must not be completely correlated. This is no limitation, since, if the inputs are totally correlated, it is actually sufficient to use a model with only one input, as the other input can then be created by letting the first input pass a linear system. Thus the model as illustrated in Figure 14.1 is invalid in that case, since we really have a single-input system.

Some additional contaminating noise, $n(t)$, is added to the sum of the signals coming from the two inputs, $u(t)$, to produce the measured output signal, $y(t)$. In the frequency domain, the output of the system then is

$$Y(f) = X_1 H_{y1} + X_2 H_{y2} + N. \tag{14.1}$$

We obtain the least squares solution to Equation (14.1) by multiplying it by the complex conjugates $X_1^*$ and $X_2^*$, respectively, and taking expected values (in a loose sense; what we do is, of course, to apply the spectral estimators from Section 10.3), which gives us the system of equations

$$\begin{cases} \mathrm{E}\left[YX_1^*\right] = G_{y1} = \mathrm{E}\left[X_1 X_1^* Hy1\right] + \mathrm{E}\left[X_2 X_1^* H_{y2}\right] + \mathrm{E}\left[NX_1^*\right] \\ \mathrm{E}\left[YX_2^*\right] = G_{y2} = \mathrm{E}\left[X_1 X_2^* Hy1\right] + \mathrm{E}\left[X_2 X_2^* H_{y2}\right] + \mathrm{E}\left[NX_2^*\right] \end{cases} \tag{14.2}$$

If we assume the noise $n(t)$ to be uncorrelated with $x_1(t)$ and $x_2(t)$, this yields

$$\begin{cases} G_{y1} = G_{11} H_{y1} + G_{21} H_{y2} \\ G_{y2} = G_{12} H_{y1} + G_{22} H_{y2} \end{cases} \tag{14.3}$$

since the cross-spectra $G_{n,x_1}$ and $G_{n,x_2}$ will approach zero as we take the expected values (i.e., we take many averages).

To solve the system of Equation (14.3) we could elaborate and find a way to diagonalize it to produce a system where each of the $H_{y1}$ and $H_{y2}$ are separated into one row. This is usually called *Gaussian elimination*, and will indeed be used in Section 14.2, as it was a historic way of solving MIMO systems. Today, however, we have good software tools for linear algebra through MATLAB/Octave and similar software. A direct matrix formulation and solution is therefore of greater interest, and we will turn to such a formulation to find the solution to Equation (14.3).

### *14.1.2 The 2-input/1-output System – matrix notation*

Equation (14.1) can be reformulated in matrix notation as

$$Y = \lfloor H \rfloor \{X\} + \{N\} \tag{14.4}$$

where $Y$ is a scalar because we have only one output (in general MIMO cases it will be a column vector, see Equation 14.11). Furthermore, with this formulation in the case of two inputs and one output, in Equation (14.4) $\lfloor H \rfloor$ is a row vector ($1 \times 2$), equal to

$$\lfloor H \rfloor = \lfloor H_{y1} \; H_{y2} \rfloor \tag{14.5}$$

and $\{X\}$ is a column vector ($2 \times 1$) equal to

$$\{X\} = \begin{Bmatrix} X_1 \\ X_2 \end{Bmatrix} \tag{14.6}$$

With these definitions Equation (14.4) is equivalent to Equation (14.1).

The least squares solution to Equation (14.4) in matrix form can now be formulated by multiplying Equation (14.4) by the matrix equivalent of the complex conjugate of $X$, namely the *Hermitian transpose*, see Appendix D. Instead of taking the complex conjugate alone, the Hermitian transpose, as the name implies, transposes *and* takes the complex conjugate. We then get

$$Y \lfloor X^* \rfloor = \lfloor H \rfloor \{X\} \lfloor X^* \rfloor + \{N\} \lfloor X^* \rfloor . \tag{14.7}$$

Averaging each term separately then yields the result

$$\lfloor G_{yx} \rfloor = \lfloor H \rfloor [G_{xx}] + \lfloor G_{nx} \rfloor . \tag{14.8}$$

The cross-spectrum matrix $[G_{nx}]$ in Equation (14.8) will of course approach zero in the averaging process. We therefore end up with the matrix equation

$$\lfloor G_{yx} \rfloor = \lfloor H \rfloor [G_{xx}] \tag{14.9}$$

which we then have to solve. Assuming the input cross-spectral matrix $[G_{xx}]$ can be inverted, the solution is the MISO $H_1$ *estimator*

$$\lfloor \hat{H}_1 \rfloor = \lfloor \hat{G}_{yx} \rfloor [\hat{G}_{xx}]^{-1} \tag{14.10}$$

where we put hats on the variables to emphasize that they are all estimates (calculated from a measurement).

There is a potential issue here: how do we know that we can invert $[G_{xx}]$? From linear algebra we know (see Appendix D) that a matrix is invertible if it has full rank, i.e., if all rows and columns are independent. This is equivalent to saying its determinant is nonzero. This is obtained in our case, if $[G_{xx}]$ is formed by averaging together several independent intermediate results $X \cdot X^*$. Two things must then be fulfilled, namely

- each element $X_m$ in the averaging process must be *independent*, and
- more than two (in this case with two inputs) independent averages must be made when forming the matrix $[G_{xx}]$.

The first point is fulfilled if we use two independent sources driving the shakers, provided the shakers are strong enough to provide forces proportional to the voltage signals by which they are fed, see Section 14.5. The second point is hardly a problem in practice, as we will always need to take more averages than we have inputs to the system, in order to make sure the cross-spectrum matrix $G_{nx}$ approaches zero.

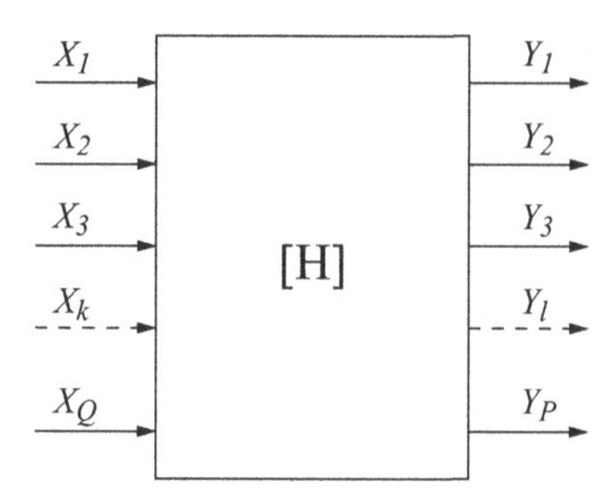


**Figure 14.2** General multiple-input-multiple-output (MIMO) system. Each output signal is caused by a linear combination of the input signals, but not by any of the other output signals

### *14.1.3 The $H_1$ Estimator for MIMO*

The MISO system (of which the 2-Input/1-Output was a small example) is easily extended into the general multiple-input/multiple-output, MIMO, system. For the MIMO system, any output signal is comprised of the contribution of a number of input signals. In structural dynamics (experimental modal analysis), for example, multiple-input models are used for the measurement of frequency response functions between input forces and response accelerations, when several dynamic shakers are mounted on the structure. Multiple-input models are also frequently used in noise source identification in acoustics, where dominant sources of disturbing noise are identified, see Chapter 15.

The general multiple-input/multiple-output (MIMO) system can be described as a general system with a number of input signals $x_q, q = 1, 2, \ldots, Q$, and a number of output signals $y_p$, $p = 1, 2, \ldots, P$, where we drop the time notation for simplicity. Each output signal is assumed to be caused only by the input signals. If there is a causal relationship between the output signals, then the model has to be redefined.

A general MIMO system is depicted in Figure 14.2. If the system is linear and time invariant, and we assume noise is contaminating only the measured output signals, we can define the system frequency response matrix $[H(f)]$ of the MIMO system so that

$$\{Y\} = [H]\{X\} + \{N\} \tag{14.11}$$

In Equation (14.11) $\{Y(f)\}$ is a column vector with dimension $(P, 1)$, $\{X(f)\}$ is a column vector with dimension $(Q, 1)$, and $[H(f)]$ is a matrix with dimension $(P, Q)$. Thus an individual element $H_{pq}(f)$ is the frequency response between input $x_q$ and output $y_p$. Row number $p$ in $[H]$ contains the frequency responses that sum up into output signal $y_p$, and a column number $q$ in $[H]$ contains the frequency responses for input signal $x_q$.

By multiplying Equation (14.11) by $\{X\}^H$, the Hermitian transpose of $\{X\}$, and take expected values of each term, we obtain

$$\left[G_{yx}\right] = [H]\left[G_{xx}\right] + \left[G_{nx}\right] \tag{14.12}$$

which is a least squares solution.

If we assume that all the noise sources in $\{N\}$ are independent of the measured input signals, the cross-spectrum matrix $[G_{nx}]$ in Equation (14.12) will be zero. Post-multiplying both sides of the equation by the inverse of $[G_{xx}]$, gives us the *MIMO $H_1$ estimator* as

$$\left[\hat{H}_1\right] = \left[\hat{G}_{yx}\right]\left[\hat{G}_{xx}\right]^{-1}. \tag{14.13}$$

Note that this equation has to be solved for each frequency of interest. The solution according to Equation (14.13) is possible if the matrix $[G_{xx}]$ is not singular, i.e., its determinant does not equal zero at any frequency. In theory, this means that no ordinary coherence between any two input signals can be equal to unity. In practice, as the ordinary coherence between two inputs reaches unity, there will be numerical problems with the matrix inversion.

As matrix inversion is rather inefficient computationally, in general the system in Equation (14.13) is better solved by Gaussian elimination. We will discuss solving Equation (14.13) in Section 14.1.5, but first we will define a suitable coherence function for MIMO estimation.

**Example 14.1.1** *Write a MATLAB/Octave script which computes a MIMO frequency response matrix based on 2 input signals and 2 output signals. The MIMO FRF matrix will be stored in a 3D matrix similar to the spectral matrices described in Section 10.7. We assume we have already computed the input cross-spectral matrix Gxx in 3D form, and the input-output cross-spectral matrix in 3D matrix Gyx, as described in Section 10.7.2.*

*The first thing we should note is that matrix inversion, using the MATLAB/Octave command* ***inv****, is not recommended. Instead we will use Gaussian elimination, which in MATLAB/Octave, in this case where the matrix to invert is on the right-hand side of the multiplication matrix, is simply accomplished by the slash, '/' operator, see Appendix D. The following code then produces the H1 estimator in a variable with the same name.*

```
% First find sizes
[Nf,D,R]=size(Gyx);
% Preallocate H1
H1=zeros(Nf,D,R);
% Loop through frequencies
for f=1:Nf
  Gxxf=squeeze(Gxx(f,:,:));
  Gyxf=squeeze(Gyx(f,:,:));
  H1(f,:,:)=Gyxf/Gxxf;
end
```

*End of example.*

### *14.1.4 Multiple Coherence*

If we look at the system in Figure 14.1 we realize that for multiple-input systems, the ordinary coherence functions calculated between each of the inputs and the output, will generally be less than unity. If, for example, the two input signals are uncorrelated and have approximately the same spectral densities, the two frequency response functions are of the same magnitude, and the contaminating noise is zero, each of the two ordinary coherence functions will be approximately 0.5. Thus the ordinary coherence function is not very useful for multiple-input systems.

For MIMO systems we define the *multiple coherence function* $\gamma^2_{y_p:x}$ for output signal $y_p$ in a similar way to which the ordinary coherence function was defined for the single-input system, namely by

$$\gamma^2_{y_p:x}(f) = \frac{G_{uu}(f)}{G_{y_p,y_p}(f)}. \tag{14.14}$$

where $u$ is the coherent output from the linear systems as in Figure 14.1.

The notation $y_p : x$ in the index is read '$y_p$ given (all inputs) $x$'. As for the single-input case, the estimation of the optimum $H$-systems, maximizes the power in $y_p$ due to the input signals and puts the remaining power in the extraneous noise signal $n_p(t)$. Thus, the multiple coherence function is interpreted, in analogy with the ordinary coherence for the single-input case, as the part of the output power spectrum $G_{y_p,y_p}$ which is linearly dependent on any of the inputs $x$. As for the ordinary coherence, evidently

$$0 \leq \gamma^2_{y_p:x}(f) \leq 1 \tag{14.15}$$

where the multiple coherence function, $\gamma^2_{y_p:x} = 1$ if and only if there is no output noise, i.e., if $n_p = 0$.

For MIMO systems, there will obviously be one multiple coherence function for each output of the system. In order to find an expression which we can use in practice to estimate the multiple coherence function for output signal $y_p$, we formulate the coherent output spectrum, $U(f)$, from Figure 14.1, taking row $p$ from the full matrix $[H]$, denoted $\lfloor H_p \rfloor$ by

$$U = \lfloor H_p \rfloor \{X\} + \{N\} \tag{14.16}$$

The Hermitian transpose of $U$, or, since it is a scalar, the complex conjugate, is

$$U^H = \lfloor X^* \rfloor \lfloor H_p \rfloor^H + \lfloor N^* \rfloor \tag{14.17}$$

Multiplying these two last equations together, and taking the expected value of each term we obtain the expression for the coherent output power spectrum $G_{uu}$ in Equation (14.18)

$$G_{uu} = \lfloor H_p \rfloor [G_{xx}] \lfloor H_p \rfloor^H . \tag{14.18}$$

Putting the result in Equation (14.18) into Equation (14.14) we obtain the equation for the multiple coherence estimator by

$$\hat{\gamma}^2_{y_p:x} = \frac{\lfloor \hat{H}_p \rfloor [\hat{G}_{xx}] \lfloor \hat{H}_p \rfloor^H}{\hat{G}_{y_p,y_p}}. \tag{14.19}$$

We can also use the expression for the $H_1$ estimator in Equation (14.13) in Equation (14.19), and after some simplification we obtain

$$\hat{\gamma}^2_{y_p:x} = \frac{[\hat{G}_{y_p,x}] \left( [\hat{G}_{xx}]^{-1} \right)^H [\hat{G}_{y_p,x}]^H}{\hat{G}_{y_p,y_p}} \tag{14.20}$$

from which it is obvious that the multiple coherence, like all other results for MIMO systems, as we have noted above, can be computed from averaged auto-and cross-spectra. The estimator in Equation (14.19) is somewhat faster to compute if the frequency response matrix $[\hat{H}]$ already exists, because it does not involve any matrix inversion. Of course, if the multiple coherence is computed simultaneously with the frequency response, the inverse of the input cross-spectral matrix could be stored as an intermediate result, and then either estimator could be used with approximately the same computation effort.

**Example 14.1.2** *Write a MATLAB/Octave script to produce the multiple coherence, using the data from Example 14.1.1.*

*We will use the first definition in Equation (14.19) and the following code produces the multiple coherence in columns in variable Cm.*

```
[Nf,D,R]=size(H);
Cm=zeros(Nf,D);
for d = 1:D    % Loop responses
    for f = 1:Nf                % Loop frequencies
        Gxxf=squeeze(Gxx(f,:,:));
        Hf=squeeze(H(f,d,:));
```

```
        Hf=Hf(:).';              % Force to row
        Cm(f,d)=real((Hf*Gxxf*Hf')/Gyy(f,d));
    end
end
```

*End of example.*

## *14.1.5 Computation Considerations for Multiple-input System*

All the discussions above assume that the input signals are not completely coherent. This in principle means that the signals used to average together the input- and output auto- and cross-spectral matrices have to be formed by independent averages. This leads to the conclusion that the data must, in general, not be periodic, since a sine wave is completely coherent with any other sine wave of the same frequency. We will present two exceptions to this later in this chapter; (i) so-called *periodic random*, see Section 14.4.3, and (ii) the multiphase stepped-sine procedure, see Section 14.4.4. For the general case, however, the input signals must not be periodic. Transient signals will in general not be valid either, except for the case where the transients consist of (independent) burst random signals.

A direct effect of the procedure of conditioning the input signals as described above, is that the number of averages used for calculation of the spectral matrices, has to be larger than the number of inputs. This fact is analogous with the single-input case, where the coherence function is not defined for the first average, as it would equal unity, even with no actual correlation between the input and output signals. Similarly, with several inputs, enough statistically independent input records have to be combined into the auto- and cross-spectra, in order for all the ordinary coherence functions between the inputs to be well defined.

This requirement is of course also necessary when using matrix inversion to solve a MIMO system as in Equation (14.13). In linear algebra terms, the rank of the input cross-spectral matrix $[G_{xx}]$ equals unity after one average, two after two averages, and so on until the number of averages is larger than or equal to the number of inputs, after which the rank of the matrix equals the number of inputs (which is also the size of the matrix $[G_{xx}]$).

## *14.1.6 The $H_v$ Estimator*

When estimating MIMO systems on mechanical systems, the $H_1$ estimator is often not optimal. In this application the MIMO system is always defined as response over force, and the $H_1$ estimator will optimize for contaminating noise in the acceleration (output) signals. Around the natural frequencies of the structure, however, the acceleration is large and thus the extraneous noise is small relative to the acceleration signal. At the same frequencies, the force signals are often low, due to limitations in the excitation system (shaker, amplifier, and stinger), which makes it difficult to excite the structure. Thus, the inputs of the MIMO system, the forces, are more prone to be contaminated by extraneous noise than the outputs, which calls for an estimator equivalent to the $H_2$ estimator for the single-input case. This proves tricky, however, as it can easily be shown that an $H_2$ estimator for multiple-input systems can be defined only for the special case where the number of response signals equals the number of forces, which is, of course, not generally the case. Because the $H_2$ estimator for MIMO systems is of limited interest, we leave the proof of this fact to Problem 14.2.

The $H_v$ *estimator* was developed in the early 1980s (Mitchell 1982; Rocklin *et al.* 1985) to yield better results in general cases with contaminating noise on both input and output. The $H_v$ estimator assumes that all the extraneous noise sources are incoherent with the true input and output signals. Then, the linear system is defined by the matrix equation

$$\{N\} = [H]\{X\} + \{Y\} \tag{14.21}$$

where $\{n\}$ is an error vector consisting of all extraneous noise. Next we form the *error autospectrum matrix*, $G_{nn}$ by multiplying Equation (14.21) by its complex conjugate and taking the expected value. This results in

$$[G_{nn}] = \left[G_{yy}\right] + [H]\left[G_{xx}\right][H]^H - [H]\left[G_{xy}\right] - \left[G_{yx}\right][H]^H . \tag{14.22}$$

Equation (14.22) can be shown to be equal to the composed matrix

$$[G_{nn}] = \left[ I \mid H \right] \begin{bmatrix} G_{yy} & G_{yx} \\ G_{xy} & G_{xx} \end{bmatrix} \begin{bmatrix} I \\ H^H \end{bmatrix} \tag{14.23}$$

Equation (14.23) is decomposed by the eigenvalue decomposition

$$[G_{nn}] = [U]^H [\Lambda] [U] \tag{14.24}$$

where the matrix $[U]$ is the eigenvector matrix, and $[\Lambda]$ is a diagonal matrix containing the eigenvalues on the diagonal. It can be shown with some advanced linear algebra, that the solution to minimizing the trace of Equation (14.24), which is a so-called *Rayleigh quotient*, is found by choosing the eigenvector corresponding to the lowest eigenvalue in $[\Lambda]$. This is the $H_v$ estimator. For the single-input/single-output case, the $H_v$ estimator reduces to the geometric mean of the $H_1$ and $H_2$ estimators

$$\hat{H}_v = \frac{G_{yx}}{\left|G_{yx}\right|} \sqrt{\frac{G_{yy}}{G_{xx}}} = \sqrt{\hat{H}_1 \hat{H}_2}. \tag{14.25}$$

### *14.1.7 Other MIMO FRF Estimators*

It can be shown (White *et al.* 2006) that the $H_v$ estimator in Section 14.1.6 is a special case of a general maximum likelihood (ML) estimator, for the case where the input and output noise at each frequency are equal. It can also be shown that the minimization problem for noise on both input and output is only soluble for general random inputs, if at least the ratio of the two noises are known, which is rarely the case in our applications. The $H_v$ estimator, is thus not a very good estimator for our purposes, since we can assume on mechanical systems with high dynamic range, that at the critical frequencies (typically the natural frequencies, and perhaps to some extent the antiresonances), we have dominant extraneous noise in *either* of the two signals (input or output), but not in both simultaneously. It therefore seems as though we are in vain trying to find any better estimator than the $H_1$ estimator, for the general case with random inputs.

In addition to the $H_1$ and $H_v$ estimators, which are the estimators commonly available in commercial software for noise and vibration analysis, several other estimators exist, which can sometimes be alternatives if dedicated software is developed. The reason to use such an alternative estimator has to be based on the fact that we want to minimize the bias error in the FRF due to noise on the input signals in addition to the noise on the output signals, because otherwise the $H_1$ is a superior estimator. We will therefore present references to some alternatives that can be used in such cases.

One alternative is the $H_c$ estimator described in Section 13.2.3, which also be defined for MIMO situations. However, the price is that one extra measurement channel per force is allocated, which makes it less attractive. A second approach is the $H_\alpha$ estimator suggested in Antoni *et al.* (2004) which makes use of theory of cyclostationarity to find an estimator which is asymptotically unbiased even in the case of input noise. A third alternative is to use time domain averaging and periodic excitation signals. This alternative will be demonstrated in Section 14.5.3.

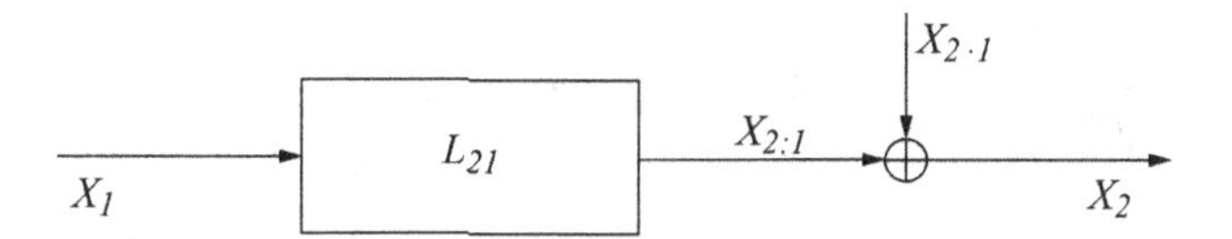


**Figure 14.3** L-system with conditioned signal $x_{2\cdot1}$ where the dependence in signal $x_2$ with signal $x_1$ is removed

## 14.2 Conditioned Input Signals

Much of the theory of estimating MIMO frequency response functions from random data was developed in early editions of Bendat and Piersol (2000). In those days matrix notation was not commonly used in this field, so instead, Bendat and Piersol used, so to speak, a manual Gaussian elimination. As part of this development, some concepts which are available in many commercial software packages for noise and vibration analysis, such as *conditioned signals* and *partial coherence,* were established. We will thus briefly discuss the approach used to develop these concepts. Although we will explain these ideas similarly to the original presentations, they can conveniently be formulated in modern matrix terminology as well, Smallwood (1995).

We will present this technique by using the 2-input/1-output system in Figure 14.1. When there is correlation between the inputs $x_1(t)$ and $x_2(t)$, it is possible and convenient to define new, uncorrelated, so-called *conditioned signals*. These 'fictive' signals are obtained by subtracting the dependence of each preceding channel from each input signal, starting with the second input signal. Thus the conditioned signals have zero correlation with each other. The concept of conditioned signals is an important tool to interpret and understand multiple-input systems. The linear systems relating one conditioned signal with another conditioned signal, or with the measured output, we term $L$-systems (Bendat and Piersol 1993, 2010). We will now deduce the relationships for conditioned signals, by first studying the system in Figure 14.3 where we look at the dependence between signal $x_1$ and $x_2$.

In Figure 14.3, we see that if we treat the two input signals as an input and an output signal to the system $L_{21}$, we get two new signals. The signal (spectrum) $X_{2:1}$ is the part of $X_2$ that is linearly dependent on $X_1$ and the 'remaining' signal, the conditioned input signal $X_{2\cdot1}$, where the index is read '2 with $x_1$ removed', is the part of $X_2$ which is uncorrelated with $X_1$. The relationship between these signals can easily be found by the simple theory for linear single-input/single-output systems discussed in Chapter 13.

From Figure 14.3 we thus have

$$X_2 = X_1 L_{21} + X_{2\cdot1}. \tag{14.26}$$

We also, using the $H_1$ estimator of Equation (13.6), that

$$L_{21} = \frac{G_{21}}{G_{11}} \tag{14.27}$$

and thus

$$X_{2\cdot1} = X_2 - \frac{G_{21}}{G_{11}} X_1. \tag{14.28}$$

The relationship for the conditioned input autospectrum $G_{22\cdot1}$ of the conditioned signal $X_{2\cdot1}$, can be found by multiplying Equation (14.28) by its complex conjugate and taking the expected value

$$\begin{aligned} G_{22\cdot1} &= E\left[X_{2\cdot1}X_{2\cdot1}^*\right] = \\ E\left[\left(X_2 - \frac{G_{21}}{G_{11}}X_1\right)\left(X_2^* - \frac{G_{12}}{G_{11}}X_1^*\right)\right] &= \ldots = G_{22} - \frac{|G_{12}|^2}{G_{11}}. \end{aligned} \tag{14.29}$$

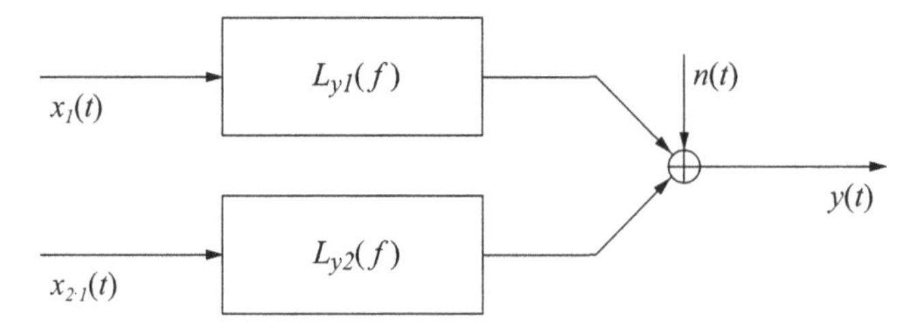


**Figure 14.4** 2-input/1-output system with uncorrelated inputs

Combining Equation (14.29) with the definition of the ordinary coherence in Equation (13.22) we obtain the simplified formulation for the conditioned input autospectrum in Equation (14.30)

$$G_{22\cdot1} = (1 - \gamma_{21}^2)G_{22}. \tag{14.30}$$

We could also have obtained this equation directly, using Equation (13.29), using the fact that our conditioned signal $x_{2\cdot1}$ is analogous to the extraneous noise in the 'standard' single-input case. It should be noted that we simplify the notation by using only the number in indexes, when the numbers stand for input signals.

It is important to note that the terms on the right-hand side of Equation (14.30) only contain functions that we already know how to calculate from the measured signals $x_1$ and $x_2$, using auto- and cross-spectra of the two signals. It is also important to realize, that the two signals $x_1$ and $x_{2\cdot1}$, are uncorrelated, that is, the ordinary coherence function between these two signals is by definition equal to zero.

Using the conditioned input signals, we can now transform the 2-input/1-output system in Figure 14.1 with possibly *correlated inputs* into a new 2-input/1-output system with *uncorrelated inputs*, as depicted in Figure 14.4. The frequency response functions in this model are also referred to as *L-systems*. The benefit of this new system is that the two unknown $L$-systems are very easily obtained using Equation (13.6), treating the two-input system as two separate single-input systems, because the input signals are now uncorrelated. Thus, we find that

$$Y = L_{y1}X_1 + L_{y2}X_{2\cdot1} + N \tag{14.31}$$

where

$$L_{y1} = \frac{G_{y1}}{G_{11}} \tag{14.32}$$

and

$$L_{y2} = \frac{G_{y2\cdot1}}{G_{22\cdot1}}. \tag{14.33}$$

In Equation (14.33) we have used the conditioned cross-spectrum $G_{y2\cdot1}$ which is found by multiplying Equation (14.31) by $X_{2\cdot1}^*$ and taking the expected value, which by using Equation (14.28) above yields

$$G_{y2\cdot1} = \mathrm{E}\left[YX_{2\cdot1}^*\right] = \mathrm{E}\left[Y\left(X_2^* - L_{21}^*X_1^*\right)\right] = \mathrm{E}\left[YX_2^*\right] - L_{21}^*\mathrm{E}\left[YX_1^*\right]. \tag{14.34}$$

Thus we have

$$G_{y2\cdot1} = G_{y2} - \frac{G_{12}}{G_{11}}G_{y1} \tag{14.35}$$

which again is a formulation where only 'standard' auto- and cross spectra are needed.

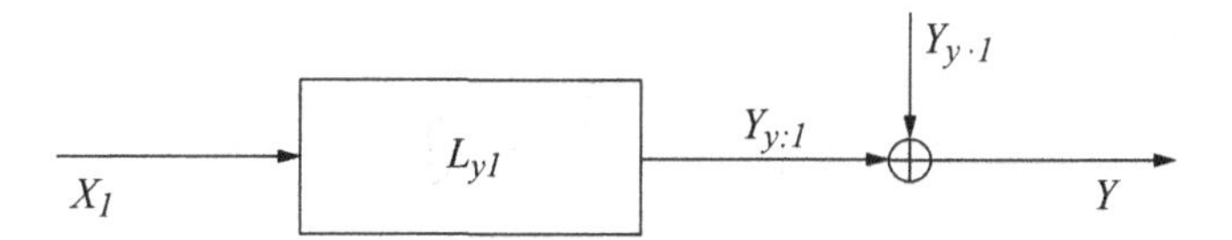


**Figure 14.5** Conditioned input/output system $L_{y1}$ obtained by looking at input and output signals $x_1$ and $y$

## 14.2.1 Conditioned Output Signals

To find relations for the multiple coherence and partial coherence functions below, we will also need a set of conditioned output signals. These are derived by using systems as depicted in Figure 14.5.

In Figure 14.5 we have put the part of the output signal, $y$, which is not dependent on the input $x_1$, into the new conditioned output signal $y_{y\cdot 1}$. The conditioned output power spectrum of $y_{y\cdot 1}$ denoted by $G_{yy\cdot 1}$ is then obtained by the following equation, which follows straight from Equation (13.29),

$$G_{yy\cdot 1} = \left(1 - \gamma_{y1}^2\right) G_{yy} \tag{14.36}$$

where $\gamma_{y1}^2$ is the ordinary coherence between input signal $x_1$ and the output $y$.

## 14.2.2 Partial Coherence

Similarly with Figure 14.5, we can form an equivalent system for the conditioned input signal $x_{2\cdot 1}$ and the conditioned output signal $y_{y\cdot 1}$ as depicted in Figure 14.6. We use the factorial symbol, !, to denote that the dependence on both input $x_1$ and $x_2$ are removed. This corresponds to the standard use of this symbol, for example $4! = 4 \cdot 3 \cdot 2 \cdot 1$.

To obtain the relationship for the output signal of the linear system in Figure 14.6, we first note that this output signal is analog with the coherent output spectrum in Equation (13.28). This gives us

$$G_{u_2u_2} = \gamma_{y2\cdot 1}^2 G_{yy\cdot 1} \tag{14.37}$$

From the definition of ordinary coherence we obtain

$$\gamma_{y2\cdot 1}^2 = \frac{\left|G_{y2\cdot 1}\right|^2}{G_{22\cdot 1} G_{yy\cdot 1}} \tag{14.38}$$

From these two relationships we obtain the final expression for the output of the linear system $L_{y2}$:

$$G_{u_2u_2} = \left|\frac{G_{y2\cdot 1}}{G_{22\cdot 1}}\right|^2 G_{22\cdot 1} = \gamma_{y2\cdot 1}^2 G_{yy\cdot 1} \tag{14.39}$$

where the new function $\gamma_{y2\cdot 1}^2$ is called the *partial coherence* of output $y_{y\cdot 1}$ with conditioned input $x_{2\cdot 1}$. Both functions $\gamma_{y2\cdot 1}^2$ and $G_{yy\cdot 1}$ can be computed from the auto- and cross-spectral functions by the equations developed above. Similarly, the partial coherence between any two conditioned signals can be

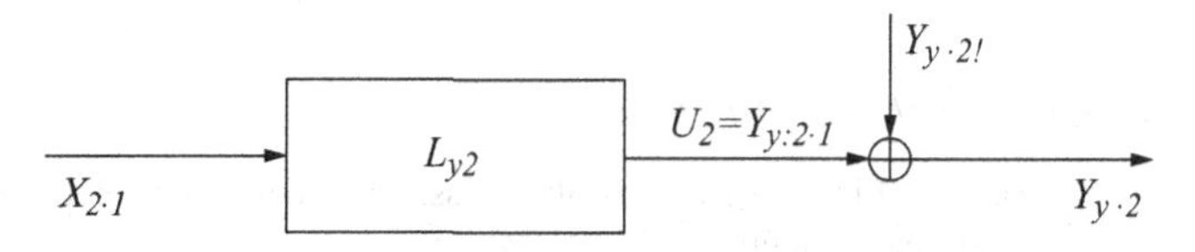


**Figure 14.6** Conditioned input and output signals for the second (conditioned) input, and for the output with the dependence with signal $x_1$ removed

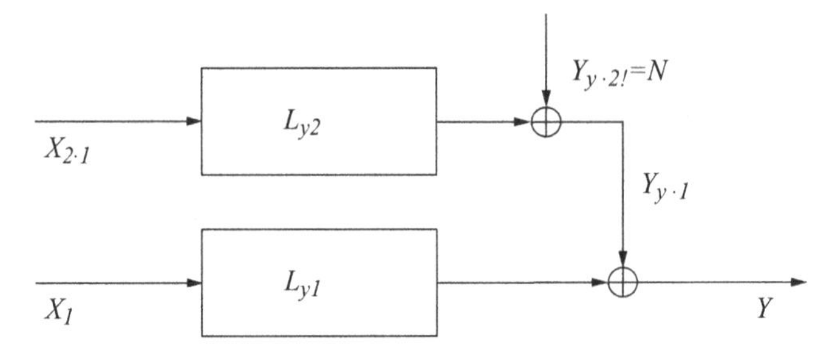


**Figure 14.7** 2-input/1-output system equivalent to the system in Figure 3.15, but using the different conditioned signals

formed. The partial coherence, furthermore, is analogous to the ordinary coherence for the two signals, but is referred to as partial when the two signals are conditioned. Note that in this context also $x_1$, although unchanged, is a conditioned signal. In many physical cases the direct interpretation of the partial coherence is difficult. In Section 15.3 on noise source identification we will discuss this in more detail.

Using the different systems described above we can now transform our 2-input/1-output system with correlated inputs, to the equivalent in Figure 14.7.

Figure 14.7 shows the principle behind the transformation of the multiple-input system ($H$-) system with correlated inputs, into the equivalent ($L$-) system with uncorrelated inputs, and with conditioned outputs. When the correlation between the inputs is removed, the $L$-systems are given by the same equations as for the single-input case. This can easily be extended to more inputs. In Section 14.2.6 we will extend this to an arbitrary number of input signals.

### *14.2.3 Ordering Signals Prior to Conditioning*

The order in which the original, correlated signals are conditioned plays an important role in the interpretation in many applications. For general multiple-input systems with more than two inputs, Bendat and Piersol (1993) recommends the inputs be ordered in one of the following ways:

1. according to the physical causal relationships (if known), or
2. so that the input with the highest ordinary coherence with the output is placed first, and then the inputs should be ordered with descending ordinary coherence functions.

A problem in ordering the inputs according to item 2. when the causal relationships are not known, is that the highest coherence value is often found for different input signals at different frequencies. In some cases it may therefore be necessary to make the conditioning and subsequent analysis using several different orders among the inputs, and trying to interpret the results accordingly. This could be done for only some frequencies in order to reduce the amount of computations.

### *14.2.4 Partial Coherent Output Power Spectra*

Using the conditioned system in Figure 14.7, it is also possible to define the partial coherent output power spectra due to the conditioned inputs $x_1$ and $x_{2\cdot 1}$ as

$$G_{y:1} = \gamma_{y1}^2 G_{yy} \tag{14.40}$$

and

$$G_{y:2\cdot 1} = \gamma^2_{y2\cdot 1} G_{yy\cdot 1} \tag{14.41}$$

which each tells how much of the power in the output power spectrum $G_{yy}$ that relates linearly to each conditioned input. These spectra were often used in noise source identification before the methods discussed in Chapter 15 were developed, but are usually replaced today with the latter methods.

### 14.2.5 Backtracking the H-systems

After having calculated the conditioned input and output spectra and the $L$-systems, it is easy to go back to the original $H$-systems. This was in fact the recommended procedure to calculate the $H$-systems, because it was considered much more computationally efficient than inverting the input spectrum matrix as discussed in Section 14.1.5. Modern software, such as MATLAB and Octave, for example, has changed this so that today there is no computational advantage in computing MIMO models like explained here. There is however, in my opinion, still an educational value in explaining it this way.

The procedure to backtrack the 'original' frequency response functions follows easily by redrawing the 2-input system as in Figure 14.8 (Allemang *et al.* 1984; Bendat and Piersol 1993, 2010). From a comparison of Figure 14.8 with Figure 14.1 and Figure 14.3, it is apparent that the $H$ system $H_{y2}$ must be equal to $L_{y2}$ since the conditioned signal with dependence on the first input signal removed, $x_{2\cdot 1}$, and the original second signal, i.e., $x_2$, must both go through the same linear system. Thus from the figure we have

$$H_{y2} = L_{y2} \tag{14.42}$$

and

$$L_{y1} = H_{y1} + L_{21} L_{y2} \tag{14.43}$$

Rewriting Equation (14.43) we then get the relationship used to calculate the system $H_{y1}$ by

$$H_{y1} = L_{y1} - L_{21} L_{y2} \tag{14.44}$$

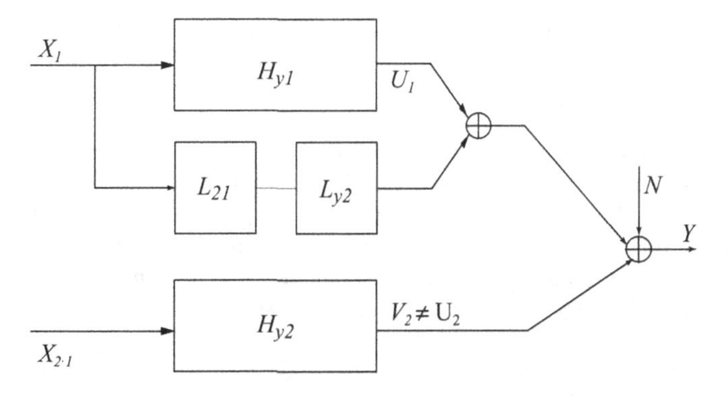


**Figure 14.8** Equivalent diagram for backtracking the $H$-systems from $L$-systems. The system in this figure is equivalent to the system in Figure 14.1, while some of the included variables are taken from the conditioned signals. The figure shows that the $L$-system $L_{y2}$ is equal to the $H$-system $H_{y2}$ since it is obvious that the conditioned signal $x_{2\cdot 1}$ goes through the same linear system as the original signal $x_2$

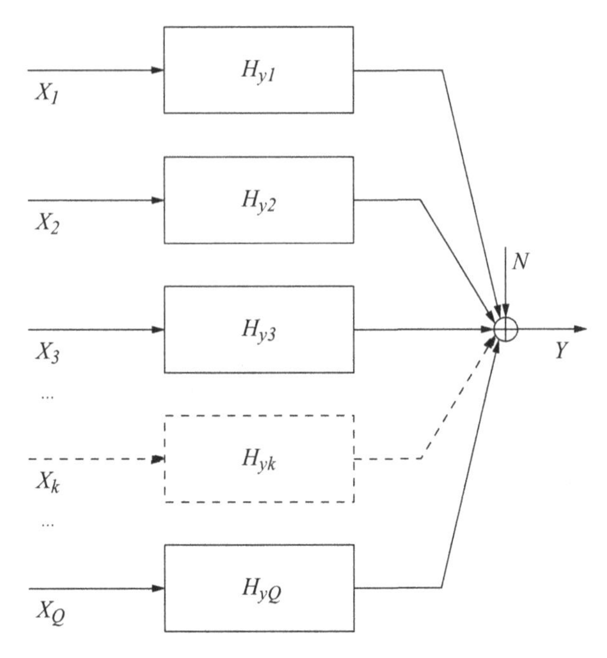


**Figure 14.9** General multiple-input/single-output (MISO) system, with (possibly) correlated inputs

### *14.2.6 General Conditioned Systems*

When analyzing systems with more than two inputs, the method we have used in the previous section is easily extended. We start with a MISO model with inputs that are correlated, as in Figure 14.9. As we discussed above, we then transform this system into a system with uncorrelated inputs as in Figure 14.10. If there is substantial correlation between some or all the inputs, the original input signals should be ordered as mentioned in Section 14.2.3.

## 14.3 Bias and Random Errors for Multiple-input Systems

We will not present any quantitative random or bias errors for the MIMO case, but restrain the discussion to some important qualitative aspects. Approximate formulas for MIMO estimates can be found in Bendat and Piersol (2010), and some errors have exact solutions developed in Antoni and Schoukens (2007, 2009).

The same arguments as we discussed for SISO systems in Section 13.5 applies to MIMO systems. The spectral densities in the auto- and cross-spectral matrices discussed in the present chapter should therefore be computed with half sine window and 67% overlap, to produce the least possible bias and random errors. The difference is small, however, compared with using the more traditional Hanning window and 50% overlap. The bias errors should be minimized by the same procedure with increasing blocksize (decreasing frequency increment) as discussed for spectral analysis and SISO FRF estimation in Chapters 10 and 13.

In general, random errors in MIMO estimates are larger than the corresponding errors for SISO systems, which is due to the fact that the input correlation has to be removed in the inversion of the input autospectral matrix. In addition, numerical difficulties with this inversion can lead to noise, as we will show in Section 14.6. It is therefore important to make sure the correlation between the inputs is sufficiently small when using multiple inputs.

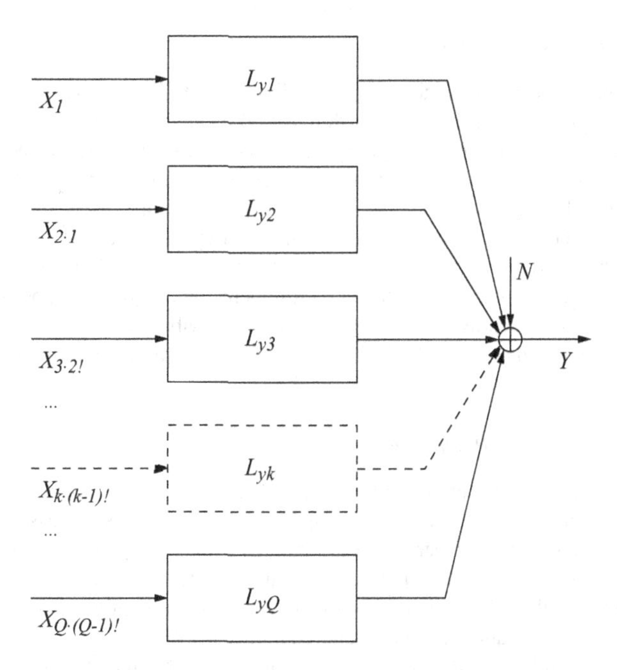


**Figure 14.10** General multiple-input/single-output (MISO) system with uncorrelated conditioned input signals

Finally, the random errors are small if the multiple coherence is unity or very close to unity. It should also be free from noise, as noise in the multiple coherence is an indication of bad signal-to-noise ratio, or possibly numerical problems.

## 14.4 Excitation Signals for MIMO Analysis

When multiple inputs are used to excite a structure in order to estimate frequency response functions by MIMO techniques, certain considerations have to be made with regard to the excitation signals used. As mentioned in Section 14.1.5, a number of independent averages have to be taken when averaging the auto- and cross-spectral matrices. Thus, the excitation signals that are used must be such that the FFT of each time block of data will be independent. This disqualifies some of the popular excitation signals used for single-input estimation, such as for example the chirp signal.

### *14.4.1 Pure Random Noise*

The simplest excitation signal which can be used for multiple-input estimation is pure random, which consists of Gaussian noise. Each input is made up of noise which is incoherent (independent) with all other input signals. The main disadvantage with this excitation signal, as in the single-input case, is that it requires relatively long blocksize and thereby long measurement time in order to reduce the bias error. This makes pure random an inconvenient excitation signal for structures with light damping. However, the pure random signal poses a minimum of requirements on the hardware, and is therefore sometimes the only available signal type. This is the case especially if external noise generators have to be used. Pure random is used with frequency domain averaging.

### 14.4.2 Burst Random Noise

The next signal type is the burst random signal. This transient signal consists of independent, pure random data for the first part of each force signal, followed by a 'silent' period, see Section 13.9.3. The burst length is chosen long enough so that the responses (outputs of the estimated system) decays to near zero, usually of the order of 20–50% of the total record time. Burst random leads to measurements with a small amount of leakage, with a proper selection of the burst-off time so that the response signals actually decay to zero. The main drawback is the poor signal-to-noise ratio. In order to synchronize the data acquisition with the excitation signals, the burst random signals usually have to be generated by the measurement system. This puts some additional requirements on the hardware, but most commercially available multi-channel data acquisition systems for noise and vibration testing today include this type of excitation signal. Averaging is made the same way as for the pure random excitation signal.

### 14.4.3 Periodic Random Noise

A disadvantage with the burst random signal is the decreased signal-to-noise ratio during the silent periods. In order to avoid this, periodic random signals can be used, as discussed for the SISO case in Sections 13.9.4 and 13.9.5. The periodic random signal for multiple-input estimation works in a somewhat different way from the single-input case. It turns out that in order for the input autospectrum matrix to be invertible in this case, it is not enough to have independent phase in each time block, but also the amplitudes at each frequency must be altered independently.

Periodic random is therefore produced by first generating a single time block of pure random data for each force signal. This block is sent to the shaker, and repeated a number of times, until the transient response of the tested structure has decayed, without any data being measured. When the structure is in its steady-state response, a time block of all signals is acquired, FFT of all channels are computed, and auto- and cross-spectra are accumulated. A new, independent time block of random data is then generated for each force signal, which is again repeated a number of times before a new time block is acquired and processed. This procedure is repeated until enough averages have been taken.

Periodic random is the best excitation signal in terms of its increased signal-to-noise ratio. As the signals are periodic within the time window, the periodic random signal has all its energy concentrated on the frequency lines of the FFT so it produces leakage-free spectra with good signal-to-noise ratio. However, a drawback with this signal is the large increase in total measurement time, since each new block has to be repeated several times, and in addition it requires special synchronization between the input channels and the output channels in the hardware. Therefore, this signal type is not very common in commercial systems.

As we will show in Section 14.5, in cases with severe input noise, the periodic random method can be used with a combination of time domain and frequency domain averaging. Although this requires more averages to yield unbiased results, it is a convenient method in this otherwise tricky case. The principle is to repeat each independent block in the periodic random sequence of blocks, and produce a time average of a large amount of such repeated blocks, using those blocks where the transient response has vanished. In order to make the input autospectrum matrix invertible, at least as many independent blocks as there are inputs to the system (in practice, say, twice that number) have to be averaged together, see Section 14.5.

### 14.4.4 The Multiphase Stepped-sine Method (MPSS)

As for the single-input case, a stepped-sine approach can be used also for the multiple-input case (Williams and Vold 1986). In this case, however, care must be taken so that the input cross-spectral matrix is indeed invertible. The special method is called the multiphase stepped-sine method (MPSS) and consists of making several 'sweeps' in frequency, between which the individual phases of the force signals are changed according to a scheme producing independent spectra. These spectra are accumulated into auto- and cross-spectral matrices which can then be solved by Equation (14.13). In order to make the input autospectrum matrix invertible, more 'sweeps' than the number of forces have to be made, making this a rather slow method. MPSS excitation has found an increased interest in the aerospace industry for so-called ground vibration tests, where the modes of whole aircraft are analyzed. Except for small fighter aircraft, which are rather stiff structures, aircraft are normally difficult to measure using broad band excitation. Recently Pauwels *et al.* (2006) presented a faster MIMO swept sine technique by replacing the usual FFT processing by digital filters.

A special consideration has to be made in order to make the forces independent when using the MPSS method. In many cases, the shakers used to excite the structure are too weak to produce exactly the same forcing function that is fed to them by the signal generator voltages. Especially around resonance, the structure responds strongly by its mode, resulting in force patterns that are dependent. In order to avoid this problem, closed-loop control of the shakers should be utilized in the measurement system.

## 14.5 Data Synthesis and Simulation Examples

After the theoretical descriptions of the MIMO estimation methods described in the preceding part of this chapter, we will now illustrate the theory by some simulation examples which will provide insight into the different issues related to MIMO FRF estimation. At the end of this section there is also an example of real data from measurements on a Plexiglas plate, to illustrate some issues which are difficult to simulate.

For all simulation examples we will use the same 2DOF system we used in Chapter 13, with the difference that we now excite both DOFs. The system has two natural frequencies at 100 and 200 Hz, and 1% damping. We use the method from Section 6.5 to produce the output due to each force (since the simulation method calculates the response in one DOF due to excitation in another DOF) and sum the two contributions.

### *14.5.1 Burst Random – Output Noise*

In our first example we use burst random excitation with uncorrelated input forces, and some contaminating extraneous output noise. In Section 13.10.2 we established that 25% burst length together with the chosen blocksize of 2048 samples was sufficient for this system to reduce the leakage to a sufficiently low degree. This will not change by the fact that we now excite two DOFs, so we choose the same burst length in this example. In Figure 14.11, some results of a MIMO estimation, using 50 frequency domain averages, of the two FRFs between both input DOFs and response in DOF 1 are plotted. In (a) and (b) we see the driving point FRF, $\hat{H}_{11}$, in DOF 1 (zoomed in around the first natural frequency in b), and in (c) and (d) we see the second FRF, $\hat{H}_{12}$. In (e) the multiple coherence is shown, and in (f) the ordinary coherence between the two forces, from which we see that the forces are uncorrelated.

The multiple coherence in Figure 14.11 (e) is very close to unity, which means that the output noise level is small enough to not seriously deteriorate the measurement. This is a representative case when measuring for example a freely supported, simple structure in a lab environment, provided appropriate force level and sensors with suitable sensitivities have been chosen. In the zoomed-in plots in (b) and (d), we see the burst random frequency lines as 'o' (rings) overlaid on the true FRF of this synthesized example. The estimates are very nearly perfect, the error remaining is the small random error due to the limited amount of averages, as the $H_1$ estimator removes the bias due to the output noise present in the measurement.

We proceed with a similar example where we add correlation between the two forces, to illustrate how this affects the FRF estimates. In Figure 14.12, the same results as in Figure 14.11 are shown for a case where the correlation between the forces is approximately 0.84 (defined by the coherence function; this means the power is correlated to 84%).

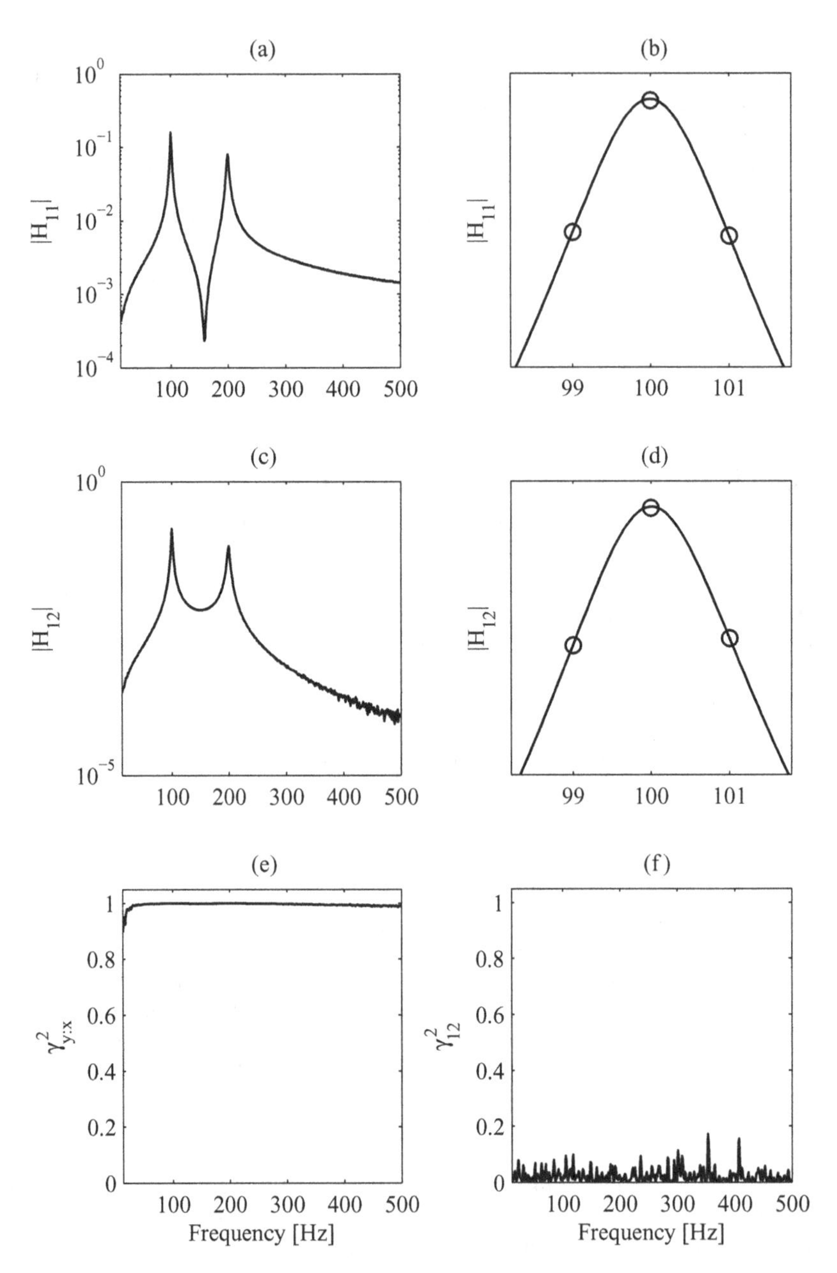


**Figure 14.11** Results from MIMO estimation of a simulation case with two FRFs on a 2DOF system, using 2048 samples blocksize, 50 averages in the frequency domain, and burst random excitation with 25% burst length and the $H_1$ estimator from Equation (14.13). In (a) the magnitude $|H_{11}|$, in (b) the magnitude of the same FRF is zoomed in around the first natural frequency (rings) overlaid on the true FRF (solid); in (c) $|H_{12}|$ is shown and in (d) the same FRF is zoomed in around the first natural frequency (rings) overlaid on the true FRF (solid). The multiple coherence, $\gamma^2_{y:x}$ is shown in (e) and the ordinary coherence between the two forces is shown in (f), in which it is seen that there is no correlation between the two forces. This example shows that burst random excitation produces virtually bias-free estimates in this ideal case. There is a small random error, although difficult to see, but the bias error is negligible. The $H_1$ estimator removes bias due to output noise

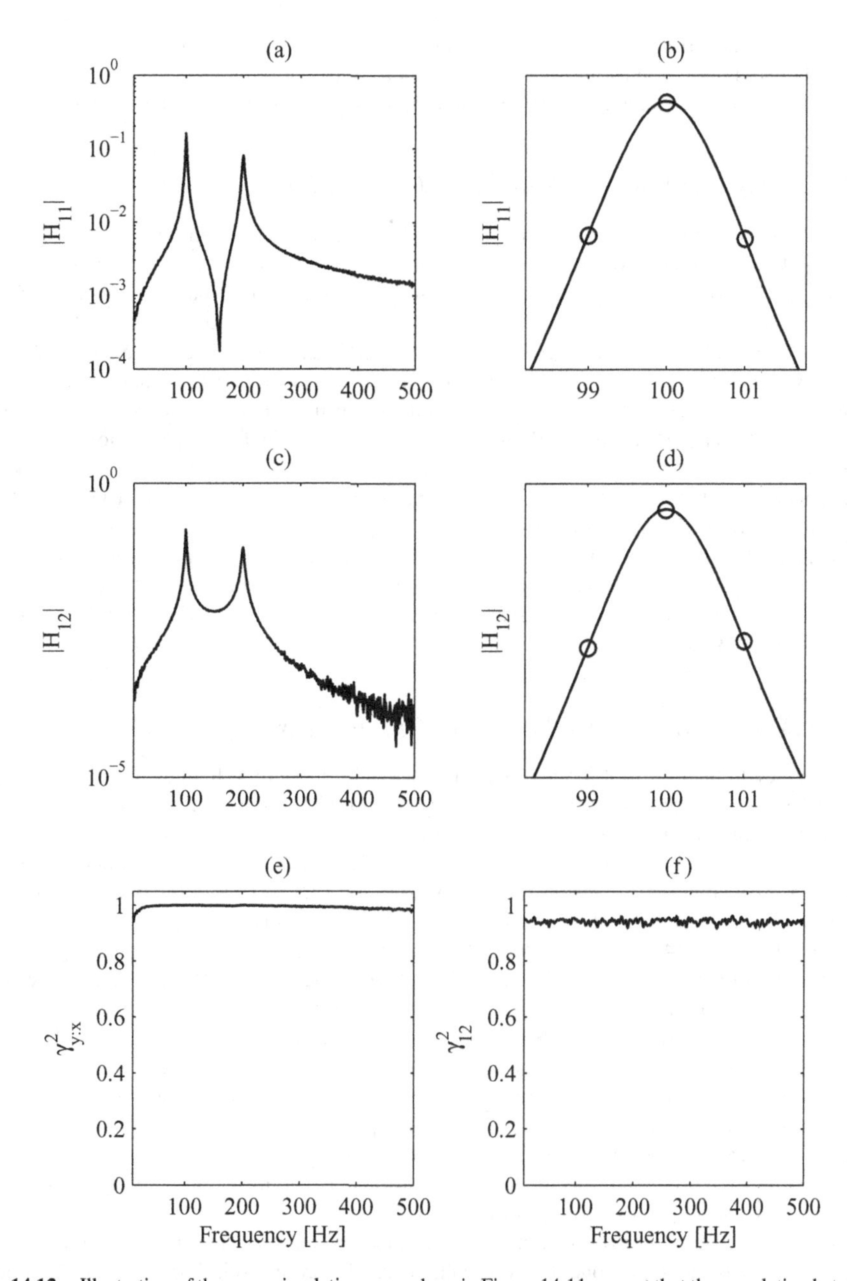


**Figure 14.12** Illustration of the same simulation example as in Figure 14.11, except that the correlation between the forces is now 0.84 as defined by the input coherence in (f). See Figure 14.11 for a description of the different plots. The difference in the results of the $H_1$ estimator in this case is an increased random error most visible in $|H_{12}|$ in (c) at higher frequencies. Even with this high correlation, the FRF estimate is relatively good, but the example shows that to obtain the best results in MIMO FRF estimation, the input correlation should be kept low

The example with correlation between the forces shows that the FRF estimates are affected when the input correlation is high. The estimates show some more sensitivity to noise when the signal-to-noise ratio is poor. Also see the example on real data in Section 14.6 for more discussion on input correlation.

### *14.5.2 Burst and Periodic Random – Input Noise*

Next we will replace the extraneous output noise used in Section 14.5.1 with input noise, which is *not* removed by the $H_1$ estimator. In Figure 14.13, results from a simulation with an input signal-to-noise ratio of approx. 40 dB (relatively poor) is shown for two input signals; burst random, and periodic random with (normal) frequency domain averaging. In Figure 14.13 (b) and (d) it is clearly seen that in this case, burst random excitation (rings) shows rather large bias, whereas the bias error is smaller, but still visible with periodic random. The reason for the smaller bias with periodic random is the increased signal-to-noise ratio with this excitation signal, most of all because it is a continuous signal, and to some extent because its energy is concentrated on the frequency lines of the DFT (see Section 13.9.1). To be sure the error we see is not seriously affected by the random error, in this simulation 400 blocks were averaged for each excitation signal, which makes the random error sufficiently small. Thus the remaining errors seen in Figure 14.13 are bias errors.

As seen in Figure 14.13, the multiple coherence for burst random excitation is clearly less than unity over the entire frequency range, whereas for periodic random it is closer to, but still slightly lower than, unity. This is a direct result of the increased signal-to-noise ratio of the periodic random signal, and follows Equation (13.36).

### *14.5.3 Periodic Random – Input and Output Noise*

As we discussed in Section 14.4.3, there is an alternative to the normal frequency domain averaging, which can be used to eliminate the bias error due to noise on both input and output signals, when we use a periodic excitation signal. The solution is to use time domain averaging to remove (attenuate) the extraneous noise on all input and output signals, prior to taking the FFT and accumulating auto and cross-spectra. An example of the results of this procedure for a severe case of input noise, and the same amount of output noise, and the $H_1$ estimator, is shown in Figure 14.14.

As seen in Figure 14.14, where the results of normal frequency domain averaging, using 400 blocks of data, and with time domain averaging (see list below for details) are compared, the time domain averaging is effective in removing the bias error. Note that the bias in the frequency domain results is a result of the $H_1$ estimator not being able to remove the error due to input noise, so it would not help to increase the number of averages in the frequency domain averaging – the error is a consistent error. In this example we have used a rather large amount of data, in total $200 \times 20$ blocks, each with 2048 samples, in the time domain averaging process, i.e., 8.192 000 samples. This has been done to reduce the random error, and to illustrate the asymptotic effect. In practice fewer averages can be used, if larger random error can be tolerated. Random errors are less serious for modal parameter curve fitting algorithms than bias errors are, so this may be acceptable in some instances.

The processing scheme for periodic random with time domain averaging is more complicated than if frequency domain averaging is done. In the simulation results presented in Figure 14.14, the following processing was followed.

1. Two random sequences, $x_1$ and $x_2$, with blocksize $N = 2048$ samples were generated as Gaussian noise, and then repeated 200 times, to generate a periodic signal with period 2048 samples.
2. These two signals, $x_1$ and $x_2$, were input to a time domain forced response algorithm, to produce output signals $y_1$ and $y_2$. These two latter signals were then summed to yield the true system output, $y$.
3. Next, the first 10 periods of all signals, $x_1$, $x_2$, and $y$, were discarded to remove the transient part of the response.

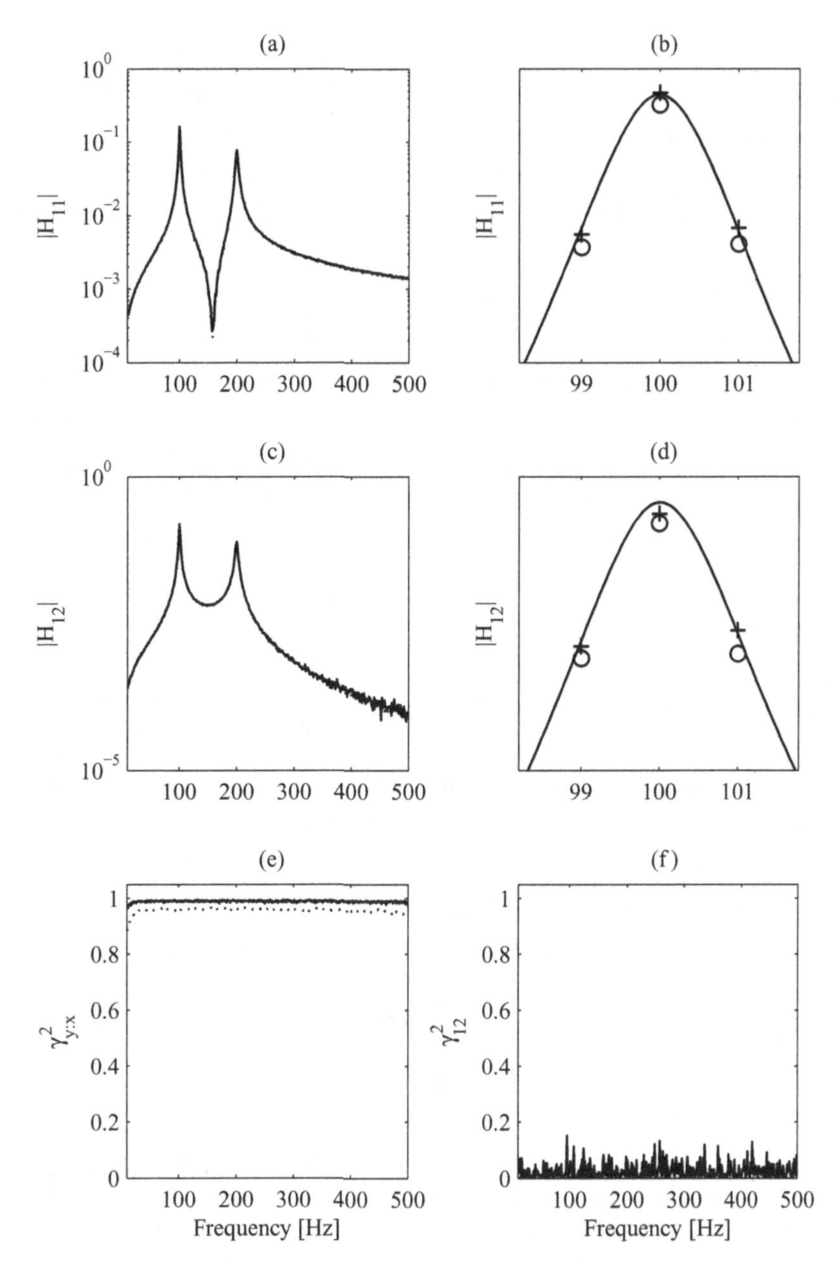


**Figure 14.13** Results of simulation using input noise on both input signals with a signal-to-noise ratio of approx. 40 dB. The results for two excitation signals are shown; burst random (rings or dotted), and periodic random (+ signs or solid). As shown, the bias error is significantly smaller with periodic random than with burst random excitation, due to the higher signal-to-noise ratio of the former. Both estimators are, however, clearly biased as the $H_1$ estimator does not remove the bias due to extraneous input noise

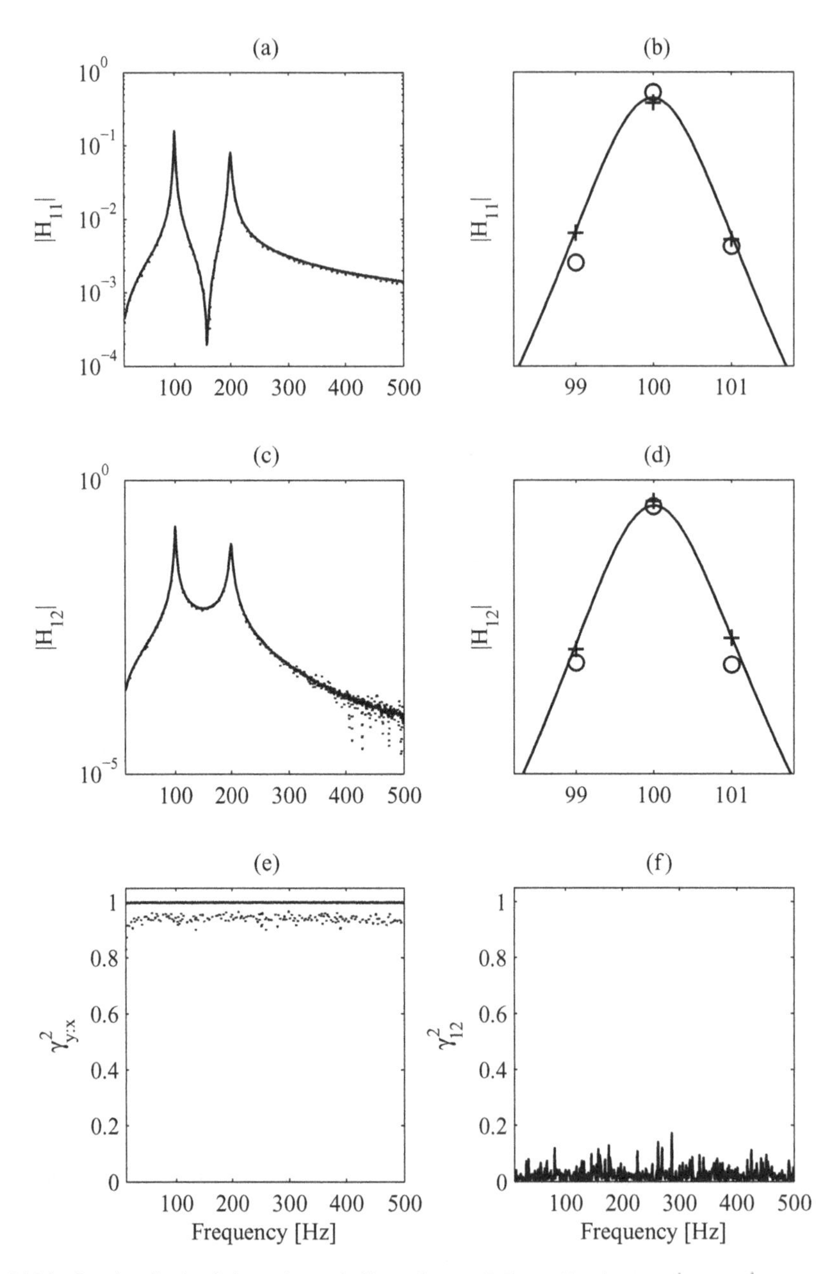


**Figure 14.14** Results of a simulation using periodic random excitation with extraneous input and output noise. The input noise had a SNR of approx. 12 dB, and the output noise approx. 40 dB. Two alternative processing techniques where used; 'normal' frequency domain averaging (dotted and rings), and time domain averaging (solid and +). As seen in (b) and (d), time domain averaging can reduce the bias error due to both input and output extraneous noise. The remaining error is predominantly random error

4. Two independent extraneous random sequences, each with an RMS level of approximately 0.25 times the RMS level of each of the input signals, $x_1$ and $x_2$, was then added to each input signal, and a random sequence with an RMS level of $10^{-4}$ times the RMS level of the output signal, $y$ was added to $y$.
5. The remaining 90 blocks, including the extraneous noise, where then time averaged to produce the pure periodic part of each signal.
6. The cleaned time signals obtained in step 5 where then used to compute an instantaneous estimate of each auto and cross-spectrum in Equation (14.13). The results were added in a cumulative calculation of the final auto- and cross-spectra.
7. Steps 1 to 6 were repeated 20 times, and the thus cumulated auto- and cross-spectra were finally used to produce an $H_1$ estimate of the frequency responses.

Note that step 7 has to be included, in order to make the input autospectral matrix invertible; time domain averaging has to be combined with, at least a few, frequency domain averages. This also makes the multiple coherence defined.

## 14.6 Real MIMO Data Case

Some properties and potential problems with MIMO shaker excitation are difficult to illustrate with simulated results as we have used in the preceding sections. We will therefore include an example where a Plexiglas plate was excited by two shakers, as illustrated in Figure 14.15. In this example we will look

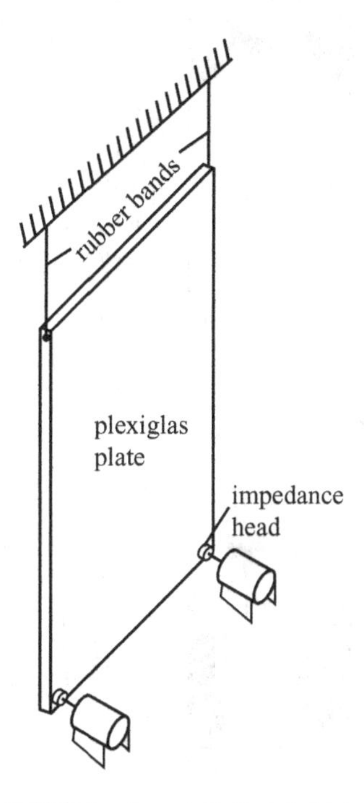


**Figure 14.15** Measurement setup for multiple-input example. Two shakers were attached through impedance heads, to a Plexiglas plate. The plate was suspended by rubber bands which were attached to a metal frame (illustrated by the rigid lines at the top of the figure), providing approximately free–free boundary conditions at frequencies above a few hertz. The two forces and the two driving point acceleration signals were measured and used to estimate a two-input/two-output system

at some of the measured functions, and study how the correlation between the input forces affects the frequency response estimation.

The Plexiglas plate was suspended by soft rubber bands, providing approximately free–free boundary conditions at frequencies of interest (above approximately 50 Hz). The driving point forces and accelerations were measured by impedance heads, ensuring that force and acceleration were measured in the same point. With this setup, the properties of the driving point frequency response functions as well as the reciprocity (see Section 6.4.2) can be verified. The frequency response functions were calculated using the $H_1$ estimator and with the procedure described in section 14.1.3.

The two electrodynamic shakers were connected to a two-channel noise generator, providing Gaussian noise with adjustable correlation. In a first measurement, the noise sources were set to be uncorrelated, and the frequency response functions were estimated. Since the noise was continuous, a Hanning window was applied in the frequency analysis.

The results of the first measurement with uncorrelated noise are found in Figure 14.16. In (a) the spectral densities of the input forces are plotted. As can be seen the forces have practically identical spectral densities. The variations are due to the mechanical impedance of the structure and the shaker characteristics. In Figure 14.16 (c), where the ordinary coherence function between the forces is plotted, it can be seen that at some frequencies (e.g., 160, 210, and 250 Hz) the correlation between the

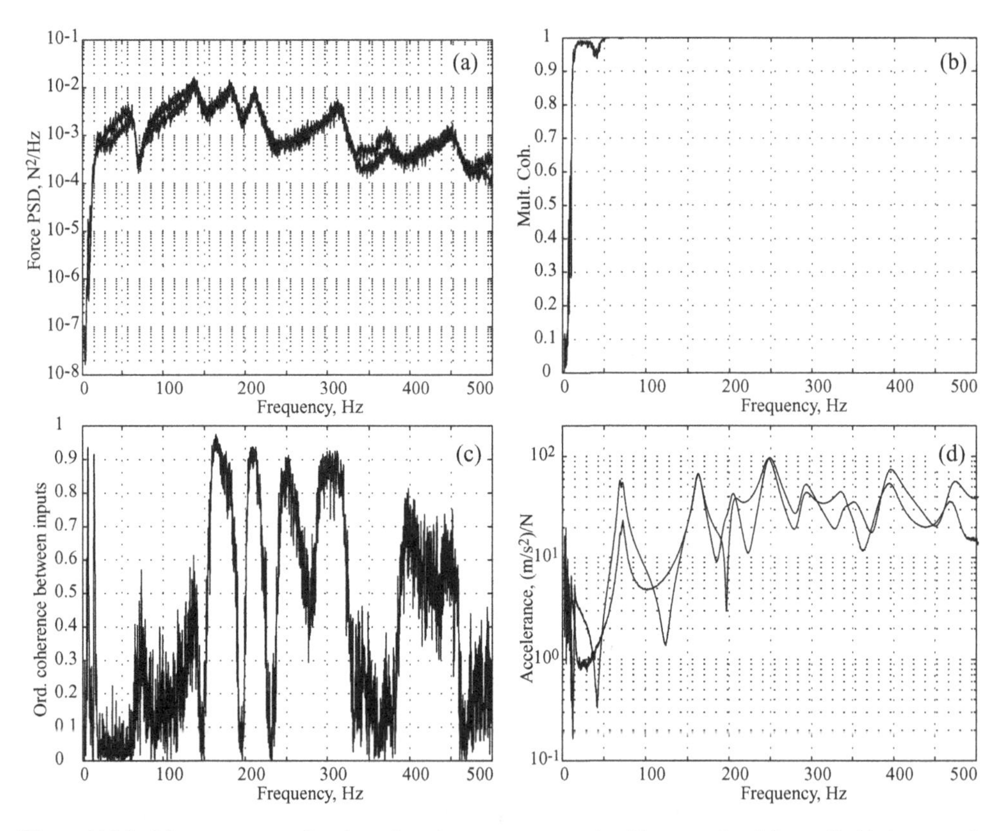


**Figure 14.16** Measurement results of two-input/two-output example with uncorrelated forces. In (a) the spectral densities of the input forces are plotted; in (b) the multiple coherence of response point 1 is plotted; (c) shows the ordinary coherence between the two forces; and (d) the frequency responses between the two forces and the response in point 1

input signals is rather high, with coherence values of 0.9 to 0.95. At other frequencies, however, the correlation is low. The increased correlation is found around antiresonances of the structure, where the stinger/amplifier/shaker combination is unable to force the structure according to the input voltage from the noise generators.

In 14.16 (b) and (d) the multiple coherence and the frequency response functions of both forces with acceleration in one of the excitation positions are plotted, respectively. From the upper plot, (b), it can be seen that for frequencies above approximately 50 Hz, the multiple coherence equals unity. Below 50 Hz, the forces are low, causing low coherence values due to contamination noise in the force transducers (and possibly accelerometers). As the first resonance frequency is approximately 130 Hz, the force input spectrum is sufficient to measure the interesting part of the FRFs. In the lower plot, (d), the suspension resonances are seen at very low frequencies, indicating that the suspension was soft enough to provide free–free conditions at the frequencies of interest.

In Figure 14.17 the results of adding some correlation between the excitation forces are shown. In Figure 14.17 (c) the coherence between the forces is plotted, and it is apparent that the correlation is very high in several frequency ranges. The multiple coherence, however, in (b) does not show any apparent degradation, which is typical. An immediate look at the plot in Figure 14.17 (a) where one of the FRFs from the uncorrelated input case is overlaid on the correlated input case, does not show any

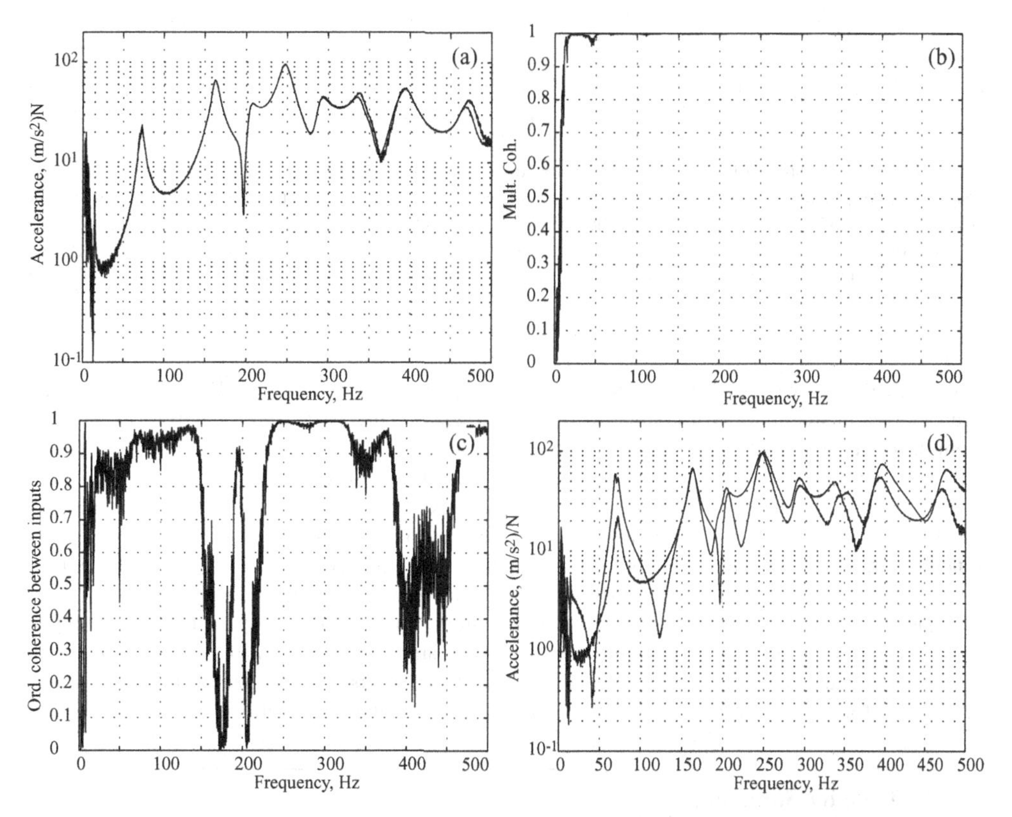


**Figure 14.17** Results from a measurement with some correlation between the input noise signals. In (a) FRF between force 2 and response 1 for uncorrelated inputs and correlated inputs are compared; in (b) the multiple coherence of response point 1 is plotted; (c) shows the ordinary coherence between the two forces; and (d) the frequency responses between the two forces and the response at point 1

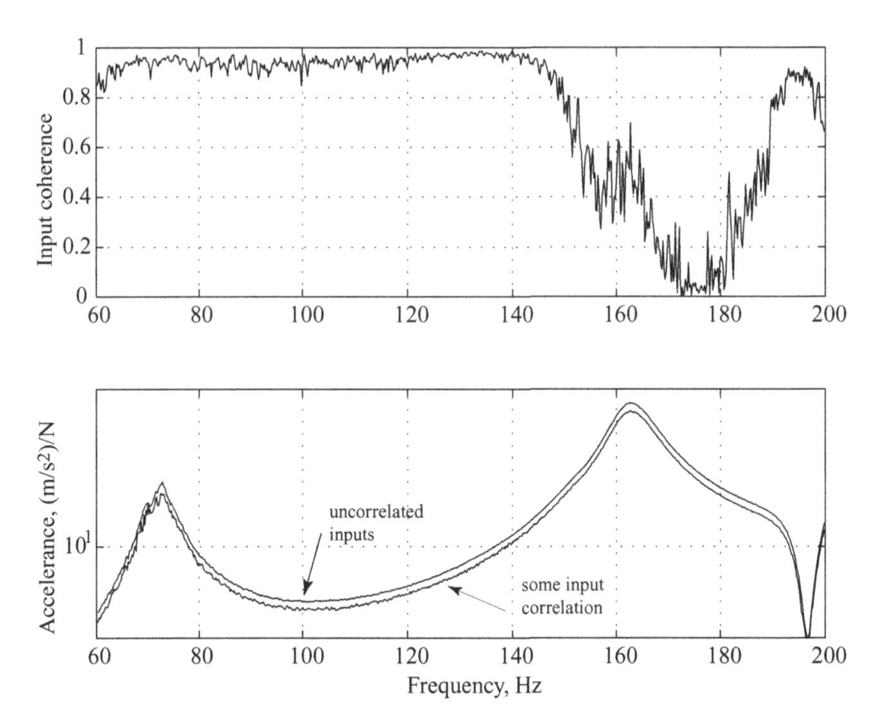


**Figure 14.18** Comparison of frequency response functions between measurements without correlation between the sources (top graph in lower plot), and with some correlation (lower graph in lower plot). The upper plot shows the coherence function between the two inputs in the case of correlated inputs, and the lower plot shows the two frequency response functions. Note that the two frequency responses are offset slightly in order for differences to show clearer. As is clear from the lower plot, the higher input correlation causes an increased variance in the estimate, seen as a 'ripple' in the lower FRF

large difference. However, as evident from Figure 14.18, where the two FRFs in Figure 14.17 (a) have been zoomed in around the first two natural frequencies, shows that there is a bias in the FRF estimate from the correlated input case. This shows how important it is to check the input correlation, as there will be no apparent visible indication in the estimates that a problem is present.

Two important conclusions can be drawn from this example. First, it is important to make sure that the input correlation is not too close to unity, when estimating multiple-input frequency response functions. In practice, a value of the ordinary coherence functions between the inputs of less than 0.8 is recommended. Second, it is important to realize, that the increased input correlation does not show in the multiple coherence, so the coherence function (or functions if more than two inputs) between the inputs must be checked. In practice, when more than two inputs are used, partial coherence functions can be used in order to avoid having to study the coherence between each pair of inputs. Principal components, see Section 15.1 is another tool which may be used to check the input correlation.

## 14.7 Chapter Summary

This chapter has presented an introduction to estimates of frequency response functions when more than one input signal is present in the system. The general multiple-input/multiple-output (MIMO) is defined such that each output is a linear combination of all the inputs, each going through an individual linear system. Thus, the outputs have no causal relations, and therefore each output can be regarded by itself, we then talk about multiple-input/single-output (MISO) systems. In general, from the point of view of understanding MIMO systems, we can view them as a number of parallel MISO systems.

For MIMO systems the $H_1$ estimator is the predominantly used estimator, since the $H_2$ estimator has the obvious drawback that it only works when the number of outputs equal the number of inputs, which is a rare case. We also introduced the $H_v$ estimator, which is an estimator attempting to minimize the error in the FRFs due to both input and output extraneous noise. However, without knowing the input and output noise properties, this estimator has to be based on an assumption that both noises are equal, which is not the case at most frequencies. It is therefore not an estimator which generally solves the problem of noise in both input and output signals.

The ordinary coherence function between an input signal and the output signal of a MISO system is not particularly useful. Instead, we defined the multiple coherence, which similarly with the ordinary coherence of a SISO system, describes how much of the output signal power is explained by all the input signals. In the case of no extraneous noise, the multiple coherence thus equals unity.

All estimators for MIMO systems include having to solve the inverse of the input autospectral matrix, $[G_{xx}]$ at every frequency. This has two important implications, namely:

- that the excitation signals have to be independent at all frequencies (i.e., not fully correlated),
- that the rank of $[G_{xx}]$ is full, which means at least as many averages has to be included in the computation of $[G_{xx}]$, as there are inputs in the system.

We discussed different excitation signals, which fulfill the demands in the list above. The most common excitation signals for MIMO estimation are pure random, burst random, and periodic random signals. Comparing these three excitation signals, we found that the periodic random signal has the best signal-to-noise ratio, and is therefore the best excitation signal in terms of measurement quality. Its drawbacks are that it requires more sophisticated hardware, and that the processing necessary for it is more complicated than is the case for the other two excitation signals. It is unfortunately, for this very reason, not very common in commercial measurement systems.

An important advantage of periodic random is also its ability to minimize the error in FRF estimates due to noise on the input and output signals simultaneously. This is not widely acknowledged but was demonstrated by an example. Therefore, in cases of severe input noise, time domain averaging with periodic random excitation can be an attractive alternative.

## 14.8 Problems

Many of the problems following are supported by the accompanying ABRAVIBE toolbox for MATLAB/ Octave and further examples which can be downloaded with the toolbox. If you have not already done so, please read Section 1.6, and follow the instructions to download this toolbox together with example files.

**Problem 14.1** *Can a MIMO system be estimated using pure random noise as excitation signals? Is it a good choice, if you can choose between pure random or burst random noise? Why, or why not?*

**Problem 14.2** *Develop an $H_2$ estimator for the MIMO case, by setting up the equations for a system with extraneous noise vector $\{M(f)\}$ on the input signals, and by multiplying by $\{Y(f)\}^H$. Explain why this estimator only works when the number of inputs is equal to the number of outputs.*

**Problem 14.3** *Write a MATLAB/Octave script which creates a mechanical 2DOF system with natural frequencies of 12.5 and 25 Hz and 2% damping, using appropriate commands from the accompanying toolbox. Set the sampling frequency to 200 Hz and use 100 (frequency domain) averages. Excite the system with pure random noise and determine which blocksize you should use as a minimum, to eliminate the bias error by running the script with larger and larger blocksize until the peaks do not become higher with a larger blocksize. Plot the magnitudes of the FRFs (with logarithmic y-axis) and coherence.*

**Problem 14.4** *Make a new script using part of the script from Problem 14.3 and replace the excitation signals by burst random noise. First adjust the burst length using a blocksize of 1024 samples until the measurement quality is good. Then check the maximum burst length (in % of the blocksize) you can have with blocksizes of 2048 and 4096, with good coherence.*

**Problem 14.5** *Use the scripts from Problems 14.3 and Problems 14.3 and add the same amount of noise to the output signal in both cases . Run the scripts with different blocksizes and note what happens with the random error. Explain why!*

## References

Allemang RJ, Rost RW and Brown DL 1984 Multiple input estimation of frequency response functions *Proc. 2nd International Modal Analysis Conference*, Orlando, FL.

Antoni J and Schoukens J 2007 A comprehensive study of the bias and variance of frequency-response-function measurements: Optimal window selection and overlapping strategies. *Automatica* **43**(10), 1723–1736.

Antoni J and Schoukens J 2009 Optimal settings for measuring frequency response functions with weighted overlapped segment averaging. *IEEE Transactions on Instrumentation and Measurement* **58**(9), 3276–3287.

Antoni J, Wagstaff P and Henrio JC 2004 H alpha – a consistent estimator for frequency response functions with input and output noise. *IEEE Transactions on Instrumentation and Measurement* **53**(2), 457–465.

Bendat J and Piersol A 1993 *Engineering Applications of Correlation and Spectral Analysis* 2nd edn. Wiley Interscience.

Bendat J and Piersol AG 2010 *Random Data: Analysis and Measurement Procedures* 4th edn. Wiley Interscience.

Mitchell L 1982 Improved methods for the FFT calculation of the frequency response function. *J. of Mechanical Design*.

Pauwels S, Michel J, M. R, Peeters B and Debille J 2006 A new MIMO sine testing technique for accelerated, high quality FRF measurements *Proc. 24th International Modal Analysis Conference*, St. Louis, Missouri Society for Experimental Mechanics.

Rocklin GT, Crowley J and Vold H 1985 A comparison of H1, H2, and Hv frequency response functions *Proc. 3rd International Modal Analysis Conference*, Orlando, FL.

Smallwood D 1995 Using singular value decomposition to compute the conditioned cross-spectral density matrix and coherence functions *Proc. 66th Shock and Vibration Symposium*, vol. 1, pp 109–120.

White PR, Tan MH and Hammond JK 2006 Analysis of the maximum likelihood, total least squares and principal component approaches for frequency response function estimation. *Journal of Sound and Vibration* **290**(3–5), 676–689.

Williams R and Vold H 1986 Multiphase-step-sine method for experimental modal analysis. *International Journal of Analytical and Experimental Modal Analysis* **1**(2), 25–34.

# 15

# Orthogonalization of Signals

The aim of conditioning of the input signals discussed in Chapter 14 was to make the input signals independent by removing any correlation between them, so that the contribution of each independent input could be added to produce the output. In linear algebra terms, this is the same as to say that the input signals are *orthogonal*. As we discussed in Chapter 14 different ordering of the input signals gave rise to different decomposed $L$-systems. This can sometimes be of use, when there is a physically motivated way of ordering the input signals. Other times, however, using conditioned signals can be difficult, especially if there is no obvious way of ordering the signals.

In this chapter we will introduce an alternative approach to orthogonalize the input signals, through the use of *principal components* and we will introduce the concept of *virtual signals* to interpret the principal components. This mathematical tool was introduced, in the form we are going to use it, by Hotelling (1933). Principal components theory has been applied to noise and vibration testing, (e.g., Otnes and Enochson 1972; Otte *et al.* 1990; Tucker and Vold 1990; Vold 1986) and is a standard tool in many current commercial measurement systems. Although the concept of principal components was originally developed on covariance matrices, we will study a corresponding formulation in the frequency domain which is more common in noise and vibration applications. A good source for more information in line with the presentation here is the dissertation thesis by (Otte 1994) who made significant contributions to the application of principal components in this field.

## 15.1 Principal Components

With $Q$ correlated input signals, we have an input cross-spectral matrix $[G_{xx}]$ with nonzero off-diagonal elements consisting of the cross-spectra between the input signals. If we could transform this matrix into a new matrix $\left\lceil G'_{xx} \right\rfloor$ which was diagonal, we would have a cross-spectral matrix of new input signals $X'_i$ which would be uncorrelated. The problem then is to diagonalize the original input spectral matrix $[G_{xx}]$ in such a manner.

The input cross-spectral matrix $[G_{xx}]$ is *Hermitian*, (sometimes *Hermitian symmetric*), i.e., it equals the complex conjugate of its transposed matrix. In mathematical terms we have

$$[G_{xx}] = \left[G^*_{xx}\right]^T = [G_{xx}]^H \tag{15.1}$$

where the superscript $^H$ is the 'Hermitian transpose' and means we transpose and take the complex conjugate, see Appendix E.

*Noise and Vibration Analysis: Signal Analysis and Experimental Procedures*, First Edition. Anders Brandt.

For a Hermitian matrix, from linear algebra theory (see Appendix D and E or any standard textbook on linear algebra, (e.g., Strang 2005) for this and much of the theory in this chapter), we know the following.

1. Its eigenvalues are real.
2. Its eigenvectors $\{u_k\}$, corresponding to the eigenvalues $\lambda_k$, can be scaled so they are *orthonormal*, i.e., they have unit length, and the dot (scalar) product of any two vectors is zero, i.e.,

$$\begin{aligned} \|\{u_k\}\|_2 &= 1 \\ \{u_k\}^H \{u_l\} &= 0, \; k \neq l. \end{aligned} \tag{15.2}$$

3. The eigenvector matrix with the eigenvectors as its columns fulfills Equation (15.3)

$$[U]^H = [U]^{-1} \tag{15.3}$$

   where $[U]$ is a matrix with each eigenvector $\{u_k\}$ in column $k$. Note: a complex matrix with orthonormal columns is called a *unitary* matrix, which the eigenvector matrix, $[U]$, thus is. (This is the complex version of orthogonal matrices.)
4. It is diagonalized by its eigenvectors through Equation (15.4)

$$[G_{xx}][U] = [U]\lceil\lambda\rfloor \tag{15.4}$$

where $\lceil\lambda\rfloor$ is a diagonal matrix with the eigenvalues of $[G_{xx}]$ on its diagonal, *in descending order*. The descending order is not necessary for Equation (15.4) to be true, but is important when we arrive at the principal components below, so we impose this restriction already here.

From the numbered points above, it is clear that we can transform our original input cross-spectral matrix into a new diagonal input spectral matrix $\lceil G'_{xx}\rfloor$ of uncorrelated 'virtual' inputs $X'_i$ using Equation (15.4) which is reformulated using Equation (15.3) into

$$\lceil G'_{xx}\rfloor = \lceil\lambda\rfloor = [U]^H [G_{xx}][U]. \tag{15.5}$$

The elements on the diagonal of $\lceil G'_{xx}\rfloor$ are called *principal components*. Note especially that the eigenvalues and corresponding eigenvectors are sorted in *descending order*, which is not necessary for the eigenvalue problem, but is an essential part of the concept of principal components. Note also that Equation (15.5) applies to one frequency at the time, and thus the principal components in our case are frequency dependent 'spectra' just as the power spectral densities.

The principal components, further, will always be positive, because input cross-spectral matrices are *positive definite*; their eigenvalues are $> 0$ (theoretically they can be equal to zero, in which case we call the matrix positive semidefinite; for matrices obtained from a measurement we will always have noise which prevents any eigenvalue from being strictly equal to zero).

An important feature of principal components is that the total power of the signals is preserved. This follows from the fact that the eigenvalue matrix $[U]$ is unitary. Thus, for each frequency, *the sum of the diagonal elements of* $G_{xx}$ *and* $\lceil G'_{xx}\rfloor$, respectively, are equal. The sum of the diagonal elements of a matrix is known as the *trace* of the matrix, i.e.,

$$\text{trace}\lceil G'_{xx}\rfloor = \text{trace}[G_{xx}] \tag{15.6}$$

**Example 15.1.1** *Use MATLAB/Octave to compute the principal components based on the input cross-spectral matrix of three measured signals,* $x_1, x_2, x_3$.

*We assume we have computed the input cross-spectral matrix* $[G_{xx}]$ *in a 3D matrix in variable* `Gxx`, *as described in Section 10.7.2. The following MATLAB/Octave code then computes the 2D matrix PC, with principal components in columns. Note that there is no need to compute the off-diagonal zeros.*

```
[N,R,dum]=size(Gxx);% Find number of inputs, R
for f=1:N
  Gxxf=squeeze(Gxx(f,:,:));% Force to 2D matrix
  PC(f,:)=eig(Gxxf);% Compute eigenvalues
  % Sort eigenvalues to produce principal components
  PC(f,:)=sort(PC(f,:),'descend');
end
```

*Note that we use the command* ***eig*** *here with only one output argument, which creates a column vector with the eigenvalues. We also have to sort the eigenvalues, because they do not necessarily come in descending order (actually, in MATLAB and Octave, they will come in ascending order). The command* ***squeeze*** *has to be used to make* `Gxxf` *a two-dimensional matrix which* ***eig*** *requires.*

*End of example.*

### *15.1.1 Principal Components Used to Find Number of Sources*

In order to illustrate a common application of principal components, we will study a simulation case of vibrations on a plate excited in two degrees of freedom. The plate model is similar to the experimental plate described in Section 14.6. The excitation signals were independent Gaussian noise. We simulate data for three accelerometers with *some added extraneous output noise*, on the plate.The autospectral densities of the simulated acceleration signals are plotted in Figure 15.1. As can be seen in the figure, the acceleration levels were of similar levels at the different points, but the actual spectra depend on the location of the accelerometers.

The principal components of the three signals were then computed using the complete input cross-spectral matrix of the three signals, i.e., in addition to the three autospectral densities in Figure 15.1, also

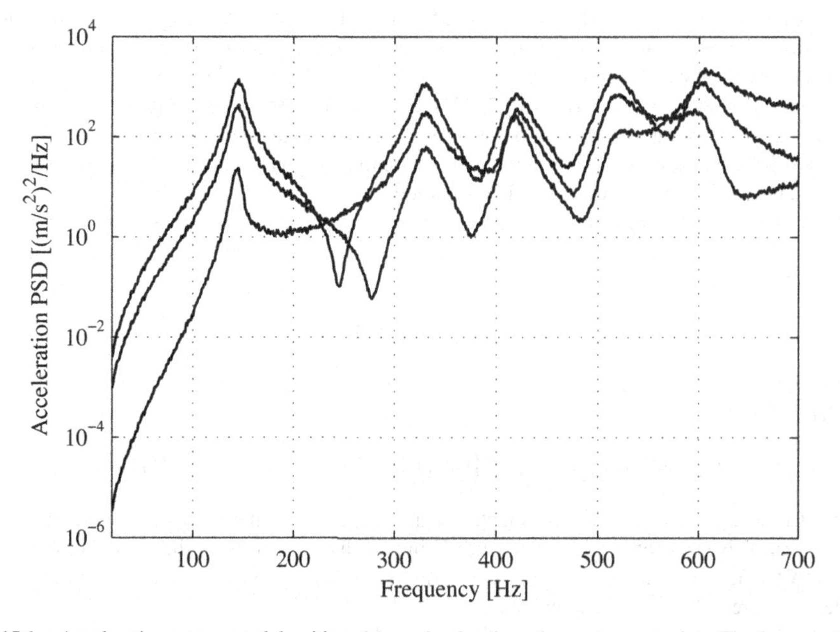


**Figure 15.1** Acceleration autospectral densities of three simulated accelerometers on a plate. The three acceleration PSDs are approximately equal level (surface under the PSDs), with local variations due to the position of each

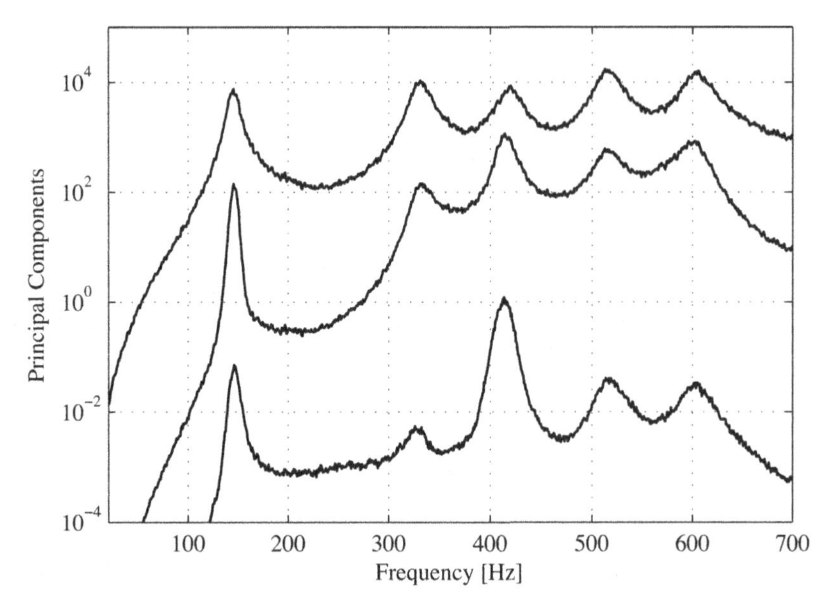


**Figure 15.2** Principal components of the three signals whose PSDs were plotted in Figure 15.1. Two principal components remain at a high level, whereas the third is reduced by approximately 50–60 dB, indicating that only two sources were present

the cross-spectral densities between each pair of accelerations were calculated. The resulting principal components are plotted in Figure 15.2. This figure clearly shows that there are two principal components with high levels relative to the third, which is some 50–60 dB lower at most frequencies. The interpretation of this is that there were only two (dominating) independent sources causing the measured vibrations, since apparently the three spectra in Figure 15.1 can be calculated as linear combinations of the two highest principal components in Figure 15.2. This is a common application of principal components in noise and vibration testing as a starting point when analyzing an unknown environment. It is of course often very helpful to know how many sources there are in a system before starting to map them; this is often called the *rank* of the system under investigation.

In order to explain the above interpretation, consider a mechanical system with a number of forces acting on it. Then, from Chapter 6 we know that the responses in column vector $\{Y\}$ given the forces in column vector $\{F\}$ satisfy

$$[H]\{F\} = \{Y\}. \tag{15.7}$$

Multiplying Equation (15.7) by its Hermitian transpose and taking the expected value of both sides yields

$$\left[G_{yy}\right] = \mathrm{E}\left[\{Y\}\{Y\}^H\right] = \mathrm{E}\left[[H]\{F\}\{F\}^H[H]^H\right] = [H]\left[G_{FF}\right][H]^H \tag{15.8}$$

Note that variable $y$ is now used for the displacement (output), as opposed to $u$ in the previous chapters on mechanics. This has been used here because $U$ in this chapter stands for the eigenvectors and we do not wish to confuse the reader too much. Furthermore, we use $F$ to denote the force instead of input variable $X$, to stress the physical application here.

Since the frequency response matrix $[H]$ is in general a matrix with full rank, the rank of $\left[G_{yy}\right]$ must clearly be equal to the rank of $[G_{FF}]$, i.e., the number of sources in the system. This gives us the first use of principal components, which is to determine the number of incoherent (uncorrelated) sources, as we demonstrated in Figure 15.2.

In the example above, we had two uncorrelated forces, generating noise on the plate. We put three accelerometers on the plate, and measured the cross-spectral matrix of these three signals. The principal components in this case, will yield the equations as follows.

$$\left[G_{yy}\right] = [U]\lceil\lambda\rfloor[U]^H = \begin{bmatrix} | & | & | \\ \{u\}_1 & \{u\}_2 & \{u\}_3 \\ | & | & | \end{bmatrix} \begin{bmatrix} \lambda_1 & 0 & 0 \\ 0 & \lambda_2 & 0 \\ 0 & 0 & 0 \end{bmatrix} \begin{bmatrix} - & \lfloor u \rfloor_1^* & - \\ - & \lfloor u \rfloor_2^* & - \\ - & \lfloor u \rfloor_3^* & - \end{bmatrix}. \tag{15.9}$$

First, we note that in Equation (15.9), the third eigenvalue is zero. This comes directly from comparing Equation (15.9) with Equation (15.8) and noting the fact that we only have two sources in the system, which means element (3,3) in $[G_{FF}]$ in Equation (15.8) is zero. From the plot of the eigenvalues as functions of frequency, thus it will immediately be clear that the number of sources is two, as was (approximately) the result plotted in Figure 15.2. The reason the third principal component in the figure is not equal to zero, is that we added some uncorrelated extraneous noise to the acceleration signals, to produce an example with a touch of reality. In general, the eigenvalues below the actual rank of the system studied will be much smaller than those due to actual sources. Another reason for the third principal component to rise is that the eigenvalue computation is sensitive to leakage, which can be seen especially around the resonances in Figure 15.2, (Otte 1994).

Principal components can also conveniently be used in multiple-input estimation of frequency response functions, by applying it to the input cross-spectral matrix of all forces. It is then easy to see if there are frequencies where one of the shakers is not contributing, by noting if one of the principal components is much lower than the remaining ones. The ideal situation in this case, as discussed in Chapter 14, is that all shakers force the structure at all frequencies, otherwise the inversion of the input cross-spectral matrix, $[G_{xx}]$, can be endangered, as described in Section 14.1.5.

### *15.1.2 Principal Components Used for Data Reduction*

From Equation (15.9) it is obvious that each column in $[U]$ is scaled by the corresponding eigenvalue to combine into a column in $\left[G_{yy}\right]$. The third column, then, does not contribute to $\left[G_{yy}\right]$, since the third eigenvalue equals zero (although there will still be an eigenvector). This illustrates the second use of principal components: it is a means of compressing the size of a system into a minimum number of independent linear combinations. By throwing away the third column of $[U]$ and the third eigenvalue in $\lceil\lambda\rfloor$, we do not loose any information. Thus, with high accuracy, we can approximate $\left[G_{yy}\right]$ by the truncated equation

$$\left[G_{yy}\right] = [U]_r\lceil\lambda\rfloor_r[U]_r^H = \begin{bmatrix} | & | \\ \{u\}_1 & \{u\}_2 \\ | & | \end{bmatrix} \begin{bmatrix} \lambda_1 & 0 \\ 0 & \lambda_2 \end{bmatrix} \begin{bmatrix} - & \lfloor u \rfloor_1^* & - \\ - & \lfloor u \rfloor_2^* & - \end{bmatrix} \tag{15.10}$$

where the index $r$ denotes the truncation to the 'approximate rank' of the eigenvalue matrix $\lceil\lambda\rfloor$.

To illustrate this method, although remotely related to noise and vibration analysis, an image is probably the best example since it is possible to see the change by the eye. A (digital) image consists of a number of picture elements, pixels, which can, of course, be put in a matrix. In most cases this matrix will be rectangular, so the approach above is not directly applicable. However, instead of using an eigenvalue decomposition, in such cases the *singular value decomposition* (SVD) can be used, see

Appendix E. In essence, the SVD decomposes any $(M \times N)$ matrix $[A]$, into three matrices, $[U]$, $[S]$, and $[V]$, so that

$$[A] = [U]\lceil S \rfloor [V]^H \tag{15.11}$$

where $[U]$ is $(M \times M)$, $\lceil S \rfloor$ is diagonal and $(M \times N)$, and contains the singular values of $A$, in descending order, and $[V]$ is $(N \times N)$. We call the left-hand matrix, $[U]$, the left singular vector matrix, and the right-hand matrix, $[V]$, is called the right singular vector matrix. For a square matrix, the singular values in matrix $\lceil S \rfloor$ equal the nonzero eigenvalues. Also note that comparing Equation (15.11) with Equation (15.5) (reorganizing the latter with $[G_{xx}]$ on the left-hand side) shows that the SVD for a (quadratic and) symmetric matrix must yield equal matrices $[U]$ and $[V]$.

For a rectangular matrix, the singular values are the square roots of the eigenvalues of both $[A][A]^H$ and $[A]^H[A]$. The columns of $[U]$ are the eigenvectors of $[A][A]^H$, and the columns of $[V]$ are the eigenvectors of $[A]^H[A]$. It is convenient, and also preferable from a computational standpoint, to use the SVD over the eigenvalue decomposition when computing principal components, so in fact it is often used instead of the eigenvalue decomposition for this purpose.

To use the SVD for data reduction, we remove all singular values below a particular threshold, and the corresponding columns in $[U]$ and $[V]$. In Figure 15.3 a picture of boats in a sunset, with 1920×2560 pixels and 256 gray scales (8 bits per pixel), is shown. A useful tool to see how much the picture can be compressed is to plot the singular values as in Figure 15.4. As can be seen in this figure, the singular values slowly drop after a rapid decrease in the first few values. This is an indication that the image in Figure 15.3 has a high rank. If the rank is low, there will be a significant drop in the singular values as the rank is exceeded. There is therefore no obvious number of singular values to keep in the compression for our picture of the boats.

In Figures 15.5 and 15.6 the same picture reduced to the 300 and 100 largest principal components, respectively, are shown. The original picture in Figure 15.3 contains information corresponding

**Figure 15.3** Original image with 1920 × 2560 pixels and 256 gray scales. [Photo: Anders Brandt]

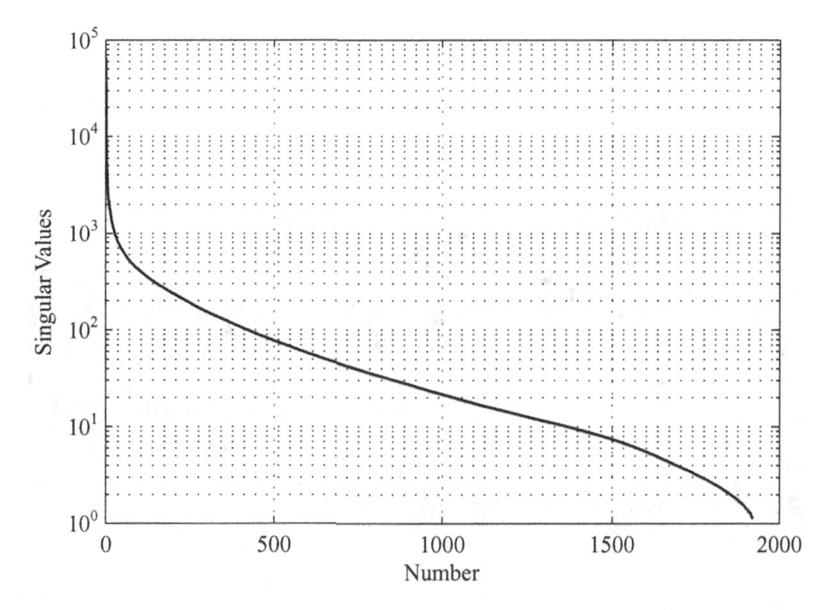


**Figure 15.4** Singular values of the picture in Figure 15.3 plotted in descending order. Although not visible in this example, sometimes there is a knee in the plot which makes it easy to select a proper number of singular values to use, see Figure 15.7. In this example the degradation of the picture is more gradual as the number of singular values is reduced

**Figure 15.5** The image in Figure 15.3 reduced to the 300 largest principal components by the method described in the text. As can be seen some detail is lost, although the picture is almost the same quality as the original picture. The amount of data is approximately 9% of the original. [Photo: Anders Brandt]

**Figure 15.6** The image in Figure 15.3 reduced to the 100 largest principal components by the method described in the text. After this much reduction, in this case the picture looks blurred. The amount of data is approximately 3% of the original. [Photo: Anders Brandt]

to $1920 \times 1920 + 1920 \times 2560 + 2560 \times 2560$ 'information units' in the sense of the SVD. The information in Figure 15.5 is condensed down to $1920 \times 300 + 300 \times 300 + 2560 \times 300$, which corresponds to approximately 9% of the original amount. In Figure 15.6 the information is condensed down to approximately 3% of the original amount of information, which in this case causes some blur in the picture. Particularly look at the bush in the upper left-hand corner, and the tree line against the sky.

In Figure 15.7 (a), an image with a limited number of levels of gray scales and a simple, periodic pattern is shown. This image clearly has a low rank . As can be seen in the plot of the singular values in Figure 15.7 (b), after the two highest singular values, the remaining values are very low. This illustrates the typical property of a singular values plot when the rank is limited.

Data compression by using SVD is very common in experimental modal analysis, where it is used to compress the frequency response matrix into a condensed size, still containing the necessary information for modal parameter extraction. In this case, a frequency response matrix obtained from a MIMO measurement can be enhanced in the sense that a large number of response degrees-of-freedom are condensed down to the minimum number needed to contain the same information as the original measured FRFs. This method was first proposed by (Lembregts 1988) and is further discussed in (Dippery *et al.* 1994). We will illustrate the method by an example.

**Example 15.1.2** *Assume we have a 3D matrix with frequency responses based on a 3-input/35-output measurement on a plate. Use MATLAB/Octave and singular value decomposition to produce enhanced frequency responses corresponding to a suitable number of singular values.*

*There are essentially two different approaches to this problem. One is to put all FRFs together into one FRF matrix, with size* $N_o \times N_i N_s$ *where* $N_o$ *is the number of outputs (responses),* $N_i$ *is the number*

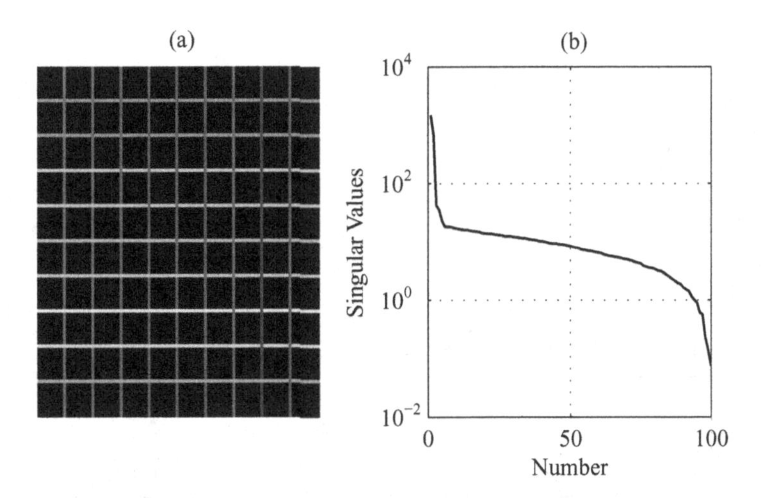


**Figure 15.7** Illustration of the drop in the singular values when the image rank is exceeded. The image in (a) has a low rank due to the limited number of gray scales and the periodicity in the graphical image. This results in a significant drop in the singular values above the first three or four singular values. Note that the lines have slightly different gray level, which increases the rank. The image is 100 × 100 pixels

*of inputs (forces), and* $N_s$ *is the number of frequency lines. This was the original proposal in (Lembregts 1988). However, this approach throws away some information we have from each shaker location. An alternative is to keep the 3D structure, and compute the SVD on the part of* $[H]$ *due to each reference separately, which is the approach we will use here. In the example, we will therefore make three singular value decompositions on, in each case, an* $N_s \times N_o$ *matrix. This will make the handling of modal participation factors (see Maia and Silva 2003) easier.*

*The following MATLAB/Octave code shows how to perform the condensation.*

```
for n=1:3
  [U,S,V]=svd(H(:,:,n));
  if n == 1
    plot(S)
    title('Select number of singular values')
    x=round(ginput(1));% Read cursor value
  end
  Ht=U(:,1:x)*S(1:x,1:x)*V(1:x,1:x)';
  He(:,:,n)=Ht;
end
```

*The results of this condensation, using the first reference, are shown in Figure 15.8, where the original FRFs and the FRFs of the condensed matrix are plotted, together with the singular values. It can be seen that the singular value plot has a 'knee' at approximately 13 singular values, after which it does not drop as fast. From the 35 FRFs originally measured, therefore, the matrix was condensed using the 13 first singular values and corresponding vectors of the singular vector matrices,* $[U]$ *and* $[V]$*. The enhanced FRFs are shown in Figure 15.8 (c). This can sometimes improve the numerical stability of modal analysis parameter extraction.*

*End of example.*

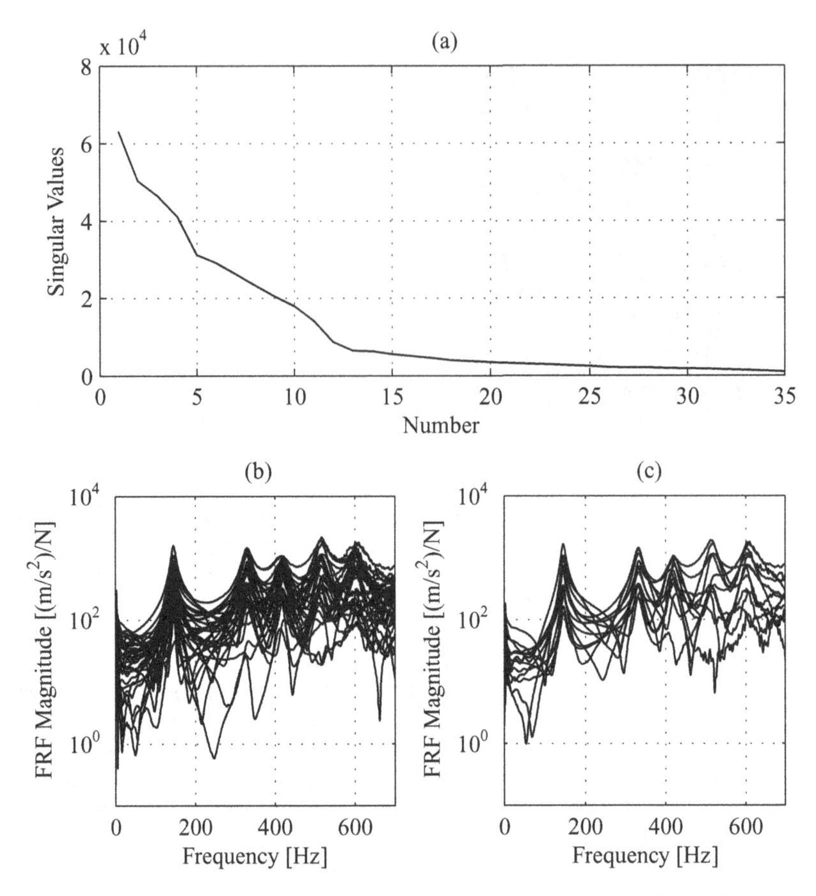


**Figure 15.8** Singular values and condensed frequency response matrix from Example 15.1.2. In (a) the singular values of a frequency response matrix from measurements on a plate with 35 measured degrees of freedom, which are plotted in (b), are shown. In (c) the 13 FRFs of the condensed FRF matrix using the 13 highest singular values and corresponding vectors of the singular vector matrices, $[U]$ and $[V]$, are shown

## 15.2 Virtual Signals

We will now go back to our reference (input) signals, and look at column vector $\{X\}$ of the spectra (one intermediate average) of the time signal $\{x(t)\}$. We now define *virtual input signals*, $\{x'(t)\}$, with spectra $\{X'\}$ as

$$\{X'\} = [U]^H \{X\}. \tag{15.12}$$

By post-multiplying Equation (15.12) by its Hermitian transpose and taking the expected value of each side of the equation we obtain

$$\lceil G'_{xx} \rfloor = E[\{X'\}\{X'\}^H] = \mathrm{E}\left[[U]^H \{X\}\{X\}^H [U]\right] = [U]^H \left[G_{xx}\right][U] \tag{15.13}$$

where we used the relationship that $(AB)^H = (B)^H(A)^H$. Equation (15.13) is equal to the principal component definition in Equation (15.5) and thus the virtual signals we have defined in Equation (15.12) are signals corresponding to the principal components (i.e., their autospectral densities are the principal components).

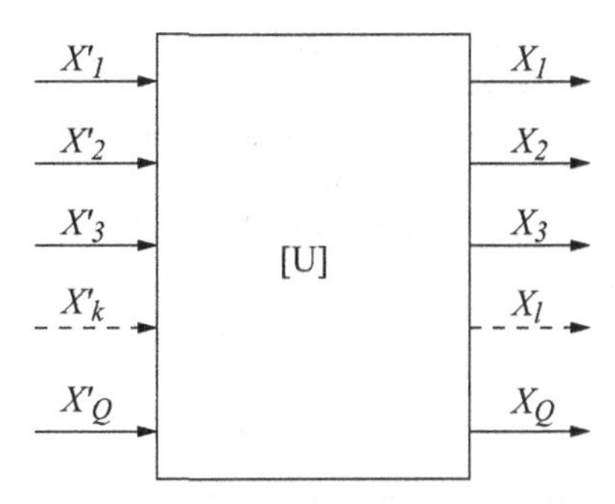


**Figure 15.9** Illustration of the linear relationship between each virtual signal in $\{X'\}$ and the original signals in $\{X\}$ being viewed as a MIMO system. The eigenvector matrix, $[U]$, is then the linear system relating virtual output signals to the original input signals

Since the spectral matrix of the virtual signals is diagonal, the virtual signals, $\{X'\}$, are apparently orthogonal (independent, uncorrelated). Although the virtual signals, like the conditioned signals in Chapter 14, cannot be measured, their auto- and cross-spectral densities can be computed from the autospectral density $[G_{xx}]$.

It is useful to note that Equation (15.12) can also be rewritten as a MIMO system with the principal components being input signals, and the original signals with spectra $\{X\}$ being the output signals. This follows directly from noting that $[U]^H = [U]^{-1}$, since the matrix is unitary, as stated in Equation (15.3). Thus, multiplying Equation (15.12) by $[U]$ leads to

$$\{X\} = [U]\{X'\} \tag{15.14}$$

which is a MIMO system with the linear system being $[U]$ and the virtual signals being the input signals. This system is depicted in Figure 15.9. From Equation (15.14) it follows that a particular element $U_{pq}$ in row $p$ and column $q$ of $[U]$, can be interpreted as the frequency response between input virtual signal $x'_q(t)$ and the measured signal $x_p(t)$.

## 15.2.1 *Virtual Input Coherence*

The concept of virtual signals is similar to the concept of conditioned signals in Chapter 14. In principle, the two methods both produce uncorrelated signals. The conditioned signals, however, become different if the order of the signals is varied, because the correlation is 'gradually' removed from the second signal and on. Virtual signals remain the same regardless of the ordering of the original signals, since they are based on the eigenvalues, which do not change if the signals are rearranged. The first virtual signal corresponds to the highest eigenvalue, the second corresponds to the second highest eigenvalue, etc. This is equivalent to ordering the virtual signals in order of their *power* at each frequency. A consequence of this is that a particular (measured) input signal does not, in general, correspond to the same virtual signal over the entire frequency range, but rather jumps between the virtual signals, depending on which of the various measured input signals are largest at each frequency.

In this section, we will show how we can compute the coherence between each virtual signal and each measured signal. This virtual coherence then shows the correlation between a particular virtual signal and a particular measured input signal, which can often show which of the measured input signals is (mostly) correlated with a virtual signal, at a particular frequency of interest.

Using Equation (15.14) the *virtual input cross-spectral density* matrix $[G_{xx'}]$ can be calculated by

$$[G_{xx'}] = \mathrm{E}[\{X\}\{X'\}^H] = \mathrm{E}\left[\{X\}\{X\}^H[U]\right] = [G_{xx}]\,[U] \tag{15.15}$$

where we again have used the relationship that $(AB)^H = (B)^H(A)^H$. The virtual input cross spectrum is useful because it allows us to define the *virtual input coherence*, which is defined as the ordinary coherence between a particular virtual input signal and a measured signal.

Using the virtual signals, analogous with the ordinary coherence function for the signals in $\{X\}$, the *virtual input coherence* functions, $g_{pq}^2$, can be defined between any original signal $x_p$ and a virtual signal $x'_q$ by

$$g_{pq}^2 = \frac{\left|G_{x_p x'_q}\right|^2}{G_{pp} G'_{qq}} \tag{15.16}$$

where, analogously with the notation in Chapter 14, we omit the $x$ and $x'$ in the indices in the denominator because we are dealing with inputs.

For each original signal, $x_p$, there will be as many virtual input coherence functions as there are input signals, i.e., there will be $Q$ virtual input coherence functions. Each virtual input coherence function tells how much of the power in $G_{pp}$ that comes from a linear relationship with the corresponding virtual signal. Thus, the virtual coherence function can be used in order to understand to what extent a particular physical signal is related to the particular virtual signal, at each frequency.

Each virtual coherence is normally not fruitful to analyze. (Otte 1994) suggested that *cumulated virtual coherences* should be used. These are computed as

$$g_{x_p:x'_{q!}}^2 = \sum_{k=1}^{q} g_{pk}^2 \tag{15.17}$$

and it is obvious that the cumulated virtual input coherences for each signal, $x_p$ sum up to unity, i.e., the last one,

$$g_{x_p:x'_{Q!}}^2 = 1. \tag{15.18}$$

The cumulated virtual input coherence functions tell how much of $G_{pp}$ that comes from the first virtual signal, the first and second virtual signals, etc.

**Example 15.2.1** *To illustrate the concept of input virtual coherence with an example we will use MATLAB/Octave to produce a 2-input-1-output system as depicted in Figure 14.1, where $H_{y1}$ and $H_{y2}$ are two SDOF systems with natural frequencies of 100 and 200 Hz, respectively. Each SDOF system is fed by a random signal, $x_1$ and $x_2$, respectively, and the SDOF output signals are summed up to an output signal y. In this example, we let the RMS level of $x_1$ be 1.5 times the RMS level of $x_2$, see Problem 15.2. To make the example more realistic, we also add some extraneous noise, $n(t)$, to the output signal, $y(t)$, with a signal-to-noise ratio of 40 dB, i.e., we let the RMS level of the extraneous noise be 0.01 times the RMS level of the output signal. We use an FFT blocksize of 2048 samples, and a sampling frequency of 2000 Hz, to have 10 times oversampling with respect to the highest natural frequency.*

*Next we compute the cumulated virtual input coherence functions for a case of two Gaussian noise signals $x_1$ and $x_2$ with 10% correlation (in the sense that the ordinary coherence between the two signals is approximately 0.1). We assume that we have already computed the input spectral matrix $G_{xx}$ as a 3D matrix as discussed in Section 10.7.2.*

*The following MATLAB/Octave code produces the 2D matrix PC with principal components of each output by using singular value decomposition, SVD. It also computes the virtual input coherence functions in variable* `VCxx`, *and cumulated virtual input coherence in* `CVCxx`.

```
[N,R,dum]=size(Gxx);
for f=1:N
  Gxxf=squeeze(Gxx(f,:,:));
```

```
  [U1,S,U2]=svd(Gxxf);
  PC(f,:)=diag(S);
  VGxxf=Gxxf*U1;             % Virtual cross spectrum
  Gxxf=real(diag(Gxxf));     % Reduce to autospectra
  for x_sig=1:R
    for pc_sig=1:R
     VCxx(f,x_sig,pc_sig)=abs(VGxxf(x_sig,pc_sig)).^2,...
     ./Gxxf(x_sig)./PC(f,pc_sig);
  end
  for x_sig=1:R
     for pc_sig=1:R
       if pc_sig == 1
         CVCxx(f,x_sig,1)=VCxx(f,x_sig,pc_sig);
       else
         CVCxx(f,x_sig,pc_sig)=VCxx(f,x_sig,pc_sig-1)+,...
         VCxx(f,x_sig,pc_sig);
       end
     end
end
```

*The cumulated virtual input coherence functions computed using this code are plotted in Figure 15.10. It shows that the main part of $G_{11}$ (approximately 95%) comes from virtual signal $x'_1$, whereas approximately 40% of $G_{22}$ comes from $x'_1$ and the remaining 60% comes from $x'_2$. Thus, it is possible that to some degree couple the different virtual input signals to the respective measured input signal, however care must be used, as we will discuss further in Example 15.2.2.*

*End of example.*

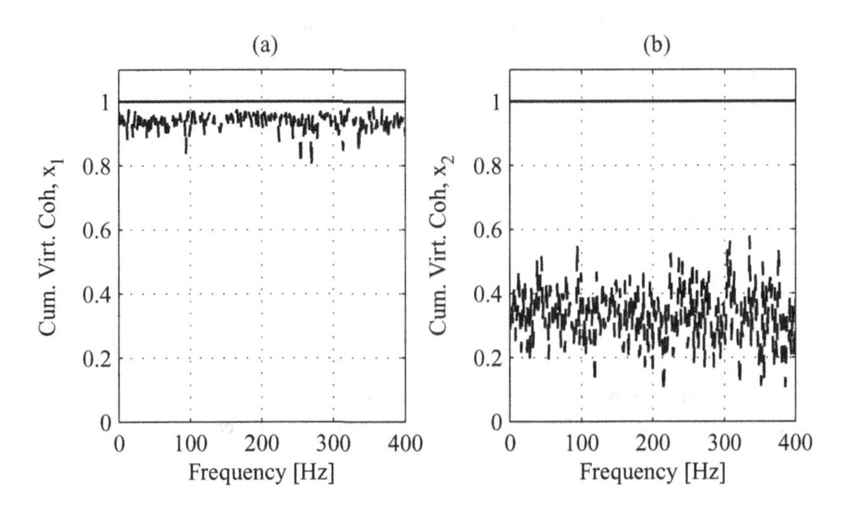


**Figure 15.10** Figure from Example 1. In (a) the cumulated virtual input coherence functions $g^2_{1:1}$ (dashed) and $g^2_{1:2!}$ for input signal $x_1$. In (b) the cumulated virtual input coherence functions $g^2_{2:1}$ (dashed) and $g^2_{2:2!}$ for input signal $x_2$. The figure shows that the first input signal is strongly correlated with the first virtual input, whereas the second input signal is a little less correlated with, but still dominated by, the second virtual input. The less correlation in the latter case is due to the fact that $x_2$ is somewhat correlated with $x_1$

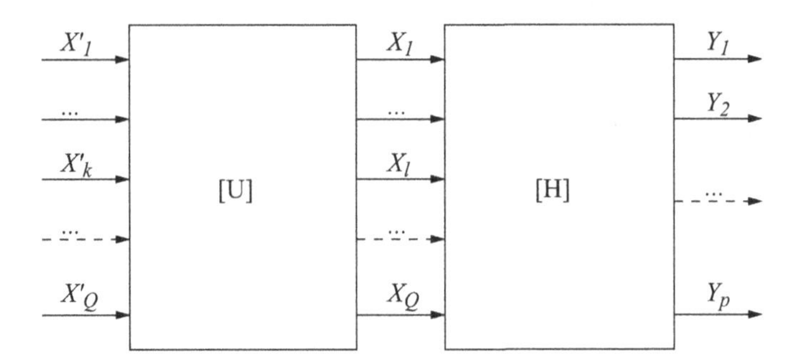


**Figure 15.11** Illustration of the concept of virtual signals on entire MIMO system

### 15.2.2 *Virtual Input/Output Coherence*

If we consider a multiple-input/multiple-output system with correlated inputs as in Chapter 14, using the concept of virtual signals, we can define *virtual input/output cross-spectral density*. This can be illustrated as in Figure 15.11, where the virtual, uncorrelated signals are the inputs, and the output signals are $y_p$. In such cases, the virtual cross-spectral densities of a particular output signal, $y_p$, and each virtual signal, $x'_q$, are of interest. The virtual input/output cross-spectral matrix elements are computed using the same principle as in Equation (15.15), which gives

$$\left[G_{yx'}\right] = \mathrm{E}[\{Y\}\{X'\}^H] = \mathrm{E}\left[\{Y\}\{X\}^H[U]\right] = \left[G_{yx}\right][U] \tag{15.19}$$

which, in the case of a single output, is a row vector, and in the case of multiple outputs, is a matrix with one row for each output.

In many noise source identification cases, it is useful to use virtual coherence signals between each virtual signal and a particular output signal $y_p$. Such virtual signals are computed consistently with the virtual coherences in Equation (15.16), replacing the original input signals with the output signals, i.e., the virtual coherence between virtual signal $x'_q$ and the output signal $y_p$ is

$$g^2_{y_p x'_q} = \frac{\left|G_{y_p x'_q}\right|^2}{G'_{qq} G_{y_p y_p}} \tag{15.20}$$

As for the input virtual coherences, it is often of more practical benefit to use the cumulated input/output virtual coherence, defined by

$$g^2_{y_p:x'_q!} = \sum_{k=1}^{q} g^2_{y_p x'_q k}. \tag{15.21}$$

Since the cumulated input/output virtual coherence sums up all contributions in $y_p$ due to the virtual inputs it must hold that, $g_{y_p \cdot x'_Q!}$, the last cumulated coherence, evidently equals the multiple coherence.

### 15.2.3 *Virtual Coherent Output Power*

In addition to the virtual coherence functions defined in Sections 15.2.1 and 15.2.2, *virtual coherent output power* spectra are very useful functions. They are defined similarly to partial coherent output power spectra for conditioned signals, as a coherence function multiplied by the output (target) spectrum. Thus, if we first consider the case of input virtual signals, for a particular measured input signal, $x_q$, we can define the virtual coherent output power of $x_q$ with virtual signal $x'_i$, which we denote $G_{qq:x'_i}$, as

$$G_{qq:x'_i} = g^2_{x_q x'_i} G_{qq}. \tag{15.22}$$

By using the virtual coherent output power spectra in comparison with the power spectral density $G_{qq}$, the individual contributions in $x_q$ to each virtual input can be plotted.

For an output signal, $y_p$, similarly the virtual coherent output power with each virtual input signal $x'_q$ can be defined as

$$G_{y_p y_p:x'_q} = g^2_{y_p x'_q} G_{y_p y_p}. \tag{15.23}$$

We will illustrate the use of virtual input/output coherence and virtual coherent output power by an example.

**Example 15.2.2** *We continue Example 15.2.1 by now looking at input/output relations. This example will show the importance of using virtual (or conditioned, see discussion below) signals instead of general MIMO techniques, when the input signals are correlated.*

*We assume that the input spectral matrix* $[G_{xx}]$*, in this case* $1025 \times 2 \times 2$*, the input/output cross-spectral matrix,* $[G_{yx}]$*, in this case* $1025 \times 1 \times 2$ *and the output autospectrum* $G_{yy}$*, in this case* $1025 \times 1$*, are all computed using the procedures from Chapter 9. They are stored in variables* Gxx*,* Gyx*, and* Gyy*, respectively in MATLAB/Octave.*

*Next, the eigenvalues and eigenvectors of the input cross-spectral matrix in* Gxx *are computed at each frequency, and the virtual input/output cross-spectral matrix, the cumulated virtual input/output coherence functions, and the virtual coherent output power, are all computed using the following MATLAB/Octave code.*

```
[Nf,R,dum]=size(Gxx);Find the size of Gxx
for f=1:Nf
  Gxxf=squeeze(Gxx(f,:,:));
  Gyxf=Gyx(f,:);
  Gyyf=Gyy(f,:);
  [U1,S,U2]=svd(Gxxf);
  PC(f,:)=diag(S);
  Gxx_p=Gxxf*U1;       % Virtual cross-spectrum
  Gxxf=diag(Gxxf);     % Reduce to autospectra
  VGyx(f,:)=Gyxf*U1;   % Virtual in/out cross-spectrum
  S=diag(S);                    % Reduce to vector
  TC=[];
  for pc_sig=1:R
    TC=[TC abs(VGyx(f,pc_sig))^2/(Gyyf*S(pc_sig))];
  end
  VC(f,:)=TC;
end
% Produce cumulated coherence and virt. coherent spectra
for r = 1:R
    if r == 1
        CVCyx(:,r)=VC(:,1);
    else
        CVCyx(:,r)=CVCyx(:,r-1)+VC(:,r);
    end
    VPyyx(:,r)=Gyy.*VC(:,r); % Virt. coherent output power
end
```

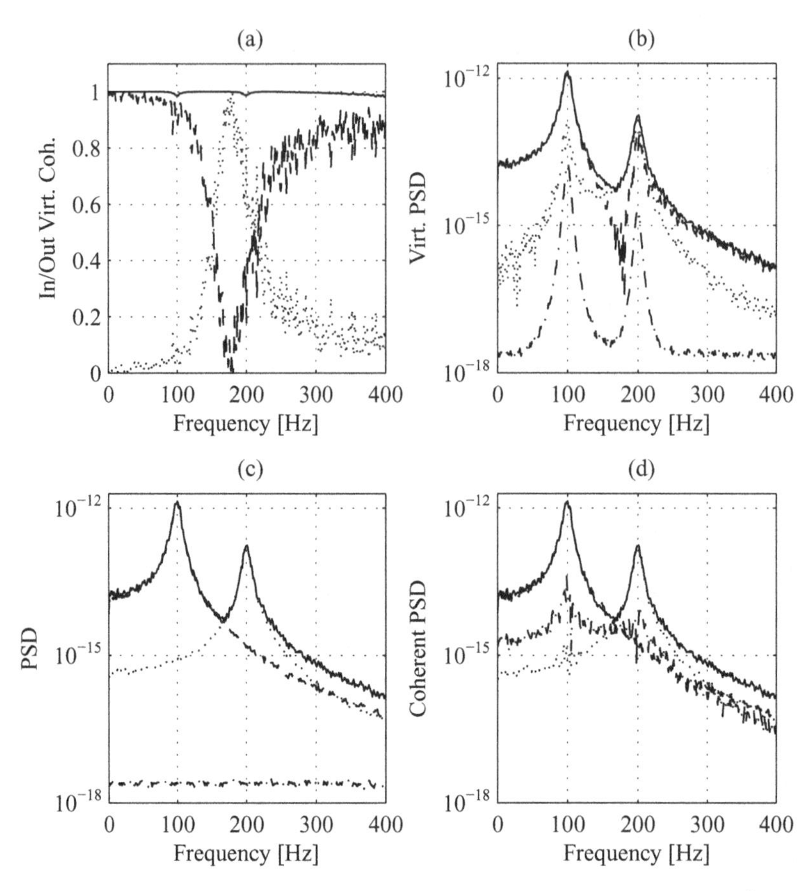


***Figure 15.12*** *Figure for Example 15.2.2. In (a) the virtual input/output coherence functions $g^2_{yx'_1}$ (dashed) and $g^2_{yx'_2}$ (dotted) are plotted together with the cumulated input/output coherence $g^2_{y:2!}$ (solid), which equals the multiple coherence. In (b) the virtual coherent output power spectra $G_{y:x'_1}$ (dashed), $G_{y:x'_2}$ (dotted), the measured output autopower spectrum $G_{yy}$ (solid), and the remaining uncorrelated spectrum (or error), $G_{nn}$ (dash-dotted). In (c) the true auto-spectral densities $G_{u_1u_1}$ (dashed), $G_{u_2u_2}$ (dotted), $G_{yy}$ (solid), and $G_{nn}$ (dash-dotted). In (d), the same spectra as in (c), but estimated using a MIMO estimation, see text for details*

*With the natural frequencies of the two SDOF systems separated, we can anticipate that one of the virtual coherence functions will dominate around the natural frequency of the first SDOF system, and the other virtual coherence should dominate around the natural frequency of the second SDOF system. In Figure 15.12 the results of the simulation with 10% correlation between the inputs is shown.*

*In Figure 15.12 the results of the code above are shown in (a) and (b). In Figure 15.12 (a), the input/output virtual coherence functions and the multiple coherence functions (also equal to the second cumulated virtual coherence) are plotted. As seen in the figure, the first virtual input/output coherence function, $g^2_{y,x'_1}$, is dominating at low frequencies, past the first natural frequency, and at high frequencies above approx. 250 Hz. This is an indication that the output, y, is caused mostly by the first virtual signal, $x'_1$. In the intermediate region around the natural frequency of the second SDOF system, the second virtual input/output coherence function, $g^2_{y,x'_2}$ is dominating, indicating that the output is here dominated by the second virtual input, $x'_2$, which is what we expect from the setup.*

*In Figure 15.12 (b) the resulting virtual coherent output power spectra are shown. What we can see here is that the PSD of the output signal, $G_{yy}$, is composed of the two virtual coherent spectra. The error, dash-dotted in the figure, is below approximately 0.01, which means that the sum of the two virtual coherent spectra add up to the measured output spectrum. Thus, using the virtual coherent output spectra, we can tell, at each frequency, how much comes from $x'_1$ and $x'_2$, respectively.*

*In Figure 15.12 (c), the autospectra of each of the signals from the simulation are shown. Most of these spectra we know only because we have simulated the system, but we could never measure them (except $G_{yy}$). This plot shows the importance of using uncorrelated signals to add up to the output spectrum! A careful examination of the frequency region (particularly) above 200 Hz, shows that the sum $G_{u_1u_1} + G_{u_1u_1} + G_{nn} \neq G_{yy}$! This is because the signals $u_1$ and $u_2$ are correlated, and in accordance with Equation (13.18), which showed that for correlated signals, we cannot add up the autospectra to produce the autospectrum of the sum. This is the main motivation for using virtual, or conditioned, signals.*

*In Figure 15.12 (d) we have included an 'experimental' version of the results in Figure 15.12 (c), in that we have produced each of the output autospectra as $\hat{G}_{u_1u_1} = G_{11}\left|H_{y1}\right|$ and $\hat{G}_{u_2u_2} = G_{22}\left|H_{y2}\right|$, where $H_{y1}$ and $H_{y2}$ where solved by a MIMO solution. Note that the multiple coherence in Figure 15.12 (a) is approximately unity, and the ordinary coherence between the two input signals was low, so the MIMO system can be solved with good precision. However, the sum of the estimated coherent output spectra using this approach do not add up to the measured output signal, so this method cannot be used when we have correlated inputs.*

*We should make one more observation from Figure 15.12 (b). Looking at the error function, which should equal the uncorrelated spectral density $G_{nn}$ shown in Figure 15.12 (c), obviously there is some problem around the natural frequencies of both SDOF systems. This is due to the leakage sensitivity of the SVD, which was mentioned in Section 15.1.1. A main conclusion from this is, that great caution should be used when trying to estimate residual spectra, as the spectrum $G_{nn}$ is really the error in the analysis, and not necessarily only the spectrum of the uncorrelated signal, $n(t)$.*

*Finally we should make a special note regarding the coherence estimates. When the coherence (of any type) is low, there is a large random error in the estimate, as is obvious from Figure 15.12. This error propagates through to some of the coherent spectra, as seen in Figure 15.12 (b).*

*End of example.*

## 15.3 Noise Source Identification (NSI)

Noise source identification is a common application of virtual and conditioned signals in applications where there is at least two random, potential sources to perceived noise. A typical example from the automotive industry is road noise and wind noise, which are both random and both contribute to the perceived sound in a car.

### *15.3.1 Multiple Source Example*

We will now apply the techniques described in the preceding part of this chapter to an experimental setup, illustrated in Figure 15.13. This setup consists of two electrical noise sources connected to a speaker and an aluminum plate, respectively, which both radiate sound. A microphone ('Mic. 1') is placed in front of the speaker to pick up this source, and an accelerometer ('Acc. 1') is placed on the plate to pick up this source. A response microphone ('Mic. 2'), is placed in the room, at a position where we want to be able to tell from which source the sound comes, at each frequency.

The first thing we want to do, is to find the number of independent sources in the room, so that we know how many reference signals we need to use for the virtual signals. The spectral densities of the voltage signals from the two microphones and the accelerometer are plotted in Figure 15.14 (a). As can be seen in the figure, the three signals have approximately similar spectral density levels. The reason we

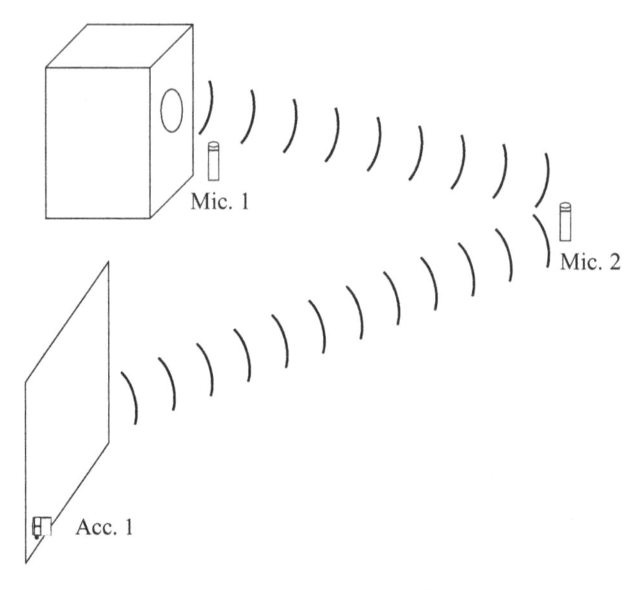


**Figure 15.13** Measurement setup for noise source identification example. Two noise sources generating approximately white Gaussian noise are fed to a speaker and (through a shaker) an aluminum plate, respectively. At some frequencies the plate will emit noise which will mix with the noise from the speaker. The noise from the speaker can be assumed not to be totally flat as the speaker is not ideal. A reference accelerometer was used on the plate to pick up the noise correlated with what was emitted by the plate, one microphone was used to record the speaker noise, and one microphone was used as the response pickup

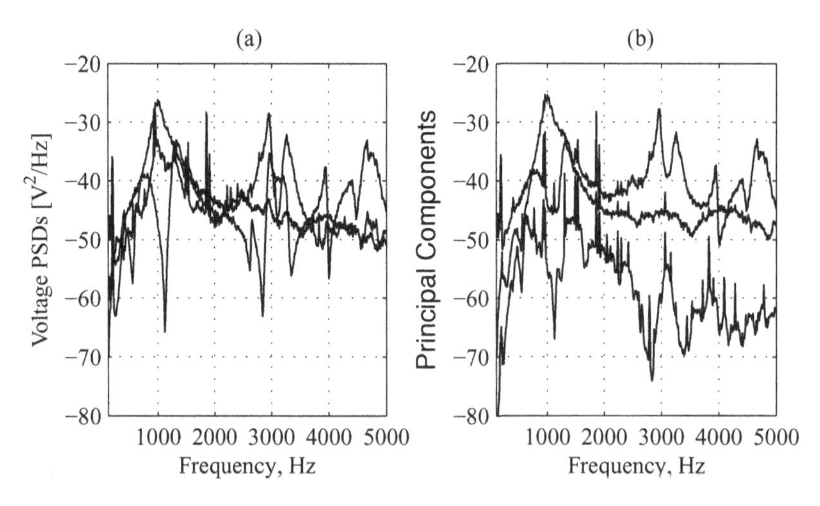


**Figure 15.14** In (a) autospectral densities of the voltage signals from the two microphones and the accelerometer in Figure 15.13. The spectral densities have approximately equal levels, with local variations due to the sound field in the room and the plate dynamics. In (b) principal components of the three signals in (a). Two principal components remain at a high level, whereas the third is reduced by approximately 10–20 dB, indicating that two sources were present

use the voltage levels is that we want the three signals scaled to approximately the same levels for the principal component calculation.

The principal components of the three signals are then computed using the complete input cross-spectral matrix of the three signals, i.e., in addition to the three autospectral densities in Figure 15.14, also the cross-spectral densities between each pair of sensors were calculated. The resulting principal components are plotted in Figure 15.14 (b). This figure shows that the two highest principal components are at least 10 dB higher than the third principal component, and more at frequencies above 2000 Hz. The interpretation of this is that there were two dominating sources in the room, although the dynamic range is not very large. Indeed, at some frequencies, for example approximately 900 and 1800 Hz, there are some narrowband peaks, most visible in the response microphone. These are tones from a fan located in the room, but these tones will not be possible to analyze as we do not have any reference related to them.

We will now use the principal components and virtual cross-spectra or virtual coherence functions to rank the sources at each frequency. The cumulated virtual input/output coherence functions are plotted in Figure 15.15. The figure shows that between 70% and close to 100 % of $G_{yy}$ can be explained by a linear relationship with one or both of the virtual signals, which indicates that there were mainly two sources during the measurements, as is expected. This is not true, however, in the frequency range from approximately 400–700 Hz, where there is apparently another source contributing considerably to the sound in the response microphone.

In Figure 15.16, the cumulated virtual coherent output spectra are plotted together with the spectral density of the response microphone, $G_{yy}$. This provides the same information as the virtual coherence functions in Figure 15.15, but this type of plot is often preferred over the former, since it shows directly

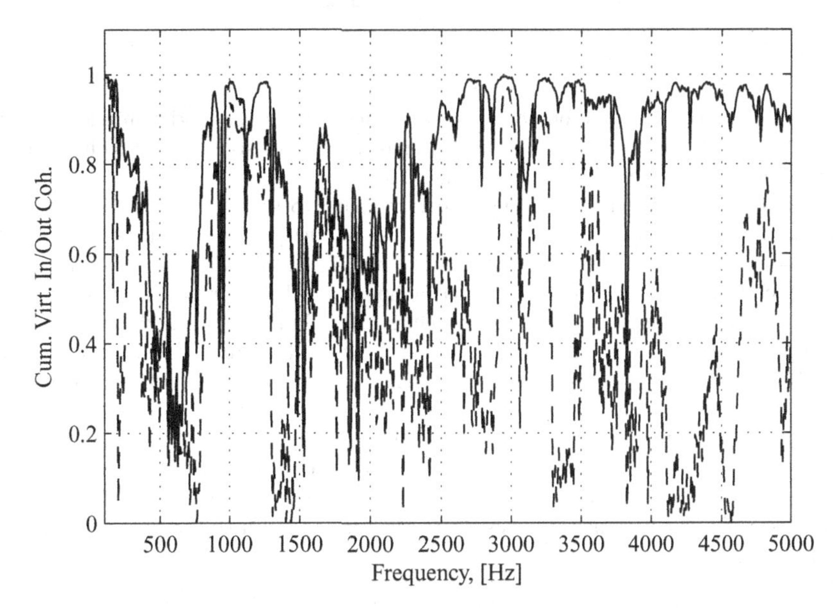


**Figure 15.15** Virtual input/output coherence functions. The functions show that (i) more than 70% of $G_{yy}$ can be explained by the two uncorrelated sources, except mainly in the frequency range 400–700 Hz where another (unmeasured) source is dominating; and (ii) the two uncorrelated sources dominate at different frequencies, which can seen by the first cumulated virtual coherence dropping, where the second cumulated coherence is still high, which means the second (not cumulated) virtual coherence is high at those frequencies. Also note that the second cumulated coherence in this case with two virtual signals, equals the multiple coherence

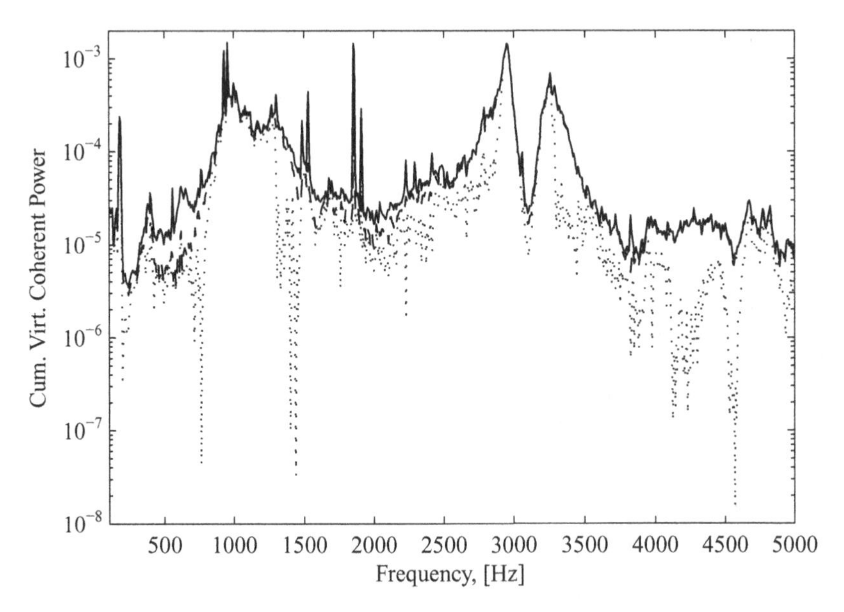


**Figure 15.16** Virtual coherent output spectra and the spectral density of the response microphone signal, $G_{yy}$. From the figure it is clear that at most frequencies the output signal is closely described by the two independent virtual signals, i.e., the sound in the response microphone *is indeed* produced by one or both of the reference signals, the speaker and the plate

the division of $G_{yy}$ into the various contributions. Although difficult to see in print, the second cumulated coherent output spectrum is very close to the measured signal at most frequencies, which can also be interpreted from Figure 15.15. Note the important difference in reading the plots, however, between the linear scale in Figure 15.15 and the logarithmic scale in Figure 15.16.

The final step may be to find which of the two sources, the plate or the speaker, is contributing to the response sound ('Mic. 2') at a particular frequency. To do this we need to compute the virtual input coherence functions, between each principal component and each input signal ('Acc. 1' and 'Mic 1', respectively). These functions are plotted in Figure 15.17. To find which of the two actual sources (plate or speaker) is dominating the response sound in 'Mic. 2' at a particular frequency, we first go to Figure 15.16, and look at the frequency range of interest. As an example we take the frequency range between approximately 1300 and 1500 Hz. In Figure 15.16 we find that in this frequency range the second virtual signal dominates the response signal (because the dotted line, which is the first virtual coherent spectrum drops). Then from Figure 15.17 (b) we find that in the same frequency range, it is the second measured signal which is correlated with the second virtual signal (the first virtual signal is plotted as dotted, and the sum of the first and second as solid). The speaker is consequently the dominating source between 1300 and 1500 Hz.

## 15.3.2 *Automotive Example*

The correlation techniques discussed in the present chapter have been used successfully in the automotive industry for many years. As an example of an application of virtual signals, we will conclude this chapter with an example of structure-borne road noise analysis from a modern car. The response sensor in this

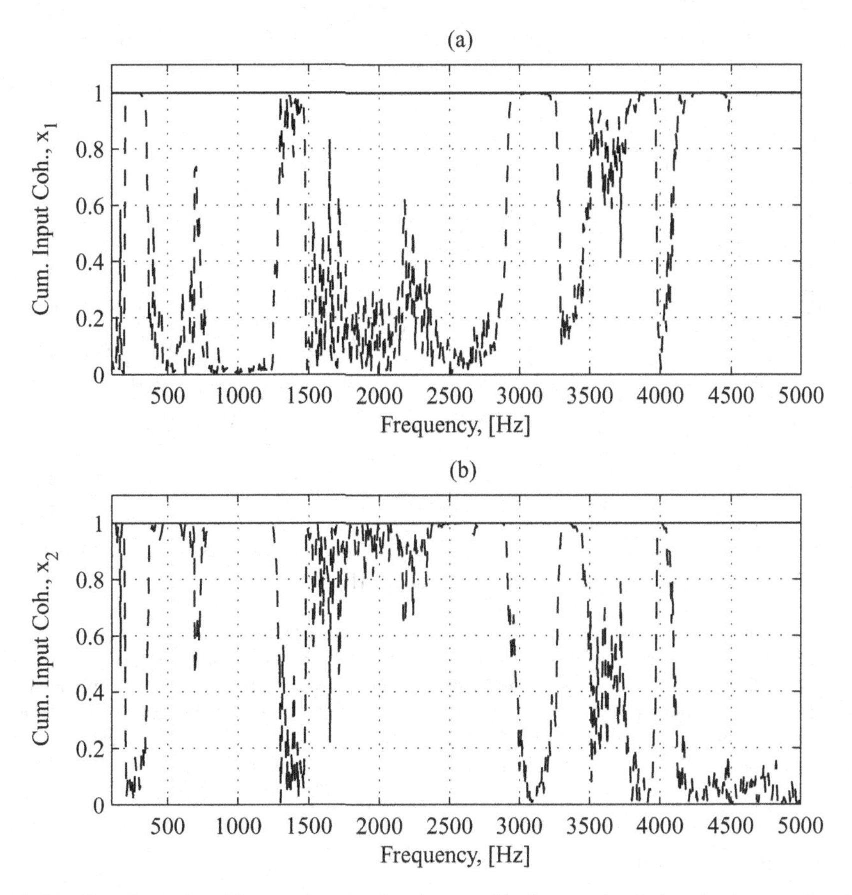


**Figure 15.17** Cumulated virtual input coherence functions. In (a), the cumulated virtual coherence functions for measured signal $x_1$, the plate acceleration, and in (b), the cumulated virtual coherence functions for measured signal $x_2$, the speaker microphone. From the plots we can conclude that, e.g., in the frequency range from approximately 1300–1500 Hz, $x_1$ is highly correlated with virtual signal $x_1'$ and $x_2$ with virtual signal $x_2'$. Because the SVD at each frequency is sorting the principal components according to the power of the measured signals, the relationship between the measured and the virtual signals swap, whenever one of the measured signals becomes higher than the other after having been smaller. Therefore, at each frequency of interest, this plot has to be investigated before concluding which of the measured signals actually correlates with a particular virtual signal

case was a microphone in the position of the driver's left ear, measured using a standard measurement microphone. The inputs of interest were the front and rear wheel axles, as they produce structure borne noise which is uncorrelated to some extent. This is usually ascribed to the fact that the left and right tires experience different road surfaces. To measure the part of the response microphone signal coherent with each of these axles, accelerometers were mounted near the center of each wheel spindle.

An analysis using principal component analysis resulted in the virtual coherent output spectral densities shown in Figure 15.18. In this plot, where the spectral densities have been A-weighted (see Section 3.3.4), it can be seen that up to approximately 500 Hz, the sound in the drivers' ear is dominated by the structure borne noise. Above 500 Hz, other sources contribute significantly to the sound in the driver's ear. A careful examination of the plot also reveals that the noise is mostly coherent with the front axle.

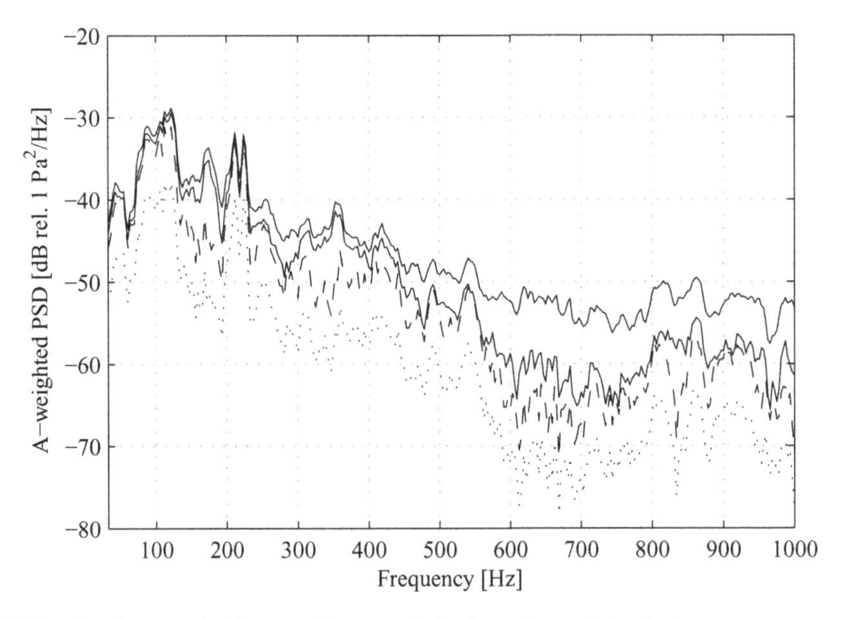


**Figure 15.18** Sound pressure in driver·s ear from an analysis of a modern car. Virtual coherent output power spectra from front axle (dash-dotted), rear axle (dotted), sum of those two (solid), and total noise spectrum (solid and highest at all frequencies). The example shows that structure-borne sound dominates the driver·s perception up to approximately 500 Hz and that the two uncorrelated sources contribute to most of the sound (within approximately 5 dB) in this frequency range. Above 500 Hz other sources contribute by more than 10 dB. Note that the y-scale is in dB, but not in sound pressure level as the curves are PSDs. [Data courtesy of SAAB Automobile AB]

## 15.4 Chapter Summary

Principal components can be seen as the autospectral densities of virtual, uncorrelated signals. To compute the principal components, the eigenvalues of the input cross-spectral matrix of all involved signals is computed at each frequency, and the eigenvalues are ordered from largest to smallest. The largest eigenvalue at each frequency is the first principal component, the second largest eigenvalue is the second principal component, etc.

The eigenvalue decomposition results in that the measured signal with highest power at a particular frequency, dominates the principal component at that frequency. Thus the principal component approach differs from the approach with conditioned signals described in Chapter 14, because the latter type of orthogonal signals is ordered manually by the user. This means that the conditioned spectra and coherence functions become different with different ordering of the signals. For principal components, the order of the input signals does not affect the result. There is also a numerical stability issue, as the SVD is very accurate also in cases with severe extraneous noise. It is therefore often preferred over the conditioned signal approach.

The virtual signals, $x_i'$, cannot be measured or estimated from measured data, just as we could not measure or estimate the conditioned signals. However, we can estimate virtual cross-spectral densities between any virtual signal (as input) and any measured input signal, $x_q$, or between any virtual signal and an output signal, $y_p$. Because we already have the auto-spectral densities of the virtual signals, the principal components, we can therefore estimate virtual coherence functions between the virtual signals and any input or output signal. these virtual coherence functions are simply ordinary coherence functions between a virtual signal and a measured signal.

The purpose of computing the virtual coherence functions is usually to split the power of an output signal (a 'target sensor') into the contribution from each virtual signal, which can be done because the virtual signals are uncorrelated. Several examples of this were given in this chapter.

We also presented an application in this chapter of how to use the SVD for data compression, where a frequency response matrix was condensed to a smaller size of frequency responses, still having the same information.

## 15.5 Problems

Many of the problems following are supported by the accompanying ABRAVIBE toolbox for MATLAB/Octave and further examples which can be downloaded with the toolbox. If you have not already done so, please read Section 1.6, and follow the instructions to download this toolbox together with example files.

**Problem 15.1** *Write a MATLAB/Octave script which creates three Gaussian signals using the* ***randn*** *command, with 102 400 samples each. Compute the cross-spectral matrix of the three signals, using a Hanning window of length 1024 samples. Then compute the principal components by using the* ***eig*** *and the* ***svd*** *commands. Compare the computed principal components and see if there are any differences. Also, clock the computation time using each of the commands for computing the principal components to see if there is any significant difference.*

*Hint: use* ***tic*** *and* ***toc****, to clock the computation time, see MATLAB/Octave help.*

**Problem 15.2** *Calculate the principal components and virtual input/output coherence functions of two Gaussian signals with equal RMS level which each pass a linear system with unity frequency response, to produce an output signal. (That is, the output signal is simply the sum of the two random input signals). Use the parameters for sampling frequency, etc. and appropriate code from Example 15.2.2. Plot the principal components and the virtual input/output coherence functions. Do they look OK? Try to explain why.*

*Hint: Think about what makes the principal components separate at each frequency, and see the discussion in Example 15.2.2.*

## References


Dippery KD, Phillips AW and Allemang RJ 1994 Condensation of the spatial domain in modal parameter-estimation. *12th International Modal Analysis Conference*, Honolulu, HI **2251**, 818–824.

Hotelling H 1933 Analysis of a complex of statistical variables into principal components. *Journal of Educational Psychology* **24**, 417–441, 498–520.

Lembregts F 1988 *Frequency Domain Identification Techniques for Experimental Multiple Input Modal Analysis* PhD Thesis Katholieke Universiteit Leuven, Leuven, Belgium.

Maia N and Silva J (eds) 2003 *Theoretical and Experimental Modal Analysis*. Research Studies Press, Baldock, Hertforsdhire, England.

Otnes RK and Enochson L 1972 *Digital Time Series Analysis*. Wiley Interscience.

Otte D 1994 *Development and evaluation of singular value analysis methodologies for studying multivariate noise and vibration problems* PhD Thesis Catholic University Leuven, Belgium.

Otte D, de Ponseele P. V and Leuridan J 1990 Operational deflection shapes in multisource environments *Proc. 8th International Modal Analysis Conference*, Kissimmee, FL.

Strang G 2005 *Linear Algebra and its Applications* 4th edn. Brooks Cole, San Diego.

Tucker S and Vold H 1990 On principal response analysis *Proc. ASELAB Conf.*, Paris, France.

Vold H 1986 Estimation of operating shapes and homogeneous constraints with coherent background noise. *Proc. ISMA 1986, International Conference on Noise and Vibration Engineering*, Catholic University, Leuven, Belgium.

# 16

# Advanced Analysis Methods

In this chapter we will discuss some signal analysis tools which have not found their right place in the preceding chapters, but which have their applications in noise and vibration analysis. Each method will be briefly discussed and references given for the reader who wants to find more information. We will also very briefly describe two main tools for vibration analysis which are the keys to solving many vibration problems: operating deflection shapes (ODS), and experimental modal analysis (EMA).

## 16.1 Shock Response Spectrum

In cases where the damaging effect of a vibration environment is of interest, various so-called *response spectra* are often used (Himmelblau *et al.* 1993; Lalanne 2002). The most common response spectrum is the *Shock Response Spectrum*, or SRS, (Greenfield 1977; ISO 18431-4: 2007; Smallwood 1981). It has been used for a long time in aerospace and military applications, especially for pyroshock signals. In recent years, it has also found a growing popularity in civilian applications, such as for example the automotive industry.

The SRS calculates the damaging potential of a signal, originally a transient signal, as the name implies. This damaging potential is of course a factor of both the frequency content of the signal, and the resonance frequencies of the structure to which the transient (vibration signal) is applied. The problem of determining the damaging potential of a vibration environment is, of course, an extremely difficult task. The shock response offers a conservative, approximative method of finding the worst damaging potential of the vibration environment in question.

While the SRS was originally developed for transients, it has proven valuable also when comparing vibration environments of very different character, such as periodic, random, and transient. Let us say that we have a sensitive electronics box, e.g., for engine control. We have to mount this box somewhere in the engine compartment of a car. In some possible mounting points, the vibrations are mainly periodic, such as on top of the engine block, whereas if we mount it on the chassis, perhaps there will be more random vibrations caused by the road. How do we compare these entirely different environments, and select the least harmful position to place our electronics box? The SRS can be used for this purpose.

Another common application of SRS is in environmental testing (see Section 1.3). It is often the case that a real-life vibration measurement has to be condensed into a test specification such as a random or shock test. SRS is a useful tool in such design (Ahlin 2006; Henderson and Piersol 2003; Lalanne 2002).

The basis for the SRS is an assumption that we are going to mount an object in a location where we know the vibration environment (usually in terms of the acceleration level), and we wish to have a measure of how dangerous this environment is to our object. A simple assumption is then that the object

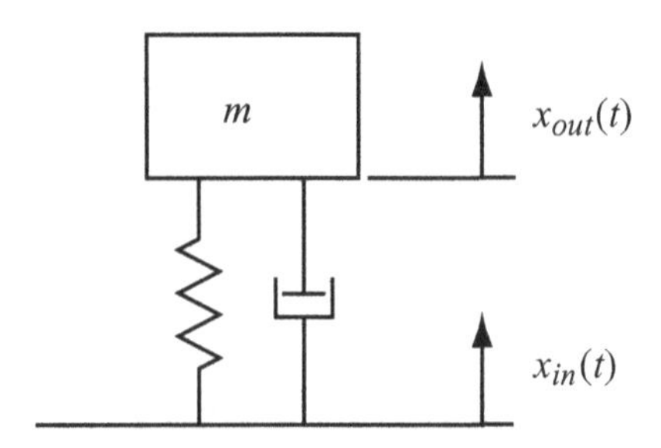


**Figure 16.1** SDOF (single-degree-of-freedom) mechanical system used for SRS (Shock Response Spectrum) calculation. For each frequency value in the SRS, the mass-spring system is tuned to that frequency. The base of the system is supposed to be excited by the input signal $x_{in}(t)$, and the SRS is defined as the maximum of the (absolute value of) the resulting response of the mass, $x_{out}(t)$. The excitation and response units are usually acceleration, but can be velocity or displacement in certain applications, or the relative displacement, velocity or acceleration, see (ISO 18431-4: 2007)

to be mounted will act as a single-degree-of-freedom system, causing a resonance amplification of the vibrations in the mounting position. This assumption does not necessarily mean that the mounted object has only one natural frequency, as each natural frequency (mode) acts as an SDOF system, as we showed in Chapter 6. As we will see, the SRS accounts for any natural frequency our object may have.

The shock response is defined as the output response of a single degree-of-freedom (SDOF) mechanical mass-spring-damper-system, to the transient, which is applied to the base of the mechanical system. For each frequency value in the SRS, the natural frequency of the mechanical SDOF system is tuned to that frequency. The vibration signal to be analyzed can arbitrarily be displacement, velocity or acceleration; however, acceleration is the most common, (ISO 18431-4: 2007). Some authors advocate the use of pseudo-velocity, (Gaberson 2003; Gaberson *et al.* 2000). For this brief introduction we will focus on acceleration output.

An illustration of an SDOF mechanical system is shown in Figure 16.1. The input signal, $x$ (the measured vibration signal) is assumed to be the acceleration of the base of the mechanical system. The output, from which the SRS is derived, is the resulting vibration (acceleration) of the mass, $m$. Denoting the (undamped) natural frequency of the system by $f_n$, and the (viscous) relative damping by $\zeta_n$, we obtain the transfer function

$$H(s) = \frac{X_{\text{out}}}{X_{\text{in}}} = \frac{1 + 2\zeta_n (s/\omega_n)}{1 + 2\zeta_n (s/\omega_n) + (s/\omega_n)^2} \tag{16.1}$$

where $\omega_n = 2\pi f_n$ is the natural angular frequency in rad/s.

Using the transfer function in Equation (16.1), the output time signal is calculated, and usually, the maximum absolute value of the output signal is taken as the SRS value. This is referred to as the *maximax* shock response. Furthermore, the output signal is divided into two parts. The part of the output signal during the time when the input signal is present is called the *primary part*, and the part of the output signal after the input signal is removed (no longer exists) is called the *residual part*. Depending on which of the two output signal parts is used, the SRS is referred to as primary SRS or residual SRS, respectively. Finally, the damping of the mechanical system has to be selected. The assumed damping of the structure for which the SRS is applied should be taken. If this value is not known, usually $\zeta_n = 0.05$ is used as a standard value. In SRS applications (and generally in environmental testing) it is more common to use the $Q$-factor, which we saw in Equation (5.27) is $Q = 1/2\zeta$.

It is obvious from the above, that the shock response spectrum is not a spectrum in general terms. It is called a spectrum only because it has frequency on its $x$-axis. The interpretation of the SRS value for a

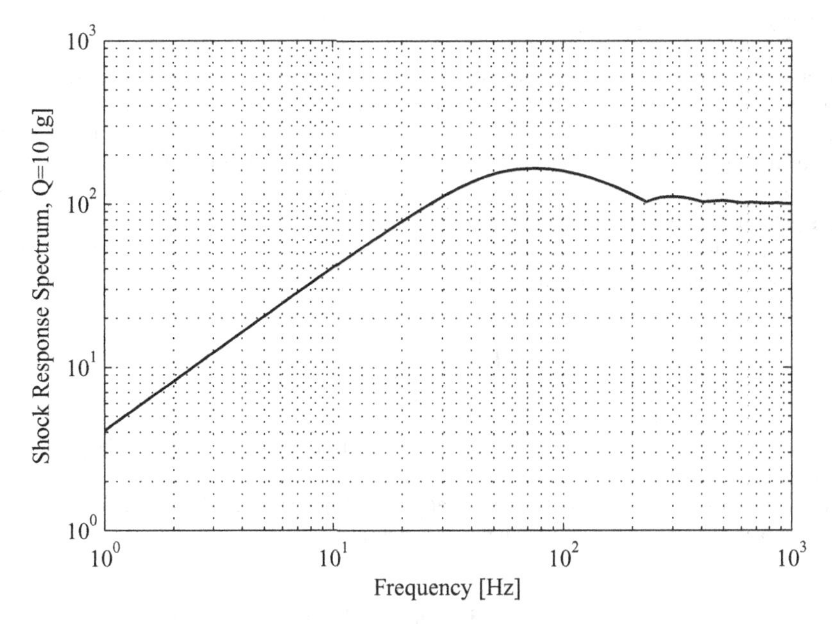


**Figure 16.2** Shock response spectrum (SRS) of a half sine of 11 ms duration, with a maximum value of 100 g, calculated with $Q = 10$. The $y$-axis as well as the frequency axis are logarithmic, and the spectrum is calculated for frequencies spaced 1/6th octave apart

particular frequency, is the maximum acceleration that will be caused by the analyzed signal, given that the structure has a natural frequency of the particular 'SRS frequency', with the damping used for the SRS calculation.

To calculate the SRS, digital filters defined in (ISO 18431-4: 2007) should be used. The SRS is usually calculated for frequencies on a logarithmic scale, often 1/6-octave bands, and for the $Q$-factor of $Q = 10$. In Figure 16.2, the maximax SRS for absolute acceleration using $Q = 10$, produced by a transient (half sine), is plotted.

When using a standard noise and vibration measurement system to acquire signals for SRS calculations, it is important to consider the oversampling rate, see Chapter 3 as well as the phase linearity of the anti-aliasing filters, see Section 11.2.2. As the SRS is calculated in the time domain, it is not sufficient to use the oversampling of 2.56 that is normally used for frequency analysis. At least a factor of 10 must be used in order for the error in the SRS calculation to be small. In many systems this can be accomplished by first sampling the data with a sufficient frequency range and a 'normal' oversampling ratio of 2.56, and in a post-processing stage digitally upsample the data by a factor of 4 using the procedure described in Section 3.2. This will limit the maximum amplitude inaccuracy due to the sampling frequency to less than 10%.

A particular characteristic of the shock response spectrum should be especially noted. At high frequencies, the SRS reaches the maximum of the absolute value of the input acceleration. This occurs at either the highest frequency contained in the analyzed acceleration signal, or the bandwidth (highest frequency) of the acquisition system, whichever is lowest. This is easily realized by considering the low-frequency part of the transfer function in Equation (16.1), which equals unity. For high SRS frequencies, the natural frequency of the SDOF system is high, and thus the bandwidth of the measured signal is low compared with the natural frequency. This means that the output of the SDOF system will be the same as the measured signal; there is no resonance amplification.

## 16.2 The Hilbert Transform

The Hilbert transform (Bendat and Piersol 2010) is a useful transform in some signal analysis applications. It is used for two different purposes; (i) for *envelope* calculations, for example when studying modulated signals, or (ii) to create so-called *analytic signals*, for which it relates the real and imaginary part. The Hilbert transform of a signal is a new signal in the same domain as the original signal. Thus the Hilbert transform of a time signal is a new time signal, and the Hilbert transform of a frequency domain signal is a new signal in the frequency domain.

The Hilbert transform, $\tilde{x}(t)$, of a real time signal $x(t)$, given that this signal exists for times $-\infty \leq t \leq \infty$, is a new, real-valued time signal which is defined by

$$\tilde{x}(t) = \int_{-\infty}^{\infty} \frac{x(u)}{\pi(t-u)} du = x(t) * \frac{1}{\pi t} \tag{16.2}$$

where $*$ denotes convolution. The Hilbert transform is a linear operator, i.e., the Hilbert transform of a sum of two signals equals the sum of the separate Hilbert transforms of each signal.

The convolution in Equation (16.2) corresponds to multiplication in the frequency domain by the Fourier transform of $1/\pi t$. It can be shown that this Fourier transform is given by

$$\mathcal{F}\left[\frac{1}{\pi t}\right] = -\mathrm{j} \cdot \mathrm{sgn}(f) = \begin{cases} -\mathrm{j}, & f > 0 \\ 0, & f = 0 \\ \mathrm{j}, & f < 0 \end{cases} \tag{16.3}$$

where $\mathcal{F}$ denotes the Fourier transform. Thus the Fourier transform of $\tilde{x}(t)$ is

$$\tilde{X}(f) = \mathcal{F}[\tilde{x}(t)] = -\mathrm{j}\mathrm{sgn}(f)X(f). \tag{16.4}$$

Equation (16.4) can alternatively be written as

$$\tilde{X}(f) = \begin{cases} \mathrm{e}^{-j\pi/2}, & f > 0 \\ 0, & f = 0 \\ \mathrm{e}^{j\pi/2}, & f < 0 \end{cases}. \tag{16.5}$$

Thus the Hilbert transform of a time domain signal equals a phase shift of $\pm 90°$ in the frequency domain. The Hilbert transform acts as an all-pass filter with uniform (flat) amplitude characteristic and with phase shift of $-90°$ for positive frequencies and $+90°$ for negative frequencies. The inverse Hilbert transform which calculates the time signal $x(t)$ from the Hilbert transform $\tilde{x}(t)$, is given by

$$x(t) = -\int_{-\infty}^{\infty} \frac{\tilde{x}(u)}{\pi(t-u)} du \tag{16.6}$$

but can also be written as

$$x(t) = \mathcal{F}^{-1}\left[\mathrm{j}\mathrm{sgn}(f)\tilde{X}(f)\right]. \tag{16.7}$$

It is important to note that the Hilbert transform does not commute with the Fourier transform, i.e.,

$$\mathcal{H}[\mathcal{F}[x(t)]] \neq \mathcal{F}[\mathcal{H}[x(t)]] \tag{16.8}$$

where we have denoted the Hilbert transform by $\mathcal{H}$.

### 16.2.1 Computation of the Hilbert Transform

To compute the Hilbert transformation, a new function $z(t)$ is defined by

$$z(t) = x(t) + j\tilde{x}(t) \tag{16.9}$$

which is called the *analytic* signal of $x(t)$. In the frequency domain, the Fourier transform of $z(t)$ will be

$$Z(f) = X(f) + j\tilde{X}(f). \tag{16.10}$$

By using Equation (16.4) with Equation (16.10), we obtain

$$Z(f) = X(f) + j(-j\operatorname{sgn}(f)X(f)) = (1 + \operatorname{sgn}(f))X(f) \tag{16.11}$$

If we define Z(0) = X(0), we obtain

$$Z(f) = \begin{cases} 2X(f), f > 0 \\ X(0), f = 0 \\ 0, f < 0 \end{cases}. \tag{16.12}$$

Using the Fourier transform $Z(f)$ we can now obtain the Hilbert transform by the inverse Fourier transform of $Z(f)$, whereby

$$\tilde{x}(t) = \mathcal{F}^{-1}[Z(f)] \tag{16.13}$$

i.e., the method to compute the Hilbert transform is by inverse Fourier transforming the Fourier transform of the analytical signal $z(t)$.

Note that the negative half-plane has been zeroed out in $Z(f)$, so that the digital formula to calculate the Hilbert transform becomes

$$\tilde{x}(n) = 2\Delta f \operatorname{Im}\left[\sum_{k=0}^{N/2} X_c(k) e^{j\frac{2\pi nk}{N}}\right]. \tag{16.14}$$

Especially note that the sum in Equation (16.14) is only carried out up to the Nyquist frequency, $N/2$. In Equation (16.14), $X_c(k)$ is the normalized discrete Fourier transform which approximates the continuous Fourier transform. Thus

$$X_c(k) = \Delta t \sum_{n=0}^{N-1} x(n) e^{-j\frac{2\pi kn}{N}}. \tag{16.15}$$

Since $f = f_s/N$ and $\Delta t = 1/f_s$ then, using the ordinary DFT $X(k)$, we can simplify Equation (16.14) into

$$\tilde{x}(n) = \frac{2}{N} \operatorname{Im}\left[\sum_{k=0}^{N/2} X(k) e^{j\frac{2\pi nk}{N}}\right] \tag{16.16}$$

which is the algorithm to compute the Hilbert transform $\tilde{x}(t) = \mathcal{H}[x(t)]$.

### 16.2.2 Envelope Detection by the Hilbert Transform

One of the common applications of the Hilbert transform in noise and vibration analysis, is to find the envelope of correlation signals, or modulated signals. Recall that for bandwidth-limited noise, the autocorrelation of a signal $x(t)$, $R_{xx}(\tau)$, becomes broadened into a sinc function, see Section 4.2.12. If we have a situation where a band-limited signal passes several paths from one point to another, as

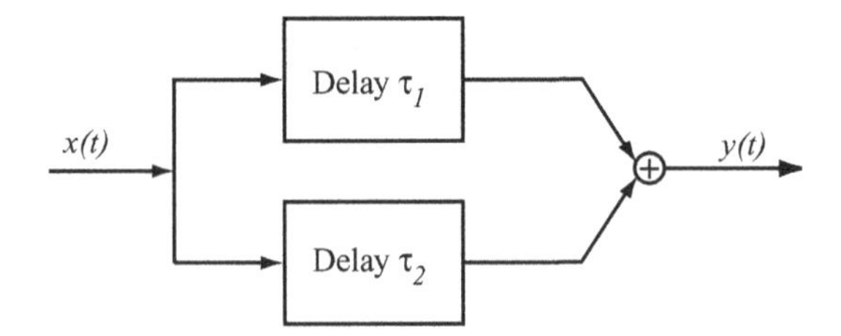


**Figure 16.3** Two-path problem. The bandwidth-limited noise $x(t)$ is passing through two paths. Each path has a different time delay, denoted $\tau_1$ and $\tau_2$. By calculating the cross-correlation between signals $x$ and $y$ and locate the two peaks in this function, potentially the time delays can be estimated

depicted in Figure 16.3, it can therefore potentially be hard to find the positions of the maxima of the cross-correlation between the two signals.

To use the Hilbert transform to calculate envelopes, the complex variable $z(t)$, called the *analytic signal* related to $x(t)$ as defined in Equation (16.9) is used. The envelope of the signal $x(t)$ is defined as the magnitude of $z(t)$, i.e., the envelope $e(t)$ is defined as

$$e(t) = |z(t)| = \sqrt{x^2(t) + \tilde{x}^2(t)}. \tag{16.17}$$

**Example 16.2.1** *We will illustrate the use of the Hilbert transform for envelope computation by an example of a two-path system as shown in Figure 16.3, where each of the two systems, in addition to a time delay, consists of an SDOF system with a natural frequency of 100 Hz and relative damping* $\zeta = 0.01$. *Both SDOF signals are excited by bandlimited white random noise with a bandwidth of 1000 Hz.*

*In Figure 16.4, the cross-correlation of the two signals,* $R_{yx}(\tau)$ *is shown. As is seen in Figure 16.4, it is hard to find exactly where the two peaks are found due to the wide correlation peaks. In Figure 16.5 the envelope calculated by using the Hilbert transform is shown. In this plot it is easy to see where the two peaks occur.*

*End of example.*

### *16.2.3 Relating Real and Imaginary Parts of Frequency Response Functions*

The second use of the Hilbert transform we will look at, is to find the relationship between the real part and the imaginary part of a frequency response function. We showed in Section 2.7.1 that any signal $h(t)$ can be divided into a sum of an even part $h_e(t)$ and an odd part $h_o(t)$.

We are especially interested at the moment in impulse response functions $h(t)$. If the impulse response function represents a physically realizable system, then it is causal, i.e.,

$$h(t) = 0, t < 0 \tag{16.18}$$

Then, from Equations (2.65) and (2.66) it follows that for times $t > 0$,

$$\begin{aligned} h_e(t) &= \frac{1}{2}h(t) \\ h_o(t) &= h_e(t) \end{aligned} \tag{16.19}$$

and for times $t < 0$

$$\begin{aligned} h_e(t) &= \frac{1}{2}h(-t) \\ h_o(t) &= -h_e(t) \end{aligned} \tag{16.20}$$

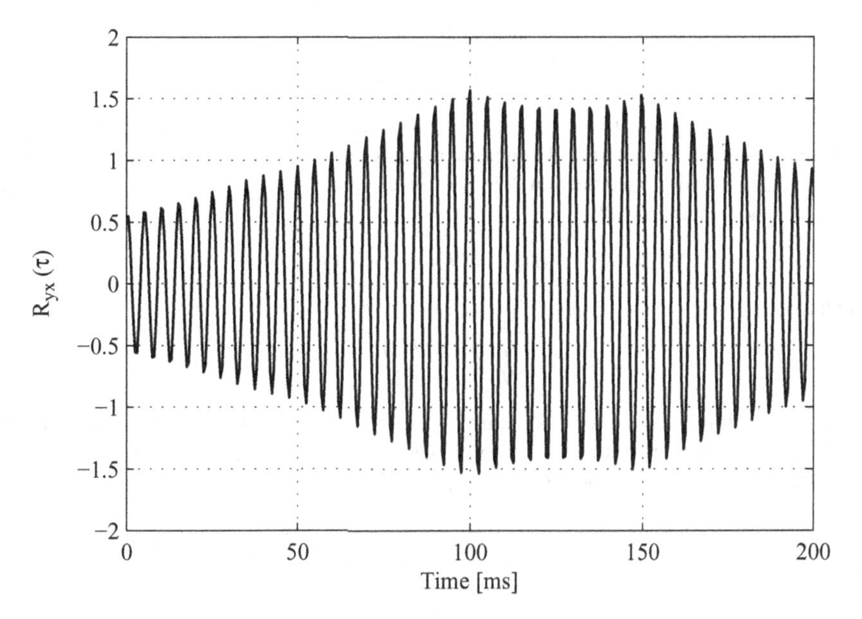


**Figure 16.4** Plot for Example 16.2.1. Cross-correlation $R_{yx}(\tau)$ of two signals with delay as illustrated in Figure 16.3, where the paths have time delays $\tau_1 = 100$ ms, and $\tau_2 = 150$ ms. In addition to the different time delays, each path consists of an SDOF system with natural frequency of 100 Hz, and 1% relative damping. Due to the band-limited nature of the noise after passing the SDOF systems, each correlation peak is broadened, making it hard to find the location of the two peaks in $R_{yx}(\tau)$

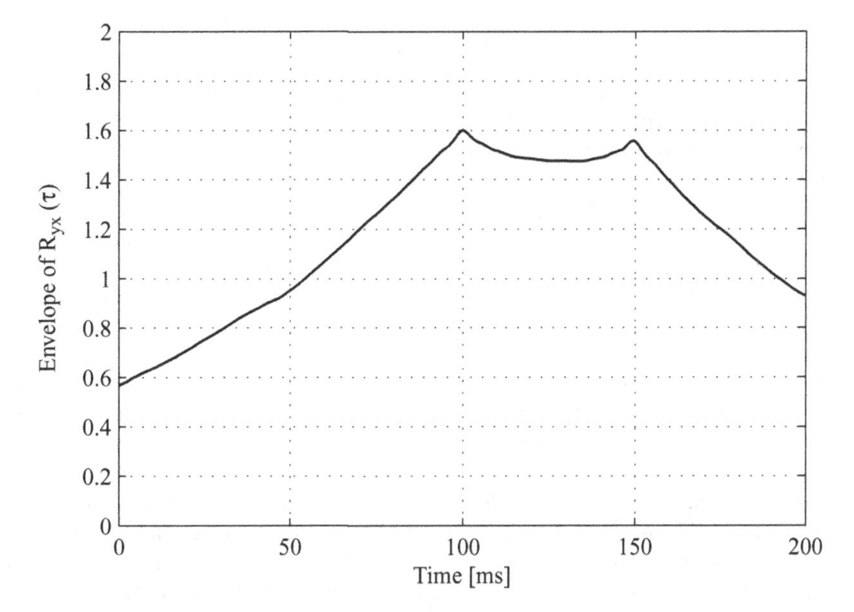


**Figure 16.5** Envelope of cross-correlation function $R_{yx}(\tau)$ in Figure 16.4. In the envelope it is easy to find the locations of the two peaks at approximately 100 and 150 ms

From Equations (16.19) and (16.20) we thus obtain

$$h_e(t) = \text{sgn}(t)h_o(t) \tag{16.21}$$

For the frequency response function of any causal system, the real part of the frequency response function comes from the even part of the impulse response, and the imaginary part comes from the odd part of the impulse response, see Section 2.7. This means that if we define the frequency response by a sum of its real and imaginary parts,

$$H(f) = H_R(f) + \mathrm{j}H_I(f) \tag{16.22}$$

then we have

$$\begin{aligned} H_R(f) &= \mathcal{F}[h_e(t)] \\ H_I(f) &= \mathcal{F}[h_o(t)] \end{aligned} \tag{16.23}$$

Now, again, multiplication in one domain, for example the time domain as in Equation (16.21), corresponds to convolution in the other domain, here the frequency domain. Using Equation (16.3) together with Equations (16.21) and (16.23), we get that

$$H_R(f) = \mathcal{F}[h_e(t)] = \mathcal{F}\left[\text{sgn}(t)h_o(t)\right] = H_I(f) * \frac{1}{\pi f} = \int_{-\infty}^{\infty} \frac{H_I(u)}{\pi(f-u)}\mathrm{d}u \tag{16.24}$$

which is the Hilbert transform of $H_I$. In other words, the real part, $H_R(f)$, of the frequency response function of a causal system, equals the Hilbert transform of the imaginary part, $H_I(f)$ of the same FRF, i.e.,

$$H_R = \mathcal{H}[H_I]. \tag{16.25}$$

Similarly, we can obtain

$$H_I(f) = -\mathcal{H}[H_R(f)] \tag{16.26}$$

which, in words, says that the imaginary part of a causal frequency response function, equals the Hilbert transform of the real part of the same FRF, with changed sign. The two statements in Equations (16.25) and (16.26) apply to all causal frequency response functions, and are called the Hilbert transform relationships between real and imaginary part.

The main application of Equations (16.25) and(16.26) in noise and vibration analysis is to investigate whether an estimated frequency response function belongs to a causal system. If that is not the case, it is a strong indication that the estimated system is nonlinear (Tomlinson and Kirk 1984).

When investigating frequency response functions by means of the Hilbert transform relationships between the real and imaginary part, it has been suggested that the functions be estimated by use of stepped-sine excitation, and not broadband excitation. The reasoning behind this is that the broadband excitation methods 'linearize' the FRF of the structure, thus yielding the best linear system between the measured input and output signals. For strong nonlinearities, however, this is not necessary, as they will result in non-causal impulse response functions.

**Example 16.2.2** *To illustrate what the Hilbert transform relations between real and imaginary parts of frequency response functions look like for a typical FRF, we will look at an FRF obtained by impact testing as described in Section 13.8. In Figure 16.6 (a) an FRF estimated by impact testing is shown. In Figure 16.6 (b) the impulse response of the same FRF is shown, and it is apparent that there is some non-causality which can be seen as a rising at the end of the block. Recall the periodicity of the discrete Fourier transform, which means that the part at the end of the block can be wrapped around to the left of time zero. This is very often seen on experimentally obtained IRFs. In Figure 16.6 (c) and (d),*

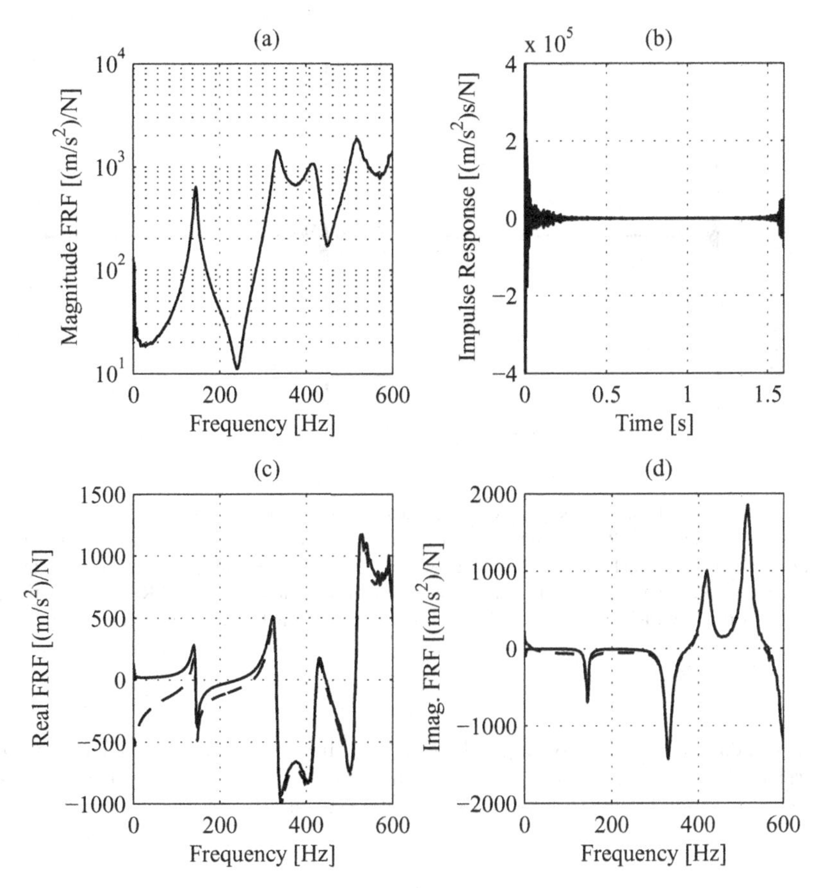


**Figure 16.6** Plots for Example 16.2.2. In (a) an FRF estimated from an impact excitation experiment is shown. Its impulse response in (b) shows non-causal behavior at the end of the block, which corresponds to negative time to the left of time zero. In (c) the real part of the FRF in (a) is shown, overlaid with the Hilbert transform of the imaginary part, and in (d) the imaginary part of the FRF in (a) is overlaid with the Hilbert transform of the real part (with changed sign)

*the real part and imaginary part of the original FRF, and the corresponding real and imaginary part obtained by the Hilbert transform relationships are plotted. As seen in the figure, in this case there is some discrepancy, particularly in the real part of the measured FRF and the real part created as the Hilbert transform of the imaginary part.*

*End of example.*

A potential use of the Hilbert transform, which has, to my best knowledge, never been published, is to 'clean up' estimated frequency responses by keeping either the real or imaginary part, and replacing the discarded part by the Hilbert transform of the kept part. This produces the result of a causal system and can potentially be more accurate than using the originally estimated FRF. In Figure 16.7 (a), the impulse response from the same FRF used for Example 16.2.2 has been computed after the FRF was processed this way, by discarding the imaginary part, and creating a new imaginary part from the real part by using Equation (16.25). In Figure 16.7 (b), the impulse response created by using the same procedure, but instead discarding the real part of the measured FRF is shown. As can be seen in the impulse response in

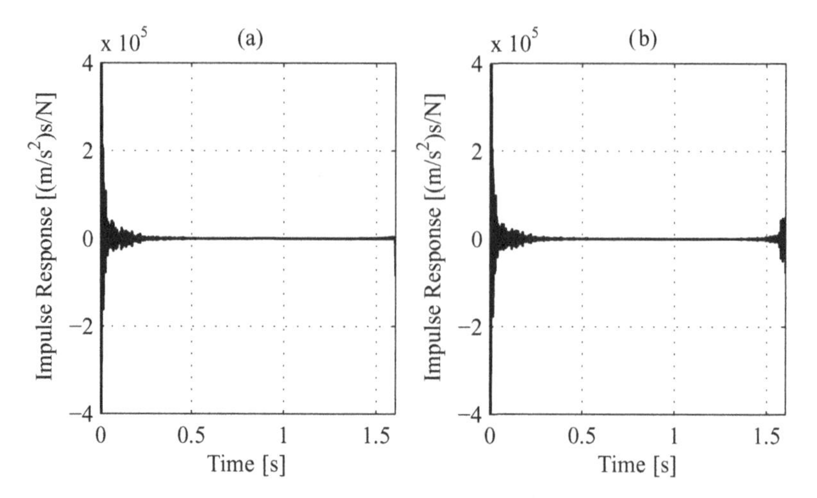


**Figure 16.7** Illustration of using Hilbert transform to 'clean up' frequency responses. In (a), an impulse response of the FRF used in Example 16.2.2 has been modified by using the real part of the FRF to ceate the imaginary part by the Hilbert relation in Equation (16.26). In (b), the impulse response using the imaginary part of the original FRF is shown. As apparent from the plots, the result is different, depending on which part of the FRF is kept. In this example, keeping the real part of the FRF and creating the imaginary part using the Hilbert transform leads to a more causal impulse response

Figure 16.7 (a), this can significantly reduce the non-causal behavior at the end of the record. However, as can be seen in Figure 16.7 (b), it matters which of the parts, the real or imaginary parts, are discarded. The impulse response in Figure 16.7 (b) shows no improvement over the original impulse response in Figure 16.6.

## 16.3 Cepstrum Analysis

Cepstrum is an analysis function with many applications in signal processing. Cepstrum analysis is used for example for echo removal, so-called deconvolution (in principle going 'backwards' through a filter), to find harmonics and hidden sidebands in signals, it is frequently used in speech recognition, etc. The term cepstrum is a paraphrase of the word spectrum, and comes from the original paper (Bogert *et al.* 1963). Since cepstrum introduces a time-related domain that is similar, but not equal to, the 'delay domain' of the autocorrelation function, the authors saw it necessary to introduce a new terminology in order to avoid confusion. Although most of their suggested terminology has not won common acceptance, a few of the terms are still used.

The originally proposed definitions of cepstra (there are several, as we will see below) are usually modified in modern texts, so some care needs to be executed when going back to the original papers. The most common definition today, is to define the cepstrum as the inverse Fourier transform of the logarithm of a spectrum. Several types of cepstra coexist, which are based on various spectra, single-sided and double-sided, etc. The two most common cepstrum functions, at least in noise and vibration analysis, the power cepstrum and the complex cepstrum will be presented here.

In noise and vibration analysis, cepstra are mostly used for diagnostics, particularly on gearboxes (Endo *et al.* 2009; Randall 1982). Lately, however, it has also received some interest for operational modal analysis applications (Gao and Randall 1996a,b; Hanson *et al.* 2007a,b; Randall and Gao 1994).

### 16.3.1 Power Cepstrum

The most common cepstrum function used in noise and vibration analysis is the *power cepstrum*, which is defined as the inverse Fourier transform of the logarithm of an autopower spectrum. In practice, the power cepstrum of a signal $x(t)$, denoted $c_{px}(\tau)$, is calculated as

$$c_{px}(\tau) = \text{IFFT}\left[\log\left(S_{xx}\right)\right] \tag{16.27}$$

where $S_{xx}$ is the double-sided autopower spectrum defined by Equation (10.1). Sometimes, the power cepstrum is instead calculated from the single-sided spectrum, which under certain circumstances will give some useful qualities to the cepstrum.

The power cepstrum has three important features:

- it is an inverse Fourier transform, which means it finds periodicities in $S_{xx}$;
- it uses $\log(S_{xx})$, which amplifies low levels, and compresses high levels. This further enhances the ability of the cepstrum to find periodicites also where the harmonics are low; and
- if the analyzed signal is a composition of an input going through a linear system (which all vibration signals are), then the cepstrum can sometimes separate the input spectrum from the linear system frequency response.

The second point unfortunately also means that the cepstrum is sensitive to noise, as the logarithm, in addition to amplifying low harmonic levels, also amplifies the lower part of the spectrum, where the background noise is normally 'hidden'.

The third point can be seen easily from the equation of an output autospectrum of a linear system which, if $x(t)$ is the input signal and $y(t)$ is the analyzed output signal, and the linear system has FRF $H(f)$, is

$$S_{yy}(f) = |H(f)|^2\, S_{xx}(f) \tag{16.28}$$

from which it is easily seen that the cepstrum $c_{py}(\tau)$ becomes

$$c_{py}(\tau) = \text{IFFT}\left[\log\left(|H|^2\right)\right] + \text{IFFT}\left[\log\left(S_{yy}\right)\right] \tag{16.29}$$

which means the cepstrum is a sum of the effect of the linear system and the effect of the input spectrum. If the linear system and the spectrum have different main frequency ranges, then it is possible to filter (lifter) out the effect of the linear system.

The time variable of the cepstrum, $\tau$, is called 'quefrency', using the paraphrasing terminology originally proposed by (Bogert *et al.* 1963). This terminology was introduced to stress the fact that, although $\tau$ has the unit of time, it is different than the lag domain of the autocorrelation function $R_{xx}$, although often the variable $\tau$ is used for both. In addition to the terms 'cepstrum' which is a paraphrase of 'spectrum', and 'quefrency' for the time variable which is a paraphrase of 'frequency', in cepstrum analysis it is also common to use the paraphrases 'liftering' for 'filtering', and sometimes 'short-pass liftering' for 'lowpass filtering' and 'long-pass liftering' for 'highpass filtering', although the two last paraphrases are rarely used.

The cepstrum calculated by Equation (16.27) is, with our definition, a real quantity, with positive and negative values, much like an autocorrelation function. In most cases, however, the magnitude of it is displayed, as there is no information to be gathered from the positive and negative values. The cepstrum is useful for analyzing signals which contain many harmonics, which is often the case in, for example, gearboxes, see (Konstantin-Hansen and Herlufsen 2010). We will illustrate its use with an example using data from a milling machine.

**Example 16.3.1** *We will look at an example with an acceleration measured on a milling machine with a four-tooth endmill, running at approximately 1655 RPM, which corresponds to 27.59 Hz. In Figure 16.8 (a), the time signal of a small segment of the data is shown. It is clearly seen that the signal is periodic with a period of approximately 9 ms, which corresponds to* $1/(4 \cdot 27.59)$ *seconds. The reason for this is, of course, that for each revolution, the four teeth grab into the material which causes the fourth harmonic of the RPM to be the main frequency.*

*In Figure 16.8 (b) and (c), linear spectra of the time signal are shown with linear y-scale in (b), and with logarithmic y-scale in (c). In the plot with linear y-scale, it is clearly seen that we have harmonics of approximately 110.3 Hz (4 times 27.59), with small peaks at every quarter of these frequencies. The harmonics rise up to approximately 1543 Hz (14th harmonic of 110.3 Hz) and then fall. In Figure 16.8 c), with logarithmic y-scale, the typical complexity of spectra from this type of machines is clearly seen. We have a 'forest' of peaks and the complexity can be startling.*

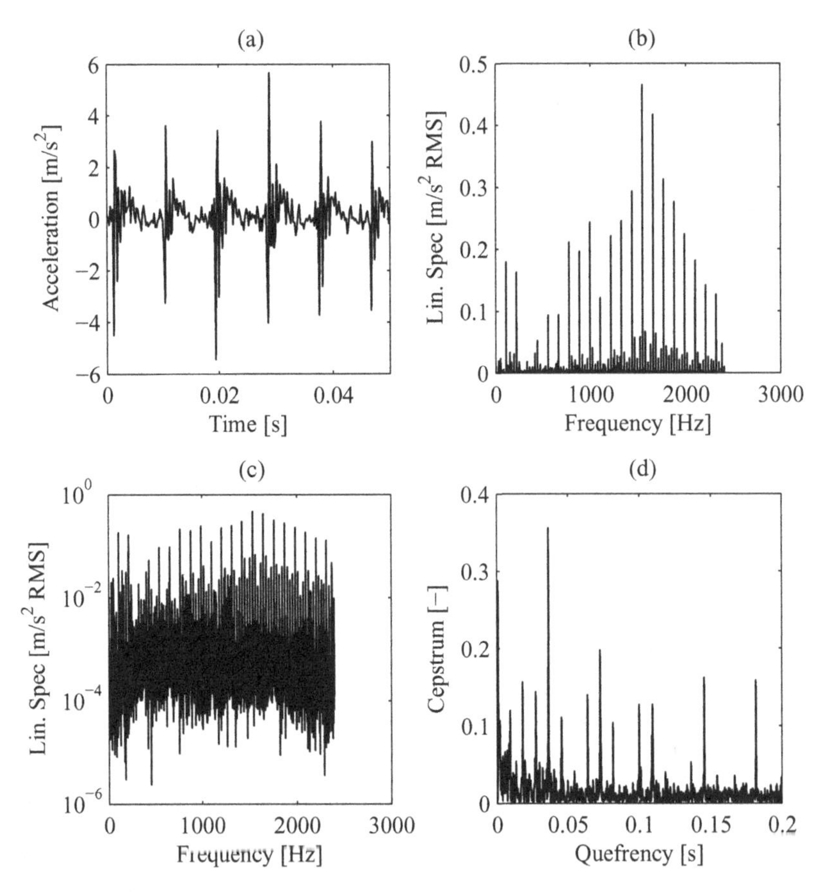


**Figure 16.8** Plot for Example 16.3.1. In (a) a small part of an acceleration time signal from a measurement on a milling machine during a milling operation is shown. In (b) and (c), the linear spectrum of the time signal is shown with linear and logarithmic amplitude axis, respectively. As can be seen particularly in the latter plot, the vibration spectrum is very complex with many harmonics. In (d), the power cepstrum of the signal is shown, in which it can be seen that the highest peak, at approximately 0.03625 s, is corresponding to the RPM of the milling tool. See Example 16.3.1 for a discussion. [Data courtesy of Prof. Kjell Ahlin]

*This is a good example of a case where the power cepstrum can concentrate the information into a more compressed form, where the interpretation can be easier. In Figure 16.8 (d), the power cepstrum of the first 200 ms is shown. The highest peak in the cepstrum is found at approximately 36.25 ms, corresponding to 27.59 Hz, the RPM of the milling tool, which is the fundamental periodic component in the 'forest' of peaks seen in Figure 16.8 (c). For diagnostic purposes, rather than monitoring the complex spectrum in (c), it might be more efficient to track the level of this peak in the cepstrum. As our purpose here is not to dwell on monitoring and diagnostic methods for milling machines, we leave the example at this.*

*End of example.*

## 16.3.2 Complex Cepstrum

Another common cepstrum function is the *complex cepstrum*, which is defined as the inverse Fourier transform of the logarithm of the (discrete) Fourier transform $X(f)$, i.e.,

$$c_c(\tau) = \text{IDFT}\left[log\left(X(f)\right)\right]. \tag{16.30}$$

The complex cepstrum is, despite its name, a real function. This is a result of the fact that if $X(f)$ is the Fourier transform of a real signal, $x(t)$, then if we write

$$X(f) = A(f)e^{-j\phi(f)} \tag{16.31}$$

the logarithm of Equation (16.31) is given by

$$\log X(f) = \log A(f) + j\phi(f). \tag{16.32}$$

Now, because $x(t)$ is real, then we know from Section 2.7.1 that:

- $A(f)$ is even,
- $\log(A(f))$ is even,
- $\phi(f)$ is odd.

This proves that $\log(X(f))$ is conjugate symmetric, i.e.,

$$\log X(-f) = \left[\log\left(X(f)\right)\right]^* \tag{16.33}$$

and thus from basic properties of the Fourier transform, the inverse Fourier transform of Equation (16.31) is a real signal.

In order to calculate the complex cepstrum, the phase $\phi(f)$ must be a continuous function, that is the phase has to be unwrapped. This can be accomplished in MATLAB/Octave by the **unwrap** command.

## 16.3.3 Inverse Cepstrum

Both the power and the complex cepstrum can be inverse transformed. This is usually done after some alteration of the cepstrum has been performed. Although today this is often referred to as filtering, the originally suggested term, liftering, is better to avoid confusion. When the inverse cepstrum is to be applied to a power cepstrum as in Equation (16.27), the double-sided autopower spectrum has to be used. The procedure to create an inverse cepstrum, is simply to forward Fourier transform the cepstrum, for example,

$$\log\left(S_{xx}(f)\right) = \text{FFT}\left[c_p(\tau)\right] \tag{16.34}$$

As is seen in Equation (16.34) the inverse cepstrum actually produces the logarithm of the autopower spectrum.

## 16.4 The Envelope Spectrum

The *envelope spectrum* is a spectrum often used in vibration monitoring, particularly to diagnose rolling element bearings. The idea of the envelope spectrum is to extract the envelope of the time signal before computing the spectrum. This is related to amplitude demodulation, and is useful in cases where a vibration phenomenon occurs modulated on top of a constant frequency, e.g., when a ball bearing has a fault which causes a periodic impact with the period of the rotation speed (or related to the rotation speed).

To understand the principle of the envelope spectrum, in Figure 16.9 we find a (zoomed-in part) of a time signal which is created by a train of impulses exciting an SDOF system. The envelope of the signal, computed by the Hilbert transform as described in Section 16.2.2, is overlaid on the signal. The envelope spectrum is the spectrum of the envelope signal in Figure 16.9, and clearly this signal has a much lower frequency than the original signal. In fact, the envelope is the *demodulated signal*, in telecommunication terminology.

Incidentally, the time plot in Figure 16.9 is relatively similar to the time plot in Figure 16.8 (a). In fact, as we will see in Example 16.4.1, the synthesized signal in Figure 16.9 is a simplified model of the phenomenon behind the vibrations on the milling machine, see also Problem 16.2.

To compute envelope spectra, it is generally necessary to bandpass filter the raw vibration signal around the frequencies of interest. Another 'trick' often used, is to take the square of the envelope prior to computing the spectrum, which makes the envelope spectrum less sensitive to higher-frequency harmonics. We will now illustrate the computation of envelope spectra with an example.

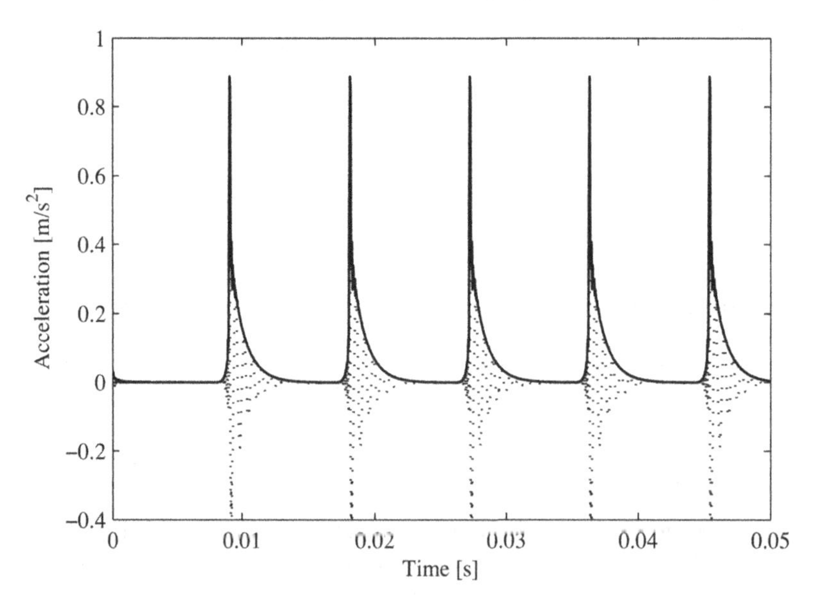


**Figure 16.9** Example of time signal where amplitude modulation occurs. The signal is synthetically generated by letting a pulse train with approximately 110.1 Hz frequency excite an SDOF system with a natural frequency of 1543 Hz and 10% relative damping. The time envelope of the signal (solid), computed by the Hilbert transform, is overlaid on the vibration signal (dotted). The envelope is a signal with low frequencies relative to the frequencies of the vibration signal

**Example 16.4.1** *We will look at the envelope spectrum of the milling machine vibration used in Example 16.3.1. Looking at the spectrum in Figure 16.8 (b) and (c), we can see that the vibrations peak at approximately 1543 Hz. This is due to a resonance in the machine or workpiece at this frequency. By setting a bandpass filter centered at* $f_c = 1543$ *Hz and with, say, 200 Hz bandwidth, we will encompass three of the harmonics on each side of the peak at 1543 Hz (because, from Example 16.3.1 we know the harmonics are spread approximately 27 Hz apart). We then compute the square of the envelope, and compute a spectrum. The following MATLAB/Octave code does all of this.*

```
% Bandpass filter the signal in variable x
fc=1543;
B=200;
flo=(fc-B/2)/(fs/2);
fhi=(fc+B/2)/(fs/2);
[b,a]=butter(4,[flo fhi]);
x=filtfilt(b,a,x);
% Compute the envelope squared
e2=abs(hilbert(x)).^2;
[E,f]=alinspec(e2,fs,ahann(8192),1);
```

*In the example code we have used the command* ***alinspec*** *from the accompanying toolbox, which computes a linear spectrum; remaining commands are standard MATLAB/Octave (signal processing toolbox) commands. The result of the code above, using the data from the milling machine, is plotted in Figure 16.10, for the first 200 Hz. The higher frequencies will not have any significant frequency content and can be thrown away. The plot clearly shows four peaks, whereof the highest is located at 110.2 Hz, which is the frequency of 4 times the RPM of the milling tool.*

*End of example.*

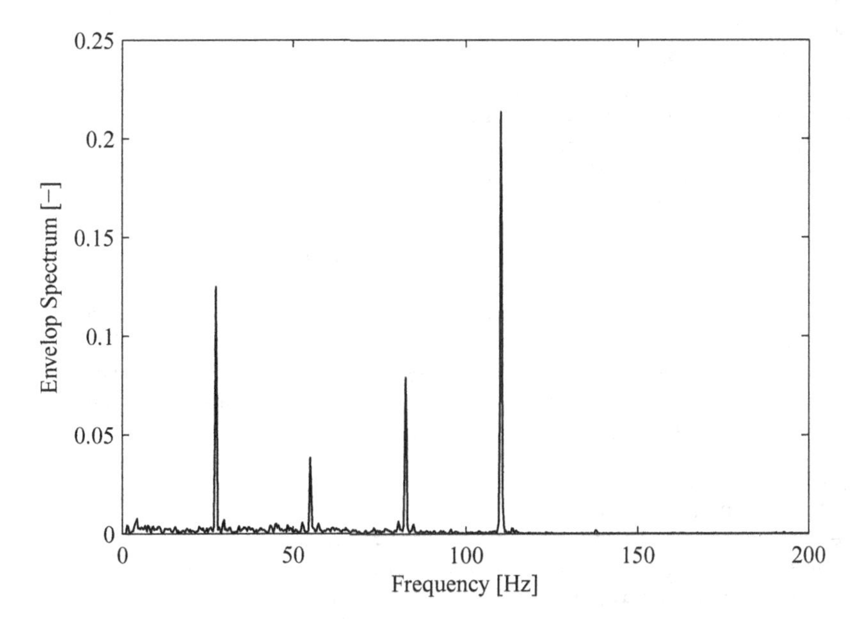


**Figure 16.10** Plot for Example 16.4.1. The plot shows the envelope spectrum of an acceleration signal from a measurement on a milling machine, after applying a bandpass filter with a center frequency of 1543 Hz and 200 Hz bandwidth. [Data courtesy of Prof. Kjell Ahlin]

## 16.5 Creating Random Signals with Known Spectral Density

It is rather common to want to create a random time signal with known properties. It can be for shaker excitation, where the spectrum is requested to compensate for stinger dynamics, for example, or it can be in environmental testing, where a test object should be excited with a particular spectrum.

It has already been mentioned, in Section 13.9.4, how to produce pseudo-random noise with a known spectrum. Here, we will limit the method to cases where the signal is short enough so that an inverse FFT can be performed on the entire signal. This is a reasonable limitation today, where FFT can be performed on several million samples of data in a matter of a second or two.

The principle of creating a pseudo-random signal, is to create an amplitude spectrum in the frequency domain with the spectrum shape (the square root of the PSD). Then, a random phase is added to each frequency, and an inverse FFT (IFFT) is performed. For the IFFT to work, we must, of course, create a double-sided spectrum prior to the IFFT, using the symmetry properties from Section 2.7. We will illustrate the method with an example using MATLAB/Octave.

**Example 16.5.1** *Assume we have a PSD in MATLAB/Octave variable* `Gxx`, *and a frequency axis in variable* `f`, *going from zero Hz to* `fs/2`. *This means the length of f and* `Gxx` *is $N/2+1$, where N was the blocksize used to compute the PSD. For this example we let $N = 2048$ samples. Create a time signal with Gaussian PDF, and with the same RMS level as that of the PSD, and with length $L > N$, in our example, $100 \cdot 1024$ samples.*

*The first thing we do is to find the sampling frequency, which is $f_s = 2\max(f)$, since f has length $N/2+1$. We should also calculate the total RMS level from the PSD. Next we interpolate the PSD in* `Gxx` *up to the length $L/2+1$, since we have a single-sided spectrum so far. The final, double-sided spectrum will then be length L. Let us show this part of the MATLAB/Octave code before proceeding.*

```
fs=2*f(end);
df=f(2);
% Compute the RMS level
R=sqrt(df*sum(Gxx));
% We need a new freq. axis length
newx=linspace(f(1),f(end),ceil(L/2)+1);
% Interpolate Gxx onto this new x-axis
P=interp1(f,Gxx,newx,'linear','extrap');
P=P(:);        % Make sure a column
```

*where the* **ceil** *is rounding upwards, in case L would be an odd number.*

*Next we will compute the amplitudes of the spectrum as the square root of the PSD. We then add a random phase betweenn $-\pi$ and $\pi$, and produce the negative frequencies with the Fourier transform symmetry properties (even amplitude, and odd phase, in this case). Then we compute the IFFT, and scale the RMS to our variable R above. This final part is done by the following code. Note that the command* **rand** *creates uniformly distributed values between zero and unity, so we make a trick to get values between –0.5 and 0.5.*

```
A=sqrt(P);
phi=2*pi*(rand(ceil(L/2)+1,1)-0.5);
% Create negative frequencies in upper half
% of the block (as FFT/IFFT wants it)
A=[A(1); 0.5*A(2:end); 0.5*A(end-1:-1:2)];
phi=[phi; -phi(end-1:-1:2)];
phi(end)=0;    % phi of fs/2 must be real
y=sqrt(2*fs*length(A))*real(ifft((A.*exp(j*phi))));
```

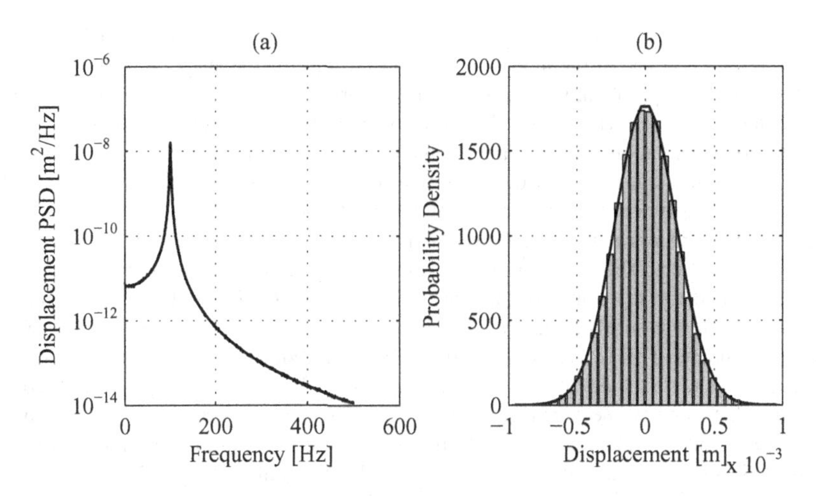


**Figure 16.11** Plots for Example 16.5.1. In (a), the PSD of a time signal synthesized to provide a known PSD is shown (solid), overlaid by the true PSD (dashed). The difference is only due to the random error in the estimated PSD and the two curves cannot be distinguished from each other on the scale presented. In (b) the probability density of the synthesized time signal is shown (bars), overlaid with a theoretical Gaussian distribution with the same mean and standard deviation (solid). The time signal is obviously Gaussian

```
y=y-mean(y); % Force mean to zero
% Scale y to proper RMS level
y=R/std(y)*y;
```

*In Figure 16.11 the resulting PSD of the time signal computed using an original PSD corresponding to the output signal of an SDOF system with natural frequency of 100 Hz and 1% damping, forced by white noise, is shown, overlaid by the true PSD, using the RMS level of the input and the magnitude squared of the SDOF frequency response.*

*End of example.*

## 16.6 Operational Deflection Shapes – ODS

One of the most useful tools for solving many vibration problems is the *operating deflection shape*. This technique is usually based on animation of the forced response of a structure in operation at a particular frequency, although it is possible also to animate time domain signals, particularly for transient analysis.

The basis for ODS is the forced response expressed by the displacement vector $\{U(f)\}$ in Equation (6.98) which is repeated here, slightly altered, for convenience

$$\{U(f)\} = [H(f)]\{F(f)\} \tag{16.35}$$

where we assume that the forces $\{F(f)\}$ are unknown. We know from Chapter 6, that the frequency response matrix $[H(f)]$ has a direct relation to the mode shapes of the structure as shown by, for example, Equation (6.106). The basis for ODS is, that in many (but not all) cases when a structure is forced, the deformation shape at a particular frequency, as described by Equation (16.35), will look very similar to one of the mode shapes. This is particularly true close to the natural frequencies, where structures are very 'unwilling' to move in any other way than by the shape of the mode in question. This is also more likely to be true the fewer forces there are driving the structure. Vibration problems naturally occur most often near the natural frequencies, because of the resonance amplification, and so, if we take a problem frequency and insert into Equation (16.35) and animate the vector $\{U\}$, in many cases we will see a motion resembling one of the modes. Knowing this deformation shape, it is then easier to find a solution to the vibration problem, without having to perform a more complex measurement such as an experimental modal analysis test. ODS is therefore a very useful trouble-shooting tool.

If we measure accelerations or velocities, Equation (16.35) needs to be multiplied by $-\omega^2$ or $j\omega$, respectively, but, since that is only a question of a number, the shape is not altered. It is therefore insignificant which entity we measure and animate, in most cases.

It remains to describe how to measure the vector in Equation (16.35), or derivatives of it if we measure with, say, accelerometers. The answer to this question is that it does not really matter, as long as we follow good practice for spectrum measurements. From Chapters 10 and 13 we know that the only possibility we have to measure the phase relation between two channels, is to measure a cross-spectrum between the two points. This is what gives us the phase of a frequency response, but we might as well obtain it directly by a cross-spectrum measurement, of course. When it comes to the amplitudes of each element in $\{U\}$, they are best measured with autospectra. The type of spectrum, of course, is determined by whether the signal is periodic or random (transients are rarely relevant in this case). A common way to perform ODS, is therefore to use the *phase spectrum* described in Section 10.2.3.

Frequency response functions (transmissibilities) could also be used for this purpose. While perfectly usable *per se*, there is, however, a disadvantage with using FRFs for ODS extraction: there will, in general, not be any peaks in transmissibilities at the frequencies of interest. Therefore, it is difficult to find the proper frequencies to extract the ODS. The phase spectrum is a better choice.

ODS analysis also requires software that allows us to generate a wire frame model of the measurement points and to animate the operating shapes. Such software is common in commercial applications and can also be implemented in, for example, MATLAB.

To make an ODS measurement is rather straightforward. The following points briefly describes the procedure.

1. After deciding which points should be measured, number each point, decide a coordinate system, and build a geometry model in the animation software.
2. Make a measurement of points on the structure during steady-state conditions, recording which sensor is in which point and direction (most commercial systems have support for this). If all requested points cannot be measured simultaneously, select one or more references which are kept in the same place in every measurement, move the other sensors around and make a measurement set of each measurement until all points have been measured. The reference or references should be selected at points with clear peaks at the frequency or frequencies of interest.
3. Compute phase spectra of all measured points, with phase reference to the reference channel.
4. If more than one set of measurements was used, each set should be scaled by the vibration level in the reference point of the same set, to equal out potential differences in the operating conditions between the sets.
5. Extract the amplitude and phase at the frequencies of interest, from all measurements, and store each frequency result in a shape vector.
6. Import the shape vectors into the animation software and animate the shapes.

### *16.6.1 Multiple-Reference ODS*

In some cases, the ODS obtained by the procedure in Section 16.6 will depend on the spatial location of the reference accelerometer, e.g., if there is a repeated root so that two (or more) modes are present at the same frequency. Then the ODS using the procedure described in Section 16.6 will show a linear combination of the modes at or near that frequency, the linear combination being dependent on the reference location. Naturally, placing a reference accelerometer on or near a node line is also a possible cause of badly defined operating shapes, see Section 6.4.5.

This situation can be solved by using more than one reference during the measurements, and computing the input/output cross-spectral matrix using the references as inputs, and all other channels as outputs. Then virtual cross-spectra of all outputs in $\{y\}$ with the $r$ highest principal components are computed using Equation (15.15). This virtual cross-spectrum matrix has one column for each virtual signal (principal component) and each of these columns is then used to extract operating shapes. Each column will give a virtual ODS. In the case of several independent operating shapes located at a particular frequency, provided at least one of the references is located off the nodal lines for each shape, the virtual shapes ,will each give one of the independent shapes (Otte *et al.* 1990; Tucker and Vold 1990). It is also possible to use this technique in the case where all points of interest cannot be measured simultaneously.

## 16.7 Introduction to Experimental Modal Analysis

Identifying the *modal parameters*, $f_r$, $\zeta_r$, and the mode shapes $[\psi]_r$ of a structure is called *modal analysis* as we discussed in Chapter 6. This type of analysis can be carried out experimentally by measuring a number of frequency responses and then extracting the parameters of, for example, the mathematical model in Equation (6.110). It is then commonly referred to as *experimental modal analysis*, EMA. If instead the mass and stiffness matrices are used, it is commonly referred to as *analytical modal analysis*.

There are several reasons for wanting to perform EMA. A common reason for determining the modal parameters experimentally is that they are desired to verify analytical results, typically from finite element models. It is particularly important to use correct damping estimates for simulations, and this alone can sometimes justify an experimental modal analysis to be performed. Damping is essentially impossible to compute analytically with current techniques but has to be estimated experimentally.

Another common reason for applying EMA is for trouble-shooting purposes when it has been established that a mode at a particular frequency, often coinciding with a high force level, for example from an engine order (see Chapter 12), is causing excessive vibration levels. In such cases, EMA can be applied by using a less rigorous measurement, as the precision of the modal parameters is of less concern. The main result of the EMA in this case is to find approximate mode shapes in order to assess some design change which will move the problem mode away from the problem frequency.

This section will by necessity be brief. There are, however, several excellent books with more comprehensive coverage of experimental modal analysis, particularly Ewins (2000) for a general and rather practical description, and Maia and Silva (2003) and Heylen *et al.* (1997) for more details about parameter extraction methods, etc.

### *16.7.1 Main Steps in EMA*

EMA is an application where the engineering experience and skills of the engineer come to their edge; some people refer to it as an art form. The degree of difficulties depends on the purpose of the test:

- For trouble-shooting purposes, in many cases it is sufficient to obtain rough estimates of the natural frequencies and mode shapes in order to find possible design changes. In this case it is usually sufficient to use impact excitation, and in many cases the structure of interest is measured in its place, i.e., with the boundary conditions it has in operation.
- If the modal parameters are going to be used for validating an analytical model, in most cases the structure has to be suspended with free–free boundary conditions as this produces measurements of the highest quality. In many cases, shakers are used for excitation, as it is usually possible to estimate FRFs of higher quality by this technique. More sophisticated modal parameter extraction methods are usually used in this case, which assure the best possible estimates of the extracted parameters.

Whereas there are many pitfalls in applying EMA, it should be noted that in most cases the key to success lies in making good-quality measurements. Following the procedures described in Chapters 13 and 14 will lead to this, if enough care is taken. When the measured FRFs are of high quality, the modal parameter extraction is often a rather simple process.

An EMA test can be decomposed into the following main steps.

1. Selection of test method: impact or shaker excitation, and in the former case, of fixed force or fixed response locations. In the latter case, single or multiple shaker excitation should be considered.
2. Selection of reference and response DOFs. At least one row or column of the frequency response matrix $[H(f)]$ has to be measured, see Section 6.4.2.
3. Selection of test structure suspension, usually free–free support if good-quality results are necessary.
4. Preliminary tests to find optimal measurement settings.
5. Check to verify correct suspension, shaker attachment, etc.
6. Data acquisition of all FRFs, usually in several 'scans', as all forces/responses cannot be measured simultaneously.
7. Data quality analysis, to verify data are good enough for good parameter extraction.
8. Modal parameter extraction.
9. Verification of obtained modal parameters.

Some of the above points will be further discussed in the following.

### *16.7.2 Data Checks*

Whether impact excitation or shaker excitation is selected usually depends on the required results. For best quality, it is most common to use shaker excitation, although for very lightly damped systems, impact excitation can be used with high quality as well (Fladung 1994; Fladung and Rost 1997). Usually, several response accelerometers should be selected as references in the latter case.

When impact excitation is selected, the most common strategy is to let one or more accelerometers be the reference sensors (the sensors common for every measurement), and the impact hammer is moved over the structure from DOF to DOF. This is usually referred to as *roving force*. In some cases, when the impact hammer cannot access some measurement DOFs, for example due to lack of space, the impact hammer has to excite the same position for every measurement, while a number of accelerometers are used to 'scan' the structure, exactly as for shaker testing.

Referring back to the definition of the dynamic flexibility (receptance) frequency responses in Equation (6.98), it is easy to see that a fixed response reference and roving force, will result in the measurement of a row of the FRF matrix $[H(f)]$. Consequently, a fixed force reference and roving response will lead to measurement of a column in $[H(f)]$. Because of the symmetry of the FRF matrix due to Maxwell's reciprocity theorem in Equation (6.111), it turns out that row $p$ in $[H(f)]$ is equal to column $p$. This means that when processing frequency response data, it is sufficient to let the software handle one case, for example fixed force reference. If a fixed response test has been done, the obtained frequency response matrix is simply transposed. Note, in this case that, using MATLAB/Octave, for example, the matrix should be transposed but *not* complex conjugated, which is done by default using the apostrophe operator. Instead, the dot apostrophe (.') operator should be used.

The selection of reference and response DOFs is a complicated matter. For references, in Section 6.4.5 we pointed out that they should be selected so that they are not on node lines, or if multiple references are used, at least one of the references is not on a node line for each mode. The response locations should be selected to obtain as good a MAC matrix as possible, which will be discussed in Section 16.7.4. This is particularly important when an analytical model is to be verified by the EMA results.

When it comes to the data acquisition, some very important points to follow are the following.

1. For high-quality results, the best choice is to suspend the test structure by free–free boundary conditions.
2. Best results are obtained by using several references, either by having several accelerometers sitting on well-selected points (away from nodal lines), or by having several shakers sitting at similar points.
3. If shakers are used, the attachment should be carefully checked to ensure that
   3.1. the driving point imaginary part must have peaks in just one directions (assuming acceleration is measured, for velocity the real part is checked), see Section 6.4.6,
   3.2. the reciprocity has to be very good; this is easily checked if two or more shakers are used, and more difficult otherwise, see Section 13.12.2, and,
   3.3. the linearity should be checked by estimating the same FRF (between same two points) with two different excitation force levels; if the structure is linear, these FRFs will be equal.
4. Great care should be exercised to ensure that the coherence is as close to unity as possible over the entire frequency range of interest. This can often require much experimentation before succeeding. On lightly damped systems, minor dips in the coherence around antiresonances may have to be accepted, but coherence values very close to unity around the natural frequencies of the structure should always be attempted.

### *16.7.3 Mode Indicator Functions*

One of the tools often used in EMA is the *mode indicator function*, MIF. Actually, there are several mode indicator functions commonly used, which can be found in Rades (1994) where a good comparison is presented. Here we will present three of the most common MIFs.

The simplest mode indicator function is a sum of the magnitude of all FRFs, usually squared, or sometimes a sum of the imaginary part squared. This type of plot exaggerates global modes, i.e., modes where most measured FRFs have a large displacement.

The 'Normal MIF', or sometimes 'MIF 1', is a one-dimensional MIF, i.e., it operates on single-reference FRF matrices. The normal MIF is based on the fact that, as we know from Chapter 6, off the natural frequencies of a structure, the FRF (if it is an accelerance or a receptance) is approximately real, whereas exactly at the undamped natural frequency, it is purely imaginary. Note that if mobilities are used, the properties of the real and imaginary parts of the FRF are swapped. Since accelerance is the most commonly used measurement function, we define the MIFs in the following text on this type of FRF.

To make the formulas a little simpler in the following, we define the real part of an FRF, by $\mathrm{Re}\left[H_{pq}\right] = H_{Rpq}$ and the imaginary part we denote by $\mathrm{Im}\left[H_{pq}\right] = H_{Ipq}$. We therefore define the normal MIF as

$$\mathrm{MIF}_1(f) = \frac{\sum\limits_p \left|H_{Rp}(f)\right|^2}{\sum\limits_p \left|H_p(f)\right|^2} \tag{16.36}$$

where $p$ sums over all measured FRFs at each frequency $f$ and we pay no regard to which reference, $q$, is used. This function will be near unity, except at frequencies where there is influence from global modes, where it will drop to near zero. An example of a normal MIF using all FRFs from a measurement on the Plexiglas plate described in Section 14.6 is shown in Figure 16.12 (a).

When multi-reference data have been obtained, more sophisticated MIFs can be computed. The most commonly used such MIF is the *multivariate MIF*, or MvMIF, described in, for example, Williams

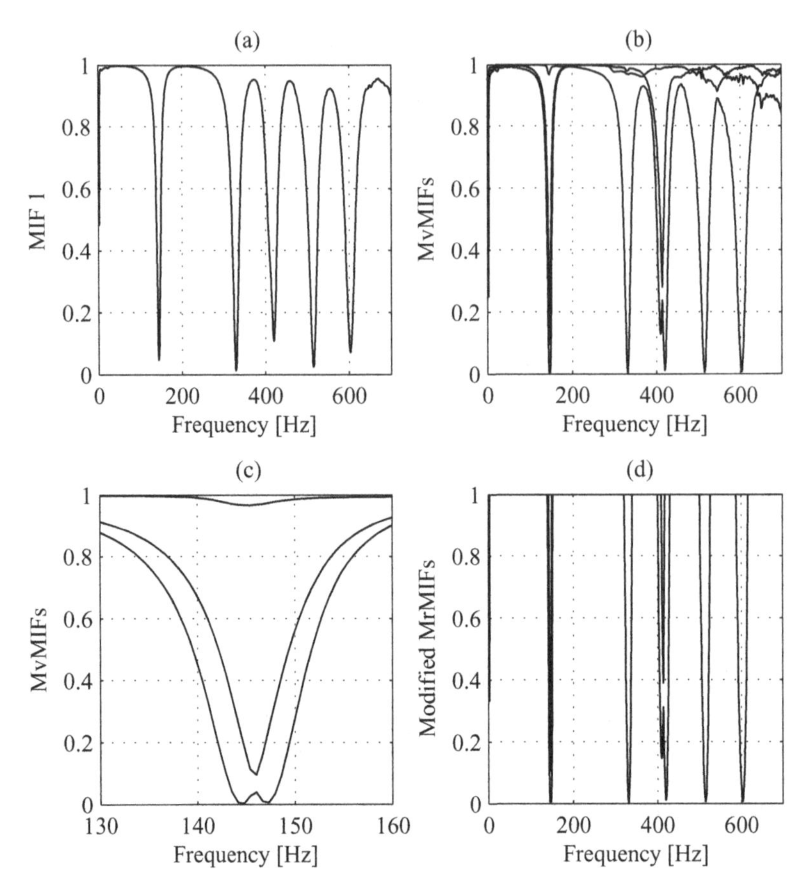


**Figure 16.12** Plots of mode indicator functions, MIFs, computed from data from measurements on a Plexiglas plate, see text. In (a), MIF 1 (normal MIF); in (b) multivariate MIF, MvMIF; in (c), zoomed in range around first modes; and in (d) the modified real MIF, MrMIF. In (c), the typical eigenvalue cross-over effect is clearly seen, see text for details. It is also seen, by comparing (b) and (d), that the MrMIF equals unity over a larger frequency range and more distinctly dips near the modes

*et al.* (1985). If we assume that the FRF matrix $[H(f)]$ has data in columns corresponding to a fixed force reference measurement, this function is based on an eigenvalue solution of

$$[H_R]^H [H_R] = \lambda_v \left( [H_R]^H [H_R] [H_I]^H [H_I] \right) \tag{16.37}$$

where the eigenvalues, $\lambda_v$, are sorted in ascending order. The MvMif functions are limited to the range zero to unity. The first (smallest) of the eigenvalues is called the first MvMIF, the second smallest eigenvalue is called the second MvMIF, and so on, for as many MvMIFs as there were references. The advantage with this mode indicator function is that it detects repeated modes, i.e., if there are two modes at a particular frequency, both the first and second MvMIF will dip towards zero, if there are three modes the first, second, and third MvMIF will all dip towards zero, etc. An example of the MvMIF for a three-reference test is shown in Figure 16.12.

In Figure 16.12 (b), at approximately 150 Hz, and also at approximately 420 Hz, the second MvMIF can be seen to dip. From a closer inspection, as shown in Figure 16.12 (c), where the plot has been zoomed in around 140 Hz, it can be seen that the first MvMIF actually has two close dips, whereas the second MvMIF dips in between these two. This phenomenon is often referred to as the *eigenvalue cross-over effect*, and should be interpreted such that there are actually only two modes, the two modes indicated by the first MvMIF. The second MvMIF dips only because of the poor frequency resolution and the cross-over effect. Some experience interpreting the MvMIF is thus necessary in order not to misinterpret the results when modes are closely spaced.

The MvMIF is actually detecting real normal modes. A similar mode indicator function, the *modified real mode indicator function*, MrMIF, is defined by the eigenvalues, $\lambda_r$, of

$$[H_R]^H [H_R] = \lambda_r \left([H_I]^H [H_I]\right) \tag{16.38}$$

where the eigenvalues are sorted in ascending order as for the MvMIF. The MrMIFs are usually plotted in logarithmic $y$-scale which produces plots similar to the MvMIFs. The MrMIFs are not bounded between zero and unity. However, since the idea of the MrMIF is that the eigenvalues $\lambda_r$ vanishes near undamped natural frequencies, I propose to set the MrMIF values which exceed unity, to unity, i.e.,

$$\text{MrMIF} = \max(\lambda_r, 1) \tag{16.39}$$

and plot the resulting MrMIF in linear $y$-scale, which produces very distinct dips at the undamped natural frequencies. An example of the MrMIFs modified as in Equation (16.39) are shown in Figure 16.12 (d). The MrMIF is rarely found in commercial modal analysis software.

It should be noted that all the MIFs are computed for each measured frequency. In cases where the frequency increment is poor, the mode indicator functions may not dip to zero, because there is no frequency value exactly at the undamped frequencies of the structure.

### 16.7.4 The MAC Matrix

It is common to want to compare the similarity between different mode shapes. The *modal assurance criterion*, MAC, is the most common tool for this purpose (Allemang and Brown 1982). The MAC value between two modes $\{\psi\}_r$ and $\{\psi\}_s$ is defined by

$$\text{MAC}_{rs} = \frac{\left|\{\psi\}_r^H \{\psi\}_s\right|^2}{\left(\{\psi\}_r^H \{\psi\}_r\right)\left(\{\psi\}_s^H \{\psi\}_s\right)} \tag{16.40}$$

which can be interpreted as the normalized correlation coefficient between the two vectors. The MAC value is a value between zero and unity and is detecting the similarity between two modes.

The MAC is usually computed as a matrix of one of two kinds: (i) the MAC between a certain set of modes and the same set of modes, which is called the auto-MAC, or (ii) the MAC between two different sets of modes, the cross-MAC.

For an interpretation of the MAC matrix, we refer to the modal orthogonality relation in Equation (6.41), which is repeated here for convenience

$$\{\psi\}_r^H [M]\{\psi\}_s = 0 \tag{16.41}$$

for any two mode vectors where $r \neq s$. Comparing Equation (16.41) with the definition of the MAC in Equation (16.40), we note that if we replace the mass matrix in the former equation by the identity matrix, we obtain the definition of the MAC. This has some important implications:

- the MAC between two different modes is not guaranteed to be zero, and
- if the MAC is zero between two modes, it is similar to replacing the mass matrix by the identity matrix in Equation 16.41. (It is not equivalent, because we are talking about experimental mode shapes, which means that we do not have the degrees-of-freedom of the mass matrix.)

An important concept of the MAC is that measurement DOFs can be chosen to minimize cross-MAC values. This is a common way of selecting measurement DOFs, particularly when the aim is to use the EMA results to verify an analytical model, where it is important to be able to separate the modes from the analytical model and the EMA.

The auto-MAC matrix is also often used to assess the quality of the parameter extraction results. In such cases high auto-MAC values off the diagonal are taken as indications that the modes have not been properly separated by the mode shape extraction process. This, however, requires that the selection of measurement DOFs has been carefully done.

### 16.7.5 Modal Parameter Extraction

We will only briefly touch the extensive field of extracting modal parameters from measured FRFs. The simplest methods used for this purpose are the SDOF extraction techniques, where one or more FRFs are processed in a narrow frequency range around a natural frequency, under the assumption that there is only one mode affecting the FRF, which is thus acting as an SDOF system, see Chapter 6. We will look at this technique assuming we have a single FRF, $H_{pq}(f)$. From Equation (5.17) we know that around the natural frequency, assuming there is only one mode, the FRF $H_{pq}(\mathrm{j}\omega)$ in the form of receptance, can be approximated by

$$H(\mathrm{j}\omega) = \frac{U(\mathrm{j}\omega)}{F(\mathrm{j}\omega)} = \frac{1/m}{-\omega^2 + \mathrm{j}2\zeta_r\omega_r\omega + \omega_r^2}. \tag{16.42}$$

One of the simplest SDOF parameter extraction techniques is found be using a least squares solution of Equation (16.42), using a few frequency lines around the natural frequency, which is found, for example, by inspecting a mode indicator function as described in Section 16.7.3. We start by rewriting Equation (16.42) at a frequency $\omega_i$ around the natural frequency, as

$$H_{pq}(\omega_i)\omega_r^2 + \mathrm{j}2\zeta_r\omega_r\omega_i H_{pq}(\omega_i) - 1/m = \omega_i^2 H_{pq}(\omega_i). \tag{16.43}$$

The mass is, of course, not going to be possible to estimate, as the FRF we are processing is not from an SDOF system, so we will replace this variable with a dummy variable, $D$. We can then rewrite Equation (16.43) in matrix form by first using a row vector times a column

$$\lfloor\, H_{pq}(\omega_i)\ \ \mathrm{j}2\omega_i H_{pq}(\omega_i)\ \ -1 \,\rfloor \begin{Bmatrix} \omega_r^2 \\ \zeta_r\omega_r \\ D \end{Bmatrix} = \omega_i^2 H_{pq}(\omega_i) \tag{16.44}$$

which is valid for any frequency around the natural frequency of the system. We now repeat Equation (16.44) by adding several frequencies, $\omega_i, i = 1, 2, \ldots, L$ and get a matrix equation

$$\begin{bmatrix} H_{pq}(\omega_1) & \mathrm{j}2\omega_1 H_{pq}(\omega_1) & -1 \\ H_{pq}(\omega_2) & \mathrm{j}2\omega_2 H_{pq}(\omega_2) & -1 \\ \cdots & \cdots & \cdots \\ H_{pq}(\omega_L) & \mathrm{j}2\omega_L H_{pq}(\omega_L) & -1 \end{bmatrix} \begin{Bmatrix} \omega_r^2 \\ \zeta_r\omega_r \\ D \end{Bmatrix} = \begin{Bmatrix} \omega_1^2 H_{pq}(\omega_1) \\ \omega_2^2 H_{pq}(\omega_2) \\ \cdots \\ \omega_L^2 H_{pq}(\omega_L) \end{Bmatrix} \tag{16.45}$$

which can be solved for example by a pseudo-inverse, see Appendix E. Typically a few frequency lines, say, 3 to 9, are used to solve for the two unknowns, $\omega_r$ and $\zeta_r$, in Equation (16.45). There are many variants of this simple method, and some include the residue, although we will look at an alternative approach, as the equation system we have used here is ill conditioned, particularly for small values for the mode shape coefficients.

An easy approach to find the mode shape coefficients once the pole has been obtained by solving Equation (16.45) is to simply insert the pole into Equation (6.108) and omit the complex conjugate part, so that

$$A_{pqr} = H_{pq}(j\omega_r)(j\omega_r - \lambda_r) \tag{16.46}$$

where the residue is defined by Equation (6.110). It should finally be noted that in order to obtain the entire residue vector, each FRF in a row or column is processed by using Equation (16.46), after which the driving point mode shape coefficient squared is located, and finally all other elements in the residue vector are divided by the driving point mode shape coefficient, see Equation (6.113) on page 136.

## 16.8 Chapter Summary

In this chapter we have presented some common signal analysis tools which have not found their proper place in other chapters, but which are important to know about. The shock response spectrum is an important tool to use for comparison of various vibration environments with respect to how harmful they can be. It is also commonly used as a tool in environmental engineering to compare a test specification with a real-life vibration environment. The shock response spectrum is based on the assumption that the worst case which can happen is that a structure, mounted in the point where the environment is measured, has a particular resonance frequency and damping. The SRS then tells how large the maximum acceleration level (if it is absolute acceleration SRS) the structure will get.

The Hilbert transform has two main applications in noise and vibration analysis; (i) it is used to compute the envelope of an oscillating function, for example a cross-correlation function; and (ii) it is used to relate the real and imaginary parts of a frequency response function. It was shown how the Hilbert transform in the latter application can also be used to 'clean up' an FRF which exhibits non-causal behavior, i.e., which has a corresponding impulse response which is not zero for negative time.

We then presented the cepstrum, which is the inverse FFT of the logarithm of a spectrum. The cepstrum measures the periodicity in the spectrum, which is useful in cases where the spectrum consists of a number of harmonics to some periodic phenomenon. Cepstra are commonly used, for example, for diagnostics of gear box vibrations.

The envelope spectrum is another common tool, particularly used for bearing diagnostics. The principle of the envelope spectrum is to bandpass filter a time signal around an important 'carrier frequency', for example the rotation speed, or a multiple of this frequency. The envelope of the bandpass-filtered signal is then computed using the Hilbert transform, and the spectrum of this envelope is computed. This essentially works as amplitude demodulation, and the resulting envelope spectrum is a cleaner version of the original spectrum, with a limited number of spectral peaks corresponding to the carrier frequency and the spectral peaks nearest to it.

We also showed how to produce random noise with Gaussian probability density and a given PSD. The procedure is to create a pseudo-random signal by setting the amplitudes of each sine in the frequency domain, and give each frequency an arbitrary phase between $-\pi$ and $\pi$ radians. The resulting time signal was proven by an example to have the correct PSD and PDF.

Finally, we presented two common tools which can help to solve many vibration problems: operating deflection shapes, ODS, and experimental modal analysis, EMA. Whereas ODS is a relatively simple analysis method, based only on spectrum measurements, EMA is a delicate technique which requires great care and quite some knowledge to succeed. One of the most important keys to successful EMA is high-quality FRFs, which can be obtained using the procedures presented in this book. Modal parameter extraction is then many times relatively easy, although the theory behind many of the algorithms used is complicated.

## 16.9 Problems

Many of the problems following are supported by the accompanying ABRAVIBE toolbox for MATLAB/Octave and further examples which can be downloaded with the toolbox. If you have not already done so, please read Section 1.6, and follow the instructions to download this toolbox together with example files.

**Problem 16.1** *Assume a sensitive electronics box which is to sit in the engine compartment of a car. There are two potential locations for mounting the box. The first place, on top of the engine, has worst vibration levels when the engine is running at 2400 RPM. Then, the dominating vibrations are orders 2, 4, and 6, with vibration levels of 2, 1, and 0.5 g respectively in the location where the box can be mounted. The other potential place is on the chassis, where the worst vibrations occur as shocks of 20 ms duration, and 20 g peak levels. The shocks can be modeled as half sine pulses.*

*Compute the shock response spectra of both vibrations, and decide which location is least harmful to the electronics box, assuming there are no long-term (fatigue) effects, but the damage will occur instantly. (Otherwise, of course, we would have to take the statistics of number of pulses, etc., into account.)*

**Problem 16.2** *Assume a rotating machine has a resonance at 400 Hz, with 5% damping. The machine is operating at 1200 RPM and a fault occurs in the drive shaft, producing one pulse per revolution. Use MATLAB/Octave to produce a simulated acceleration signal, if each pulse is 100 N and the machine can be modelled as an SDOF system with a mass of 200 kg. Use a sampling frequency of 10 000 Hz and 5 seconds of data. Compute the envelope spectrum, using a bandpass filter with center frequency of 400 Hz, and try different bandwidths.*

**Problem 16.3** *Use the time signal from Problem 16.2 and compute a power cepstrum.*

**Problem 16.4** *Compute a time signal with a frequency range from 0 to 1000 Hz and a PSD which is constant between 0 and 400 Hz, then 10 times higher from 400 to 600 Hz, and then as the same level as between 0 and 400 Hz up to 1000 Hz. The total RMS of the signal should be 20 g, and the signal should be Gaussian. After generating the signal, verify its properties by computing a PSD and a PDF.*

## References


Ahlin K 2006 Comparison of test specifications and measured field data. *Sound And Vibration* **40**(9), 22–25.

Allemang R and Brown D 1982 A correlation coefficient for modal vector analysis *Proc. 1st International Modal Analysis Conference*, Orlando, FL.

Bendat J and Piersol AG 2010 *Random Data: Analysis and Measurement Procedures* 4th edn. Wiley Interscience.

Bogert B, Healy M and Tukey J 1963 The quefrency alanysis of time series for echoes: Cepstrum, pseudo-autocovariance, cross-cepstrum and saphe cracking In *Proc. Symposium on time Series Analysis* (ed. Rosenblatt M), pp. 209–243.

Endo H, Randall RB and Gosselin C 2009 Differential diagnosis of spall vs. cracks in the gear tooth fillet region: Experimental validation. *Mechanical Systems and Signal Processing* **23**(3), 636–651.

Ewins DJ 2000 *Modal Testing: Theory, Practice and Application* 2nd edn. Research Studies Press, Baldock, Hertfordshire, England.

Fladung W 1994 *The development and implementation of multiple reference impact testing* Master's Thesis University of Cincinnati.

Fladung W and Rost R 1997 Application and correction of the exponential window for frequency response functions. *Mechanical Systems And Signal Processing* **11**(1), 23–36.

Gaberson HA 2003 Using the velocity shock spectrum to predict shock damage. *Sound And Vibration* **37**(9), 5–6.

Gaberson HA, Pal D and Chapler RS 2000 Classification of violent environments that cause equipment failure. *Sound And Vibration* **34**(5), 16–23.

Gao Y and Randall RB 1996a Determination of frequency response functions from response measurements .1. extraction of poles and zeros from response cepstra. *Mechanical Systems and Signal Processing* **10**(3), 293–317.

Gao Y and Randall RB 1996b Determination of frequency response functions from response measurements .2. regeneration of frequency response functions from poles and zeros. *Mechanical Systems and Signal Processing* **10**(3), 319–340.

Greenfield J 1977 Dealing with the shock environment using the shock response spectrum analysis. *J. Soc. Environmental Engineers* (9), 3–15.

Hanson D, Randall RB, Antoni J, Thompson DJ, Waters TP and Ford RAJ 2007a Cyclostationarity and the cepstrum for operational modal analysis of MIMO systems – part i: Modal parameter identification. *Mechanical Systems and Signal Processing* **21**(6), 2441–2458.

Hanson D, Randall RB, Antoni J, Waters TP, Thompson DJ and Ford RAJ 2007b Cyclostationarity and the cepstrum for operational modal analysis of mimo systems – part ii: Obtaining scaled mode shapes through finite element model updating. *Mechanical Systems and Signal Processing* **21**(6), 2459–2473.

Henderson GR and Piersol AG 2003 Evaluating vibration environments using the shock response spectrum. *Sound and Vibration* **37**(4), 18–21.

Heylen W, Lammens S and Sas P 1997 *Modal Analysis Theory and Testing* 2nd edn. Catholic University Leuven, Leuven, Belgium.

Himmelblau H, Piersol AG, Wise JH and Grundvig MR 1993 *Handbook for Dynamic Data Acquisition and Analysis*. Institute of Environmental Sciences and Technology, Mount Prospect, Illinois.

ISO 18431-4: 2007 *Mechanical Vibration and Shock – Signal Processing – Part 4: Shock spectrum analysis*. International Organization for Standardization.

Konstantin-Hansen H and Herlufsen H 2010 Envelope and cepstrum analyses for machinery fault identification. *Sound and Vibration* **44**(5), 10–12.

Lalanne C 2002 *Mechanical Vibration & Shock – Specification Development, Vol. 5*. CRC Press.

Maia N and Silva J (eds) 2003 *Theoretical and Experimental Modal Analysis*. Research Studies Press, Baldock, Hertforsdhire, England.

Otte D, de Ponseele P. V and Leuridan J 1990 Operational deflection shapes in multisource environments *Proc. 8th International Modal Analysis Conference*, Kissimmee, FL.

Rades M 1994 A comparison of some mode indicator functions. *Mechanical Systems and Signal Processing* **8**(4), 459–474.

Randall RB 1982 Cepstrum analysis and gearbox fault-diagnosis. *Maintenance Management International* **3**(3), 183–208.

Randall RB and Gao Y 1994 Extraction of modal parameters from the response power cepstrum. *Journal of Sound and Vibration* **176**(2), 179–193.

Smallwood D 1981 An improved recursive formula for calculating shock response spectra. *Shock and Vibration Bulletin* (51), 4–10.

Tomlinson G and Kirk N 1984 Modal analysis and identification of structural non-linearity *Proc. 2nd Int. Conf. on Rec. Adv. in Struct. Dyn.*, 1984, University of Southampton, pp 495–510.

Tucker S and Vold H 1990 On principal response analysis *Proc. ASELAB Conf.*, Paris, France.

Williams R, Crowley J and Vold H 1985 The multivariate mode indicator function in modal analysis *Proc. 3rd International Modal Analysis Conference*, Orlando, FL.

# Appendix A

# Complex Numbers

Complex numbers are frequently used in signal analysis. A complex number, $c$ is defined as

$$c = a + \mathrm{j}b \tag{A.1}$$

where the real numbers $a$ and $b$ are called the real part and imaginary part, respectively, of $c$. The number j, the imaginary number, also sometimes denoted i, is equal to the square root of $-1$. Of course, this does not (at least immediately) provide any insight into the use of complex numbers, so we shall here show some fundamental use of complex numbers.

First we define the *complex conjugate*, $c^*$, of $c$, by

$$c^* = a - \mathrm{j}b. \tag{A.2}$$

A useful picture of complex numbers is obtained if we plot the real and imaginary parts of $c$ as $x$ and $y$ coordinates, respectively, in a coordinate system as in Figure A.1. Complex numbers represented by Equation (A.1) is often called *rectangular form*, or the *Euclidian* form.

From Figure A.1 it directly follows that the complex number, $c$, may be written using trigonometric functions, as

$$c = A[\cos\phi + \mathrm{j}\sin\phi] \tag{A.3}$$

from which it follows that

$$A = \sqrt{a^2 + b^2} \tag{A.4}$$

and

$$\phi = \arctan\left(\frac{b}{a}\right). \tag{A.5}$$

The expression of the complex number, $c$, in Equation (A.3) is often called the *trigonometric form*. The factor $A$, is also the square root of the *amplitude squared* of the complex number, $c$, which is obtained by

$$|c|^2 = cc^* = (a + \mathrm{j}b)(a - \mathrm{j}b) = a^2 + b^2. \tag{A.6}$$

*Noise and Vibration Analysis: Signal Analysis and Experimental Procedures,* First Edition. Anders Brandt.

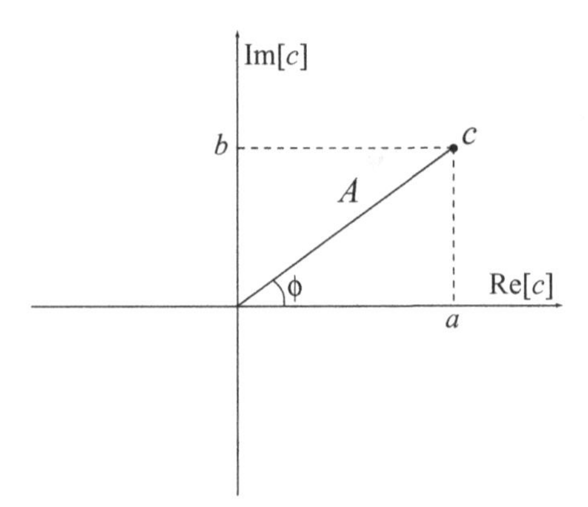


**Figure A.1** The complex plane

There is a third common notation for expressing $c$; the *Euler form* or *polar form*. Here, $c$ is written as in Equation (A.7), which can readily be seen as a special notation.

$$c = A\mathrm{e}^{\mathrm{j}\phi} \tag{A.7}$$

where $A$ and $\phi$ are equal to those in Equation (A.3).

The polar form also has a simplified notation commonly used in, for example, electrical engineering. Here, $c$ is written as

$$c = A\angle\phi \tag{A.8}$$

where the symbol $\angle$ is read 'angle.'

When we use complex numbers in signal analysis, there are mainly two operations of interest. The first is a summation of two complex numbers, say $c_1 = a_1 + \mathrm{j}b_1$ and $c_2 = a_2 + \mathrm{j}b_2$. An example of this case is when we have two sound waves with a certain common frequency, and the two sounds are added together at a certain point. Since the sound information contains both amplitude and phase, it becomes a complex addition, see also below where we describe how complex numbers are used to describe sinusoids. With the addition of two complex numbers, the rectangular form is most suitable and the sum, $c$, of the two numbers is

$$c = c_1 + c_2 = (a_1 + a_2) + \mathrm{j}(b_1 + b_2) \tag{A.9}$$

that is, the real and imaginary parts are summed separately. This is equivalent to vector addition.

The other important operation is multiplication of two complex numbers. An example of this is if we let a sinusoidal force excite a structure for which we know the frequency response between force and response at a certain point. The response at this point may be obtained by multiplying the complex sinusoid by the (complex) value of the frequency response at the frequency of the sinusoid. When we multiply two complex numbers, we prefer to use the polar form of Equation (A.7) and the product then becomes

$$c = c_1 \cdot c_2 = A_1 A_2 \mathrm{e}^{\mathrm{j}(\phi_1+\phi_2)} \tag{A.10}$$

that is, with multiplication, the amplitudes are multiplied and the phase angles are summed.

The most important reason for using complex numbers in signal analysis (noise and vibration analysis) is that when we have sinusoids, it is quite effective to replace them with their complex analogs. Assume first that we have a real, time-dependent signal, $x(t)$, e.g., a measured acceleration signal of a certain frequency

$$x(t) = A\cos(\omega t) \tag{A.11}$$

A complex sinusoid is now defined as

$$\tilde{x}(t) = A\mathrm{e}^{\mathrm{j}(\omega t+\phi)} = C\mathrm{e}^{\omega t} \tag{A.12}$$

where

$$C = A\mathrm{e}^{\mathrm{j}\phi} \tag{A.13}$$

Using this notation, our actual (original) signal can be written as

$$x(t) = \mathrm{Re}\,[\tilde{x}(t)]. \tag{A.14}$$

By introducing the complex signal, $\tilde{x}(t)$ , we are able to easily change both the amplitude and phase of our signal, for example, passing through a frequency response. The resulting signal is then obtained by taking the real part of the calculated complex signal. We achieve the same result as if we had used the real signal the whole time, but without the complicated trigonometric rules. The imaginary part of the complex signal sometimes also has interpretations which we shall not delve into here, but basically we can say that it simply follows along as a complement to the calculations.

**Example A.0.1** *As an example of using complex numbers, assume that we have a sinusoidal force with amplitude 30 N and frequency 100 Hz. The force passes through an SDOF system with a natural frequency of 100 Hz, where we let the frequency response of accelerance type be* $0.1\angle\pi/2$ *[(m/s2)/N]. We let the phase of our force be the reference, that is, 0 radians. What is the resulting acceleration? Our force signal,* $F(t)$*, can be written in complex form as*

$$F(t) = C\mathrm{e}^{\mathrm{j}2\pi f_0 t} \tag{A.15}$$

*where* $C = 10\mathrm{e}^{\mathrm{j}0} = 10$ *[N], and* $f_0 = 100$ *[Hz]. Furthermore, the frequency response at 100 Hz is*

$$H(100) = 0.1\mathrm{e}^{\mathrm{j}\pi/2}. \tag{A.16}$$

*We thus obtain from Equation (A.10) that the resulting acceleration is*

$$a(t) = F(t)H(100) = 10\cdot 0.1\mathrm{e}^{\mathrm{j}(2\pi f_0 t+0+\pi/2} = \mathrm{e}^{\mathrm{j}(2\pi f_0 t+\pi/2} \tag{A.17}$$

*or, if we write the actual, real acceleration, that is, the real part of Equation (A.17), then*

$$a(t) = \cos(2\pi f_0 t + \pi/2) \tag{A.18}$$

*End of example.*

# Appendix B

# Logarithmic Diagrams

Logarithmic (log) scales are often used when displaying spectra. There are two reasons for this:

1. the compression that occurs when changing to log scale (usually on the $y$-axis) reveals details in the curve that are not as obvious using a linear scale, and
2. many curves become straight lines on a log–log scale (where both axes are logarithmic).

Logarithms can be defined with an arbitrary base. The logarithm we most often use within noise and vibration is the base 10 logarithm, or 'log-base-10.' For this logarithm, if

$$x = 10^y \tag{B.1}$$

then

$$y = \log_{10}(x) \tag{B.2}$$

which is read as 'the log-base-10 of $x$ is equal to $y$.' For example, the log-base-10 of 1000 is equal to 3. A simple algebraic rule for logarithms when multiplying, which follows from the definition, is that

$$\log_{10}(a \cdot b) = \log_{10}(a) + \log_{10}(b) \tag{B.3}$$

Especially useful, as we will discover in Appendix C, dealing with decibels, is that

$$\log_{10}\left(a^2\right) = 2 \cdot \log_{10}(a) \tag{B.4}$$

The log-base-10 as a function of $x$ is shown in Figure B.1 (a).

A common reason why log scales are used is that many curves become straight lines on a log–log scale (when both the $x$-scale and $y$-scale are logarithmic). More specifically, this is valid for curves which are exponential expressions, that is, where $y = x^a$. In Figure B.1 (b), two such functions are plotted with linear and log–log scales. In this case we must use log scales on both the $x$-axis and the $y$-axis in order

*Noise and Vibration Analysis: Signal Analysis and Experimental Procedures,* First Edition. Anders Brandt.

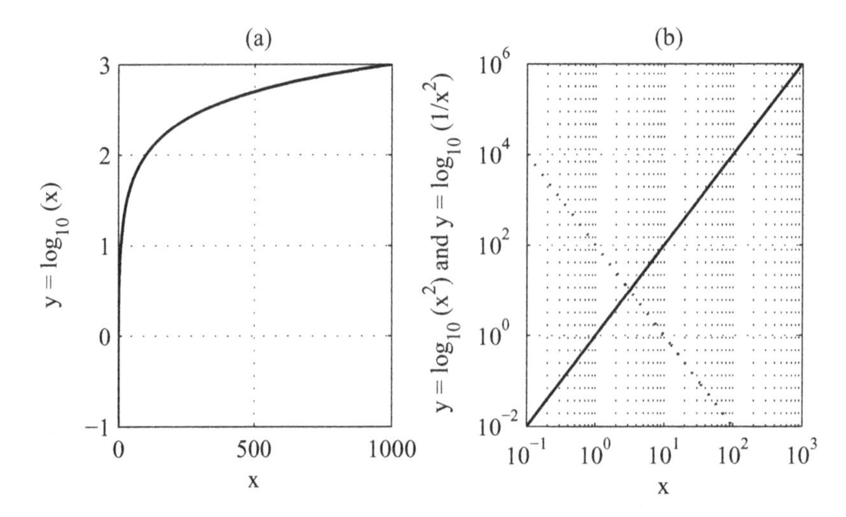


**Figure B.1** In (a), $y = \log_{10}(x)$ is shown as a function of $x$, for $x$-values from 0.1 to 1000. As seen in the diagram, taking the log results in a strong compression, that is, a large difference in $x$-values give only small differences in $y$-values ('logged' values). In (b), the functions $y = x^2$ (solid) and $y = 100/x^2 = 100x^{-2}$ (dotted), are plotted with log–log scales. We see in (b) how exponential functions of the form $y = x^a$ become straight lines

to obtain straight lines. Log–log plots are common for plotting filter characteristics, and spectra in, e.g., environmental testing.

The compression effect of the logarithm is often used, for example, when we plot the frequency response of mechanical structures; otherwise we lose details because of the large dynamic range of such frequency responses, as illustrated in Figure B.2.

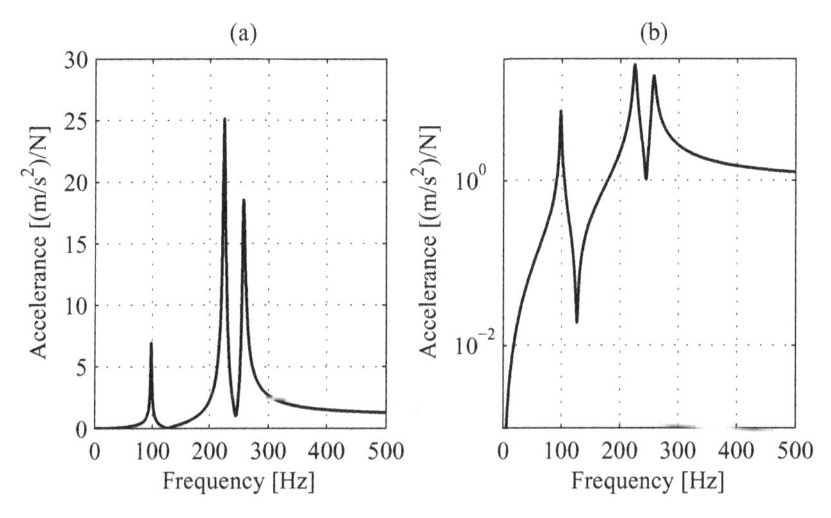


**Figure B.2** Frequency response plotted with (a) linear $y$-scale and (b) logarithmic $y$-scale. Comparing the two formats shows that many details of the curve are only visible when using a log $y$-scale

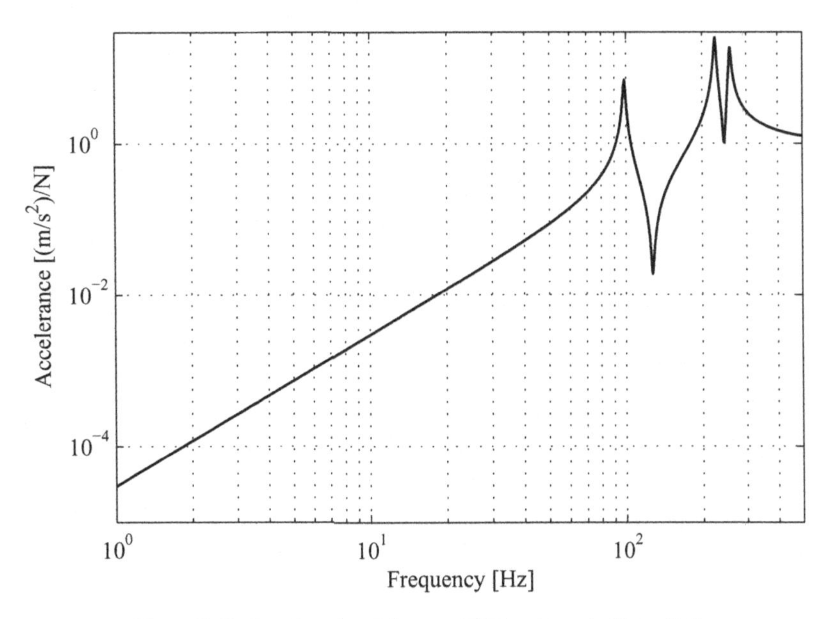


**Figure B.3** Log–log plot of the same FRF as shown in Figure B.2

Some people also prefer to plot frequency responses of mechanical systems on a log–log scale, although that is not the case with the author of this book. The reason is shown in Figure B.3. Although there are some good points for plotting FRFs on a log–log scale, for example the fact that resonance bandwidths are constant relative bandwidths, the natural frequencies have a tendency to be 'packed' in the upper part of the $x$-axis when using a log–log scale. As apparent throughout this book, I therefore prefer logarithmic $y$-axis and linear $x$-axis format for FRF plots.

# Appendix C

# Decibels

The concept of decibels is central to noise and vibration analysis. It is primarily used within acoustics, where the concept is related to the logartihmic sensitivity of the human ear. It is also used for plotting, for example, frequency responses and filter characteristics. Thus, it is essential to understand how the decibel expression is calculated.

The decibel was invented (be people at the Bell Telephone Labs) for use in telecommunications, and was invented to make expressions independent of the context, that is, if amplitude or power is used. Or, maybe we should say the unit Bell, was invented; the Bell unit is so large, however, that in most fields it is most common to use the *deci*Bell, a tenth of a Bell. To obtain this desired quality, the decibel is defined as a relative measure using a power ratio. For a power, $P$, which is to be converted to decibels relative to a reference power, $P_0$, the resulting power, $P_{\mathrm{dB}}$, in decibels is

$$P_{\mathrm{dB}} = 10 \cdot \log_{10}\left(\frac{P}{P_0}\right) \tag{C.1}$$

If we, for example, have a power of 100 watts and the reference power is 1 watt, then we obtain a power of 20 dB relative to 1 watt.

Of course, we often measure entities with linear units, and not powers. We can therefore use an analog with an electrical circuit, where the power consumed by a circuit component, for example a resistor, is proportional to the product of the current through the resistor, and the voltage across it. That is, if the resistance is $R$, then the power consumed by the resistor is

$$P = UI = U \cdot \frac{U}{R} = \frac{U^2}{R}. \tag{C.2}$$

If our reference power is $P_0$, which corresponds to a reference voltage $U_0$, then

$$P_0 = \frac{U_0^2}{R}. \tag{C.3}$$

We now express the power, $P$ in decibels relative to $P_0$, and obtain

$$P_{\mathrm{db}} = 10 \cdot \log_{10}\left(\frac{P}{P_0}\right) = 10 \cdot \log_{10}\left(\frac{U}{U_0}\right)^2 \tag{C.4}$$

*Noise and Vibration Analysis: Signal Analysis and Experimental Procedures*, First Edition. Anders Brandt.

If we now use the relation $\log_{10}(a^2) = 2log_{10}(a)$ from Appendix B, we obtain an alternative formula for calculating the decibel expression for linear quantities, namely

$$P_{dB} = 20 \cdot \log_{10}\left(\frac{U}{U_0}\right). \tag{C.5}$$

Equation (C.5) thus expresses a voltage ratio in decibels. Note that the decibel value is the same if we use a voltage ratio as if we use the power ratio. That is the point with decibels!

Of course, in noise and vibration analysis we rarely express electrical voltages directly in decibels. For arbitrary units we use the following rules:

1. If the unit of the entity we wish to convert is linear (not squared) we use Equation (C.5). This is true, for example, for an acceleration in $m/s^2$.
2. If the unit is quadratic, we use Equation (C.1) and replace $P$ with our measured entity. This is true, for example, for a PSD of an acceleration in $[(m/s)^2)^2/Hz]$.

Finally, we must observe that a decibel value is only meaningful when the reference that has been used is indicated! This may seem confusing if you are used to hearing, for example, sound levels given in dB without any reference. In acoustics, however, standard reference values are commonly used which are often not reported explicitly. For sound pressure levels (which sound levels usually are), the reference 20 μPa is used.

# Appendix D

# Some Elementary Matrix Algebra

I have assumed the reader to be acquainted with some basic linear algebra in this book. In this appendix we will summarize some important matrix algebra relations, and define the nomenclature for matrices used throughout the book. For a complete coverage see for example Strang (2004).

First, let us define the nomenclature used in this book. A *column vector* is denoted by $\{x\}$, by which we mean

$$\{x\} = \begin{Bmatrix} x_1 \\ x_2 \\ \cdots \\ x_M \end{Bmatrix} \tag{D.1}$$

if we assume the vector has $M$ elements. A *row vector* is denoted by $\lfloor y \rfloor$, by which we mean

$$\lfloor y \rfloor = \lfloor y_1 \; y_2 \; \ldots \; y_N \rfloor \tag{D.2}$$

if we assume the vector has length $N$. A regular *matrix*, $[A]$, is denoted by brackets, i.e.,

$$[A] = \begin{bmatrix} a_{11} & a_{12} & \ldots & a_{1N} \\ a_{21} & a_{22} & \ldots & a_{2N} \\ \ldots & \ldots & \ldots & \ldots \\ a_{M1} & a_{M2} & \ldots & a_{MN} \end{bmatrix} \tag{D.3}$$

and we call this matrix an $M$-by-$N$ matrix, or we say that it has size $M \times N$. A *diagonal matrix* is denoted by, e.g., $\lceil S \rfloor$, and it has no nonzero off-diagonal elements, i.e.,

$$\lceil S \rfloor = \begin{bmatrix} s_{11} & 0 & \ldots & 0 \\ 0 & s_{22} & \ldots & 0 \\ \ldots & & & \ldots \\ 0 & 0 & \ldots & s_{MM} \end{bmatrix} \tag{D.4}$$

and, of course, this matrix has to be square, $M \times M$. The most common diagonal matrix is perhaps the *identity matrix*, $\lceil I \rfloor$, which is

$$\lceil I \rfloor = \begin{bmatrix} 1 & 0 & \ldots & 0 \\ 0 & 1 & \ldots & 0 \\ \ldots & & & \ldots \\ 0 & 0 & \ldots & 1 \end{bmatrix} \tag{D.5}$$

*Noise and Vibration Analysis: Signal Analysis and Experimental Procedures,* First Edition. Anders Brandt.

and which has the important property that, for any matrix, $[A]$,

$$[A]\lceil I\rfloor = [A]. \tag{D.6}$$

We also have the important property that the inverse of the identity matrix equals itself, i.e.,

$$\lceil I\rfloor^{-1} = \lceil I\rfloor. \tag{D.7}$$

We denote the *transpose* of a real vector or matrix by the superscript $^T$. For complex vectors and matrices we usually replace the transpose by the *Hermitian transpose*, $[A]^H$, which is equal to the complex conjugate (see Appendix A), of the transposed matrix, i.e.,

$$[A]^T = \begin{bmatrix} a_{11} & a_{21} & \dots & a_{M1} \\ a_{12} & a_{22} & \dots & a_{M2} \\ \dots & \dots & \dots & \dots \\ a_{1N} & a_{2N} & \dots & a_{MN} \end{bmatrix} \tag{D.8}$$

and

$$[A]^H = \begin{bmatrix} a^*_{11} & a^*_{21} & \dots & a^*_{M1} \\ a^*_{12} & a^*_{22} & \dots & a^*_{M2} \\ \dots & \dots & \dots & \dots \\ a^*_{1N} & a^*_{2N} & \dots & a^*_{MN} \end{bmatrix} \tag{D.9}$$

The 'standard' matrix equation $[A]\{x\} = \{b\}$ has the solution

$$\{x\} = [A]^{-1}\{b\} \tag{D.10}$$

where we call $[A]^{-1}$ the *inverse* of $[A]$. Of course, the solution to Equation (D.10) may or may not exist. If the matrix $[A]$ is square, we have the situation of an equation system with the same number of unknowns (in $\{x\}$) as we have equations. Then, the inverse exists if the *determinant* of $[A]$, denoted $|A|$, is nonzero. We leave the details of determinants to the text, see for example Section 6.3.1 where we use the determinant of a system matrix.

If, on the other hand, there are more lines than unknowns, i.e., $M > N$, then we have to find some other solution. The most common solution in that case is the *least squares* solution,

$$\{x\} = \left([A]^T[A]\right)^{-1}[A]^T\{b\}. \tag{D.11}$$

The inverse of a matrix is a numerically unstable entity and should be avoided in computations. In MATLAB/Octave, the best way of solving standard, square equations, is therefore to use the *slash* and the *backslash* operators. These work so that, if we have an equation of three matrices $A$, $B$, and $C$, so that the solution is

$$[A] = [B][C]^{-1} \tag{D.12}$$

then the solution in MATLAB/Octave is best obtained by the code

```
A = B / C
```

and if the equation has the solution

$$[A] = [C]^{-1}[B] \tag{D.13}$$

then the solution in MATLAB/Octave is best obtained by the code

```
A = C \ B
```

where, in both codes, you should note that the inverse entity is, so to speak, below the division sign; in 'B/C', C is below the sign, indicating the inverse is taken of C. In 'C \ B', C is also below the division sign. In MATLAB/Octave these two operators are called 'right division' (/) and 'left division' (\), respectively, which points to which side of the non-inverted matrix ($[B]$), the inverse matrix ($[C]^{-1}$) is standing.

## Reference

Strang G 2005 *Linear Algebra and its Applications* 4th edn. Brooks Cole, San Diego.

# Appendix E

# Eigenvalues and the SVD

Modern noise and vibration analysis applications involve many advanced linear algebra concepts, and we have used some in this book, particularly in Chapters 6, 14 and 15. Some of the linear algebra theory we have used is not included in most curricula, even at graduate level. We will therefore summarize some of the most important concepts here. More details can be found in some textbooks on linear algebra, see for example Strang (2004).

## E.1 Eigenvalues and Complex Matrices

The concept of eigenvalues and eigenvectors is very important. Generally, the eigenvalue problem is related to the equation

$$([A] - \lambda \lceil I \rfloor)\{x\} = \{0\} \tag{E.1}$$

which is also sometimes written as

$$[A]\{x\} = \lambda \{x\}. \tag{E.2}$$

The solutions, $\lambda_n$, to any of these two equations are called the *eigenvalues* of $[A]$. Furthermore, if we put such an eigenvalue, $\lambda_n$, into Equation (E.1), then there is only a particular vector, $\{x_r\}$, which satisfies the equation. Such vectors are called the *eigenvectors* of $[A]$ and correspond to each eigenvalue. Naturally, there are infinitely many eigenvectors for each eigenvalue, because we can scale it with a factor, and it will still satisfy Equation (E.1), but there is only one unique vector times a scale factor.

The eigenvalues in the solution to Equation (E.1) are found be finding the values, $\lambda_n$, satisfying

$$|[A] - \lambda \lceil I \rfloor| = 0 \tag{E.3}$$

where $|\ |$ denotes the determinant. Thus the solution is found by finding the values $\lambda_n$ which makes the determinant in Equation (E.3) equal zero. This is also called the *characteristic equation* of $[A]$.

Eigenvalues and eigenvectors are closely related to complex matrices, particularly those matrices which are *Hermitian* (sometimes 'Hermitian symmetric'), i.e., complex matrices $[A]$ which are equal to their Hermitian transpose, i.e.,

$$[A]^H = [A]. \tag{E.4}$$

Hermitian matrices are the equivalent of real symmetric matrices, for which $[A]^T = [A]$. There are some fundamental properties of the eigenvalues of symmetric matrices, which are of particular interest to us. We will therefore state some of these briefly.

---

*Noise and Vibration Analysis: Signal Analysis and Experimental Procedures*, First Edition. Anders Brandt.

- The eigenvalues of any symmetric matrix are real.
- The eigenvalues of any Hermitian matrix are real.
- If the eigenvalues are all larger than zero, the matrix is called *positive definite*, and if some eigenvalues are allowed to be equal to zero, the matrix is referred to as *positive semidefinite*.
- The eigenvectors of any symmetric or Hermitian matrix can be chosen orthonormal, i.e., with unity length, and orthogonal to each other.
- For an orthonormal matrix, $[Q]$, $[Q]^T = [Q]^{-1}$.

We now come to the important *diagonalization* properties of symmetric matrices. Any symmetric matrix, $[A]$, can be diagonalized by its eigenvectors into

$$[A] = [Q] \lceil \Lambda \rfloor [Q]^T \tag{E.5}$$

where the matrix $[Q]$ has each eigenvector as a column, and $\lceil \Lambda \rfloor$ is a diagonal matrix with the eigenvalue $\lambda_n$ in its $(n, n)$ element, corresponding to the eigenvector in column $n$ of $Q$. If the matrix $[Q]$ is chosen so that it is orthonormal, then

$$\lceil \Lambda \rfloor = [Q]^T [A] [Q] \tag{E.6}$$

and it should be noted that MATLAB/Octave produces orthonormal eigenvectors by default, using the **eig** command.

For complex matrices, for example the input cross-spectral matrix $[G_{xx}]$ discussed in Section 10.7 and in Chapters 14 and 15, the above equations turn into equivalent equations if we replace the transpose with Hermitian transpose. Thus, for any Hermitian (and therefore complex) matrix, $[A]$, then

$$[A] = [U] \lceil \Lambda \rfloor [U]^H \tag{E.7}$$

and

$$\lceil \Lambda \rfloor = [U]^H [A] [U] \tag{E.8}$$

where the matrix $[U]$ is a matrix with each eigenvector corresponding to the element $(n, n)$ in $\Lambda$ located in column $n$ of $[U]$. The matrix $[U]$ is called a *unitary* matrix, in that it is complex and has orthonormal columns. For this matrix, of course, it is true that $[U]^H = [U]^{-1}$.

## E.2 The Singular Value Decomposition (SVD)

The singular value decomposition, SVD, is perhaps one of the most powerful decompositions used in modern engineering. It resembles the eigenvalue decomposition in Equations (E.6) and (E.7), but it works on *any* (possibly rectangular) matrix $[A]$, whereas the eigenvalue decompositions, of course, only work for square matrices. The SVD of any matrix $[A]$ is

$$[A] = [U_1] \lceil S \rfloor [U_2]^H \tag{E.9}$$

where the columns in $[U_1]$ are called the *left singular vectors*, the values in the diagonal matrix $[S]$ are called the *singular values*, and the columns in $[U_2]$ are the *right singular vectors*. If $[A]$ is an $M \times N$ matrix, then

- $[U_1]$ is an $M \times M$ matrix, whose columns are the eigenvectors of $[A][A]^H$,
- the matrix $\lceil S \rfloor$ is $M \times N$ and the singular values on the diagonal equal the square roots of the eigenvalues of $[A]^H [A]$ and $[A][A]^H$,
- the singular values are always real and non-negative, and sorted in descending order,
- $[U_2]$ is an $N \times N$ matrix, whose columns are the eigenvectors of $[A]^H [A]$, and

- the columns of both the left and the right singular vector matrices, $[U_1]$ and $[U_2]$, are orthonormal (orthogonal and unity length).

The matrix $\lceil S \rfloor$ is an $M \times N$ matrix, so we need to define how such a matrix can be diagonal. It is simply diagonal in such a way that each diagonal element starting from element $(1, 1)$ contains a (perhaps) nonzero value, and all remaining values in the matrix are zero. If $M > N$, then $\lceil S \rfloor$ is

$$\lceil S \rfloor = \begin{bmatrix} s_{11} & 0 & \ldots & 0 \\ 0 & s_{22} & \ldots & 0 \\ 0 & 0 & \ldots & 0 \\ \ldots & & \ldots & s_{NN} \\ 0 & 0 & \ldots & 0 \\ \ldots & & & \ldots \\ 0 & 0 & \ldots & 0 \end{bmatrix} \tag{E.10}$$

and if $M < N$ then $\lceil S \rfloor$ instead looks like

$$\lceil S \rfloor = \begin{bmatrix} s_{11} & 0 & \ldots & 0 & \ldots & 0 \\ 0 & s_{22} & \ldots & 0 & \ldots & 0 \\ 0 & \ldots & \ldots & 0 & \ldots & 0 \\ 0 & 0 & \ldots & s_{MM} & \ldots & 0 \end{bmatrix}. \tag{E.11}$$

An important property of the SVD of a Hermitian matrix, is that, for a Hermitian matrix $[A]$, $[U_1] = [U]$ and $[U_2] = [U]$, and the singular values are equal to the eigenvalues $\lceil S \rfloor = \lceil \Lambda \rfloor$.

A concept closely related to the SVD is the *pseudo-inverse*, $[A]^+$ of any $M \times N$ matrix $[A]^+$. The pseudo-inverse is the solution to

$$[A]\{x\} = \{b\} \tag{E.12}$$

for rectangular matrices, where the pseudo-inverse solution is

$$\{x\} = [A]^+ \{b\}. \tag{E.13}$$

The solution to Equation (E.13) is of particular interest in many measurement situations where typically $M > N$, i.e., we have an *overdetermined* set of equations (more equations than unknowns). The pseudoinverse can be shown to be equal to

$$[A]^+ = [U_2] \lceil S^+ \rfloor [U_1]^H \tag{E.14}$$

where the inverse singular value matrix $\lceil S^+ \rfloor$ contains the reciprocal of the singular values, i.e., $s_{nn}^+ = 1/s_{nn}$.

The pseudoinverse is closely related to the least squares solution in Equation (D.11) and its main advantage is its very robust numerical performance. At a slight increase in computational effort, the pseudo-inverse has proven to perform extremely well in cases where the matrices involved are noisy or ill-conditioned, and is therefore often preferred over the former.

Finally it should be mentioned that eigenvalues, the SVD, and pseudo-inverse are all integral parts of MATLAB/Octave. They are performed by the commands **eig**, **svd**, and **pinv**, respectively.

For applications of the SVD, see Section 15.1.

# Reference

Strang G 2005 *Linear Algebra and its Applications* 4th edn. Brooks Cole, San Diego.

# Appendix F

# Organizations and Resources

For the newcomer to the field of noise and vibrations it is desirable to find good resources for information. To facilitate this need, this appendix gives an overview of some useful places to search for more information. It has not been my intention to exclude any particular organization or resource. However, by necessity, the information in this appendix will be an incomplete list. See it as examples, and if you go to these places to look for more information, you will be 'in the loop' and you will be able to nest your way through the vast range of information available, particularly with the easy access to Internet today.

I should also point out that there are domestic organizations in most countries which are involved in (usually) either acoustics, vibrations, or environmental engineering vibration testing.

For scientific journals, the reader is suggested to look in the Bibliography section. The (main) journals published by the organizations below, are, however, mentioned in conjunction with each organization.

The **Acoustical Society of America**, ASA, is a scientific organization, as its name implies predominantly in acoustics. Apart from publishing the *Journal of the Acoustical Society of America*, it organizes a number of scientific meetings.

The **Catholic University of Leuven**, K.U. Leuven, in Belgium, organizes a biannual conference, *International Conference on Noise and Vibration Engineering*, also known as ISMA, which is the largest international conference on noise and vibration engineering in Europe. Information about ISMA can be found at http://www.isma-isaac.be.

The **Institute of Environmental Sciences and Technology** is a professional organization not only for vibrations, but for all kinds of environmental testing (climate, etc.). It organizes the annual ESTECH conference and its website is http://www.iest.org.

**The International Institute of Acoustics and Vibration**, IIAV, is a worldwide scientific organization which, among other things, organize an annual conference, the *International Conference on Sound and Vibration*, ICSV. The IIAV website is http://www.iiav.org.

**International Institute of Noise Control Engineering**, I-INCE, is an international consortium of organizations in this field, and organizes the annual INTER-NOISE conference. Its website is http://www.i-ince.org.

**The International Operational Modal Analysis Conference**, IOMAC, is a relatively new, biannual, conference specializing in operational modal analysis, OMA. Information about the conference can be found at http://www.iomac.dk.

The **Shock and Vibration Information Analysis Center**, SAVIAC, is an American organization which, among other things, organize the annual *Shock and Vibration Symposium* in the U.S. It also publishes the *Shock and Vibration Journal*. Its website is http://www.saviac.org.

*Noise and Vibration Analysis: Signal Analysis and Experimental Procedures*, First Edition. Anders Brandt.

The **Society for Experimental Mechanics**, SEM, is a worldwide organization. It organizes, among other things, the annual *International Modal Analysis Conference*, IMAC, which is, despite its name, a conference with topics presented from many fields of noise and vibration engineering. SEM also publishes *Experimental techniques*, and has a website at http://www.sem.org.

**SAE International** is an organization which, among many things, has sections on vibrations in, predominantly, automotive and aerospace applications. SAE organize a biannual conference on vibrations, and the website is http://www.sae.org.

**Sound and Vibration** is a practical engineering publication that is free to U.S. subscribers. There is a $60/year fee for all other countries, except $25/year for Canada. It contains a variety of articles from various noise and vibration applications. It also has a large number of advertisements, and is a good source for the reader who wants to find vendors of measurement and analysis equipment. Its website, http://www.sandv.com contains a variety of useful information, including articles from past issues.

# Bibliography

Ahlin K 2006 Comparison of test specifications and measured field data. *Sound and Vibration* **40**(9), 22–25.

Ahlin K, Magnevall M and Josefsson A 2006 Simulation of forced response in linear and nonlinear mechanical systems using digital filters *Proc. ISMA2006, International Conference on Noise and Vibration Engineering*, Catholic University, Leuven, Belgium.

Allemang R and Brown D 1982 A correlation coefficient for modal vector analysis *Proc. 1st International Modal Analysis Conference*, Orlando, FL.

Allemang RJ, Rost RW and Brown DL 1984 Multiple input estimation of frequency response functions *Proc. 2nd International Modal Analysis Conference*, Orlando, FL.

ANSI S1.11 2004 *Specification for Octave-Band and Fractional-Octave-Band Analog and Digital Filters*. American National Standards Institute.

Antoni J and Schoukens J 2007 A comprehensive study of the bias and variance of frequency-response-function measurements: Optimal window selection and overlapping strategies. *Automatica* **43**(10), 1723–1736.

Antoni J and Schoukens J 2009 Optimal settings for measuring frequency response functions with weighted overlapped segment averaging. *IEEE Transactions on Instrumentation and Measurement* **58**(9), 3276–3287.

Antoni J, Wagstaff P and Henrio JC 2004 H alpha – a consistent estimator for frequency response functions with input and output noise. *IEEE Transactions on Instrumentation and Measurement* **53**(2), 457–465.

Bendat J and Piersol A 1993 *Engineering Applications of Correlation and Spectral Analysis* 2nd edn. Wiley Interscience.

Bendat J and Piersol AG 2010 *Random Data: Analysis and Measurement Procedures* 4th edn. Wiley Interscience.

Blackman RB and Tukey JW 1958a The measurement of power spectra from the point of view of communications engineering .1. *Bell System Technical Journal* **37**(1), 185–282.

Blackman RB and Tukey JW 1958b The measurement of power spectra from the point of view of communications engineering .2. *Bell System Technical Journal* **37**(2), 485–569.

Blough J 1998 *Improving the Analysis of Operating Data on Rotating Automotive Components* PhD Thesis University of Cincinnati, College of Engineering.

Bogert B, Healy M and Tukey J 1963 The quefrency alanysis of time series for echoes: Cepstrum, pseudo-autocovariance, cross-cepstrum and saphe cracking In *Proc. Symposium on time Series Analysis* (ed. Rosenblatt M), pp 209–243.

Brandt A and Ahlin K 2003 A digital filter method for forced response computation *Proc. 21st International Modal Analysis Conference*, Kissimmee, FL.

Brandt A and Brincker R 2010 Impact excitation processing for improved frequency response quality *Proc. 28th International Modal Analysis Conference*, Jacksonville, FL Society for Experimental Mechanics.

Brandt A, Lagö T, Ahlin K and Tuma J 2005 Main principles and limitations of current order tracking methods *Sound and Vibration* **39**(3), 19–22.

Brownlee K 1984 *Statistical Theory and Methodology*. Krieger Publishing Company.

Carlsson B 1991 Maximum flat digital differentiator. *Electronics Letters* **27**(8), 675–677.

Cooley J, Lewis P and Welch P 1967 Historical notes on the fast fourier transform. *IEEE Trans. on Audio and Electroacoustics* **15**(2), 76–79.

---

*Noise and Vibration Analysis: Signal Analysis and Experimental Procedures*, First Edition. Anders Brandt.

Cooley JW and Tukey JW 1965 An algorithm for machine calculation of complex fourier series. *Mathematics of Computation* **19**(90), 297–301.

Cooley JW and Tukey JW 1993 On the origin and publication of the FFT paper – a citation-classic commentary on an algorithm for the machine calculation of complex Fourier-series – Cooley, J.W., Tukey, J.W. *Current Contents/ Engineering Technology and Applied Sciences* (51–52), 8–9.

Cooley JW, Lewis PAW and Welch PD 1970 The application of the fast Fourier transform algorithm to the estimation of spectra and cross-spectra. *Journal of Sound and Vibration* **12**(3), 339–352.

Craig RR and Kurdila AJ 2006 *Fundamentals of Structural Dynamics*. John Wiley & Sons, Inc.

Daniell PJ 1946 Discussion of 'on the theoretical specification and sampling properties of autocorrelated time-series'. *Journal of the Royal Statistical Society* **8 (suppl.)**(1), 88–90.

Den Hartog JP 1985 *Mechanical Vibrations*. Dover Publications Inc.

Dippery KD, Phillips AW and Allemang RJ 1994 Condensation of the spatial domain in modal parameter-estimation. *12th International Modal Analysis Conference*, Honolulu, HI 818–824.

(ed. Maia N and Silva J) 2003 *Theoretical and Experimental Modal Analysis*. Research Studies Press, Baldock, Hertforsdhire, England.

Endo H, Randall RB and Gosselin C 2009 Differential diagnosis of spall vs. cracks in the gear tooth fillet region: Experimental validation. *Mechanical Systems and Signal Processing* **23**(3), 636–651.

Ewins DJ 2000 *Modal Testing: Theory, Practice and Application* 2nd edn. Research Studies Press, Baldock, Hertfordshire, England.

Fladung W 1994 *The development and implementation of multiple reference impact testing* Master's Thesis University of Cincinnati.

Fladung W and Rost R 1997 Application and correction of the exponential window for frequency response functions. *Mechanical Systems and Signal Processing* **11**(1), 23–36.

Fladung W, Zucker A, Phillips A and Allemang R 1999 Using cyclic averaging with impact testing *Proc. 17th International Modal Analysis Conference*, Kissimmee, FL.

Fyfe KR and Munck EDS 1997 Analysis of computed order tracking. *Mechanical Systems and Signal Processing* **11**(2), 187–205.

Gaberson HA 2003 Using the velocity shock spectrum to predict shock damage. *Sound and Vibration* **37**(9), 5–6.

Gaberson HA, Pal D and Chapler RS 2000 Classification of violent environments that cause equipment failure. *Sound and Vibration* **34**(5), 16–23.

Gao Y and Randall RB 1996a Determination of frequency response functions from response measurements .1. extraction of poles and zeros from response cepstra. *Mechanical Systems and Signal Processing* **10**(3), 293–317.

Gao Y and Randall RB 1996b Determination of frequency response functions from response measurements .2. regeneration of frequency response functions from poles and zeros. *Mechanical Systems and Signal Processing* **10**(3), 319–340.

Goyder H 1984 Foolproof methods for frequency response measurements *Proc. 2nd International Conference on Recent Advances in Structural Dynamics*, Southampton.

Greenfield J 1977 Dealing with the shock environment using the shock response spectrum analysis. *J. Soc. Environmental Engineers* (9), 3–15.

Halvorsen WG and Brown DL 1977 Impulse technique for structural frequency-response testing. *Sound and Vibration* **11**(11), 8–21.

Hannig J and Lee TCM 2004 Kernel smoothing of periodograms under Kullback–Leibler discrepancy. *Signal Processing* **84**(7), 1255–1266.

Hanson D, Randall RB, Antoni J, Thompson DJ, Waters TP and Ford RAJ 2007a Cyclostationarity and the cepstrum for operational modal analysis of MIMO systems – part i: Modal parameter identification. *Mechanical Systems and Signal Processing* **21**(6), 2441–2458.

Hanson D, Randall RB, Antoni J, Waters TP, Thompson DJ and Ford RAJ 2007b Cyclostationarity and the cepstrum for operational modal analysis of mimo systems – part ii: Obtaining scaled mode shapes through finite element model updating. *Mechanical Systems and Signal Processing* **21**(6), 2459–2473.

Harris FJ 1978 Use of windows for harmonic-analysis with discrete fourier-transform. *Proceedings of The IEEE* **66**(1), 51–83.

Haykin S 2003 *Signals and Systems* 2nd edn. John Wiley & Sons, Inc.

Heidemann MT, Johnson DH and Burrus CS 1985 Gauss and the history of the fast Fourier-transform. *Archive For History of Exact Sciences* **34**(3), 265–277.

Henderson GR and Piersol AG 2003 Evaluating vibration environments using the shock response spectrum. *Sound and Vibration* **37**(4), 18–21.

Heylen W, Lammens S and Sas P 1997 *Modal Analysis Theory and Testing* 2nd edn. Catholic University Leuven, Leuven, Belgium.

Higgins RJ 1990 *Digital Signal Processing in VLSI*. Prentice Hall.

Himmelblau H, Piersol AG, Wise JH and Grundvig MR 1993 *Handbook for Dynamic Data Acquisition and Analysis*. Institute of Environmental Sciences and Technology, Mount Prospect, Illinois.

Håkansson B and Carlsson P 1987 Bias errors in mechanical impedance data obtained with impedance heads. *J. Sound and Vibration* **113**(1), 173–183.

Hotelling H 1933 Analysis of a complex of statistical variables into principal components. *Journal of Educational Psychology* **24**, 417–441, 498–520.

IEC 61260 1995 *Electroacoustics – Octave-Band and Fractional-Octave-Band Filters*. International Electrotechnical Commission.

IEC 61672-1 2005 *Electroacoustics – Sound level meters – Part 1: Specifications*. International Electrotechnical Commission.

IEEE 1451.4 2004 *A Smart Transducer Interface for Sensors and Actuators – Mixed-mode Communication Protocols and Transducer Electronic Data Sheet (TEDS) Formats*. IEEE Standards Association.

Inman D 2007 *Engineering Vibration* 3rd edn. Prentice Hall.

ISO 18431-1: 2005 *Mechanical Vibration and Shock – Signal Processing – Part 1: General Introduction*. International Organization for Standardization, Geneva, Switzerland.

ISO 18431-4: 2007 *Mechanical Vibration and Shock – Signal Processing – Part 4: Shock spectrum analysis*. International Organization for Standardization, Geneva, Switzerland.

ISO 2631-1: 1997 *Mechanical vibration and shock – Evaluation of human exposure to whole-body vibration – Part 1: General requirements*. International Organization for Standardization, Geneva, Switzerland.

ISO 2631-5: 2004 *Mechanical vibration and shock – Evaluation of human exposure to whole-body vibration – Part 5: Method for evaluation of vibration containing multiple shocks*. International Organization for Standardization, Geneva, Switzerland.

ISO 2641: 1990 *Vibration and shock – Vocabulary*. International Organization for Standardization, Geneva, Switzerland.

ISO 8041: 2005 *Human response to vibration – Measuring instrumentation*. International Organization for Standardization, Geneva, Switzerland.

Kay SM and Marple SL 1981 Spectrum analysis – a modern perspective. *Proceedings of the IEEE* **69**(11), 1380–1419.

Kennedy C and Pancu C 1947 Use of vectors in vibration measurement and analysis. *J. of the Aeronautical Sciences* **14**(11), 603–625.

Konstantin-Hansen H and Herlufsen H 2010 Envelope and cepstrum analyses for machinery fault identification. *Sound and Vibration* **44**(5), 10–12.

Kozin F and Natke HG 1986 System-identification techniques. *Structural Safety* **3**(3-4), 269–316.

Kumar B and Roy SCD 1988 Coefficients of maximally linear, FIR digital differentiators for low-frequencies. *Electronics Letters* **24**(9), 563–565.

Lalanne C 2002 *Mechanical Vibration and Shock – Specification Development, Vol. 5*. CRC Press.

Lebihan J 1995 Maximally linear FIR digital differentiators. *Circuits Systems and Signal Processing* **14**(5), 633–637.

Lembregts F 1988 *Frequency Domain Identification Techniques for Experimental Multiple Input Modal Analysis* PhD Thesis Katholieke Universiteit Leuven, Leuven, Belgium.

Mansfield NJ 2005 *Human Response to Vibration*. CRC Press.

Mitchell L 1982 Improved methods for the FFT calculation of the frequency response function. *J. of Mechanical Design*.

Mitchell L and Cobb R 1987 An unbiased frequency response function estimator *Proc. 5th International Modal Analysis Conference*, London, UK, pp 364–373, London, U.K.

Newland DE 2005 *An Introduction to Random Vibrations, Spectral, and Wavelet Analysis* 3rd edn. Dover Publications Inc.

Newton I 1687 *Philosophiæ Naturalis Principia Mathematica*. London.

Norfield D 2006 *Practical Balancing of Rotating Machinery*. Elsevier Science.

Nuttall A and Carter C 1982 Spectral estimation using combined time and lag weighting. *IEEE* **70**(9), 1115–1125.

Nuttall AH 1981 Some windows with very good sidelobe behavior. *IEEE Transactions on Acoustics Speech and Signal Processing* **29**(1), 84–91.

Nyquist H 2002 Certain topics in telegraph transmission theory (reprinted from *Transactions of the AIEE*, February 1928 617–644). *Proceedings of the IEEE* **90**(2), 280–305.

Oppenheim AV and Schafer RW 1975 *Digital Signal Processing*. Prentice Hall.

Oppenheim AV, Schafer RW and Buck JR 1999 *Discrete-Time Signal Processing*. Pearson Education.

Otnes RK and Enochson L 1972 *Digital Time Series Analysis*. Wiley Interscience.

Otte D 1994 *Development and evaluation of singular value analysis methodologies for studying multivariate noise and vibration problems* PhD Thesis Catholic University Leuven, Belgium.

Otte D, de Ponseele P. V and Leuridan J 1990 Operational deflection shapes in multisource environments *Proc. 8th International Modal Analysis Conference*, Kissimmee, FL.

Pan MC and Wu CX 2007 Adaptive Vold–Kalman filtering order tracking. *Mechanical Systems and Signal Processing* **21**(8), 2957–2969.

Pan MC, Liao SW and Chiu CC 2007 Improvement on gabor order tracking and objective comparison with Vold–Kalman filtering order tracking. *Mechanical Systems And Signal Processing* **21**(2), 653–667.

Papoulis A 2002 *Probability, Random Variables, and Stochastic Processes* 4th edn. McGraw-Hill.

Parks TW and McClellan J 1972 Chebyshev approximation for nonrecursive digital filters with linear phase. *IEEE Transactions on Circuit Theory* **CT19**(2), 189–194.

Pauwels S, Michel J, M. R, Peeters B and Debille J 2006 A new MIMO sine testing technique for accelerated, high quality FRF measurements *Proc. 24th International Modal Analysis Conference*, St. Louis, Missouri Society for Experimental Mechanics.

Pelant P, Tuma J and Benes T 2004 Vold–Kalman order tracking filtration in car noise and vibration measurements *Proc. Proc. 33rd International Congress and Exposition on Noise Control Engineering, INTER-NOISE*, Prague, Czech Republic.

Phillips AW and Allemang RJ 2003 An overview of MIMO-FRF excitation/averaging/processing techniques. *Journal of Sound and Vibration* **262**(3), 651–675.

Pintelon R and Schoukens J 1990 Real-time integration and differentiation of analog-signals by means of digital filtering. *IEEE Transactions on Instrumentation And Measurement* **39**(6), 923–927.

Pintelon R, Peeters B and Guillaume P 2008 Continuous-time operational modal analysis in the presence of harmonic disturbances. *Mechanical Systems and Signal Processing* **22**(5), 1017–1035.

Potter R 1990a A new order tracking method for rotating machinery. *Sound and Vibration* **24**(9), 30–34.

Potter R 1990b Tracking and resampling method and apparatus for monitoring the performance of rotating machines.

Proakis JG and Manolakis DG 2006 *Digital Signal Processing: Principles, Algorithms, and Applications* 4th edn. Prentice Hall.

Qian S 2003 Gabor expansion for order tracking. *Sound and Vibration* **37**(6), 18–22.

Rabiner LR and Schafer RW 1974 On the behavior of minimax relative error FIR digital differentiators. *Bell System Technical Journal* **53**(2), 333–361.

Rades M 1994 A comparison of some mode indicator functions. *Mechanical Systems and Signal Processing* **8**(4), 459–474.

Randall RB 1982 Cepstrum analysis and gearbox fault-diagnosis. *Maintenance Management International* **3**(3), 183–208.

Randall RB and Gao Y 1994 Extraction of modal parameters from the response power cepstrum. *Journal of Sound and Vibration* **176**(2), 179–193.

Rao S 2003 *Mechanical Vibrations* 4th edn. Pearson Education.

Reljin IS, Reljin BD and Papic VD 2007 Extremely flat-top windows for harmonic analysis. *IEEE Transactions on Instrumentation and Measurement* **56**(3), 1025–1041.

Rimell AN and Mansfield NJ 2007 Design of digital filters for frequency weightings required for risk assessments of workers exposed to vibration. *Industrial Health* **45**(4), 512–519.

Rocklin GT, Crowley J and Vold H 1985 A comparison of H1, H2, and Hv frequency response functions *Proc. 3rd International Modal Analysis Conference*, Orlando, FL.

Saavedra PN and Rodriguez CG 2006 Accurate assessment of computed order tracking. *Shock and Vibration* **13**(1), 13–32.

Schmidt H 1985a Resolution bias errors in spectral density, frequency response and coherence function measurements, i: General theory. *Journal of Sound and Vibration* **101**(3), 347–362.

Schmidt H 1985b Resolution bias errors in spectral density, frequency response and coherence function measurements, iii: Application to second-order systems (white noise excitation). *Journal of Sound and Vibration* **101**(3), 377–404.

Schoukens J, Rolain Y and Pintelon R 2006 Analysis of windowing/leakage effects in frequency response function measurements. *Automatica* **42**(1), 27–38.

Serridge M and Licht T 1986 *Piezoelectric Accelerometer and Vibration Preamplifier Handbook*. Brüel & Kjær, Nærum, Denmark.

Shannon CE 1998 Communication in the presence of noise (reprinted from the proceedings of the IRE, vol 37, pp 10–21, 1949). *Proceedings of the IEEE* **86**(2), 447–457.

Shao H, Jin W and Qian S 2003 Order tracking by discrete Gabor expansion. *IEEE Transactions on Instrumentation and Measurement* **52**(3), 754–761.

Sheskin D 2004 *Handbook of Parametric and Nonparametric Statistical Procedures* 3rd edn. Chapman & Hall.

Smallwood D 1981 An improved recursive formula for calculating shock response spectra. *Shock and Vibration Bulletin* (51), 4–10.

Smallwood D 1995 Using singular value decomposition to compute the conditioned cross-spectral density matrix and coherence functions *Proc. 66th Shock and Vibration Symposium*, vol. 1, pp 109–120.

Stoica P and Sundin T 1999 Optimally smoothed periodogram. *Signal Process.* **78**(3), 253–264.

Strang G 2005 *Linear Algebra and its Applications* 4th edn. Brooks Cole, San Diego.

Thrane N 1979 The discrete fourier transform and FFT analyzers. Brüel & Kjær Technical Review 1.

Tomlinson G and Kirk N 1984 Modal analysis and identification of structural non-linearity *Proc. 2nd Int. Conf. on Rec. Adv. in Struct. Dyn.*, 1984, University of Southampton, pp 495–510.

Tran T, Claesson I and Dahl M 2004 Design and improvement of flattop windows with semi-infinite optimization *Proc. The 6th International Conference on Optimization: Techniques and Applications*, Ballarat, Australia.

Tucker S and Vold H 1990 On principal response analysis *Proc. ASELAB Conf.*, Paris, France.

Tuma J 2004 Sound quality assessment using Vold–Kalman tracking filtering *Seminar, Instruments and Control*, Ostrava, Czech Republic.

Tuma J 2005 Setting the passband width in the Vold–Kalman order tracking filter *Proc. 12th ICSV*, Lisbon, Portugal.

Vold H 1986 Estimation of operating shapes and homogeneous constraints with coherent background noise *Proc. ISMA 1986, International Conference on Noise and Vibration Engineering*, Catholic University, Leuven, Belgium.

Vold H and Leuridan J 1993 High resolution order tracking at extreme slew rates *Proc. SAE Noise and Vibration Conference*, Traverse City, MI Society of Automotive Engineers.

Vold H, Crowley J and Nessler J 1988 Tracking sine waves in systems with high slew rates *Proc. 6th International Modal Analysis Conference*, Kissimmee, FL, pp 189–193.

Vold H, Mains M and Blough J 1997 Theoretical foundation for high performance order tracking with the Vold–Kalman tracking filter *Proc. 1997 Noise and Vibration Conference, SAE*, vol. 3, pp 1083–1088.

Welch P 1967 The use of fast Fourier transform for the estimation of power spectra: A method based on time averaging over short, modified periodograms. *IEEE Trans. on Audio and Electroacoustics* **AU-15**(2), 70–73.

White PR, Tan MH and Hammond JK 2006 Analysis of the maximum likelihood, total least squares and principal component approaches for frequency response function estimation. *Journal of Sound and Vibration* **290**(3–5), 676–689.

Wicks A and Vold H 1986 The Hs frequency response estimator *Proc. 4th International Modal Analysis Conference*, Los Angeles, CA.

Williams R and Vold H 1986 Multiphase-step-sine method for experimental modal analysis. *International Journal of Analytical and Experimental Modal Analysis* **1**(2), 25–34.

Williams R, Crowley J and Vold H 1985 The multivariate mode indicator function in modal analysis *Proc. 3rd International Modal Analysis Conference*, Orlando, FL.

Wirsching PH, Paez TL and Ortiz H 1995 *Random Vibrations: Theory and Practice*. Wiley Interscience.

Wise J 1983 The effects of digitizing rate and phase distortion errors on the shock response spectrum *Proc. Institute of Environmental Sciences, Annual Technical Meeting, 29th, April 19–21*, Los Angeles, CA.

Wowk V 1991 *Machinery Vibration: Measurement and Analysis*. McGraw-Hill.

Zwillinger D (ed.) 2002 *CRC Standard Mathematical Tables and Formulae* 31st edn. Chapman & Hall.

# Index

Accelerance, 96
Accelerometer, 152
    base strain sensitivity, 156
    calibration of, 159
    mass calibration, description of, 160
    mass calibration, illustration, 161
    mass calibration, results, 164
    mass loading, 155
    mounted resonance frequency, 154
    mounting of, 155
    temperature sensitivity, 156
    transverse sensitivity, 156
Acoustic A-weighting
    frequency domain, 236–7
    time domain, 50
Acoustic C-weighting
    frequency domain, 236–7
    time domain, 50
Aliasing, 37
    by circular convolution, 200
    preventing, 38, 250
Analytic signal, 380
Angle domain, *see* Order domain
Angular frequency, 9
Anti-alias filter, 250
Antiresonance, 140
Autocorrelation, *see* Correlation function
Average
    ensemble, 64
    time, 64
Averaging
    exponential, 261
    frequency domain, 260
    interrupted, 261
    linear, 260
    of rotating machinery signals, 276
    peak hold, 261
    stable, 260
    time domain, 260

Bandwidth
    of signal, 36
    resonance, 94
Bandwidth–time product, 69, 178
    discussion for order tracking, 264, 279
Bartlett window, 196
Beating
    on SDOF system, 95
    principle, 11
Bias error
    definition of, 65
    normalized, 65
    of average estimate, 69
    of frequency response estimate, 292–3
    of smoothed periodogram estimator, 223
    of Welch estimator, 212–17
Blocksize
    choosing for impact testing, 303, 306
    choosing for linear spectrum, 231
    choosing for shaker excitation, 312, 314, 320
    choosing for spectral density estimation, 233
    definition of, 180
Burst random, *see* Shaker excitation

Central moment, 70
Cepstrum, 384–8
    complex cepstrum, 387
    inverse, 388

*Noise and Vibration Analysis: Signal Analysis and Experimental Procedures,* First Edition. Anders Brandt.

Cepstrum (*cont.*)
  liftering, 385
  power cepstrum
    definition of, 385
    example of, 386
    features, 385
  quefrency, 385
Charge amplifier, 148
  cable for, 149
Circular convolution, 199–200
Coherence function
  estimator, 291
  multiple, 329
    estimator, 328
  random error, 296
Coherent output power, 295
  virtual, *see* Virtual signals
Complex
  number, 403
  sine wave, 10
Condenser microphone, *see* Microphone
Conditioned signals, 331–6
  L-systems, 331
  partial coherence, 333
  partial coherent output spectra, 334
  rules for ordering inputs, 334
Confidence interval, 66
Consistent estimator, 66
Convolution
  as aliasing, 200
  by time window, 189, 194, 198
  circular, 199–200
  description of, 22–5
  integral, 21
Correlation function
  auto-
    definition of, 71
    estimator, 225
  cross-
    definition of, 71
    estimator, 225
  properties of, 72
Cosine, *see* Sine wave
Crest factor, 71
Cross-correlation, *see* Correlation function
CSD, *see* Spectral density, cross-
Cyclic frequency, 9

Damping
  determining, 103–4, 398
  effect on frequency response, 92
  hysteretic, 106
  modal, 129
  models of, 106
  nonproportional, 130–3
    eigenvalue problem, 131
  proportional, 128–9
  relative, of SDOF system, 89
  structural, 106
  viscous, 88
Data quality assessment, 84
Data quality assessment, 81
Decibel, 411
Degree-of-freedom, 121
DFT, *see* Discrete Fourier transform
Differentiation
  in frequency domain, 96, 238
  of time signals, 55–8
Dirac unit impulse, 22
Discrete Fourier transform, 180
  comparison with continuous, 181
  definition of, 180
  inverse, 180
  leakage, *see* Leakage
  periodicity of, 183
  periodogram, 209
  picket-fence effect, 189
  principle of, 183
  properties of, 186
  relation with continuous, 186
  scaling of, 181
  smearing, 269
  symmetry properties, 186
  table of pairs, 186
  time windows, *see* Time windows
  zoom, 201
DOF, *see* Degree-of-freedom
Driving point, *see* Frequency response
Dynamic flexibility, 95

Eigenvalue
  from characteristic equation, 123
  problem, example of, 124–5
  problem, for nonproportional damping, 131
  relation to pole, 122–3
  theory of, 417–18
Eigenvector
  of nonproportionally damped MDOF system, 130
  of undamped MDOF system, 123

Eigenvectors
  of proportionally damped MDOF system, 128
Energy spectral density
  definition of, 173
  estimator, 226
Enhanced frequency response, 358
Ensemble
  average of spectra, 205
  statistical, 64
Envelope
  by Hilbert transform, 380
  spectrum, *see* Spectrum
Equivalent noise bandwidth, 194
Equivalent number of averages, 219
Equivalent sound pressure, 49
Ergodic random process, 64
Error function, erf, 73
ESD, *see* Energy spectral density
Excitation signal, *see* Shaker excitation
Expected value
  definition of, 65
  estimate of, 68
  of general function, 68
Experimental modal analysis, *see* Modal analysis
Exponential window
  compensating effect of, 303–5
  for impact excitation, 300
  for transient, 227

Fast Fourier transform, 181
FFT, *see* Fast Fourier transform
FFT Analyzer, *see* Measurement system
Filters
  1/1-octave, 47
  1/3-octave, 47
  1/3-octave, example, 48
  1/n-octave, 48
  acoustic weighting, 49
  analog, 43
  anti-aliasing, 250
  Butterworth, 43
  differentiation, 55–8
  digital, 45
  first-order HP, 43
  first-order LP, 43–4
  for analog integration, 49
  for frequency weighting, 49
  fractional octave, 48
  ideal characteristics, 42
  integration, 51–5
  linear-phase
    for recording transients, 257
    importance of, 51
    producing, 46
  octave center frequencies, 47
  octave time constant, 179
  phase distortion of, 46
  smoothing, 46
  time delay of, 45
Flattop window, 195, 269, 276
Folding, *see* Aliasing
Force transducer, 157
Force window, *see* Time windows
Forced response
  of SDOF system, 95
  relation with ODS, 391
  time domain simulation of, 141–3
Fourier series
  complex, 168
  definition of, 168
  with amplitude and phase, 168
Fourier Transform
  properties of, 27
Fourier transform
  definition of, 26
  description, 25–9
  discrete, *see* Discrete Fourier Transform
  inverse, 26
  properties of, 29
  table of, 26
Frequency
  analysis
    nonparametric, 177
    nonparametric principle, 178
    parametric, 177
    principle of, 177
  angular, 9
  cyclic, 9
  damped natural, of SDOF, 90
  damped resonance, 93
  Nyquist, 36
  undamped, *see* Natural frequency
Frequency resolution, 198
Frequency response
  bias error, 292–3, 295
  definition of, 29
  driving point properties, 141
  enhanced, 358
  from poles and residues, 135–6
  from transfer function, 30

Frequency response (*cont.*)
  $H_1$ estimator
    of MIMO system, 326
    of SISO system, 287
  $H_2$ estimator
    of MIMO system, 329, 349
    of SISO system, 288
  $H_c$ estimator, 290
  $H_v$ estimator, 329–30
  imaginary part of, 100
  impact excitation, *see* Impact excitation
  $[M]$, $[C]$, $[K]$, 134
  $[M]$, $[K]$, and $\zeta_r$, 138
  magnitude of, 97
  multiple-input
    optimizing measurements, 339–48
  names of, 97
  Nyquist plot of, 100
  of SDOF system, 92
  phase of, 97
  plot formats, 97–103
  random error, 293
  real part of, 100
  shaker excitation, *see* Shaker excitation
  single-input
    optimizing measurements, 312–18
  using modal damping, 138

Gaussian, *see* Probability distribution
GNU Octave
  arranging spectra, 239
  using, 5

$H_1$ estimator
  of MIMO system, 326
  of SISO system, 287
$H_2$ estimator
  of MIMO system, 329, 349
  of SISO system, 288
$H_c$ estimator, 290
$H_v$ estimator, 329–30
Half sine window, 196
Half-power bandwidth, 94
Hamming window, 195
Harmonic removal, 223
Hermitian
  matrix, 417
  transpose, 325, 414
Hilbert transform, 378–84
  analytic signal, 380
  computation of, 379
  definition of, 378
  for computing envelope, 380
  to clean up FRF estimates, 383
  to compute imaginary part of FRF, 382
  to compute real part of FRF, 382
Histogram, 67
Hypothesis test
  for normality, 77
  for stationarity, 79–81
  theory of, 74

IEPE transducers
  current supply, 149
  electrical model, 149
  TEDS, 152
  time constants of, 151
Impact excitation, 297–306
  alternative processing, 306
  compensating effect of exponential window, 303–5
  default settings, 302
  double impact, 305
  exponential window, 300
  force signal, 298–9
  force window, 299
  optimizing settings for, 300–3
  setting blocksize, 303
  triggering, 301
  when appropriate, 298
Impact hammer, *see* Impulse hammer
Impedance head, 158
Impulse hammer, 159, 297
Impulse response
  definition of, 21
  from poles and residues, 141
  of SDOF system, 90
Input cross-spectral matrix
  definition of, 239
  inversion of, 325
Input/output cross-spectral matrix, 239
Integration
  in frequency domain, 238
  of time signals, 51–5
Interpolation
  for synchronous resampling, 274
  of time signal, 40
Inverse DFT, *see* Discrete Fourier transform
Inverse Fourier transform, *see* Fourier transform
Inverse Laplace transform, *see* Laplace transform

Kurtosis
definition of, 70
excess, 71
for data quality assessment, 82

Laplace transform, 17–21
definition, 17
inverse, 17
of derivative, 18
table of, 18
transfer function, 20
Leakage, 187–91
explanation of, 187, 194
illustration of, 187
in spectrum of periodic signal, 188
in spectrum of random signal, 199
Liftering, *see* Cepstrum
Linear spectrum, *see* Spectrum
Linear system
coherent output power, 295
definition, 16
noise on in- and output, 287
$H_c$ estimator, 290
noise on input, 288
$H_2$ estimator, 288
noise on output, 287
$H_1$ estimator, 287
theory of, 16–25
Lumped parameter model, 121

MAC matrix, 397–8
definition, 397
Mass calibration, 160
Mass loading, 155
MATLAB
arranging spectra, 239
using, 5
Maxwell·s reciprocity, 136
MDOF system
eigenvalues of, 123
eigenvectors of, 123
frequency response, 134–5
impulse response, 141
matrix equations for, 121
nonproportional damping
natural frequency, 131
poles, 131
normal modes of, 123
poles of, 122–3, 128
state-space equations of, 130
undamped natural frequency of, 122
Measurement system
absolute accuracy, 255
analog-to-digital conversion, 247
sigma-delta ADC, 252
anti-alias filter, 250
anti-alias protection, 255
averaging options, 260
block processing, 258
cross-channel match, 256
cross-channel talk, 257
data scaling, 259
dynamic range, 247, 251, 256
FFT parameters, 261
optimizing input range, 248
overload, 250, 253
overview, 246
quantization, 247
real-time bandwidth, 258
recording signals, 258
recording transients, 257
sample-and-hold, 255
sampling requirements, 250
signal conditioning, 247
triggering, 259
Microphone
calibration of, 162
types, 161
MIF, *see* Mode indicator function
MIMO system
bias error discussion, 336
computation considerations, 329
$H_1$ estimator, 326
$H_2$ estimator, 329, 349
$H_v$ estimator, 329–30
illustration of, 326
noise on input and output
unbiased estimation, 342
noise on output, 326
principle of, 323
random error discussion, 336
Mobility, 96
Modal A
definition of, 131
relation with modal mass, 139
Modal analysis
analytical, 139, 393
experimental, 393–9
data checks, 394–5
driving point FRF check, 346

Modal analysis (*cont.*)
mode indicator function, *see* Mode indicator function
overview, 394
parameter extraction, 398–9
reciprocity check, 346
reference DOFs, 394
response DOFs, 394
Modal assurance criterion, *see* MAC matrix
Modal B, 131
Modal coordinates
for undamped system, 127
of proportionally damped system, 128
Modal damping, *see* Damping
Modal mass
of proportionally damped system, 128
of undamped system, 126
relation with Modal A, 139
Modal scale constant, 136
Modal stiffness
of proportionally damped system, 128
of undamped system, 126
Modal superposition, 135, 141
Mode from wave equation, 120
Mode indicator function, 395
modified real, 397
multivariate, 395
normal, 395
Mode shape
determining experiementally, 398
scaling of, 126, 138–9
scaling to unity modal A, 139
scaling to unity modal mass, 139
weighted mass orthogonality, 125
weighted orthogonality of, 125–6
weighted stiffness orthogonality, 125
Multi-channel data
spectrum of, *see* Spectrum
Multiple coherence, *see* Coherence function
Multiple-input/multiple-output system, *see* MIMO system

Natural frequency
damped, of SDOF system, 90
determining, 103–4, 398
of nonproportionally damped MDOF system, 131
undamped, of MDOF system, 122
undamped, of SDOF system, 89

Newton's
equation, state-space form, 130
equations for MDOF system, 121
Newton·s
laws, 87
equation for SDOF system, 88
Node line, 139
Noise source identification, 367
Nonproportional damping, *see* Damping
Normal distribution, *see* Probability distribution
Normal mode, 123
Nyquist
frequency, 36
plot format, 100

Octave (software), *see* GNU Octave
Octave filter, *see* Filters
ODS, 391–3
definition of, 391
multiple reference, 392
procedure, 392
recommended measurement function, 392
One-sided spectral density, *see* Spectral density, single-sided
Operating deflection shape, *see* ODS
Order, *see* Rotating machinery analysis
Order domain, 273
Order track
fixed sampling frequency, 272
synchronous sampling frequency, 276
Ordinary coherence, *see* Coherence function
Orthogonality
between input signals, 351
between sines, 13
Overlap processing
effect, 218–20
illustration of, 206
Oversampling
definition of, 38
for shock response, 377

Parseval's theorem
discrete, 186, 189
Parseval's theorem
continuous, 27
Partial coherence, *see* Conditioned signals
Partial coherent output spectra, *see* Conditioned signals

Partial fraction expansion
definition of, 19
of SDOF system, 90
Period, 9
Periodic chirp, *see* Shaker excitation
Periodic signal, *see* Signal
Periodogram, 209
Phase angle
definition of complex, 403
of sine wave, 9
Phase distortion, *see* Filters
Phase spectrum, *see* Spectrum, linear
Picket-fence effect, 189
Piezoelectric
accelerometers, principle of, 152
effect, 147
sensor, electrical models of, 148
Poles
of nonproportionally damped MDOF system, 131
of proportionally damped MDOF system, 128
of SDOF system, 89
of undamped MDOF system, 123
Polynomial multiplication
by convolution, 25
of filter coefficients, 46
Power cepstrum, *see* Cepstrum
Principal components, *see* alsoVirtual signals, 351–9, 369
finding number of independent sources, 353
Principal coordinates, *see* Modal coordinates
Probability density
calculation of, 68
definition of, 66
Probability distribution
definition of, 66
Gaussian, 72
normal, 72
test of normality, 77
Proportional damping, *see* Damping
PSD, *see* Spectral density, auto
Pseudo-random, *see* Shaker excitation
Pure random, *see* Shaker excitation

Q-factor
definition of, 94
for shock response, 376
Quefrency, *see* Cepstrum

Random error
definition of, 66
normalized, 66
of frequency response, 293–5
of smoothed periodogram estimator, 224
of standard deviation, 69
of variance, 69
Welch estimator, 217–21
Random signal, *see* Signal
Receptance, *see* Dynamic flexibility
Reciprocity
check for, 346
definition of, 136
Recording signals, *see* Measurement system
Rectangular window, 188
Relative damping, *see* Damping
Resampling
of time signal, 41
synchronously with RPM, 272–5
Residue
definition of, 19
from mode shape coefficients, 136
Resonance bandwidth, *see* Bandwidth
Resonance frequency, *see* Natural frequency
Reverse arrangements test, 78
Rigid body mode, 123
RMS level
calculation of, 69
computing from spectrum, 234–6
definition, 15
random error of, 178
weighted spectrum, 236
Root mean square, *see* RMS level
Rotating machinery analysis
averaging, 276
bandwidth–time product, 264, 279
color map, 268
maximum order, 276
nonparametric methods, 281
order, 263
order domain, 273
order resolution, 276
order track
fixed sampling frequency, 272
synchronous sampling frequency, 276
RPM, 267
RPM map, 267
selecting time window, 271
smearing, 269
synchronous sampling, 272–5

Rotating machinery analysis (*cont.*)
DFT parameters, 276
tachometer processing, 265–7
time–frequency analysis, 264
waterfall plot, 268

Sampling
frequency, 35
synchronous, *see* Rotating machinery analysis
theorem, 36
SDOF system
forced response, 95
frequency response of, 92
illustration of, 88
impulse response of, 90
poles of, 89
transfer function of, 89
Shaker
description, 162
Shaker excitation
checking shaker attachment, 318
multiple-input, 337–9
burst random, 338–9, 342
checking input correlation, 348, 355
multiphase stepped sine, 338
optimizing measurements, 339–48
periodic random, 338, 342
pure random, 337
single-input, 306–12
burst random, 310, 312
optimizing measurements, 312–18
periodic chirp, 311, 314
pseudo-random, 310, 314
pure random, 308, 312
stepped-sine, 311
SNR of, 307
sources of error, 319
Shock response spectrum, 375–7
maximax, 376
oversampling rate for, 377
primary, 376
Q-factor, 376
residual, 376
Signal
classes, description of, 167
period of, 9
periodic, 8
random, 13
sine wave, 8
transient, 14
recording requirements, 257
Sine wave, 8
complex, 10
multiplication of two, 12
sum of two, 11
Single degree-of-freedom, *see* SDOF system
Single-input/single-output system, *see* SISO system, 286
Singular value decomposition, 418
Sinusoid, *see* Sine wave
SISO system
illustration of, 286
noise on input, 288
noise on input and output, 289
noise on output, 287
Skewness
definition of, 70
use for data quality assessment, 83
Smearing, 269
Smoothed periodogram PSD, *see* Spectral density
Smoothing
filter, 46
of periodogram for PSD, 223
Sound pressure level, 49, 412
Spectral density
creating time signal with known PSD, 390
double-sided auto
definition of, 171
double-sided cross-
definition of, 172
guidelines for computing, 231–3
of mixed property signal, 233
single-sided auto-
definition of, 172
property of, 172
Welch estimator, 211
single-sided cross-
definition of, 172
property of, 172
Welch estimator, 211
smoothed periodogram, 221–4
advantages, 223
bias error, 223
estimator, 223
for harmonic removal, 223
random error, 224
Welch
bias error, 212–17

bias error for SDOF resonance, 215
normalized random error, 220
random error, 217–21
random error with Hanning and 50% overlap, 221
scaling factor, 212

Spectral lines
usable, 199
vs blocksize, 199

Spectrum
autopower
estimator, 207
envelope, 388
computation of, 388
estimation of
guidelines, 228–34
interpretation of, 173–4
linear
definition of, 169
estimator, 208
guidelines, 229–31
phase spectrum, 170, 208
of multi-channel data, 238–40
of periodic signal, *see* Spectrum, linear
of random signal, *see* Spectral density
of transient
definition of, 173
estimator, 226

Standard deviation
definition of, 69
random error, 69

State-space, 130

Stationarity
of random process, 64
test for, 78–80

Steady-state response
definition of, 30

Stepped-sine, *see* Shaker excitation

Stinger
checking, 318
for force sensor, 157

SVD, *see* Singular value decomposition

Synchronous sampling, 272–5
DFT parameters, 276

2-input/1-output system, 324

2DOF system
introduction, 110–12
matrix equations for, 121

Time Windows
for Random Signals, 198

Time windows
amplitude correction, 193
Bartlett, 196
comparison of, 191
equivalent noise bandwidth, 194
normalized, 194
exponential, 300
flattop, 195, 269, 276
for periodic signals, 191–6
for random signals, 199
for transient signals, 227
force, 299
half sine, 196
Hanning, 191, 195
rectangular, 188
resolution of, 198
smearing, 269

Time–frequency, *see* alsoBandwidth–time product
analysis limitations, 277
analysis principle, 264–5
illustration of, 170

Transfer function
definition of, 20
of SDOF system, 89

Transient signal, *see* Signal

Transient spectrum
definition of, 173
estimator, 226

Triboelectric effect, 149

True random, *see* Shaker excitation, pure random

Tuned damper, 113–14

Two-sided spectral density, *see* Spectral density, double-sided

Undamped natural frequency, *see* Natural frequency

Uniform window, *see* Rectangular window

Variance
definition of, 69
estimate of, 69
random error, 69

Vibration isolation, 107–9

Virtual signals, 360–7
cumulated virtual coherence, 362
cumulated virtual coherent output power, 365, 369

Virtual signals (cont.)
cumulated virtual input/output coherence, 364, 369
definition of, 352, 360
number of independent sources, 353
principal components, 351
virtual coherence, 362
virtual coherent output power, 365
virtual input coherence, 362, 370
virtual input cross-spectrum, 362

Vold–Kalman filter, 281

Wave equation, 120
Welch estimator, *see* Spectral density
Wiener–Khinchine relations, 172
Window, *see* Time windows

Zero padding, 200–1
Zoom FFT, 201